Lecture Notes in Computer Sci

Commenced Publication in 1973
Founding and Former Series Editors:
Gerhard Goos, Juris Hartmanis, and Jan van Leeuwen

Editorial Board

David Hutchison
Lancaster University, UK

Takeo Kanade
Carnegie Mellon University, Pittsburgh, PA, USA

Josef Kittler
University of Surrey, Guildford, UK

Jon M. Kleinberg
Cornell University, Ithaca, NY, USA

Friedemann Mattern
ETH Zurich, Switzerland

John C. Mitchell
Stanford University, CA, USA

Moni Naor
Weizmann Institute of Science, Rehovot, Israel

Oscar Nierstrasz
University of Bern, Switzerland

C. Pandu Rangan
Indian Institute of Technology, Madras, India

Bernhard Steffen
University of Dortmund, Germany

Madhu Sudan
Massachusetts Institute of Technology, MA, USA

Demetri Terzopoulos
New York University, NY, USA

Doug Tygar
University of California, Berkeley, CA, USA

Moshe Y. Vardi
Rice University, Houston, TX, USA

Gerhard Weikum
Max-Planck Institute of Computer Science, Saarbruecken, Germany

Jiannong Cao   Wolfgang Nejdl
Ming Xu (Eds.)

Advanced Parallel Processing Technologies

6th International Workshop, APPT 2005
Hong Kong, China, October 27-28, 2005
Proceedings

 Springer

Volume Editors

Jiannong Cao
Hong Kong Polytechnic University
Hung Hom, Kowloon, Hong Kong, China
E-mail: csjcao@comp.polyu.edu.hk

Wolfgang Nejdl
University of Hannover, Information Systems Institute
Knowledge Based Systems (KBS), L3S Research Center
Appelstr. 4, 30167 Hannover, Germany
E-mail: nejdl@l3s.de

Ming Xu
National University of Defense Technology
Department of Networking Engineering, Computer College
Changsha, Hunan 410073, China
E-mail: xuming64@public.cs.hn.cn

Library of Congress Control Number: 2005934413

CR Subject Classification (1998): D, B, C, F.1-3, G.1-2

ISSN   0302-9743
ISBN-10   3-540-29639-5 Springer Berlin Heidelberg New York
ISBN-13   978-3-540-29639-3 Springer Berlin Heidelberg New York

This work is subject to copyright. All rights are reserved, whether the whole or part of the material is
concerned, specifically the rights of translation, reprinting, re-use of illustrations, recitation, broadcasting,
reproduction on microfilms or in any other way, and storage in data banks. Duplication of this publication
or parts thereof is permitted only under the provisions of the German Copyright Law of September 9, 1965,
in its current version, and permission for use must always be obtained from Springer. Violations are liable
to prosecution under the German Copyright Law.

Springer is a part of Springer Science+Business Media

springeronline.com

© Springer-Verlag Berlin Heidelberg 2005
Printed in Germany

Typesetting: Camera-ready by author, data conversion by Scientific Publishing Services, Chennai, India
Printed on acid-free paper   SPIN: 11573937   06/3142   5 4 3 2 1 0

Preface

Welcome to the proceedings of APPT 2005: the 6th International Workshop on Advanced Parallel Processing Technologies. APPT is a biennial workshop on parallel and distributed processing. Its scope covers all aspects of parallel and distributed computing technologies, including architectures, software systems and tools, algorithms, and applications. APPT originated from collaborations by researchers from China and Germany and has evolved to be an international workshop. APPT 2005 was the sixth in the series. The past five workshops were held in Beijing, Koblenz, Changsha, Ilmenau, and Xiamen, respectively.

The Program Committee is pleased to present the proceedings for APPT 2005. This year, APPT 2005 received over 220 submissions from researchers all over the world. All the papers were peer reviewed by two to three Program Committee members on their relevance, originality, significance, technical quality, and presentation. Based on the review result, 55 high-quality papers were selected to be included in the proceedings. The papers in this volume represent the forefront of research on parallel processing and related fields by researchers from China, Germany, USA, Korea, India, and other countries. The papers accepted cover a wide range of exciting topics, including architectures, software, networking, and applications.

The excellent program was the result of the hard work and the collective effort of many people and organizations. We would like to express our special thanks to the Architecture Professional Committee of the China Computer Federation (APC-CCF), the Hong Kong Polytechnic University, the National University of Defense Technology, China, and the Harbin Institute of Technology, China. We would like to thank the general chair, Prof. Xingming Zhou, and the general co-chairs, Prof. Xiaodong Zhang and Prof. David Bader, for their great support. Thanks to all members of the Program Committee and all the other reviewers for the time and hard work they put into the thorough reviewing of the large number of papers. We appreciate the keynote speakers, Prof. Francis C.M. Lau and Prof. Kurt Rothermel, for their strong support of the program. We would also like to express our gratitude to Springer for its assistance in putting the proceedings together. Last but not least, our thanks go to the Local Organizing Committee for the great job it did in making the local arrangements and organizing an attractive social program. Without their dedicated help and diligent work, the workshop would not have been so successful.

We would like to take this opportunity to thank all the authors, many of whom traveled great distances to participate in this workshop and make their valuable contributions. We hope that all participants enjoyed the program and found it worthwhile. We warmly welcome any comments and suggestions to improve our work.

August 2005   Jiannong Cao
                                                               Wolfgang Nejdl
                                                                    Ming Xu

Organization

APPT 2005 was organized mainly by the Department of Computing, Hong Kong Polytechnic University and the National University of Defense Technology, China.

Executive Committee

General Chair   Xingming Zhou
(Member of Chinese Academy of Sciences, National Lab for Parallel and Distributed Processing, China)

General Vice Co-chairs   Xiaodong Zhang
(College of William and Mary, USA)
David A. Bader
(Georgia Institute of Technology, USA)

Program Co-chairs   Jiannong Cao
(Hong Kong Polytechnic University, China)
Wolfgang Nejdl
(University of Hannover, Germany)

Publicity Chair   Cho-Li Wang
(University of Hong Kong, China)

Publication Chair   Laurence T. Yang
(St. Francis Xavier University, Canada)

Local Organization Chair   Allan K.Y. Wong
(Hong Kong Polytechnic University, China)

Finance/Registration Chair   Ming Xu
(National Lab for Parallel and Distributed Processing, China)

Sponsoring Institutions

Architecture Professional Committee of the China Computer Federation, China
Hong Kong Polytechnic University, China
National University of Defense Technology, China
Association for Computing Machinery, Hong Kong Chapter
Springer

Program Committee

Srinivas Aluru	Iowa State University, USA
Jose Nelson Amaral	University of Alberta, Canada
Wentong Cai	Nanyang Technological University, Singapore
Yiu-Keung Chan	City University of Hong Kong, China
Tarek El-Ghazawi	George Mason University, USA
Binxing Fang	Harbin Institute of Technology, China
John Feo	Cray Inc., USA
Guang Gao	University of Delaware, USA
Ananth Grama	Purdue University, USA
Manfred Hauswirth	EPFL, Switzerland
Bruce Hendrickson	Sandia National Laboratory, USA
Mehdi Jazayeri	Technical University of Vienna, Austria
Zhenzhou Ji	Harbin Institute of Technology, China
Ashfaq Khokhar	University of Illinois, Chicago, USA
Ajay Kshemkalyani	University of Illinois, Chicago, USA
Francis Lau	University of Hong Kong, China
Xiaoming Li	Peking University, China
Xinsong Liu	University of Electronic Sciences and Technology of China, China
Yunhao Liu	Hong Kong University of Science and Technology, China
Xinda Lu	Shanghai Jiao Tong University, China
Siwei Luo	Northern Jiao Tong University, China
Beth Plale	Indiana University, USA
Bernhard Plattner	Swiss Federal Institute of Technology, Switzerland
Sartaj Sahni	University of Florida, USA
Nahid Shahmehri	Linköping University, Sweden
Chengzheng Sun	Griffith University, Australia
Zhimin Tang	Institute of Computing, CAS, China
Bernard Traversat	Sun Microsystems, USA
Peter Triantafillou	University of Patras, Greece
Xingwei Wang	Northeastern University, China
Lars Wolf	Technical University Braunschweig, Germany
Jie Wu	Florida Atlantic University, USA
Li Xiao	Michigan State University, USA
Chengzhong Xu	Wayne State University, USA
Weimin Zheng	Tsinghua University, China

Table of Contents

Keynote Speech

Architecture

Algorithm and Theory

System and Software

Grid Computing

Networking

Applied Technologies

Research Issues in Adapting Computing to Small Devices

Francis C.M. Lau

Department of Computer Science, The University of Hong Kong, China
fcmlau@cs.hku.hk

Abstract. Advances in pervasive and mobile technologies are making computing available to us at anytime anywhere. Availability however does not automatically mean it is in a form that implies ease of use. Usability in the mobile world amounts to a set of problems that are not so much precedented in the history of computing. Handheld mobile devices that are thin-lean-mean for instance present challenges that require fundamental changes in the way computation is carried out, its architecture, or its supporting environment. A practical goal is to minimize these changes, which calls for automatic or semi-automatic adaptation of existent computing to the small devices. We discuss the issues and research challenges of "X adapting to Y", where X includes content, data, code, computation, GUI, and so on, and the changes in semantics and/or syntax due to the adaptation are to satisfy the constraints of Y. Some experiments we have carried out for content and code adaptation provide some useful illustration.

J. Cao, W. Nejdl, and M. Xu (Eds.): APPT 2005, LNCS 3756, p. 1, 2005.
© Springer-Verlag Berlin Heidelberg 2005

Mobile Context-Aware Systems – Linking the Physical and Digital World

Kurt Rothermel

Institute of Parallel and Distributed Systems and Centre of Excellence Nexus,
Universität Stuttgart, Germany
Kurt.Rothermel@informatik.uni-stuttgart.de

Abstract. The rapid miniaturisation and decline in prices of computer, communication and sensor technology give rise to a number of interesting developments, such as multifunctional mobile devices, sensor platforms embedded into everyday things, and sensor nodes organised in a wireless network. Those systems can capture and process sensory data and communicate this information to other peers in their proximity or to an existing server infrastructure. The sensory data are fed into spatio-temporal models of the physical world, which build the basis for the promising class of context-aware applications. Based on these developments it can be anticipated that there will be billions of sensor systems in our physical environment near future. Consequently, we envision most of the future applications to be context-aware, sharing highly dynamic digital world models offered by a large number of content providers. Obviously, the realisation of this vision will cause a number of both technological and social challenges. Some of these challenges are subject to the research of Nexus, a Centre of Excellence established at University of Stuttgart in the year 2003. In this talk, we present the vision and objectives of Nexus. Moreover, we will discuss some aspects of scaleable context management.

J. Cao, W. Nejdl, and M. Xu (Eds.): APPT 2005, LNCS 3756, p. 2, 2005.
© Springer-Verlag Berlin Heidelberg 2005

A Data Transformations Based Approach for Optimizing Memory and Cache Locality on Distributed Memory Multiprocessors[*]

Xia Jun and Xue-Jun Yang

School of Computer Science, National University of Defense Technology,
Changsha 410073, Hunan, China
ddk@nudt.edu.cn

Abstract. Data locality is one of the key factors in affecting the performance of parallel programs running on distributed memory multiprocessors. This paper presents an approach for optimizing memory locality and cache locality of perfect or non-perfect loop nests using linear data transformations on distributed memory multiprocessors. The approach optimizes memory locality with the data space fusion technique and cache locality with the projection-delamination technique, and combines the both techniques effectively to make the overheads of remote memory accesses and local memory accesses as low as possible. We conduct experiments with nine programs and the results show the approach is effective in optimizing memory locality and cache locality simultaneously.

1 Introduction

Data locality has an important affection on the performance of parallel programs running on distributed memory multiprocessors. Generally, the locality optimization problem on distributed memory multiprocessors can be divided into two subproblems. One is the memory locality optimization problem. For distributed memory multiprocessors, the memory access time from a processor to its own local memory is generally much faster than the time to local memory of the other processors. Hence, an efficient parallel executing program requires programmers or compilers to distribute code and data carefully to reduce remote memory access overheads. The other is the cache locality optimization problem. When a processor accesses its local memory, good cache locality can improve cache hit rate and reduce local memory access overheads.

Over the last decade, a great number of researchers have paid attention to the memory locality optimization problem for distributed memory multiprocessors [1-3]. Chen and Chang [1] present a skewed alignment instead of traditional dimension-order alignment techniques to align arrays. Chang et al. [2] present two new alignment functions for loop iteration space and arrays with linear subscripts in three loop index variables or quadratic subscripts. Xia and Yang [3] give an approach of effec-

---

[*] This research is supported by **NNSF** (National Natural Science Foundation grant No. 69825104).

J. Cao, W. Nejdl, and M. Xu (Eds.): APPT 2005, LNCS 3756, pp. 3 – 12, 2005.
© Springer-Verlag Berlin Heidelberg 2005

tive alignment of computation and data for a sequence of perfect loop nests. All the above researchers only consider memory locality optimization and their approaches are effective to reduce remote memory access overheads on distributed memory multiprocessors. As they don't consider cache locality optimization, there may exist many local memory access overheads because of bad cache locality, which may prevent the whole performance of parallel programs from improving further.

Over the last decade, many researchers also used loop transformations [4-6], data transformations [7-9] and combined loop and data transformations [10-12] to optimize cache locality. We discuss the most related of their work in the following.

Loop Transformations. Wolf and Lam [4] show how to use unimodular loop transformations followed by tiling loops that carry some form of reuse to improve locality. McKinley et al. [5] present a method that considers loop fusion, distribution, permutation, and reversal for improving locality. Bik et al. [6] present a method that can simultaneously reshape the access patterns of several occurrences of multi-dimensional arrays along certain desired access directions.

Data Transformations. Clauss et al. [7] use the parameterized polyhedra theory and Ehrhart polynomials to provide a new array reference evaluation function to the compiler, such that the data layout corresponds exactly to the utilization order of these data. Kandemir et al. [8] present a hyperplane based approach for optimizing spatial locality in loop nests. Xia et al. [9] present a projection-delamination technique for optimizing spatial locality and a data transformation framework based on it.

Combined Loop and Data Transformations. Cierniak and Li [10] use loop and array dimension permutations in an exhaustive search to determine the appropriate loop and data transformations. Kandemir et al. [11] give a matrix-based approach for optimizing the global locality using loop and data transformations. Kandemir et al. [12] use integer linear programming and the memory layout graph to find the best combination of loop and data layout transformations for optimizing the global locality.

The above researchers only consider cache locality optimization and their approaches are effective to improve cache hit rate and reduce local memory access overheads. As they don't consider memory locality optimization, their approaches are more suitable for use in uniprocessors or shared memory multiprocessors than in distributed memory multiprocessors because of remote memory access overheads.

In this paper, we present an approach for simultaneously optimizing memory locality and cache locality through linear data transformations on distributed memory multiprocessors. Through much research, we find the rows in linear data transformation matrices have two different effects on improving locality. Some rows have the effect of optimizing memory locality. They partition data space effectively and put dependent data together to reduce remote memory access overheads. We call these rows as *memory locality optimizing rows* (MLORs). The other rows have the effect of optimizing cache locality. They reshape the access patterns of array references along columns to reduce local memory access overheads (we assume the default memory layout is column-major, but the approach can be applied to row-major memory layout too). We call these rows as *cache locality optimizing rows* (CLORs). In this paper, we first present a theoretical framework of data space fusion, which is used to partition data space effectively, and determine MLORs based on it. Then under the condition of not affecting memory locality, we use the approach presented in [9] to reshape the access patterns along columns for cache locality and determine CLORs accordingly. At last, we combine MLORs and CLORs to form the final data transformation matrices. The approach can handle not only perfect loop nests but also non-perfect loop nests. It can simultaneously optimize memory and cache locality and can be naturally integrated with data replication and offset alignment. Therefore, our approach can reduce remote and local memory access overheads as much as possible. We conduct experiments with nine programs and the results show our approach is effective.

2 Technical Preliminaries

The program model used here is a single perfect or non-perfect loop nest and contains explicit information about which loop can be parallelized. An m-dimensional array X defines an m-dimensional polyhedron, each point of which can be denoted by an $m \times 1$ column vector. Assume the number of all the loops enclosing a reference of X is n, then the iteration space of this n-deep loop nest can be viewed as an n-dimensional polyhedron where each point is denoted by an $n \times 1$ column vector $\bar{I} = (i_1, i_2, \cdots, i_n)^T$; here, each i_k denotes a loop index. We call $\bar{I}$ as the *iteration vector*. We assume all loop bounds and subscript expressions are affine functions of enclosing loop indices and symbolic constants. Then, the reference can be denoted by $A\bar{I} + \bar{o}$, $m \times n$ matrix A is called as the *access matrix*, and $m \times 1$ column vector $\bar{o}$ is called as the *offset vector* [4]. Moreover, we assume at least one of the loops enclosing each reference of each array can be parallelized.

We use $span\{\bar{b}_1, \cdots, \bar{b}_l\}$ to denote the space spanned by vectors $\bar{b}_1, \cdots, \bar{b}_l$, Q to denote rational number field, Q^n to denote the space composed of all the $n \times 1$ rational number vectors, and $\dim(\Psi)$ to denote the dimension of vector space Ψ. We define $\bar{b} + \Psi = \{\bar{q} \mid \bar{q} = \bar{b} + \bar{p}, \bar{p} \in \Psi\}$, where $\bar{b}$ is an $n \times 1$ column vector, and Ψ is a vector space composed of $n \times 1$ column vectors.

3 Memory Locality Optimization

3.1 Data Space Decompositions

Given a reference of an m-dimensional array X, we assume the loop indices of all the loops enclosing this reference are $i_1, \cdots, i_n$ from outermost to innermost respectively. Moreover, we assume loops $i_{p_1}, \cdots, i_{p_v}$ can be parallelized. We denote the general form of the iteration space decomposition of this n-deep loop nest as

$B_X(i_{p_1},\cdots,i_{p_v})=i_{p_1}\bar{e}_{p_1}+\cdots+i_{p_v}\bar{e}_{p_v}+span\{\bar{e}_{p_{v+1}},\cdots,\bar{e}_{p_n},\bar{0}\}$, where $\bar{e}_{p_j}$ is an $n\times1$ unit column vector. We call $\{B_X(i_{p_1},\cdots,i_{p_v})|i_{p_1},\cdots,i_{p_v}\in Q\}$ as a *linear computation decomposition* of the iteration space of this n-deep loop nest and define its *parallelism* as v. We also call $B_X(i_{p_1},\cdots,i_{p_v})$ as the *iteration partition*.

Definition 1. Given an m-dimensional array X, a set of $m\times1$ column vectors $\bar{\delta},\bar{\gamma}_1,\cdots,\bar{\gamma}_v$, and a vector space Ω_X composed of $m\times1$ column vectors. Let $D_X(k_1,\cdots k_v)=\bar{\delta}+k_1\bar{\gamma}_1+\cdots+k_v\bar{\gamma}_v+\Omega_X$ ($v\geq0$). Assume $\dim(\Omega_X)=u$ and $\bar{\eta}_1,\cdots,\bar{\eta}_u$ are a basis of Ω_X. If $v=0$ or $v\neq0$ and $\forall 1\leq j\leq v$, $\bar{\gamma}_j$ can not be expressed as a linear combination of $\bar{\gamma}_1,\cdots,\bar{\gamma}_{j-1},\bar{\gamma}_{j+1},\cdots,\bar{\gamma}_v,\bar{\eta}_1,\cdots,\bar{\eta}_u$, then we call $\{D_X(k_1,\cdots,k_v)|k_1,\cdots,k_v\in Q\}$ as a *linear data decomposition* of the data space of array X and the *parallelism* of this linear data decomposition is v. We also call $D_X(k_1,\cdots k_v)$ as the *data partition*.

Assume the reference of array X is $A_X\bar{I}_X+\bar{o}_X$, where $\bar{I}_X=(i_1,\cdots,i_n)^T$, and Let $\bar{\gamma}_{p_j}=A_X\bar{e}_{p_j}$, then the data space accessed by iteration partition $B_X(i_{p_1},\cdots,i_{p_v})$ is $D_X(i_{p_1},\cdots,i_{p_v})=\bar{o}_X+i_{p_1}\bar{\gamma}_{p_1}+\cdots+i_{p_v}\bar{\gamma}_{p_v}+span\{\bar{\gamma}_{p_{v+1}},\cdots,\bar{\gamma}_{p_n},\bar{0}\}$.

If $\forall 1\leq j\leq v$, $\bar{\gamma}_{p_j}$ can not be expressed as a linear combination of $\bar{\gamma}_{p_1},\cdots,\bar{\gamma}_{p_{j-1}}$, $\bar{\gamma}_{p_{j+1}},\cdots,\bar{\gamma}_{p_n}$, then from definition 1 we know $\{D_X(i_{p_1},\cdots,i_{p_v})|i_{p_1},\cdots,i_{p_v}\in Q\}$ is a linear data decomposition of array X's data space and its parallelism is v. As the data partitions accessed by any two different iteration partitions are different, each iteration partition with its accessed data partition can be scheduled to the same processor to make the reference local.

If $\exists 1\leq j\leq v$, $\bar{\gamma}_{p_j}$ can be expressed as a linear combination of $\bar{\gamma}_{p_1},\cdots,\bar{\gamma}_{p_{j-1}}$, $\bar{\gamma}_{p_{j+1}},\cdots,\bar{\gamma}_{p_n}$. Let $0\leq v'<v$. Without loss of generality, we assume $\forall 1\leq j\leq v'$, $\bar{\gamma}_{p_j}$ can not be expressed as a linear combination of $\bar{\gamma}_{p_1},\cdots,\bar{\gamma}_{p_{j-1}},\bar{\gamma}_{p_{j+1}},\cdots,\bar{\gamma}_{p_n}$ while $\forall v'+1\leq j\leq v$, $\bar{\gamma}_{p_j}$ can. Let $D_X'(i_{p_1},\cdots,i_{p_{v'}})=\bar{o}_X+i_{p_1}\bar{\gamma}_{p_1}+\cdots+i_{p_{v'}}\bar{\gamma}_{p_{v'}}+span\{\bar{\gamma}_{p_{v'+1}},\cdots,\bar{\gamma}_{p_n},\bar{0}\}$, and then $\{D_X'(i_{p_1},\cdots,i_{p_{v'}})|i_{p_1},\cdots,i_{p_{v'}}\in Q\}$ is a linear data decomposition of array X's data space and its parallelism is v'. As $\forall i_{p_1},\cdots,i_{p_v}\in Q, D_X(i_{p_1},\cdots,i_{p_v})\subseteq D_X'(i_{p_1},\cdots,i_{p_{v'}})$ and $v'<v$, there exist the circumstances that more than one iteration partition accesses the same data partition, and therefore the data partitions have to be replicated to make the reference local.

Definition 2. Given a linear computation decomposition of the iteration space of a loop nest and a linear data decomposition of an array referenced in the loop nest that is got by the linear computation decomposition, we assume the parallelism of the linear computation decomposition is v and the parallelism of the linear data decomposition is f. Then we call the *replication degree* of the linear data decomposition relative to the linear computation decomposition is $v - f$.

We always hope the replication degree is as low as possible, because it will make the amount of data needed to be replicated as small as possible and will reduce the runtime overhead of maintaining the consistency of the replicated data.

3.2 Data Space Fusion

Given a loop nest with an m-dimensional array X's q references $A_{X^j} \bar{I}_{X^j} + \bar{o}_{X^j}$ ($1 \le j \le q$), assume $\bar{I}_{X^j}$ contains n_j elements and the parallelizable loops enclosing all references of array X are all same. Let the loop indices of these parallelizable loops are $i_1, \cdots, i_v$. $\forall 1 \le j \le q$, we define a position function ps_{X^j} for $\bar{I}_{X^j}$. The definition domain Δ_j of ps_{X^j} is composed of all the elements in $\bar{I}_{X^j}$, and ps_{X^j} takes the value of $1, \cdots, n_j$. $\forall i \in \Delta_j$, ps_{X^j} returns the position of i in $\bar{I}_{X^j}$. Let

$$B_{X^j}(i_1, \cdots, i_v) = i_1 \bar{e}^j_{ps_{X^j}(i_1)} + \cdots + i_v \bar{e}^j_{ps_{X^j}(i_v)} + span\{\{\bar{e}^j_1, \cdots, \bar{e}^j_{n_j}, \bar{0}\} - \{\bar{e}^j_{ps_{X^j}(i_1)}, \cdots, \bar{e}^j_{ps_{X^j}(i_v)}\}\}.$$

According to Section 3.1, we can know $\{B_{X^j}(i_1, \cdots, i_v) | i_1, \cdots, i_v \in Q\}$ is a linear computation decomposition of the iteration space of the n_j-deep loop nest, which is composed of all the loops enclosing reference $A_{X^j} \bar{I}_{X^j} + \bar{o}_{X^j}$. Therefore for each reference of array X, we can use the method presented in Section 3.1 to get the corresponding linear data decomposition of array X's data space. If $q > 1$, there will be more than one linear data decomposition of array X's data space. As the linear data decomposition of X should be unique (for we only consider static data decompositions in the paper and don't consider data redistribution), therefore we have to fuse all those linear data decompositions into a unique one. Assume the fused unique linear data decomposition is $\{D_X(i_{h_1}, \cdots, i_{h_{v'}}) | i_{h_1}, \cdots, i_{h_{v'}} \in Q\}$, where $0 \le v' \le v$ and $1 \le h_1, \cdots, h_{v'} \le v$. Then it should satisfy the condition: $\forall i_1, \cdots, i_v \in Q, 1 \le j \le q$, there has $\bar{o}_{X^j} + A_{X^j} B_{X^j}(i_1, \cdots, i_v) \subseteq D_X(i_{h_1}, \cdots, i_{h_{v'}})$. It is possible that many linear data decompositions satisfy the condition, and we want to acquire the one with the highest parallelism, which will make the replication degree the lowest. We give the algorithm of finding the fused unique linear data decomposition that satisfies the condition and has the highest parallelism in Fig. 1.

From the algorithm we can see that as long as the loop indices of the parallelizable loops enclosing all references of given arrays are same, the algorithm can be used to get the fused unique linear data decompositions for these arrays respectively, otherwise we can make them same by aligning parallelizable loops. As the limitation of the

Input: a loop nest with array X's q references $A_{X^1} \bar{I}_{X^1} + \bar{o}_{X^1}, \cdots, A_{X^q} \bar{I}_{X^q} + \bar{o}_{X^q}$. The parallelizable loops enclosing all references of array X are all same.

Output: array X's fused unique linear data decomposition.

Assume $\bar{I}_{X^j}$ contains n_j elements, the loop indices of these parallelizable loops are $i_1, \cdots, i_v$
and $\bar{\gamma}_{jk}$ is the kth column of A_{X^j}, where $1 \le j \le q, 1 \le k \le n_j$.

$\theta = \{\bar{0}\}$;

DO $j = 1, q$

$\quad \theta = \theta \cup (\{\bar{\gamma}_{j1}, \cdots, \bar{\gamma}_{jn_j}\} - \{\bar{\gamma}_{j(ps_{X^j}(i_1))}, \cdots, \bar{\gamma}_{j(ps_{X^j}(i_v))}\})$

ENDDO

DO $j = 2, q$

$\quad$ DO $k = 1, v$

$\quad\quad \theta = \theta \cup \{\bar{\gamma}_{j(ps_{X^j}(i_k))} - \bar{\gamma}_{1(ps_{X^1}(i_k))}\}$

$\quad$ ENDDO

ENDDO

DO $j = 2, q$

$\quad \theta = \theta \cup \{\bar{o}_{X^j} - \bar{o}_{X^1}\}$

ENDDO

$\Omega_X = span\{\theta\}$; $D_X(i_1, \cdots, i_v) = \bar{o}_{X^1} + i_1 \bar{\gamma}_{1(ps_{X^1}(i_1))} + \cdots + i_v \bar{\gamma}_{1(ps_{X^1}(i_v))} + \Omega_X$;

Let $\dim(\Omega_X) = u$ and $\bar{\eta}_1, \cdots, \bar{\eta}_u$ are a basis of Ω_X ;

Let $0 \le v' \le v$, without loss of generality, assume $\forall 1 \le j \le v'$, $\bar{\gamma}_{1(ps_{X^1}(i_j))}$ can not be expressed

as a linear combination of $\bar{\gamma}_{1(ps_{X^1}(i_1))}, \cdots, \bar{\gamma}_{1(ps_{X^1}(i_{j-1}))}, \bar{\gamma}_{1(ps_{X^1}(i_{j+1}))}, \cdots, \bar{\gamma}_{1(ps_{X^1}(i_v))}, \bar{\eta}_1, \cdots, \bar{\eta}_u$, while

$\forall v' + 1 \le j \le v$, $\bar{\gamma}_{1(ps_{X^1}(i_j))}$ can; let $\Omega'_X = span\{\Omega_X \cup \{\bar{\gamma}_{1(ps_{X^1}(i_{v'+1}))}, \cdots, \bar{\gamma}_{1(ps_{X^1}(i_v))}\}\}$ and

$D'_X(i_1, \cdots, i_{v'}) = \bar{o}_{X^1} + i_1 \bar{\gamma}_{1(ps_{X^1}(i_1))} + \cdots + i_{v'} \bar{\gamma}_{1(ps_{X^1}(i_{v'}))} + \Omega'_X$;

Return $(\{D'_X(i_1, \cdots, i_{v'}) | i_1, \cdots, i_{v'} \in Q\})$

Fig. 1. The algorithm of finding the fused unique linear data decomposition

space, the detailed steps of the alignment of parallelizable loops are omitted here. In
the following discussions, we assume the loop indices of the parallelizable loops enclosing all references of all arrays are same.

3.3 Determining Memory Locality Optimizing Rows

Given an m-dimensional array X's multiple references, let array X's fused unique
linear data decomposition gotten by the algorithm given in Fig. 1. is
$\{D_X(i_1, \cdots, i_v) = \bar{o}_X + i_1 \bar{\gamma}_1 + \cdots + i_v \bar{\gamma}_v + \Omega_X | i_1, \cdots, i_v \in Q\}$. The parallelism of this
linear data decomposition is v. We determine the MLORs for X in the following.

If $v = 0$, array X has to be replicated over all dimensions of the processor space to make all the accesses to X local. Therefore, we determine X's MLORs are null. In the following, we will discuss the circumstance of $v \neq 0$.

Assume $\dim(\Omega_X) = u$ and $\overline{\eta}_1, \cdots, \overline{\eta}_u$ are a basis of Ω_X. If $u + v < m$, we add $m - u - v$ column vectors $\overline{\eta}_{u+1}, \cdots, \overline{\eta}_{m-v}$ to make $\overline{\gamma}_1, \cdots, \overline{\gamma}_v, \overline{\eta}_1, \cdots, \overline{\eta}_{m-v}$ a basis of Q^m. Let $\Omega'_X = span\{\overline{\eta}_1, \cdots, \overline{\eta}_{m-v}, \overline{0}\}$. We can find an orthogonal basis of Ω'_X's orthogonal space, which is assumed as $\overline{\beta}_1, \cdots, \overline{\beta}_v$. Assume $\overline{\gamma}_k = z_{k1} \overline{\beta}_1 + \cdots + z_{kv} \overline{\beta}_v + z_{k(v+1)} \overline{\eta}_1 + \cdots + z_{km} \overline{\eta}_{m-v}$. Let $C = (\overline{\beta}_1, \cdots, \overline{\beta}_v)$,

$$P = \begin{pmatrix} 1/\|\overline{\beta}_1\|^2 & \cdots & 0 \\ \vdots & \ddots & \vdots \\ 0 & \cdots & 1/\|\overline{\beta}_v\|^2 \end{pmatrix}, \quad H = \begin{pmatrix} z_{11} & \cdots & z_{v1} \\ \vdots & \ddots & \vdots \\ z_{1v} & \cdots & z_{vv} \end{pmatrix} \text{ and } \overline{d}_j \text{ be the } j \text{ th row of}$$

$H^{-1}P(C)^T$. $\forall 1 \leq j \leq v$, we multiply $\overline{d}_j$ by a smallest positive integer to make all the elements in $\overline{d}_j$ become integers, and we finally determine array X's MLORs are $\overline{d}_1, \cdots, \overline{d}_v$ and the offset needed by the affine data transformation is $\overline{o}_X$.

4 Cache Locality Optimization and the Determination of the Final Linear Data Transformation Matrices

We use the approach presented in [9] to optimize cache spatial locality and continue to use the notations used in Section 3.3 in the following discussion. Given a reference of array X, assume the innermost loop index occurring in the subscript expressions of this reference is ζ. Without loss of generality, we determine the CLORs and the final linear data transformation matrix N for this reference in the following.

1. if $v = 0$, use the approach presented in [9] to find the data transformation matrix M that can optimize the spatial locality of this reference. Let the rows of M from the first to the last be $\overline{p}_1, \cdots, \overline{p}_m$ respectively. We then determine $\overline{p}_1, \cdots, \overline{p}_m$ are the CLORs and N's rows from the first to the last are $\overline{p}_1, \cdots, \overline{p}_m$ respectively with its corresponding affine data transformation Nx.

2. if $v \neq 0$ and $\forall 1 \leq j \leq v$, $\zeta \neq i_j$, find $m - v$ integer row vectors $\overline{b}_1, \cdots, \overline{b}_{m-v}$ to make $(\overline{d}_1)^T, \cdots, (\overline{d}_v)^T, (\overline{b}_1)^T, \cdots, (\overline{b}_{m-v})^T$ a basis of Q^m and use them to compose a non-singular square matrix E, where its first $m - v$ rows from the first to the last are $\overline{b}_1, \cdots, \overline{b}_{m-v}$ respectively and its last v rows are $\overline{d}_1, \cdots, \overline{d}_v$. Use $E(x - \overline{o}_X)$ to do affine transformation on the reference and let $\overline{\omega}$ be the array composed of the first $m - v$ dimensions of the transformed reference. We use the approach presented in [9] to find the $(m - v) \times (m - v)$ non-singular data transformation matrix R that can

optimize the spatial locality of array ϖ. Let $M = R\left((\bar{b}_1)^T \quad \cdots \quad (\bar{b}_{m-v})^T\right)^T$, and the rows of M from the first to the last be $\bar{\rho}_1, \cdots, \bar{\rho}_{m-v}$ respectively. We determine $\bar{\rho}_1, \cdots, \bar{\rho}_{m-v}$ are the CLORs and N's rows from the first to the last are $\bar{\rho}_1, \cdots, \bar{\rho}_{m-v}$, $\bar{d}_1, \cdots, \bar{d}_v$ respectively with its corresponding affine data transformation $N(x - \bar{o}_X)$.

3. if $v \neq 0$ and $\exists 1 \leq j \leq v$, $\zeta = i_j$, find $m - v$ integer row vectors $\bar{b}_1, \cdots, \bar{b}_{m-v}$ to make $(\bar{d}_1)^T, \cdots, (\bar{d}_v)^T, (\bar{b}_1)^T, \cdots, (\bar{b}_{m-v})^T$ a basis of Q^m and use them to compose a non-singular square matrix E, where its first $m - v + 1$ rows from the first to the last are $\bar{d}_j, \bar{b}_1, \cdots, \bar{b}_{m-v}$ respectively and its last $v - 1$ rows are $\bar{d}_1, \cdots, \bar{d}_{j-1}, \bar{d}_{j+1}, \cdots, \bar{d}_v$. Use $E(x - \bar{o}_X)$ to do affine transformation on the reference and let ϖ be the array composed of the first $m - v + 1$ dimensions of the transformed reference. We use the approach presented in [9] to find the $(m - v + 1) \times (m - v + 1)$ non-singular data transformation matrix that can optimize the spatial locality of array ϖ, and get matrix R by replacing the first row of the above data transformation matrix with a $1 \times (m - v + 1)$ row vector where its first element is numeral one and all the others are zeros. Let $M = R\left((\bar{d}_j)^T \quad (\bar{b}_1)^T \quad \cdots \quad (\bar{b}_{m-v})^T\right)^T$, and the rows of M from the second to the last be $\bar{\rho}_1, \cdots, \bar{\rho}_{m-v}$ respectively. We determine $\bar{d}_j, \bar{\rho}_1, \cdots, \bar{\rho}_{m-v}$ are the CLORs and N's rows from the first to the last are $\bar{d}_j, \bar{\rho}_1, \cdots, \bar{\rho}_{m-v}$, $\bar{d}_1, \cdots, \bar{d}_{j-1}, \bar{d}_{j+1} \cdots, \bar{d}_v$ respectively with its corresponding affine data transformation $N(x - \bar{o}_X)$. In this circumstance, $\bar{d}_j$ is both the MLOR and the CLOR.

5 Experimental Results

We will present performance results for the following nine programs: *matmult* is a matrix-multiplication routine; *syr2k* is a banded matrix update routine from BLAS; *stencil* is a five-point stencil computing code; *htribk* is a test program from Eispack; *mxm*, *cholsky* and *cfft2d1* are three test programs from Spec92/NASA benchmark suite; *mxmxm* is a routine from [10] that multiplies three matrices; *transpose* is a routine from a large computational chemistry application [13]. We conduct experiments with FORTRAN versions of these programs. For each program, we experiment with four different versions: the version with parallelization analysis only (denoted by o); the version with parallelization analysis and memory locality optimization using the method presented in Section 3 (denoted by o+m); the version with parallelization analysis and cache locality optimization using the approach presented in [9] (denoted by o+c); and the version with parallelization analysis and memory and cache locality optimization using the approach presented in this paper (denoted by o+mc).

We report speedups for up to 64 processors on some distributed memory machine. This machine has 32 nodes and each node has two processors. Shared memory architecture is adopted inside each node while distributed memory architecture is adopted among nodes. Table 1 gives speedups of the test programs in different versions.

Table 1. The speedups of the test programs in different versions

pn	matmult				syr2k				stencil			
	o	o+m	o+c	o+mc	o	o+m	o+c	o+mc	o	o+m	o+c	o+mc
2	1.69	1.66	24.3	25.4	3.27	2.15	35.6	40.3	1.20	1.27	5.34	7.39
4	3.32	3.32	46.5	50.4	6.39	4.26	57.3	75.8	2.09	2.40	6.20	12.5
8	6.61	6.63	84.9	101	11.9	8.38	69.6	136	2.97	4.26	5.79	22.7
16	12.9	12.9	134	201	19.1	16.3	57.3	224	2.91	7.14	3.96	33.9
32	24.4	25.5	151	394	25.1	31.2	45.6	322	2.33	10.9	2.69	43.5
64	40.5	49.8	128	544	22.2	58.6	27.6	427	1.52	14.9	1.59	50.7

pn	htribk				mxm				cholsky			
	o	o+m	o+c	o+mc	o	o+m	o+c	o+mc	o	o+m	o+c	o+mc
2	1.89	1.91	11.3	11.5	2.05	2.03	10.9	10.1	0.17	2.23	0.24	12.2
4	3.64	3.69	19.5	28.5	4.05	4.03	15.7	19.9	0.13	4.30	0.15	21.1
8	7.65	7.84	58.5	82.3	7.68	8.14	27.5	44.2	0.08	7.03	0.09	33.1
16	13.6	12.4	63.1	161	13.3	15.9	38.4	84.7	0.04	10.1	0.05	46.2
32	17.0	23.8	49.3	302	18.8	31.3	37.0	156	0.02	15.5	0.02	57.8
64	18.4	44.9	26.5	514	18.0	59.8	21.9	289	0.01	28.9	0.01	57.8

pn	cfft2d1				mxmxm				transpose			
	o	o+m	o+c	o+mc	o	o+m	o+c	o+mc	o	o+m	o+c	o+mc
2	0.27	3.32	0.72	3.32	8.07	2.07	26.6	27.4	0.50	1.47	0.71	11.8
4	0.24	4.39	0.50	4.24	5.90	4.03	49.2	54.6	0.57	2.81	0.69	23.5
8	0.19	5.59	0.30	5.59	8.83	8.24	82.7	108	0.52	5.05	0.56	41.7
16	0.13	6.47	0.17	9.46	14.9	16.2	108	212	0.38	9.00	0.39	92.2
32	0.07	7.23	0.09	15.4	30.1	32.3	108	412	0.24	14.0	0.24	179
64	0.04	7.23	0.05	20.5	39.2	64.6	75.3	761	0.12	19.8	0.13	342

pn denotes processor number. [matmult-2048×2048 matrices; syr2k-1024×1024 matrices; stencil-2048×2048 matrices; htribk-1024×1024 matrices; mxm-1024×1024 matrices; cholsky-the size parameters are set to 1000; cfft2d1-1024×1024 matrices and 2×1024 arrays; mxmxm-1024×1024 matrices; transpose-4096×4096 matrices]

As the o versions have not been optimized for memory and cache locality, their performance is bad for high remote and local memory access overheads. As the o+m versions have been optimized for memory locality, the remote memory access overheads will be low. But as they have not been optimized for cache locality, high local memory access overheads may prevent their performance from improving further. Although the o+c versions have been optimized for cache locality to make local memory access overheads low, there may exist high remote memory access overheads for bad memory locality. Moreover, the remote memory access overheads may increase with the increase of processor number. When the negative effect of remote memory access overheads on the performance exceeds the positive effect of cache locality optimization, the performance will get worse instead. As the o+mc versions have been optimized for memory and cache locality simultaneously, there will be low remote and local memory access overheads and therefore their performance should be better than the other three versions' performance. From Table 1 we can see that in all nine programs, the o+mc versions' performance is the best in four versions and for most of the nine programs, with the increase of processor number, the gap of optimizing effect between the o+mc versions and the other three versions becomes larger.

6 Conclusions

To solve the problems of memory and cache locality optimization on distributed memory multiprocessors, we present an approach of optimizing memory and cache locality simultaneously using data transformations in this paper. We first determine MLORs for memory locality, then determine CLORs for cache locality and combine MLORs and CLORs in some order to form the final linear data transformation matrices that can optimize memory and cache locality simultaneously at last. The experimental results show our approach is effective on distributed memory multiprocessors.

References

1. T.-S. Chen and C.-Y. Chang. Skewed data partition and alignment techniques for compiling programs on distributed memory multicomputers. The Journal of Supercomputing, 21(2): 191-211, 2002.
2. W.-L. Chang, C.-P. Chu and J.-H. Wu. Communication-free alignment for array references with linear subscripts in three loop index variables or quadratic subscripts. The Journal of Supercomputer, 20(1): 67-83, 2001.
3. XIA Jun, YANG Xue-Jun and DAI Hua-Dong. Data space fusion based approach for effective alignment of computation and data. In Proc. of 5th International Workshop on Advanced Parallel Processing Technology, Xiamen, China, pp. 215-225, 2003.
4. M. Wolf and M. Lam. A data locality optimizing algorithm. In SIGPLAN91 Conference on Programming Language Design and Implementation, Toronto, Canada, pp. 30-44, 1991.
5. K. McKinley, S. Carr and C.W. Tseng. Improving data locality with loop transformation. ACM Transactions on Programming Languages and Systems, 18(4): 424-453, 1996.
6. A.J.C. Bik and P.M.W. Knijnenburg. Reshaping Access Patterns for Improving Data Locality. In Proc. of 6th Workshop on Compilers for Parallel Computers, 1996.
7. P. Clauss and B. Meister. Automatic memory layout transformations to optimize spatial locality in parameterized loop nests. ACM SIGARCH Computer Architecture News, 28(1): 11-19, 2000.
8. M. Kandemir, A. Choudhary, N. Shenoy, P. Banerjee and J. Ramanujam. A hyperplane based approach for optimizing spatial locality in loop nests. In Proc. of 1998 ACM International Conference on Supercomputing (ICS'98), Melbourne, Australia, pp. 69-76, 1998.
9. XIA Jun, YANG Xue-Jun, ZENG Li-Fang and ZHOU Hai-Fang. A projection-delamination based approach for optimizing spatial locality in loop nests. Chinese Journal of Computers, 26(5):539-551, 2003.
10. M. Cierniak and W. Li. Unifying data and control transformations for distributed shared memory machines. In SIGPLAN95 Conference on Programming Language Design and Implementation, La Jolla, CA, pp. 205-217, 1995.
11. M. Kandemir, A. Choudhary, J. Ramanujam and P. Banerjee. A matrix-based approach to global locality optimization. Journal of Parallel and Distributed Computing, 58:190-235, 1999.
12. M. Kandemir, P. Banerjee, A. Choudhary, J. Ramanujam and E. Ayguade. Static and dynamic locality optimizations using integer linear programming. IEEE Transactions on Parallel and Distributed Systems, 12(9): 922-940, 2001.
13. High Performance Computational Chemistry Group. NWChem: A computational chemistry package for parallel computers, version 1.1. Pacific Northwest Laboratory, 1995.

A Fetch Policy Maximizing Throughput and Fairness for Two-Context SMT Processors

Caixia Sun, Hongwei Tang, and Minxuan Zhang

College of Computer, National University of Defense Technology,
Changsha 410073, Hunan, P.R. China
{cxsun_nudt, hwtang_nudt}@yahoo.com.cn
mxzhang@nudt.edu.cn

Abstract. In Simultaneous Multithreading (SMT) processors, co-scheduled threads share the processor's resources, but at the same time compete for them. A thread missing in L2 cache may hold a large number of resources which other threads could be using to make forward progress. And as a result, the overall performance of SMT processors is degraded. Currently, many instruction fetch policy focus on this problem. However, these policies are not perfect, and each has its own disadvantages. Especially, these policies are designed for processors implementing any ways simultaneous multithreading. The disadvantages of these policies may become more serious when they are used in two-context SMT processors.

In this paper, we propose a novel fetch policy called RG-FP (Resource Gating based on Fetch Priority), which is specially designed for two-context SMT processors. RG-FP combines reducing fetch priority with controlling shared resource allocation to prevent the negative effects caused by loads missing in L2 cache. Simulation results show that our RG-FP policy outperforms previously proposed fetch policies for all types of workloads in both throughput and fairness, especially for memory bounded workloads. Results also tell that our policy shows different degrees of improvement over other fetch policies. The increment over PDG is greatest, reaching 41.8% in throughput and 50.0% in Hmean on average.

1 Introduction

Simultaneous Multithreading (SMT) processors [1,2,3] improve performance by allowing running instructions from several threads simultaneously at a single cycle. Co-scheduled threads share some resources, such as issue queues, physical registers, and functional units. The way of allocating shared resources among the threads will affect the overall performance of SMT processors. Currently, shared resources allocation in SMT processors is dynamically decided by the instruction fetch policy.

In SMT processors, the number of shared resources is limited, so if a thread holds critical resources for a long time, other threads may run slower than they could or even stall because of lack of resources. A load missing in L2 cache usually causes this

J. Cao, W. Nejdl, and M. Xu (Eds.): APPT 2005, LNCS 3756, pp. 13–22, 2005.
© Springer-Verlag Berlin Heidelberg 2005

happen. Many instruction fetch policies have been proposed to address this problem, some of which are well known, such as STALL, FLUSH [4], DG and PDG [5]. A newly proposed fetch policy called DWarn [6] is also very efficient to handle L2 cache misses. However, these policies are not perfect, and each has its own disadvantages. Especially, these policies are very general, that is to say, they are designed for processors implementing any ways simultaneous multithreading. The disadvantages of these policies may become more serious when being used in two-context SMT processors. For example, FLUSH deallocates all the resources allocated to the thread with L2 cache misses and makes them available to the other threads. It will produce resource under-use when the resources deallocated are not required by any other thread. For two-context SMT processors, the pressure on shared resources is not high, so resource under-use will be more serious.

It is well known that many commercial processors implement two-way simultaneous multithreading, like Intel Xeon [7] and IBM Power5 [8]. Furthermore, in SMT processors with more than two hardware contexts, it is very common that only two threads are running together because there are not enough thread-level parallelisms. Therefore, we believe it is very important to specially design a fetch policy for two-context SMT processors. Certainly, when there are more than two hardware contexts, such a fetch policy can also be combined with other policies to achieve better performance.

In this paper, we propose a novel fetch policy called RG-FP, which is specially used in two-context SMT processors. RG-FP is built on top of ICOUNT, and combines reducing fetch priority with controlling shared resource allocation to prevent the effects of loads that miss in L2 cache. In out policy, a thread with cache misses would not be stalled immediately, but executes at a lower fetch priority. Thus, resource underutilization and idle cycles are reduced greatly. Furthermore, resources are allocated between threads based on the fetch priority, and a thread with cache misses will be gated when it attempts to exceed its assigned resources. In this way, we can prevent resource monopolization.

This paper is organized as follows. Section 2 introduces the related work. In Section 3 we detail our policy. Sections 4 and 5 present the methodology and the results. Finally, concluding remarks are given in Section 6.

2   Related Work

ICOUNT [2] prioritizes threads with few instructions in decode, rename, and the instruction queues. It presents good results for threads with high ILP(Instruction Level Parallelism). However, ICOUNT could not address the problems caused by L2 cache misses. As long as the fetch priority of a thread is the highest, ICOUNT will fetch instructions from this thread, even if this thread is experiencing L2 cache misses. The reason that we introduce ICOUNT here is the following policies that handle L2 cache misses are all based on it.

STALL [4] attempts to prevent the threads with L2 cache misses occupying most of available resources. It detects that a thread has a pending L2 miss and stalls fetching from this thread to avoid resource abuse. However, L2 miss detection may be so late that shared resources have been clogged. Furthermore, it is possible that the resources allocated to a thread are not required by any other thread, so stalling fetching from this thread will produce resource under-use. Obviously, for two-context SMT processors, it is easier to cause resource abuse or produce resource under-use.

FLUSH [4] is an extension of STALL. It tries to correct the case in which an L2 miss is detected too late by deallocating all the resources of the offending thread, making them available to the other executing threads. However, compared to STALL, it is more likely to produce resource under-use for FLUSH. Furthermore, extra fetch and power are required to redo the work for the flushed thread. For two-context SMT processors, the pressure on shared resources is not high, so resource under-use will be more serious.

Data Gating (DG) [5] attempts to reduce the effects of loads missing in the L1 data cache by stalling threads on each L1 data miss. However, there is not resource abuse when an L1 miss does not cause an L2 miss. Thus, to stall a thread every time it experiences an L1 miss may be too severe. For two-context SMT processors, another problem is that it is very easy to produce idle cycles, because the probability of two threads simultaneously experiencing L1 cache misses is high.

Predictive Data Gating (PDG) [5] works like STALL, that is, it prevents a thread from fetching instructions as soon as a cache miss is predicted. By using a miss predictor, they avoid detecting the cache miss too late, but resource under-use still exists. Furthermore, cache misses prove to be hard to predict accurately [9]. Like DG, it is also very easy to produce idle cycles for two-context SMT processors.

DCache Warn (DWarn) [6] uses a hybrid mechanism. When less than two threads run, the priority of the thread missing in L1 data cache miss is reduced. After that, if the L1 miss turns to an L2 miss, its thread is gated. When the number of execution threads is higher than 2, DWarn only reduces the priority of the threads with cache misses. For two-context SMT processors, the problem of DWarn is there exist idle cycles of the processor when two threads are all gated. Furthermore, although the fetch priority of threads with cache misses is reduced, these threads are stalled until L2 miss is declared, which may still be too late to prevent shared resources being clogged.

The main problems of fetch policies previously introduced are summed up as follows: First, not to effectively prevent shared resources being monopolized by threads with pending L2 misses; Second, to produce resource under-use when preventing a thread occupying resources which are not required by any other thread; Third, to produce idle cycles of the processor when all threads are stalled because of cache misses; Fourth, the most important one, the three problems above may become more serious when these policies are used in two-context SMT processors.

3   RG-FP Fetch Policy

RG-FP is specially designed for two-context SMT processors. It attempts to prevent the effects of loads that miss in L2 cache, and at the same time to avoid the problems of the fetch policies above.

3.1 Basic Idea

RG-FP is built on top of ICOUNT. Furthermore, it is based on the combination of two ideas, namely, reducing fetch priority and controlling shared resources allocation.

Reducing Fetch Priority. RG-FP supports three priority levels for each thread, from Level 1 to Level 3. Level 1 is the highest and Level 3 is the lowest. At the beginning, all threads are at the highest priority, Level 1. If a thread is experiencing an L1 data cache miss, its fetch priority is reduced to Level 2. After that, if the L1 cache misses finally turns to an L2 cache miss, the fetch priority of the thread is reduce further to Level 3. The transition of fetch priority is detailed in Figure 1. The threads at the same priority level are sorted by ICOUNT.

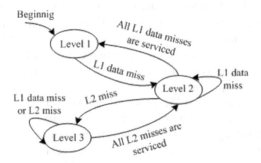

Fig. 1. Transition of the fetch priority

Controlling Shared Resource Allocation. The number of resources allocated to threads with cache misses is restricted to a certain value and threads with no outstanding cache misses can use as many resources as that are available. If a thread with cache misses is exceeding its assigned resources, it is gated until it releases some of the allocated resources or it is allowed to occupy more resources.

By reducing the fetch priority of threads with cache misses instead of stalling fetching from them, on one hand, the opportunity of keeping the fetch bandwidth fully used is given the thread with no outstanding cache misses, which implies the improvement of throughput; on the other hand, threads with pending cache misses are not stalled, so resource under-use and idle cycles of the processor are reduced greatly.

However, resources monopolization may still happen if we only reduce the fetch priority. Therefore, the idea of controlling resource allocation [10] is added to our policy. Because the resources allocated to threads with cache misses are limited, it is impossible that shared resources are monopolized by these threads. Furthermore, controlling resource allocation between threads may achieve a better throughput-fairness balance [10,11].

3.2 Resource Allocation Mechanism

We can use two methods to allocate shared resources between threads: static resource allocation (SRA) and dynamic resource allocation (DRA). In our simulations, we implement these two methods respectively. In the rest of this paper, we call RG-FP using static resource allocation RG-FP-S, and RG-FP using dynamic resource allocation RG-FP-D.

In static resource allocation, the number of resources allocated to threads at Level 2 and Level 3 is fixed, all equal to T divided by N, where T is the total number of some resource and N is the number of running threads. Because we only talk about two-context SMT processors, N is two.

In dynamic resource allocation, shared resources are dynamically allocated to threads based on each thread's fetch priority. The number of some resource allocated to thread i (i=0, 1) M_i is given in the equation (1), where T is the total number of some resource and PL_i is the fetch priority level of thread i.

$$M_i = \frac{PL_i}{PL_i + PL_{1-i}} * T \tag{1}$$

Table 1 shows an example of the resource allocation for all cases. We can see that in static resource allocation, the thread with cache misses can only occupy one half of some resource at most. In dynamic resource allocation, the thread with lower priority can borrow resources from the thread with higher priority, and as a result, the former can use the resources not required by the latter to reduce resource under-use.

Table 1. Resource allocation values for a 32-entry resource. "-" represents the thread can use as many resources as that are available.

		SRA		DRA	
PL_0	PL_1	M_0	M_1	M_0	M_1
1	1	-	-	-	-
1	2	-	16	-	21
1	3	-	16	-	24
2	1	16	-	21	-
2	2	16	16	16	16
2	3	16	16	13	19
3	1	16	-	24	-
3	2	16	16	19	13
3	3	16	16	16	16

3.3 Implementation

To implement RG-FP, each thread requires two cache miss counters, which are used to track L1 data misses L2 misses respectively, and 5 resources usage counters. Each instruction occupies an active list entry and maybe a physical register before committing. It uses an entry in the issue queues if its operands are not ready, and also require a functional unit. But each thread can have its own active list and functional units are pipelined. Therefore we only need to restrict the usage of issues queues and physical registers by threads with cache misses. There are three kinds of issue queues: integer, fp and load/store, so each thread requires three issue queues usage counters. Two more resource usage counters are required to track physical registers (integer and fp) per thread. The additional complexity required to introduce these counters depends on the particular implementation, but we do not expect it to be more complex than other hardware counters already present in most processors [10].

L1 data miss counters are incremented every time a thread experiences an L1 cache miss and decremented when the data caches fill occurs. L2 miss counters are incremented every time an L1 miss turns to an L2 miss and decremented when L2 cache fill occurs. If the L2 miss counter of a thread is nonzero, this thread is at Level 3, otherwise if the L1data miss counter is nonzero, it is at Level 2. Only when the L1 data miss counter and the L2 miss counter are all zero, is the thread at Level 1. Issue queues usage counters are incremented in the decode stage and decremented when an instructions is issued for execution. Physical registers usage counters are incremented in the decode stage and decremented in the commit stage.

In RG-FP, we use ICOUNT2.8 as the basic fetch policy. Each cycle, RG-FP fetches as many instructions as possible from the thread with higher priority, then fill in with instructions from the other thread, up to eight totally. If a thread at Level 2 or Level 3 is allocating more resources, it is fetch-stalled.

Now, we concern how to implement resource allocation between threads. For SRA, the number of resources allocated threads with cache misses is fixed, so extra circuit is not needed. For DRA, simple control logic is needed. DRA can be implemented in two ways. One is using combinational circuit to implement equation (1). The inputs of this circuit include fetch priority of each thread and the total number of some resource. This circuit gives the number of resources allocated to each thread with cache misses. The second way is using a direct-mapped table indexed by fetch priority of each thread. This table has 9 entries, as shown in Table 1. By searching this table, we can acquire the number of resources allocated to each thread with cache misses.

4 Methodology

Table 2 summarizes the benchmarks used in our simulations. All benchmarks are taken from the SPEC2000 suite [12] and use the reference data sets as inputs. It is time-consuming to simulate the complete SPEC benchmark suit. So we follow the idea proposed in [13] to run the most representative 300 million instruction segment of each benchmark. Benchmarks are divided into two groups based on their cache behaviors: those experiencing between 0.02 and 0.12 L2 cache misses per instruction, on average, over the simulated portion of the code are considered memory-intensive applications, and the rest have lower miss rates and higher inherent ILP. Table 3 lists the multithreaded workloads used in our simulations. All of the simulations in this paper either contain threads all from the first group (the MEM workloads in Table 3), or all from the second group (ILP), or an equal mix from each group (MIX).

Execution is simulated on an out-of-order superscalar processor model derived from SMTSIM [14]. The simulator models all typical sources of latency, including caches, branch mispredictions, TLB (Translation Lookaside Buffer) misses, etc. It also carefully models execution down the wrong path between branch misprediction and branch misprediction recovery. The baseline configuration of our simulator is shown in Table 4.

Table 2. Benchmarks used

Type	Benchmark
MEM	mcf, twolf, vpr, parser, ammp, applu, art, swim
ILP	aspi, fma, eon, gcc, gzip, vortex, crafty, bzip2

Table 3. Multithreaded Workloads used

Type	Applications
ILP	{gzip, bzip2}, {gcc, aspi}, {vortex, fma}, {eon, crafty}, {gzip, vortex}, {aspi, bzip2}, {gcc, crafty}, {fma, eon}
MIX	{gzip, vpr}, {gcc, ammp}, {art, vortex}, {fma, parser}, {aspi, twolf}, {crafty, art}, {bzip2, swim}, {eon, applu}
MEM	{mcf, vpr}, {ammp, parser}, {twolf, art}, {mcf, swim}, {vpr, applu}, {ammp, art}, {parser, twolf}, {swim, applu}

Table 4. Baseline configuration of the simulator

Parameter	Value
Fetch Width	8 instructions per cycle
Basic Fetch Policy	ICOUNT2.8
Instruction Queues	32 int, 32 fp, 32 load/store
Functional Units	6 int, 3 fp, 4 load/store
Renaming Physical Registers	100 int, 100 fp
Active List Entries	256 per thread
Branch Predictor	2K gshare
Branch Target Buffer	256 entries, 4-way associative
RAS	256 entries
L1I cache, L1D cache	64KB, 2-way, 64-bytes lines, 1 cycle access
L2 cache	512KB, 2-way, 64-bytes lines, 10 cycles latency
Main Memory Latency	100 cycles

We use two metrics to make a fair comparison: IPC and the Harmonic Mean (Hmean). Just as stated in [4], IPC may be a questionable metric if a fetch policy favors high IPC threads. Hmean is the harmonic mean of the relative IPC of the threads in a workload and it attempts to avoid artificial improvements achieved by giving more resources to threads with high ILP.

5   Results

Because we implement two mechanisms to allocate shared resources in our policy, we will first compare RG-FP-S with RG-FP-D. After that, we compare our policy with some fetch policies used widely, including STALL, FLUSH, DG, PDG and DWarn.

5.1 RG-FP-S vs. RG-FP-D

Figure 2 shows the throughput/Hmean results of RG-FP-D compared to RG-FP-S. We can see that RG-FP-D outperforms RG-FP-S in both throughput and fairness, by 3.3% and 4.2% on average, respectively. This is because when deciding the number of resources that a thread can use, static resource allocation only examines the cache behaviors of this thread, and ignores the other one. While dynamic resource allocation examines the two threads simultaneously. Therefore, DRA can make better use of the shared resources.

The results also show that the improvement of RG-FP-D over RG-FP-S is higher for MIX workloads, especially in fairness. The key point is that from Table 1, we can see that only when the fetch priority of two threads is not equal, are there differences between SRA and DRA. Table 5 gives how often threads in two-thread workloads are either at the same fetch priority or at different fetch priority. We can see that for MIX workloads, it is more common that two threads are at different priority levels than ILP and MEM workloads.

Table 5. Distribution of threads in fetch priority for two-thread workloads

Workload Type	The same level	Different Level
ILP	53.8	46.2
MIX	24.8	75.2
MEM	64.4	35.6

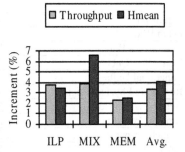

Fig. 2. Throughput/Hmean results of RG-FP-D compared to RG-FP-S

5.2 RG-FP vs. Other Policies

Figure 3(a) shows the throughput improvement of RG-FP over other policies. We only give the results of RG-FP-S. Combining the results in Figure 2, we can easily acquire the improvement of RG-FP-D over other policies.

Results show that RG-FP-S outperforms each of the other policies, especially PDG, by 41.8% on average, and DG takes second place, by 20.7% on average. This is because in DG, a thread is gated on each L1 data miss. In fact, the pressure on shared resources is not high in two-thread workloads. Therefore, resource under-use is very serious in DG. In our policy, a thread with cache misses would not be stalled immediately, but executes at a lower priority, and is gated only when this thread attempts to exceed its assigned resources. Thus, resource underutilization is reduced greatly. For PDG, it suffers the same problem as DG. In addition, cache misses prove to be hard to predict accurately, which reduces further the advantage of this technique.

Figure 3(b) depicts the Hmean improvement of RG-FP-S over other policies. Similarly, RG-FP-S outperforms all the other policies for all types of workloads. The key point is our policy never stalls or squashes a thread with cache misses directly, but lets this thread run at a lower priority. As a result, the thread with cache misses can use the resources that are not required by the other thread. Therefore, under the condition of not significantly affecting a thread, our policy tries to improve the performance of the other thread as highly as possible.

From Figure 3 (a) and (b), we can also observe that our policy outperforms other policies mainly for MIX and MEM workloads. Recall that the difference between RG-FP and previously proposed policies is RG-FP can avoid such problems as resource under-use, resource monopolization, and idle cycles of the processor. We know that these problems are produced when cache misses take place. Therefore, compared to other policies, our policy works better for memory-bounded applications.

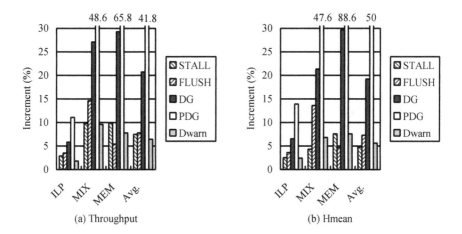

Fig. 3. The improvement of RG-FP-S over STALL, FLUSH, DG, PDG and DWarn

6 Conclusions

In SMT processors, a thread experiencing a miss in L2 cache may hold a large number of resources which other threads could be using to make forward progress. As a result, the overall performance of processors is degraded. Currently, many instruction fetch policies focus on this problem, such as STALL, FLUSH, DG, PDG and DWarn. However, these policies are not perfect, and each has its own disadvantages, mainly including resource monopolization, resource under-use and idle cycles of the processor. The disadvantages of these policies will become more serious when being used in two-context SMT processors.

Our contribution is that we propose a novel fetch policy called RG-FP. Our policy is specially designed for two-context SMT processors to prevent the negative effects of loads missing in L2 cache, and at the same time to avoid the problems of previously proposed fetch policies. Simulation results show that:

1. RG-FP outperforms previously proposed policies for all types of workloads in both throughput and fairness. Especially for **MIX** and **MEM** workloads, the improvement is more obvious.
2. RG-FP shows different degrees of increment over other policies. The increment over PDG is most significant, 41.8% in throughput, and 50.0% in Hmean, on average.
3. When using DRA to allocate resources between threads, RG-FP can outperform other policies further. Compared with SRA, the additional improvement achieved by using DRA in throughput is 3.8% for ILP workloads, 3.9% for MIX workloads and 2.3% for MEM workloads; in harmonic mean is 3.4% for ILP workloads, 6.6% for MIX workloads and 2.5% for MEM workloads.

Acknowledgements

This work was supported by "863" project No. 2002AA110020, Chinese NSF No. 60376018, No. 60273069 and No. 90207011. The authors would like to thank Peixiang Yan and Yi He for their work on the simulator.

References

1. D. Tullsen, S. Eggers and H. Levy: Simultaneous multithreading: Maximizing on-chip parallelism. In Proc. ISCA-22(1995)
2. D. Tullsen, S. Eggers, et al.: Exploiting choice: Instruction fetch and issue on an implementable simultaneous multithreading processor. In Proc. ISCA-23(1996)
3. S. J. Eggers, J. S. Emer, et al.: Simultaneous Multithreading: a Platform for next-generation processors. IEEE Micro, 17(5):12-19(1997)
4. D. Tullsen and J. Brown: Handling long-latency loads in a simultaneous multithreaded processor. In Proc. MICRO-34(2001)
5. A. El-Moursy and D. Albonesi: Front-end policies for improved issue efficiency in SMT processors. In Proc. HPCA-9(2003)
6. F. J. Cazorla, A. Ramirez, et al.: DCache Warn: an I-Fetch policy to increase SMT efficiency. In Proc. IPDPS-18(2004)
7. D. Koufaty and D. T. Marr: Hyperthreading technology in the Netburst microarchitecture. IEEE Micro (2003)
8. R. Kalla, B. Sinharoy and J. Tendler: IBM POWER5 chip: a dual-core multithreaded processor. IEEE Micro (2004)
9. A. Yoaz, M. Erez et al.: Speculation techniques for improving load related instruction scheduling. In Proc. ISCA-26(1999)
10. F. J. Cazorla, A. Ramirez, et al.: Dynamically controlled resource allocation in SMT processors. In Proc. MICRO-37(2004)
11. F. J. Cazorla, et al.: Implicit vs. explicit resource allocation in SMT processors. In Proceedings of the Euromicro Symposium on Digital System Design (2004)
12. The standard performance evaluation corporation, WWW cite: http://www.specbench.org.
13. T. Sherwood, E. Perelman and B. Calder: Basic block distribution analysis to find periodic behavior and simulation points in applications. In Proceedings of the International Conference on Parallel Architectures and Compilation Techniques (2001)
14. D. Tullsen: Simulation and modeling of a simultaneous multithreading processor. In Proceedings of 22nd Annual Computer Measurement Group Conference (1996)

A Loop Transformation Using Two Parallel Region Partitioning Method

Sam Jin Jeong and Jung Soo Han

Division of Information and Communication Engineering, Cheonan University,
Anseo-dong 115, Cheonan City, Korea 330-704
{sjjeong, jshan}@cheonan.ac.kr

Abstract. Loop parallelization is an important optimization issue in the execution of scientific programs. This paper proposes loop transformation techniques for finding parallel regions within nested loops with non-uniform dependences in order to improve parallelism. By parallelizing anti dependence region using variable renaming, there remains only flow dependence in the loop. We then divide the iteration space into FDT (Flow Dependence Tail set) and FDH (Flow Dependence Head set). By two given equations, we show how to determine whether the intersection of FDT and FDH is empty or not. So, we can find two parallel regions for doubly nested loops with non-uniform dependences. In the case that FDT does not overlap FDH, we will divide the iteration space into two parallel regions by a line.

1 Introduction

Computationally expensive programs spend most of their time in the execution of DO-loops. Therefore, an efficient approach for exploiting potential parallelism is to concentrate on the parallelism available in loops in ordinary programs and has a considerable effect on the speedup [1]. Current parallelizing compilers pay much of their attention to loop parallelization.

Some techniques, based on Convex Hull theory [6], are the minimum dependence distance tiling method [4], [5], the unique set oriented partitioning method [3], and the three region partitioning method [2], [8].

Fig. 1(a) shows the dependence patterns of Example 1 in the iteration space.

This paper will focus on loop transformation techniques of perfectly nested loops with non-uniform dependences. Especially, it shows us how to find two parallel regions in doubly nested loops with non-uniform dependences by two given equations.

Example l.

```
do i = 1, 10
  do j = 1, 10
    A(i+j, 3*i+j+3) = . . .
              . . . = A(i+j+1, i+2*j+4)
  enddo
enddo
```

J. Cao, W. Nejdl, and M. Xu (Eds.): APPT 2005, LNCS 3756, pp. 23 – 30, 2005.
© Springer-Verlag Berlin Heidelberg 2005

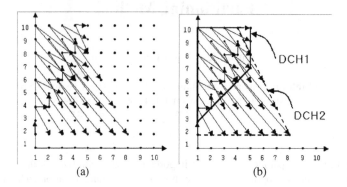

Fig. 1. (a) Iteration Spaces (b) CDCH of Example 1

The rest of this paper is organized as follows. Chapter two describes our loop model, and introduces the concept of Complete Dependence Convex Hull (CDCH). In chapter three, we define the properties of FDT (Flow Dependence Tail set) and FDH (Flow Dependence Head set). We show how to find FDT and FDH, to determine the intersection of FDT and FDH, and to divide iteration space into two parallel regions by a line. Chapter four shows comparison with related works. Finally, we conclude in chapter five.

2 Program Model and Dependence Analysis

The loop model considered in this paper is doubly nested loop program of the form shown in Figure 2. For the given loop, l_1 (l_2) and u_1 (u_2) indicate the lower and upper bounds respectively, and should be known at compile time. We also assume that the program statements inside these nested loops are simple assignment statements of arrays. The dimensionality of these arrays is assumed to be equal to the nested loop depth. To characterize the coupled array subscripts, the array subscripts, $f_1(I, J)$, $f_2(I, J)$, $f_3(I, J)$, and $f_4(I, J)$, are linear functions of the loop index variables.

$$
\begin{aligned}
&\text{do } I = l_1, u_1 \\
&\quad \text{do } J = l_2, u_2 \\
&\qquad A(f_1(I, J), f_2(I, J)) = \ldots \\
&\qquad \qquad \ldots = A(f_3(I, J), f_4(I, J)) \\
&\quad \text{enddo} \\
&\text{enddo}
\end{aligned}
$$

Fig. 2. A doubly nested loop model

The most common method to compute data dependences involves solving a set of linear diophantine equations with a set of constraints formed by the iteration boundaries. The loop in Fig. 2 carries cross iteration dependences if and only if there exist four integers (i_1, j_1, i_2, j_2) satisfying the system of linear diophantine equations given by (1) and the system of inequalities given by (2). The general solution to these equations can be computed by the extended GCD or the power test algorithm [7] and forms a **DCH** (Dependence Convex Hull).

$$f_1(i_1, j_1) = f_3(i_2, j_2) \text{ and } f_2(i_1, j_1) = f_4(i_2, j_2) \tag{1}$$

$$l_1 \le i_1, i_2 \le u_1 \text{ and } l_2 \le j_1, j_2 \le u_2 \tag{2}$$

There are two approaches to solve the system of Diophantine equations of (1). One way is to set i_1 to x_1 and j_1 to y_1 and get the solution of i_2 and j_2.

$$a_{21}i_2 + b_{21}j_2 + c_{21} = a_{11}x_1 + b_{11}y_1 + c_{11}$$
$$a_{22}i_2 + b_{22}j_2 + c_{22} = a_{12}x_1 + b_{12}y_1 + c_{12}$$

We have the solution as

$$i_2 = \alpha_{11}x_1 + \beta_{11}y_1 + \gamma_{11}$$
$$j_2 = \alpha_{12}x_1 + \beta_{12}y_1 + \gamma_{12}$$

where

$$\alpha_{11} = (a_{11}b_{22} - a_{12}b_{21})/(a_{21}b_{22} - a_{22}b_{21})$$
$$\beta_{11} = (b_{11}b_{22} - b_{12}b_{21})/(a_{21}b_{22} - a_{22}b_{21})$$
$$\gamma_{11} = (b_{22}c_{11} + b_{21}c_{22} - b_{22}c_{21} - b_{21}c_{12})/(a_{21}b_{22} - a_{22}b_{21})$$
$$\alpha_{12} = (a_{21}a_{12} - a_{11}a_{22})/(a_{21}b_{22} - a_{22}b_{21})$$
$$\beta_{12} = (a_{21}b_{12} - a_{22}b_{11})/(a_{21}b_{22} - a_{22}b_{21})$$
$$\gamma_{12} = (a_{21}c_{12} + a_{22}c_{21} - a_{21}c_{22} - a_{22}c_{11})/(a_{21}b_{22} - a_{22}b_{21})$$

The solution space is the set of points (x, y) satisfying the equations given above. The set of inequalities can be written as

$$l_1 \le i_1 \le u_1 \text{ and } l_2 \le j_1 \le u_2 \text{ and}$$
$$l_1 \le \alpha_{11}x_1 + \beta_{11}y_1 + \gamma_{11} \le u_1 \text{ and } l_2 \le \alpha_{12}x_1 + \beta_{12}y_1 + \gamma_{12} \le u_2 \tag{3}$$

where (3) defines a DCH denoted by DCH1.

Another approach is to set i_2 to x_2 and j_2 to y_2 and solve for the solution of i_1 and j_1.

$$a_{11}i_1 + b_{11}j_1 + c_{11} = a_{21}x_2 + b_{21}y_2 + c_{21}$$
$$a_{12}i_1 + b_{12}j_1 + c_{12} = a_{22}x_2 + b_{22}y_2 + c_{22}$$

We have the solution as

$$i_1 = \alpha_{21}x_2 + \beta_{21}y_2 + \gamma_{21}$$
$$j_1 = \alpha_{22}x_2 + \beta_{22}y_2 + \gamma_{22}$$

where

$$\alpha_{21} = (a_{21}b_{12} - a_{22}b_{11})/(a_{11}b_{12} - a_{12}b_{11})$$
$$\beta_{21} = (b_{12}b_{21} - b_{11}b_{22})/(a_{11}b_{12} - a_{12}b_{11})$$
$$\gamma_{21} = (b_{12}c_{21} + b_{11}c_{12} - b_{12}c_{11} - b_{11}c_{22})/(a_{11}b_{12} - a_{12}b_{11})$$
$$\alpha_{22} = (a_{11}a_{22} - a_{12}a_{21})/(a_{11}b_{12} - a_{12}b_{11})$$
$$\beta_{22} = (a_{11}b_{22} - a_{12}b_{21})/(a_{11}b_{12} - a_{12}b_{11})$$
$$\gamma_{22} = (a_{11}c_{22} + a_{12}c_{11} - a_{11}c_{12} - a_{12}c_{21})/(a_{11}b_{12} - a_{12}b_{11})$$

The solution space is the set of points (x, y) satisfying the solution given above. In this case the set of inequalities can be written as

$$l_1 \le i_2 \le u_1 \text{ and } l_2 \le j_2 \le u_2 \text{ and}$$

$$l_1 \le \alpha_{21}x_2 + \beta_{21}y_2 + \gamma_{21} \le u_1 \text{ and } l_2 \le \alpha_{22}x_2 + \beta_{22}y_2 + \gamma_{22} \le u_2 \tag{4}$$

where (4) defines another DCH, denoted by DCH2.

The above two sets of solutions are both valid. Each of them has the dependence information on one extreme. For some simple cases, for instance, since there is only one kind of dependence, either flow or anti dependence, one set of solution (*i.e.* DCH) should be enough. Punyamurtula and Chaudhary used (3) for their technique [4], while Zaafrani and Ito used (4) for their technique [8]. For those more complicated cases, where both flow and anti dependences are involved and dependence patterns are irregular, we need to use both sets of solutions.

If iteration (i_2, j_2) is dependent on iteration (i_1, j_1), then we have a dependence distance vector $d(x, y)$ with

$$d_i(x, y) = i_2 - i_1, \; d_j(x, y) = j_2 - j_1 \tag{5}$$

For DCH1, we have

$$d_i(x_1, y_1) = (\alpha_{11} - 1)x_1 + \beta_{11}y_1 + \gamma_{11}, \; d_j(x_1, y_1) = \alpha_{12}x_1 + (\beta_{12} - 1)y_1 + \gamma_{12} \tag{6}$$

For DCH2, we have

$$d_i(x_2, y_2) = (1 - \alpha_{21})x_2 - \beta_{21}y_2 - \gamma_{21}, \; d_j(x_2, y_2) = -\alpha_{22}x_2 + (1 - \beta_{22})y_2 - \gamma_{22} \tag{7}$$

Clearly if we have a solution (x_1, y_1) in DCH1, we must have a solution (x_2, y_2) in DCH2, because they have been solved from the same set of linear Diophantine equations (1). The union of DCH1 and DCH2 is called Complete Dependence Convex Hull (**CDCH**), and all dependences lie within the CDCH. Fig. 1(b) shows the CDCH of Example 1.

We can write these dependence distance functions in a general form as

$$d(i_1, j_1) = (d_i(i_1, j_1), d_j(i_1, j_1)), \; d(i_2, j_2) = (d_i(i_2, j_2), d_j(i_2, j_2))$$

$$d_i(i_1, j_1) = p_1 * i_1 + q_1 * j_1 + r_1, \; d_j(i_1, j_1) = p_2 * i_1 + q_2 * j_1 + r_2 \tag{8}$$

$$d_i(i_2, j_2) = p_3 * i_2 + q_3 * j_2 + r_3, \; d_j(i_2, j_2) = p_4 * i_2 + q_4 * j_2 + r_4$$

where p_i, q_i, and r_i are real values and i_1, j_1, i_2, and j_2 are integer variables of the iteration space. The properties of DCH1 and DCH2 can be found in [3].

The set of inequalities and dependence distances of the loop in Example 1 are computed as follows.

DCH1 : $1 \le i_1 \le 10$ DCH2 : $1 \le j_2/2 \le 10$

$\quad\quad 1 \le j_1 \le 10$ $\quad\quad 1 \le i_2 + j_2/2 + 1 \le 10$

$\quad\quad 1 \le -i_1 + j_1 - 1 \le 10$ $\quad\quad 1 \le i_2 \le 10$

$\quad\quad 1 \le 2i_1 \le 10$ $\quad\quad 1 \le j_2 \le 10$

$\quad\quad d_i(i_1, j_1) = -2i_1 + j_1 - 1$ $\quad\quad d_i(i_2, j_2) = i_2 - j_2/2$

$\quad\quad d_j(i_1, j_1) = 2i_1 - j_1$ $\quad\quad d_j(i_2, j_2) = -i_2 + j_2/2 - 1$

3 Two Parallel Region Partitioning Method

We define the flow dependence tail set (FDT) and the flow dependence head set (FDH) as follows.

Definition 1. *Let L be a doubly nested loop with the form in Fig. 2. If line $d_i(i_1, j_1) = 0$ intersects DCH1, the flow dependence tail set of the DCH1, namely **FDT**(L), is the region H, where H is equal to*

$$DCH1 \cap \{(i_1, j_1) \mid d_i(i_1, j_1) \geq 0 \text{ or } d_i(i_1, j_1) \leq 0 \} \tag{9}$$

Definition 2. *Let L be a doubly nested loop with the form in Fig. 2. If line $d_i(i_2, j_2) = 0$ intersects DCH2, the flow dependence head set of the DCH2, namely **FDH**(L), is the region H, where H is equal to*

$$DCH2 \cap \{(i_2, j_2) \mid d_i(i_2, j_2) \geq 0 \text{ or } d_i(i_2, j_2) \leq 0 \} \tag{10}$$

Property 1. *Suppose line $d_i(i, j) = p*i+q*j+r$ passes through CDCH. If $q > 0$, FDT(FDH) is on the side of $d_i(i_1, j_1) \geq 0$ ($d_i(i_2, j_2) \geq 0$), otherwise, FDT(FDH) is on the side of $d_i(i_1, j_1) \leq 0$ ($d_i(i_2, j_2) \leq 0$).*

Fig. 3(b) shows FDH and FDT of the loop in Example 1 after variable renaming.

In our proposed algorithm, Algorithm Region_Partition, we can determine whether the intersection of FDT and FDH is empty by position of two given lines $d_i(i_1, j_1) = 0$ and $d_i(i_2, j_2) = 0$, and two real values q_1 and q_3 given in (8). If the intersection of FDT and FDH is not empty, we divide the iteration space into two parallel regions and one serial region by two appropriate lines as given in the three region partitioning method [5], [7]. If the intersection of FDT and FDH is empty, we divide the iteration space into two parallel regions by the line $d_i(i_1, j_1) = 0$ or $d_i(i_2, j_2) = 0$.

From property 1, we know that the real value $q_1(q_3)$ determines whether the position of FDT(FDH) is on side of the line $d_i(i_1, j_1) \geq 0 (d_i(i_2, j_2) \geq 0)$ or not. The line is the bounds of two parallel loops.

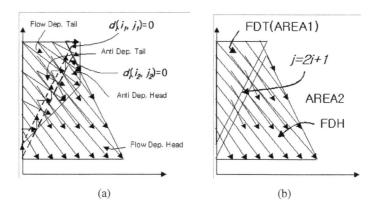

(a)   (b)

Fig. 3. (a) D ependence and Anti Dependence unique set, (b) FDT and FDH of Example 1

```
Algorithm Region_Partition
INPUT: two lines (d_i(i_1, j_1) = 0, d_i(i_2, j_2) = 0) and two real values
(q_1, q_3)
OUTPUT: two parallel regions
BEGIN
IF (line d_i(i_1, j_1) = 0 is on the left side of line d_i(i_2, j_2) =0)
    IF (q_1 > 0 and q_3 < 0){
        /* AREA1 does not overlap AREA2 */
        AREA1: {(i_1, j_1) | d_i(i_1, j_1) • 0}
        AREA2: {(i_1, j_1) | d_i(i_1, j_1) < 0} }
ELSE IF (d_i(i_1, j_1) = 0 is on the right side of d_i(i_2, j_2) = 0)
    IF (q_1 < 0 and q_3 > 0) {
        /* AREA1 does not overlap AREA2 */
        AREA1: {(i_1, j_1) | d_i(i_1, j_1) • 0}
        AREA2: {(i_1, j_1) | d_i(i_1, j_1) > 0} }
ELSE Use Three Region Partitioning Method
END Region_Partition
```

Fig. 4. Algorithm of determining the intersection of FDT and FDH

In this algorithm, the line $d_i(i_1, j_1) = 0$ is expressed by $j = Ai+B$, where $A = (1-\alpha_{11})/\beta_{11}$, $B = -\gamma_{11}/\beta_{11}$, which are derived from (5). We know that the line can be the upper or lower bound in the transformed loops based on the corresponding region of the loop technique. The line $d_i(i_1, j_1) = 0$ is the upper boundary in AREA2 and lower boundary in AREA1 in Example 1. In this case, the iteration space is divided into two parallel regions, AREA1 and AREA2, by line $j = 2i_1 +1$ as shown in Fig 3(b). The execution order is AREA1 $\rightarrow$ AREA2. Transformed loops are given as follows.

/* AREA1 */

do $i = l_1, u_1$

   do $j = \max(l_2, \lceil 2*i+1 \rceil)$, u_2
      A(i+j, 3*i+j+3) = . . .
         . . . = A(i+j+1, i+2*j+4)
   enddo
enddo

/* AREA2 */

do $i = l_1, u_1$

   do $j = l_2, \min(u_2, \lceil 2*i+1 \rceil)$
      A(i+j, 3*i+j+3) = . . .
         . . . = A(i+j+1, i+2*j+4)
   enddo
enddo

Fig. 5. Transformation of the loop by two parallel region partitioning method in Example 1

4 Performance Analysis

Theoretical speedup for performance analysis can be computed as follows. Ignoring the synchronization, scheduling and variable renaming overheads, and assuming an unlimited number of processors, each partition can be executed in one time step. Hence, the total time of execution is equal to the number of parallel regions, N_p, plus the number of sequential iterations, N_s. Generally, speedup is represented by the ratio of total sequential execution time to the execution time on parallel computer system as follows:

$$Speedup = (N_i * N_j)/(N_p + N_s)$$

where N_i, N_j are the size of loop i, j, respectively

By using an example given in Example 1, we compare the performance of our proposed method with that of related works.

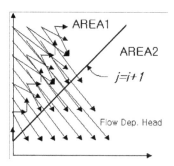

Fig. 6. Regions of the loop partitioning by the unique sets oriented partitioning in Example 1

Applying the unique set oriented partitioning to this loop illustrates case 4 of this technique [3], which is the case that there are two kinds of dependence and DCH1 overlaps with DCH2, and there is at least one isolated unique set. Fig. 6 shows one possible partitioning case of this method. AREA2 contains only flow dependence heads, and AREA1 contains flow dependence tails and anti dependence heads and tails by the line $j = i+1$. So, the speedup for this method is $(10*10)/(45+2) = 2.1$.

Applying the minimum dependence distance tiling method to this loop illustrates case 2 of this technique [4], which is the case that line $d_i(i, j) = 0$ and $d_j(i, j) = 0$ pass through the IDCH. The minimum values of $d_i(i, j)$, d_{imin}, and $d_j(i, j)$, d_{jmin}, occur at the extreme point $(1, 1)$ and both $d_{imin} = 1$ and $d_{jmin} = 1$. There is only serial region, and no speedup for this method.

This example is the case which FDT does not overlap the FDH. In this case, we apply our proposed method - two parallel region partitioning method. After variable renaming, there remains flow dependence tail set (FDT) and flow dependence head set (FDH) separately. Thus, a line $d_i(i, j) = 0$ between two sets divides the iteration space into two parallel areas as shown in Fig. 4(b). The iterations within each area can be fully executed in parallel. The speedup for this method is $(10*10)/2 = 50$.

In the above comparisons, our proposed partitioning method exploits more parallelism than the other related methods.

5   Conclusions

In this paper, we have studied loop transformation techniques for finding parallel regions within nested loops with non-uniform dependences in order to maximize parallelism.

When there are both flow and anti dependence sets within the doubly nested loop, we eliminate anti dependence sets by variable renaming. After variable renaming, there remains only flow dependence in the nested loop. We then divide the iteration space into FDT (Flow Dependence Tail set) and FDH (Flow Dependence Head set).

In our proposed algorithm, we can determine whether the intersection of FDT and FDH is empty by position of two given lines $d_i(i_1, j_1) = 0$ and $d_i(i_2, j_2) = 0$, and two real values. If the intersection of FDT and FDH is not empty, we divide the iteration space into two parallel regions and one serial region by two appropriate lines. If the intersection of FDT and FDH is empty, we divide the iteration space into two parallel regions by the line $d_i(i_1, j_1) = 0$ or $d_i(i_2, j_2) = 0$. The iterations within each area can be fully executed in parallel.

In comparison with some previous partitioning methods, our proposed method leads to better speedup compared with other methods such as minimum tiling method and unique set oriented partitioning method.

References

1. D. Kuck, A. Sameh, R. Cytron, A. Polychronopoulos, G. Lee, T. McDaniel, B. Leasure, C. Beckman, J. Davies, and C. Kruskal, "The effects of program restructuring, algorithm change and architecture choice on program performance," in *Proceedings of the 1984 International Conference on Parallel Processing*, pp. 129-138, August 1984.
2. C. K. Cho and M. H. Lee, "A loop parallization method for nested loops with non-uniform dependences", in *Proceedings of the International Conference on Parallel and Distributed Systems,* pp. 314-321, December 10-13, 1997.
3. J. Ju and V. Chaudhary, "Unique sets oriented partitioning of nested loops with non-uniform dependences," in *Proceedings of International Conference on Parallel Processing*, vol. III, pp. 45-52, 1996.
4. S. Punyamurtula and V. Chaudhary, "Minimum dependence distance tiling of nested loops with non-uniform dependences," in *Proceedings of Symposium on Parallel and Distributed Processing*, pp. 74-81, 1994.
5. S. Punyamurtula, V. Chaudhary, J. Ju, and S. Roy, "Compile time partitioning of nested loop iteration spaces with non-uniform dependences," *Journal of Parallel Algorithms and Applications*, October 1996.
6. T. Tzen and L. Ni, "Dependence uniformization: A loop parallelization technique," *IEEE Transactions on Parallel and Distributed Systems*, vol. 4, no. 5, pp. 547-558. May 1993.
7. M. Wolfe and C. W. Tseng, "The power test for data dependence," *IEEE Transactions on Parallel and Distributed Systems*, vol. 3, no. 5, pp. 591-601, September 1992.
8. A. Zaafrani and M. R. Ito, "Parallel region execution of loops with irregular dependences," in *Proceedings of the International Conference on Parallel Processing*, vol. II, pp. 11-19, 1994.

Criticality Based Speculation Control for Speculative Multithreaded Architectures

Rahul Nagpal and Anasua Bhowmik

Department of Computer Science and Automation,
Indian Institute of Science, Bangalore, India
{rahul, anasua}@csa.iisc.ernet.in

Abstract. Unending quest for performance improvement coupled with the advancements in integrated circuit technology have led to the development of new architectural paradigm. Speculative multithreaded architecture (SpMT) philosophy relies on aggressive speculative execution for improved performance. However, aggressive speculative execution comes with a mixed flavor of improving performance, when successful, and adversely affecting the performance (and energy consumption) because of useless computation in the event of mis-speculation. Dynamic instruction criticality information can be applied to control and guide such an aggressive speculative execution.

In this paper, we propose a model to determine the dynamic instruction criticality of SpMT execution. We have also developed two novel techniques, utilizing the criticality information, namely delaying the non-critical loads and the criticality based thread-prediction for reducing useless computations. Our experiments with criticality based speculation control show a significant reduction in useless computation with little reduction in speedup.

1 Introduction

Speculative multithreaded (SpMT) execution paradigm [2] takes a step ahead in aggressive ILP based execution. In SpMT execution model, a sequential program is divided into threads by combination of hardware and software techniques and the threads are executed in parallel on multiple out-of-order superscalar processing elements (PEs). SpMT processors allow threads to run speculatively in presence of ambiguous control and data dependencies and recover upon the detection of dependency violations. Control speculation allows the future threads to be predicted and speculatively started. Data speculation allows memory operations to be speculatively performed on a buffer or cache and later committed to memory, if found successful. However, the thread mis-prediction leads to squashing of the mis-predicted thread as well as all the subsequent threads. Similarly data dependence violation leads to squashing and re-execution of threads or instructions violating the dependence. Thus aggressive speculative execution comes with a mixed flavor of improving the performance when correctly speculated and adversely affecting the performance and energy consumption by doing useless computation in the event of mis-speculation.

J. Cao, W. Nejdl, and M. Xu (Eds.): APPT 2005, LNCS 3756, pp. 31–40, 2005.
© Springer-Verlag Berlin Heidelberg 2005

As energy consumed by the processor is fast becoming an important performance parameter, it is necessary that the speculation should be done judiciously in order to reduce the energy wastage while maintaining the program performance. Identification of the dynamic critical instructions and performing the speculation depending on the instruction criticality can reduce the risk of redundant computation. This can help to achieve a balance between power and performance. Whereas the earlier works on determining and exploiting dynamic critical paths have targeted out-of-order superscalar processor models, in this paper, we propose an analytical model for determining the dynamic critical path of programs for SpMT execution paradigm. We have analyzed the dynamic critical path information in detail to get the insight about the behavior of the programs under SpMT execution model and then used that knowledge to perform both the control speculation and the data speculation in an aggressive SpMT processor.

In order to restrict the control speculation, we do not speculatively execute the next thread when that thread is dependent on low confidence non-critical branch. Our experiments with this scheme has resulted in 40.44% reduction in useless computation while the IPC degraded by only 6.52%. Similarly, data speculation is controlled by delaying the non-critical speculative loads and the experimental evaluation reveals 26.5% savings in redundant execution with 8% reduction in IPC. These results also validate our criticality model.

The rest of the paper is organized as follows. We discuss the related work in section 2. Section 3 presents our graph model for determining the dynamic critical path for SpMT processors. Section 4 presents the detail measurement and analysis of critical path information. In section 5, we describe the speculation control techniques using the critical path information and also present the experimental results. Section 6 contains the conclusions and future works.

2   Related Work

Critical path analysis, which is traditionally used by compiler, has been recently adopted by architectural community to control aggressive optimization. Tune et al.[3] have proposed various heuristics based on micro-execution events to decide on the criticality of instructions. Their heuristic based critical instruction prediction have been shown to be effective for driving various optimization techniques such as instruction steering in a clustered architecture, value prediction, and reducing power consumption. Fields et al.[1] follow a modeling based approach for predicting critical instructions. They proposed a graph model to capture hardware constraints such as finite reorder buffer, and branch mis-prediction apart from traditional data dependencies. They have used this model to develop a token passing based predictor and used it for selective value prediction and instruction steering for a clustered architecture. They have shown that the model based approach is generic and more accurate than a heuristic based approach[1]. Though our model is based on the model proposed of Fields et al. [1], it is significantly more challenging to accurately model and determine critical paths in

SpMT processors, since there exist many parallel paths of execution through different threads at the same time.

Reducing energy in microprocessors is an active area of research and many power optimization techniques have been proposed. The works that are closest to our work are [4], [7]. Manne et. al.[4] introduced the idea of pipeline gating and used branch confidence estimator to reduce wrong-path instructions in the pipeline to save energy and [7] reduces processor power dissipation by throttling the different pipeline stages selectively based on branch confidence estimation.

3 Modeling the Critical Path for SpMT Processor

The central idea behind SpMT is to execute multiple threads obtained from a sequential program in parallel. SpMT architectures generally supports both control speculation and data speculation as well as recovery mechanism to handle incorrect speculations. This enables parallelization of applications, despite any compile-time uncertainty about (control or data) dependences that may exist between the threads running in parallel.

A typical SpMT processor consists of a collection of simpler processing elements (PEs) that are connected by an interconnection network. Each PE has its own fetch, decode, and execution units. Each individual PE follows the out-of-order execution. A thread spawns the next successor thread speculatively based on thread level prediction. A spawned thread becomes active after a PE becomes available. At any point of time only one thread executes non-speculatively (*the head thread*). Although instructions are executed out-of-order, they are committed in program order. A speculative thread (and all the subsequent threads) can be squashed before committing because of mis-prediction or dependence violation.

To determine the critical path in a program, we capture the micro-execution of the program in the SpMT processor by building a dynamic dependence graph (DDG) at run-time. The critical path is the longest path in the dynamic dependence graph and the nodes lying on the critical path constitute the critical instructions. Each instruction in the DDG is represented by three nodes namely **F, E,** and **C** nodes. A **F** node represent the event of an instruction fetched, decoded, and put in the reorder buffer (ROB). Instructions in the same thread are fetched in order subject to limited fetch bandwidth and ROB space. Instructions in the different currently active threads are fetched in parallel. A **E** node represents the event of execution of an instruction. Decoded instructions are executed in parallel in each active PE subject to the availability of input operands, functional units, and under the constraint on limited issue width. A **C** node represents the commit event of an instruction. Instructions are committed in program order across all the threads. An edge between two nodes depict the dependence between them and the edge weight represent the resultant delay. The edges between the nodes in the instruction execution model are listed in Table 1. There are inter-thread as well as intra thread edges depicting the

Table 1. List of edges for instruction execution in SpMT execution model, Xj,i refer to node X of ith instruction in j^{th} thread

Edge type	No.	Constraint modeled	Name	Edge
Intra-thread	1.	In-order Fetch	FF_i	$F_{j,i-1} \rightarrow F_{j,i}$
	2.	Finite Size ROB	CF_i	$C_{j,i-R} \rightarrow F_{j,i}$, R is the size of ROB buffer
	3.	Control Dependence	EF_i	$E_{j,i-1} \rightarrow F_{j,i}$, if instr $i-1$ is the mis-predicted branch
Inter-thread	4.	Next Thread activation	FF_I	$F_{j-1,1} \rightarrow F_{j,1}$
	5.	Finite PE	CF_I	$C_{j-N,l} \rightarrow F_{j,i}$, N is the number of PEs
	6.	Control/data mis-speculation	EF_I	$E_{j,i} \rightarrow F_{j+1,1}$
Intra-thread	7.	Execute Follow Fetch	FE_i	$F_{j,i} \rightarrow E_{j,i}$
	8.	Data Dependency	EE_i	$E_{j,m} \rightarrow E_{j,n}, \quad m < n$
Inter-thread	9.	Data Dependency	EE_I	$E_{k,m} \rightarrow E_{j,n}, \quad k < j$
Intra-thread	10.	Commit Follow Execution	EC_i	$E_{j,i} \rightarrow C_{j,i}$
	11.	In-order Commit	CC_i	$C_{j,i-1} \rightarrow C_{j,i}$
Inter-thread	12.	In-order Commit	CC_I	$C_{j-1,l} \rightarrow C_{j,1}$

interaction between the threads running in parallel. Our critical path model takes into consideration micro-architectural constraints of dynamic execution in the SpMT processor. Next we precisely describe the constraint modeled by each of the edge.

1. FF_i models the constraint of in order fetch within a thread. So for all threads j there is an edge between the fetch nodes of successive instructions in the thread. If the two instructions can be fetched in the same cycle then the edge weight is zero.

2. CF_i models the limited size ROB within each PE, since the fetch of the instruction i is possible only after commit of instruction i-R within the PE where R is the size of ROB buffer.

3. EF_i models the intra-thread branch mis-prediction where fetch of an instruction following a branch is not possible till the outcome of the branch is known.

4. FF_I models the activation of successive threads in program order. A thread can not start fetching instructions until its immediate predecessor have started fetching the instructions. This is modeled by an edge between the fetch nodes of the instructions of successive threads.

5. The number of threads simultaneously active are limited by number of available PEs. CF_I represent this constraint as an edge between the commit of the last instruction of thread $j - N$ to first instruction of the thread j, where N is the number of available PEs. Only after commit of the last instruction of thread $j - N^{th}$ thread the jth thread can be activated.

6. EF_I constraint models the thread mis-prediction that limit the fetch of instructions in the next thread. In case of thread mis-prediction by j^{th} thread, the first instruction from the $(j+1)^{th}$ thread can be fetched only after executing the instruction i [1] in thread j, that detects the thread mis-prediction.

---

[1] The instruction could be *branch, jump, call or return.*

7. FE_i models that an instruction can be executed only after it is fetched decoded and put into RUU.
8. EE_i models intra-thread data dependency.
9. EE_I models inter-thread data dependency.
10. EC_i represents commit follow execution constraint.
11. CC_i models in order commit within the thread.
12. CC_I models the in-order commit across threads The first instruction of thread j can be committed only after the last instruct l of thread $(j-1)$ is committed. This is represented by $C_{j-1,l} \rightarrow C_{j,1}$.

3.1   Example of Graph Model

In this subsection we give a detailed example of our dependence graph model for finding the critical path. Figure 1(a) shows a dynamic program segment consisting of four threads T1, T2, T3, and T4. Figure 1(b) shows the corresponding dependence graph model assuming two PEs and a fetch width of 2 and ROB buffer size of 3 instructions. Note that in the graph we have shown all the edges but not all the weights. The critical path in that graph is shown with the thick arrows.

Let us consider that the correct execution order of threads be T1, T2, and T4. T1 is allocated in PE1 and fetched there. The edges between the **F** nodes are shown and since the issue width is 2, instructions $I1$ and $I2$ are fetched in the same cycle so the delay of the $F_{1,I1}F_{1,I2}$ is 0. The edges $E_{1,I1}E_{1,I2}$ and $E_{1,I1}E_{1,I3}$ show the intra-thread data dependency due to register **r1**. The edge $C_{1,I1}F_{1,I4}$ shows the ROB buffer size constraint, i.e., since the buffer size is only 3, instruction $I4$ can not be fetched until $I1$ commits. The intra-thread edges between commit nodes model the in-order commit and the edge weights are always 0, since there is no restriction on number of commits possible in one cycle.

The inter-thread edge $F_{1,I1}F_{2,I1}$ models the fact that thread T2 is predicted from T1 and it takes 2 cycles to start execution of T2 in PE2. Similarly T2 predicts T3. The edge $C_{1,I4}F_{3,I1}$ models the limited number of PEs. Thread T3

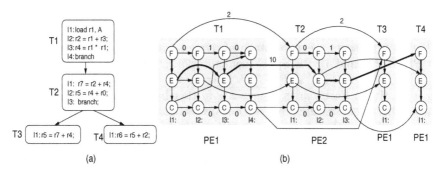

(a)   (b)

Fig. 1. Example of critical path modeling (a)Example of program threads (b)The dependence graph for two PEs

can not start execution unless the last instruction of T1 commits and PE1 becomes available. The edges $E_{1,I2}E_{2,I1}$ and $E_{1,I3}E_{2,I2}$ represent the inter-thread data dependency. The weight of $E_{1,I3}E_{2,I2}$ is 10 assuming that multiplication operation needs 10 cycles.

In this example, T4 is the correct thread to be executed after T2 and not T3. The branch instruction at the end of T2 detects the thread mis-prediction and T3 gets squashed and T4 starts executing in PE1. Since the first instruction of T4 can be fetched only after the branch instruction of T2 gets executed there is an edge from $E_{2,I3}$ to $F_{4,I1}$. Note that there is no out going inter-thread edge from T3 and this ensures that no nodes from T3 will come into the critical path. If the number of PEs in the system be n, the out going inter-thread edges from the threads Tk to be squashed can only span at most next $n - 2$ threads[2] and all these $n - 2$ threads will be squashed along with Tk since there existence is dependent on Tk. This ensures that no nodes from the squashed threads come into critical path.

4   Detection and Analysis of Critical Path

In this section, we present a detailed analysis of the critical path that is necessary in order to evaluate the potential and effective mechanism for speculation control under the SpMT execution model. We have implemented the dynamic dependence graph model in a cycle accurate simulator of the multiscalar[2] processor developed at University of Wisconsin that simulates an SpMT processor. The parameter of the simulated architecture are given in table 2.

Table 2. Hardware parameters used in experimental evaluation

Component	Description
PEs	2-way issue, 32-entry ROB, 2 integer, 1 FP, 1 branch, 1 memory
Intra-task prediction	gshare with 16-bit history, 64K-entry table of 2-bit counters
Inter-task prediction	path-based with 16-bit history, 64K-entry table of 2-bit counters
Register Ring	2 values per cycle, bypass same cycle between adjacent PEs
Memory Buffer	32 entries/PE, 32 x No. of PE bytes/entry fully associative, 2 cycle hit
L1 I-cache	16 * No. of PE KB , 2-way associative, 32 byte blocks, 1cycle hit
L1 D-cache	16 * No. of PE KB, 2-way associative, 32 byte blocks, 2 cycle hit

Figure 2 presents the percentage of dynamic instructions that are critical in 4 PE, 8 PE, and 12 PE SpMT configurations. We have found that on the average 40.87%, 21.14%, and 17.52% instructions are critical for 4, 8, and 12 PEs respectively. Each bar in Figure 2 is further broken down into three parts showing the *fetch, execute,* and *commit critical* instructions. An instruction is *fetch critical* if its **F** node comes into the critical path. From Figure 2, we see that approximately half of the critical instructions in all three configurations are *fetch critical*. The percentage of *execute critical* instructions are 7.35%, 5.89%, and 5.63% for 4, 8, and 12 PEs respectively.

---

[2] When Tk is the thread following the head thread.

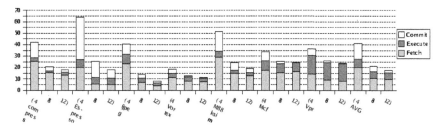

Fig. 2. % breakup of dynamic instructions found critical

From Figure 2 we also observe that the number of critical instructions decrease with increasing number of PEs. With more number of PEs, more operations are executed in parallel, decreasing the number of critical instructions. Most of the reduction is due to effective increase in fetch bandwidth that reduces the number of *fetch critical* instructions. Increasing number of PEs in effect relaxes the CF_I constraint described in Table 1, i.e. more instructions could be fetched and executed without waiting for commit. We also observe that thread misprediction increases marginally with increase in the number of PEs. and hence the EF_I constraint does not offset the benefits of relaxing the CF_I constraint.

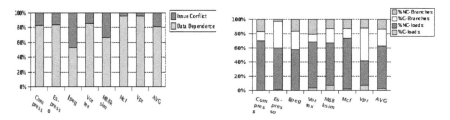

Fig. 3. % breakup of execute critical instructions

Fig. 4. % breakup of squash loss time

Figure 3 presents the breakup of execute critical instructions based on the reason for their criticality. An instruction could be execute critical if it needs to wait for the operands (data dependency) or gets delayed due to limited issue width despite the availability of data values. From the Figure 3 we see that the average percentage of instructions that become *execute critical* due to data dependency in a 8 PE SpMT model is 81.68%. Due to space constraints, we do not show the results for 4 and 12 PE model. We have observed that the data dependence fraction of *execute critical* instructions increase with more number of PEs due to the increased effect of inter-thread communication. With increasing number of PEs, threads that are far ahead in the execution get free PEs and then wait for the earlier threads to supply the data. Therefore, the data dependence fraction of *execute critical* instructions are increased with increasing number of PEs. The existence of significant amount of issue conflict in Figure 3 implies the scope for performance improvement by giving priority to critical instructions while scheduling.

Our experiments have found that for 8 PE model on the average 18.77% of load instructions and 12.76% of store instructions are critical and the average numbers of mispredicted branches that are on the critical path is 32.98%. To gain further insight into the cause of useless computation, we have also measured the time spent in useless computation due to thread squashing. The results for 8 PE configuration are shown in Figure 4. where *squash loss time* (SLT) due to the memory dependence violation and branch mis-prediction are presented under each bar separately. The instructions causing memory dependence violation (i.e. load/store instructions) and branch mis-prediction are again classified into critical and non-critical instructions. From Figure 4 we see that on the average, critical load-store contributes for only 3.03% of the squashing where as 59.42% of squashing is caused due to non-critical loads. The figure also shows that on the average 24.0% of squashing is caused due to critical branch type instructions that are mis-predicted and 13.52% of squashing is attributed to non-critical branches. The high percentage of non-critical load-store instructions in Figure 4 points toward the possibility of avoiding speculative execution of such instructions in order to reduce power consumption due to unnecessary computation. Similarly the high percentage of non-critical mis-predicted branches points toward the possibility of avoiding speculation across these branches.

5   Using Instruction Criticality for Speculation Control

In this section we propose and evaluate two speculation control techniques using the instruction criticality information to reduce the useless computation. We present results only for the 8 PE configuration. We observe similar results for the 4 PE and the 12 PE configurations.

5.1   Reducing Useless Computation by Delaying Non-critical Loads

In SpMT processors a load instruction is speculatively executed assuming that the store on which the load is dependent has already taken place. The processor loads the value from either the memory or the intermediate buffer and continues execution. If an earlier store to the same location is executed after the load, the processor detects a memory dependence violation and squashes the violating load and its dependent instructions and this causes dynamic energy wastage. Although, speculative loads may lead to dynamic energy wastage, they are necessary to speedup the execution. Our experiments have shown that nearly 18% of all load instructions lie on the critical path. Therefore speculative loads can not be removed altogether. However, from Figure 4, we see that in all the programs a significant percentage (avg. 59.42%) of squashing is due to non-critical loads. Since the total execution time is not likely to depend on the non-critical loads, we delay the non-critical loads (we call this scheme DL) in order to reduce dynamic energy wastage without affecting the speedup, We determine the average time between issue of loads and resultant squashing (average time to squash or ATS) and the non-critical loads are delayed by this duration.

From Figures 6 and 5 we see that the average reduction of squash loss time is 26.43% with IPC reduction of 8.11% for delayed load scheme. The degradation in IPC is happening mainly because at present we delay all the non-critical loads in a program by a fixed ATS and this is not very efficient, since different non-critical loads should be delayed by different amount. We are currently experimenting with a on line time-to-squash predictor that predicts the required load delay cycles for each load individually and we expect that this we will further improve the IPC.

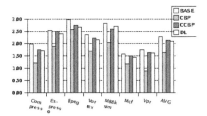 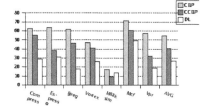

Fig. 5. IPC values   **Fig. 6.** % reduction in squash loss time

5.2 Reducing Useless Computation Due to Thread Mis-prediction

In SpMT processor a thread mis-prediction leads to the squashing of the mis-speculated thread and all the subsequent threads thus causing huge wastage of dynamic energy. From Figure 4 we can see that a significant portion of squash loss is due to branch mis-prediction. Earlier works have used confidence estimator for speculation control[5,7]. However, our experiments show that confidence based prediction alone is not sufficient for speculation control in SpMT thread prediction. Therefore we have combined criticality information with the confidence estimation to perform speculation control. In confidence based prediction mechanism, speculation is done only for branches with high confidence value whereas the combined criticality and confidence based prediction mechanism speculates across branches with low confidence as well if its found on the critical path of the program. We have implemented a 5 bit JRS[6] predictor with resetting counter for confidence estimation. We experiment with both purely confidence based predictor (CBP) and a combined confidence and criticality based predictor(CCBP) and compare the result with the base line performance (i.e., no confidence estimator).

From Figure 6 we see that a purely confidence based predictor is able to get a maximum reduction in squash loss time (on the average 55%) but suffers from intolerably high performance penalty (29.81% reduction in IPC). The confidence estimator identifies many correct predictions as low confidence branches and by not speculating on those predictions the processor misses parallelism opportunities. This result is in agreement with the earlier studies[5] done in the context of superscalar processors. On the other hand, our combined criticality and confidence based predictor is able to reap most of the benefits of purely confidence based predictor in terms of reducing SLT (on the average, reduction is 40.44% for CCBP compared to 54.44% of CBP) with much less performance penalty (6.52% for CCBP compared to 29.81% of CBP).

6   Conclusions and Future Work

Whereas the earlier work on critical path analysis of program is limited to out-of-order superscalar processors, we have developed a model to identify the dynamic critical instructions in SpMT execution. We proposed two novel techniques that use the criticality information for speculation control. Our experiments show significant reduction in useless computation with little performance degradation. The future extension of this work involves development of on-line criticality predictor and estimation of the exact energy savings gained by criticality based speculation.

References

1. B. Fields, S. Rubin, and R. Bodik. Focusing Processor Policies via Critical-path Prediction. In *Proc. of Intl. Symp. on Computer Architecture*, 2001.
2. M. Franklin. Multiscalar Processors. *Kluwer Academic Publishers*, 2002.
3. E. Tune, D. M. Tullsen, and B. Calder. Quantifying Instruction Criticality. In *Proc. of Intl. Conf. on Parallel Architectures and Compilation Techniques*, 2002.
4. S. Manne, A. Klauser, and D. Grunwald. Pipeline Gating: Speculation Control For Energy Reduction. In *Proc. of Intl. Symp. on Computer Architecture*, 1998.
5. D. Grunwald, A. Klauser, S. Manne, and A. Pleszkun. Confidence Estimation for Speculation Control. In *Proc. of Intl. Symp. on Computer Architecture*, 1998.
6. E. Jacobsen, E. Rotenberg, J. E. Smith. Assigning Confidence to Conditional Branch Predictions. In *Proceedings of Intl. Symp. on Microarchitecture*, 1996.
7. J. L. Aragon, J. Gonzalez, and A. Gonzalez. Power-Aware Control Speculation Through Selective Throttling. In *Proc. of Intl. Symp. on High Performance Computer Architecture*, 2003.
8. G. Ascia, V. Catania, M.Palesi, and D. Patti. A System-level Framework for Evaluating Area/Performance/Power Trade-offs of VLIW-based Embedded Systems. In *Asia and South Pacific Design Automation Conference*, 2005.

Design and Implementation of Semantic Caching Coherency Control Scheme Toward Distributed Environment

Hai Wan[1,2,*] and Lei Li[2]

[1] Department of Computer Science and Technology, SUN YAT-SEN University,
Guangzhou 510275, PRC
[2] Software Research Institute of SUN YAT-SEN University, Guangzhou, 510275, PRC
whwanhai@163.com

Abstract. Semantic caching is very attractive for use in distributed computing environments based on historical queries and their descriptions, one of whose important issues is how to best maintain semantic caching using a coherency control scheme. With the object of applying semantic caching into practice, the cache coherency problems including the data between the server and its caching as well as the cached data and their semantic descriptions are analyzed. This paper presents conflicts existing in semantic caching and their formal definitions, proposes the semantic caching model, and coherency control scheme, meanwhile derives update list optimization algorithm adopted in server and coherency control algorithm used in clients. Finally, the performance of the semantic caching coherency control scheme is examined and analyzed through a simulation study in detail.

1 Introduction

Semantic caching is a kind of cache technology, which is based on historical queries and their descriptions. Because of the low bandwidth and high expense of wireless network in distributed systems and mobile computing environments, caching of frequently accessed data at clients plays an important role due to the reduced network traffic and the improved response time. As one of the most important caching technique, semantic caching can maintain the cache in client both semantic descriptions and results of previous queries. Unlike other caching such as: page or tuple caching whose clients only cache data without semantic descriptions, since semantics about the cached items are stored in a semantic cache, the client is able to reason from the local cache to determine whether a query can be totally answered, how much it can be answered, and what data are missing [1]. Hence, Semantic caching is very attractive and important for ensuring efficient access to widely distributed data in mobile, narrow bandwidth, frequent disconnection environments.

Because the essence of caching technology maintains many duplicates of server in cache nodes, how to maintain those coherencies is a key technology. Furthermore,

---

* Corresponding author.

J. Cao, W. Nejdl, and M. Xu (Eds.): APPT 2005, LNCS 3756, pp. 41–51, 2005.
© Springer-Verlag Berlin Heidelberg 2005

that not only the results are cached in semantic caching but also their semantic descriptions make semantic caching control more complex, which is also the bottleneck of applying semantic caching technique. Accordingly, an important issue to semantic caching is the maintenance of coherency, including the data between the server and its caching as well as the cached data and their semantic descriptions.

Semantic caching has been studied in previous literature, including research on semantic cache coherency. Qun Ren presented a formal semantic caching definition, explored the semantic caching query processing strategies[1]. In [2,3],Wu tingting presented QPID algorithm which can find cache items related to the query at first, classify them, and then process data. Considering the factors of bandwidth, overload, and reliability, the client may allow the cache to maintain weak coherency [4]. Paper [5] proposed a semantic cache coherency asynchronous scheme. On the other hand, data caching has been widely researched, paper [6] presented a scheme which supports for speculative update propagation; paper [7] proposed validation-based reprocessing scheme for updating spatial data in mobile environments; Huang analyzed distributed data cache systems and presented coherency control scheme[8].

From what we have discussed so far, it is clear that research on semantic caching coherency is very important and necessary. According to the practicality request of semantic caching, we present conflicts existing in semantic caching and their formal definitions, propose the semantic caching model and its coherency control scheme, and meanwhile derive update list optimization algorithm adopted in server and coherency control algorithm used in clients.

The rest of this paper is organized as follows: in section 2, we analyze previous the cache coherency problems on semantic caching as well as other related issues and present some formal definitions which can help us study these problems. A formal semantic cache model and its coherency control scheme are proposed in section 3. Section 4 discusses semantic caching coherency control algorithm adopted in server and coherency control algorithm used in clients. Section 5 proposes semantic caching interfaces and protocol. Performance of coherency control scheme and optimization Update list are examined and analyzed through a simulation study in section 6. Finally, we summarize our work and discuss future research in section 7.

2   Semantic Caching Coherency Problems

Like traditional cases, semantic caching can also become obsolete because tuples have been *updated* either in the clients or in the server. On the other hand, when some newly *inserted/modified/updated* tuples satisfy the semantic descriptions, they must be added to the caching even though they were not in it before, which also causes the caching to be out-dated. In brief, the situation of semantic caching coherency control is complicated. We argue that this complexity can be reduced significantly by classifying those situations as *data error conflict* and *data non-integrity conflict*. We have termed our classifications in order to define semantic cache coherency.

Definition 1. Data error conflict and data non-integrity conflict.

(1) Data error conflict: is that some tuples or results do not belong to the data set represented by the cached semantic descriptions; **(2) Data non-integrity conflict:** is that the data set in client represented by the cached semantic descriptions does not contain the same data obtained in server by executing the same semantic descriptions.

We present a coherency control approach to avoid these two kinds of conflicts.

Theorem 1. If there are no *data error conflict* and *data non-integrity conflict* semantic caching, its coherency is satisfiable.

Proof. Consider t is a tuple cached in semantic caching DC_{rs}, DC_p is the semantic description of DC_{rs}, and satisfies $DC_{rs}= \sigma_{DCp}(DC_R)$; S_P is the same description as DC_p, but executing in server S_R and satisfies $S_{rs}= \sigma_{Sp}(S_R)$. If there is no data error conflict, for any $t \in DC_{rs}$, it has $t \in \sigma_{Sp}(S_R)$, so $DC_{rs} \subset \sigma_{Sp}(S_R)$. On the other hand, if there is no data non-integrity conflict, for any $t \in \sigma_{DCp}(DC_R)$, it has $t \in S_{rs}$, so $DC_{rs} \supset \sigma_{Sp}(S_{rs})$. Accordingly, we can conclude $DC_{rs}=\sigma_{Sp}(S_R)$ and this theorem can be proved.   ■

From Theorem 1, we can define *Semantic Caching Coherency* as follows.

Definition 2. Semantic caching coherency.

Given a database server S, and S has $<S_R,S_P,S_{rs}>$, which satisfies $S_{rs}= \sigma_{Sp}(S_R)$, $S_R \subset S$; Consider DC is a semantic caching of S, and has $<DC_R,DC_P,DC_{rs}>$,which satisfies $DC_{rs}= \sigma_{DCp}(DC_R),DC_R \subset S$.

If DC can satisfy: **(1)** Cached data in semantic caching can maintain coherency whenever *inserting/modifying/updating* in S or DC, i.e. $\forall$ query q, $S_{rsq}= \sigma_{Sq}(S_R)$ is satisfiable in S, and meanwhile $DC_{rsq}= \sigma_{DCq}(DC_R)$ is satisfiable in DC; **(2)** Cached data in semantic caching can match the semantic descriptions whenever *inserting/modifying/updating* in S or DC, i.e. $\forall$query q, $DC_{rsq}= \sigma_{DCq}(DC_R)$ is satisfiable;

We say semantic caching coherency is satisfiable.

For the semantic caching coherency, maintaining the semantic description firstly and then updating the cached data is usually adopted. Nevertheless, we can demonstrate this method is unsatisfiable through the following examples.

Table 1. A relation in server   **Table 2.** Cached data in semantic caching

ID	Name	Sex	Salary	EduLevel	Age
001	Smith	Male	8000	Master	25
002	Victor	Male	3000	Bachelor	22
003	Rose	Female	4000	None	23
004	Cathy	Female	5000	Docto	27
005	Tom	Male	6000	Master	25
006	Susan	Female	7000	Bachelor	22

ID	Name	Sex	Salary	EduLevel	Age
002	Victor	Male	3000	Bachelor	22
006	Susan	Female	7000	Bachelor	22

Consider a relation R in the server (shown in table 1), and the client has cached data DC_T (shown in table 2) and its semantic description T: " EduLevel ='Bachelor' ".

(1) When the client obtain an *update* operation:

Update R Set EduLevel= 'master' where EduLevel=' bachelor'

T_{new}: "EduLevel = master", and T_{new} is relevant with T, SC_T can also *update* along with the server; however after updating, the *Data error conflict* will happen.

(2) When the client obtain an *update* operation:

Update R Set Age=24 where Age=22

T_{new}: "Age=22", However, T_{new} is irrelevant with T, so DC_T can not *update* along with the server, which will lead to *Data non-integrity conflict*.

So we can conclude in order for the semantic caching coherency *update* the data firstly and then maintain their semantic descriptions should be adopted. As Definition 2 denotes the operations of *update, delete* and *insert* will initiate the *Data error conflict* or *Data non-integrity conflict* situation. We argue that the complexity of those operations can be reduced significantly by substituting *update* with the operations of *delete* and *insert*. To test this hypothesis satisfying the semantic coherency, we developed specification 1 and Theorem 2 as follows.

Specification 1. (1) Substitute *update* with the operations of *delete* and *insert*; (2) When some tuples cached in the client are *deleted* in the server, we *delete* the cached tuples directly regardless of their semantic descriptions; (3) When *insert* happens in the server, all cached semantic descriptions should be traversed; because the cached semantic descriptions DCp are irrelevant with each other, if the *insert* semantic description match one of the DCp, the tuples should be inserted into DC_{rs}.

Let us reprocess the above *update* situations using specification 1:

1) *Update* in the server, and the results are shown in table 3.

Table 3. An updated relation in server

ID	Name	Sex	Salary	EduLevel	Age
001	Smith	Male	8000	Master	25
002	Victor	Male	3000	Bachelor	24
003	Rose	Female	4000	None	23
004	Cathy	Female	5000	Docto	27
005	Tom	Male	6000	Master	25
006	Susan	Female	7000	Bachelor	24

Table 4. Updated semantic caching

ID	Name	Sex	Salary	EduLevel	Age
002	Victor	Male	3000	Bachelor	24
006	Susan	Female	7000	Bachelor	24

In the client, Substitute *"Update R Set Age=24 where Age=22"* with:

Step 1: "Delete from R where Age=22"; *Step 2*: "Insert into R (001,'Victor','Male', 3000,'Bachelor',24)" and "Insert into R (006,'Susan', 'Female',7000,'Bachelor',24)".

Do Step 1 firstly, and compare T:"EduL evel='Bachelor'" with tuples to be inserted, Because they are relevant, do Step 2 and the updated DC_T is shown in table 4.

2) *Update* in the server, and the results are shown in table 5.

Table 5. An updated relation in server

ID	Name	Sex	Salary	EduLevel	Age
001	Smith	Male	8000	Master	25
002	Victor	Male	3000	Master	22
003	Rose	Female	4000	None	23
004	Cathy	Female	5000	Docto	27
005	Tom	Male	6000	Master	25
006	Susan	Female	7000	Master	22

In client, Substitute *"Update R Set EduLevel= 'master' where EduLevel=' Bachelor'"* with: *Step1*:"Delete from R where EduLevel=' Bachelor'"; *Step2*:"Insert into R (001,'Victor','Male', 3000,' Master', 22)" and "Insert into R (006,'Susan', 'Female', 7000,'Master', 22)"

Do Step 1 firstly, and compare T: "EduLevel ='Bachelor'" with the tuples to be inserted. Because they are irrelevant, do not process Step 2 and the updated DC_T is φ.

Both of processes can stay coherency, and we can prove the specification 1 by theorem 2 given as follows.

Theorem 2. After the operations of *delete* or *insert,* the semantic coherency of DC is satisfiable.

Proof. We first prove the operation of *delete* is satisfiable. Consider t is the tuple to be *deleted,* and $t \in DC_{rs}$; after deleting t, DC_{rs} will change to $<DC_R, DC_P, DC_{rs}-\{t\}>$. Consider A is the data set in the server before the operation of *delete,* i.e. $A=S_{rs}=\sigma_{Sp}(S_R)$, and $t \in \sigma_{Sp}(S_R)$; after deleting t, S_{rs} will change to $<S_R, S_P, S_{rs}-\{t\}>$. Because the coherency of SC_{rs} is satisfiable, we have $A=SC_{rs}$, then $B=A-\{t\}=SC_{rs}-\{t\}$. So the semantic coherency after *delete* operation can be proved.

Next, we prove the operation of *insert* is satisfiable. Consider t is the tuple to be *inserted* in the server, if abandoning the operation of insert, i.e. t can not satisfy any semantic descriptions cached in client, it has $t \notin DC_{rs}$, and $t \notin \sigma_{Sp}(S_R)$. If so, never are semantic descriptions affected by *insert* t to S_R in the server. If t is the tuple to be inserted in the caching, according to the *insert* specification, t must match one of the semantic descriptions of DC_R. Suppose DC is $<DC_R, DC_P, DC_{rs}>$ before inserting, then DC will change to $<DC_R, DC_P, DC_{rs} \cup \{t\}>$ after that operation. Consider $A=DC_{rs}$, because t satisfy the semantic descriptions of DC and $t \in \sigma_{DCp}(DC_{rs})$, then $B=\sigma_{Sp}(S_R)= A \cup \{t\}$. Because SC satisfy coherency before inserting, $B=A \cup \{t\}= SC_{rs} \cup \{t\}$. So the semantic coherency after insert operation can be proved.   ■

Because of the complexity of semantic caching coherency and so many factors that have to be considered, some constraint conditions are given as following to simplify the problems. (**1**) We argue that the semantic caching model and processing can be reduced significantly by using *Embedded Database* to store data and their semantic descriptions. (**2**) Either the server or its clients should process the operations of *query/update/insert/delete*, in which we should term out semantic caching coherency control scheme to do with *update/insert/delete*. To simplify, we argue if the operations of *update/insert/delete* happen in client, they should be sent to server and committed. If the server commits successfully, it is the *updated/inserted/deleted* data in the server which should be considered how to maintenance coherency between the server and its clients. (**3**) The cached semantic descriptions should stay original without being changed by coherency control scheme. (**4**) We should optimize the committed processes in the server when the network is disconnected.

3   Semantic Caching Model

According to we have discussed, to construct semantic caching model, the following objectives and goals should be achieved: (**1**) Both results of previous queries and their semantic descriptions should be stored and processed; meanwhile, the semantic caching coherency defined in definition 2 should be satisfiable; (**2**) It is easy to process the operations of trimming, evaluation, combination, and replacement. Fig 1 illustrates the semantic caching model consists of client model and server model.

Because we argue that it is easy to process projection in clients, we simplify the model by caching the whole tuples without *projection*, where F is the semantic

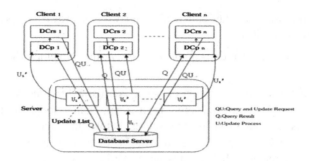

Fig. 1. Frame of semantic caching model

descriptions discussed in this article and we only consider one relation or view, although the proposed model can be easily extended to handle other more complicated queries. Unlike other semantic caching models, *failure* query is cached as well in our client model, which could further reduce the accessing to the database and simplify the sub expressions in *Where* clause.

Definition 3. Query Q is $<Q_R, Q_P, Q_{rs}>$, where Q_R is the relation or view to be queried; Q_P is the semantic description, i.e., F; $Q_{rs} = \sigma_{Qp}(Q_R)$, $Q_R \subset SA$; where SA is the semantic caching and query Q does not process projection.

Definition 4. Query Result Caching DC

DC is previous query results and their semantic descriptions *queried/trimmed /combined/replaced* at one time. Every DC_i has the unique corresponding semantic description DC_{Pi}; DC is $<DC_R, DC_P, DC_{rs}>$. where $DC_{rs} = \sigma_{DCp}(DC_R)$, $DC_R \subset SA$.

Definition 5. Result Caching SC

SC is one of the server relation or view cached in client; SC is $<SC_R, SC_P, SC_{rs}>$, where $SC_R = \cup DC_{iR}$; $SC_P = \vee DC_{ip} (1 \le i \le n)$; $SC_{rs} = \sigma_{SCp}(SC_R)$; and SC is composed of DC_i with the same relation or view.

Fig 2 shows the SC made up of 3 DC_i, obviously, $SC_R = \cup DC_{iR}$ (i=1, 2, 3).

We design and implement semantic caching based on *Embedded Database*. Consequently, we show how to implement it in detail in Fig 3.

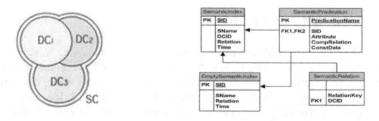

Fig. 2. Frame of semantic caching model   **Fig. 3.** E-R diagram of Semantic caching in clients

The objective to design server model is to determine which client is relevant to the *query/delete/insert* processes in the server and optimize *query/delete/insert* processes when network is disconnected.

Definition 6. Client Index CI

CI is <S, Q, C>, where S is all relations or views in the server, Q is a query, C is the client which is the results queried by R, i.e., $C = \sigma_Q(S)$;

Definition 7. Update List U_i

U_i is the update queue made up of the operations of *update, delete* and *insert* corresponding to one client.

U_i is $U_i = \{U_{1i}, U_{2i}, ..., U_{ji},\}$, and any two U_{hi}, U_{ki} ($1 \le h, k \le j, h \ne k$) are irrelevant. That is to say,

1) If U_{hi}, U_{ki} is *Insert*, i.e. Insert into R tuples A_{hi} (or A_{ki}), then $A_{hi} \ne A_{ki}$ is satisfiable.

2) If U_{hi}, U_{ki} is *Delete*, i.e. Delete from R_{hi} (or R_{ki}) where T_{hi}, T_{ki}, then

①$R_{hi} \ne R_{ki}$; or② when $R_{hi} = R_{ki}$, $T_{hi} \wedge T_{ki} = \emptyset$ is satisfiable.

3) If U_{hi}, U_{ki} is *Update*, i.e.Update R_{hi} (or R_{ki}) set A_{hi}(or A_{ki}) where $T_{hi}(T_{ki})$, then

①$R_{hi} \ne R_{ki}$; or②when $R_{hi} = R_{ki}$, $A_{hi} \wedge A_{ki} \ne \emptyset$; or ③when $R_{hi} = R_{ki}$ and $A_{hi} \wedge A_{ki} = \emptyset$, then $T_{hi} \wedge T_{ki} = \emptyset$ is satisfiable.

4   Semantic Caching Coherency Control Scheme

Based on semantic caching model given in section 3, we propose formal definitions and control scheme associated with semantic caching coherency, which consist of update list control scheme, client coherency control scheme, etc. Our scheme is belong to weak coherency control scheme [4].

4.1   Update List Control Scheme

As discussed in section 2.3 and 3.2, when there is an operation (*update/delete /insert*) committed successfully in the server, we would use update list control scheme to optimize the update list. When the network connection is reliable, every successfully committed operation will be sent to corresponding clients directly. However, when the network is disconnected, all of those operations will store in *update* list to wait for the network reconnected again, meanwhile, we can optimize the update operation by the *Update List Optimization Algorithm* as follows, with which we can combine and replace relevant *update* lists to save bandwidth cost.

Algorithm 1. Update List Optimization Algorithm

   (1) $\exists U_{new}$, U_{new} is ①when there is an operation update committed successfully in the server, we can know which client i is associated with U_{new} by retrieving client index-CI defined in definition 6; or②in update list and to be sent to client I, if the network is disconnected, U_{new} should be added into update list.

   (2) If U_i is empty, U_{new} will be added into U_i and execute (4); if not, go on (2);

(3) If $\forall U_{hi}$ ($U_{hi} \in U_i, 1 \leq h \leq j$), ①$R_{hi} \neq R_{new}$ is satisfiable; or ②if $R_{hi}=R_{new}$, $A_{hi} \neq A_{new}$ is satisfiable; or ③if $R_{hi}=R_{new}$ and $A_{hi}=A_{new}, T_{hi} \bigwedge T_{new} = \emptyset$ is satisfiable, then execute $U_i=U_i+U_{new}$; and go on (4).

(4) If $\exists\ U_{hi}$ ($U_{hi} \in U_{temp}, 1 \leq h \leq j$), when $R_{hi}=R_{new}$, $A_{hi}=A_{new}$ $T_{hi} \bigwedge T_{new} \neq \emptyset$ is satisfiable, then suppose $U_{temp}=U_{new}$, and go on executing:

while ($U_{temp} \neq \emptyset$, $\ulcorner U_{hi}$ ($U_{hi} \in U_{temp}, 1 \leq h \leq j$))

$\{$if (($R_{hi} \neq R_{new}$) $\bigvee$ ($R_{hi}=R_{new}, \boxminus A_{hi} \neq A_{new}$) $\bigvee$ ($R_{hi}=R_{new}$, $A_{hi}=A_{new} \boxminus T_{hi} \bigwedge T_{new} = \emptyset$)) then

$\{U_i=U_i+U_{new}$; $U_{temp}= U_{temp}-U_{hi}$; $\}$

else if($R_{hi}=R_{new}$, $A_{hi}=A_{new} \boxminus T_{hi}=T_{new}$) then

$\{A_{hi}= A_{new}$; $U_i=U_i-U_{hi}$; $U_{temp}= U_{temp}+U_{hi}$;$\}\}$

if ($U_{temp} \neq \emptyset$) then k=the number of U_{hi} ($U_{hi} \in U_{temp}$)

for (i=1;i=i+1;i<k)

$\{T_{hi}= T_{hi}-T_{hi} \bigwedge T_{new}$; $T_{new}= T_{new}+T_{hi} \bigwedge T_{new}$; Update U_{hi};$\}$

(5) If the network is reconnected again, send *update* items to corresponding client i and *update*.

By Algorithm 1, we can the *update* irrelevant items, which can save the network bandwidth and optimize *update* operation in clients.

4.2 Client Update Control Scheme

When the network connection is reliable, the committed operations including *update/delete/insert* in the server will be sent to corresponding clients. According to Specification 1 and Theorem 2, we should process *Client Update Control Algorithm*:

Algorithm 2. Client Update Control Algorithm
 (1) If the received operation is *delete*, process it directly in SC_R;
 (2) If the received operation is *insert*, analyze it into atom *insert* operation; i.e.
   Insert into R tuples (A_i (i=1,2,..., n));
   ①Traverse cached semantic description and compare atom *insert* with every SC_P;
   ②If $\forall\ DC_{Pj}$, $DC_{Pj} \subseteq SC_P$; meanwhile $\forall A_i$, and $A_i \subseteq DC_{Pj}$, then process *insert*;
   ③ If not, abandon the operation.
 (3) If the received operation is *update*,
   ① Analyze and substitute it with *delete* and *insert*;
   ② Process *delete* firstly as (1) shows;
   ③ And then do with *insert* as (2) describes.

5 Semantic Caching Interfaces and Protocol

We design semantic caching interface and protocol. The update list component is built in the server or agents, which communicates with the client through interface. We design SCP (Semantic caching Protocol) to support the communication between the servers or agents and the clients, belonging to the application layer of TCP/IP protocol.

SCP Data Pack is composed of HEADER (the length of message, 4 bytes) and DATA. Field Meanings of DATA are showed in table 6 and 7:

Table 6. Field meanings of data

LENGTH	Is the length of DATA, 2 bytes.
OPCODE	Is the operation type, which is showed in detail in table 7, has 3 bits, represents 8 operations, and has used 6 in SCP.
LOC	Specifies where the message comes from, 0 represents client and 1 represents server.
PRENUM	Specifies how many predicts in this message, 3 bits.
RR	Specifies the message is request or response, 0 represents request and 1 represent response.
MESG-ID	Uniquely Identifies a SCP message combined with IP address, 32 bits.
OP-DATA	Operation content, maximum 64K.

Table 7. Opecode and Op-data

OPCODE	Instruction	The meaning of Instruction	LOC	OP-DATA
0	NOP	No action, only test	0 : client; 1 : server	None
1	RT_C2A	Return result to caching	0	Result
2	REQ_C2S	Send request to server	0	Predict, R, Result
3	REQ_A2C	Send request to client	0	Predict, R, Result
4	REPLY_S2C	Return result to client	1	Predict, R, Result
5	REQ_S2S	Send request to the component in server	1	Predict, R, Result

6 Performance Study

We design a simulation study to examine the performance of the semantic caching coherency control scheme, especially effects analysis of optimization *Update* list and coherency control scheme. **(1) Semantic caching model:** we use *Ebase* database researched and designed by *Software Research Institute of SUN YAT-SEN University* as the embedded database in semantic caching model. The system is modeled to be composed of a server, a client and a network connecting them, in which the server's OS is Sun Solaris7 and database is Oracle9i, the client's OS is windows 2000 in which semantic caching is built.**2) Simulation Environment:** the simulator is implemented in C++, which simulates a simple but typical client-server model and is composed of *Query* Generator module, *Update* Generator module, Network Manager Module and Statistic Analysis module. **3) Experiment Data:** we use a modified Wisconsin Benchmark[9] to examine the performance. The benchmark database contains one single relation, R, with 5,000 tuples; each tuple has a size of 256 bytes, all of which are stored in the server (as shown in table 8).

Table 8. Main parameters of the simulation experiments

Parameter	Value	Meaning
DBSize	5000	Number of tuples in the sample database
TupleSize	256	Tuple size of the database(Bytes)
R_{update}	500	Rate of update at database(second)
P_{ins}	0.5	Proportion of insert operation in update at database
P_{del}	0.5	Proportion of delete operation in update at database

We compare the performance of semantic caching coherency control scheme with the case when there is no *Update* list in the server, and then demonstrate the effects optimization *Update* list

1) Effects analysis of coherency control scheme (as Fig 4 shows)
For examining the performance of semantic caching service, we compare the response time of queries from a client to retrieve in a replication: one is with semantic caching service, while the other is without. With 60 kinds of different query types and numbers, 11 sets of test tuples, and the results of the experiments are shown in Fig 4.

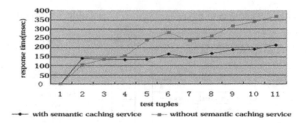

Fig. 4. Effects analysis of semantic caching service

Obviously, the change in *Query* complexity and the number of *Query* items have impact on the both response time, but semantic caching service can improve retrieving process, save response time and make replication more efficient.

2) Effects analysis of optimization *Update* list (as Fig 5 shows)
As we can see, with the help of optimization *Update* list, we can optimize the *Update* list and process efficiently, even though the growth of *update* items.

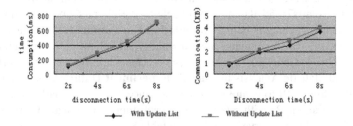

Fig. 5. Effects analysis of optimization *Update* list

7 Conclusions and Future Work

We have examined and analyzed one critical aspect of the semantic caching technology – semantic caching coherency control scheme from the respects of definition, conflicts analysis, semantic model, coherency control scheme, interface and protocol, and performance evaluations.

For the future research, we plan to extend our semantic caching coherency control scheme to include more complicated *update* such as multiple R or materialized views, etc, which increase the complexity in coherency control processing.

References

1. Qun Ren, Margaret H. Dunham, Semantic Caching and Query Processing, IEEE transactions on knowledge and data engineering,vol.15,No.1,January/February 2003 192-210
2. Wu Tingting, Zhou Xingming, Extracting Query Results from Semantic Cache, Chinese Computer Journal, 2002, 10:1104-1110

3. Wu Tingting, Zhang Wensong, Zhou Xingming Answering Query through Cache during Disconnection, Chinese Computer Journal,2003,10: 1393-1399
4. Wu Tingting, Zhang Wensong, Zhou Xingming, Weak Cache Coherency in Mobile Environments, Computer Engineering & Science Vol.26,No.4,2004:59-61
5. Wu Hengshan, Deng Zhifeng, A semantic Cache coherency Schemes in Mobile computing Environments, Computer Engineering, May 2003:126-127
6. U.Cetintemel, P.J. Keleher Support for Speculative Update Propagation and Mobility in Deno, Proc.IEEE Int'1 Conf. Distributed Computing Systems, 2001.
7. Dong Hyun Kim Validation-Based Reprocessing Scheme for Updating Spatial Data in Mobile Computing Environments Proceedings of the17th International Conference on Advanced Information Networking and Applications (AINA'03)
8. Huang Shi-neng, Xi Jian-qing, distributed data cache systems Huang Shi-neng, Xi Jian-qing, Journal of software 2001,12(7):1094-1100
9. J. Gray, The Benchmark Handbook. Morgan Kaufmann, 1993.

Energy Efficient United L2 Cache Design with Instruction/Data Filter Scheme

Zhiqiang Ma, Zhenzhou Ji, Mingzeng Hu, and Yi Ji

Department of Computer Science and Technology of Harbin Institute of Technology,
150001 Harbin, China
{mzq, jzz, mzhu, jiyi}@pact518.hit.edu.cn

Abstract. The on-chip caches usually consume a significant amount of energy in modern microprocessors. This paper presents an I/D filter scheme to reduce the energy consumption of united L2 caches shared by instructions and data. By adding an I/D indicator bit, the cache block is classified into I-block and D-block. For instruction and data accesses, only the corresponding blocks instead of all the blocks in the same set selected are accessed. By this method, we can easily filter the unnecessary way activities and save the energy consumption. This technique uses a small amount of additional hardware without increasing the cache access latency, and the area overhead is negligible. Simplescalar simulator and CACTI were used to evaluate the performance of our proposed architecture, the results shows that the I/D filter scheme is energy efficient for set-associative caches.

1 Introduction

The need in low-power processor design is growing due to the reliability problem for high frequency, high temperature processor chip and expanding market for battery powered mobile devices. The on-chip cache memory is one significant source of power consumption to which many researchers have paid attention. For example, the 21264 DEC Alpha chip dissipates 21% [1] and the ARM 920T dissipates 44% [2] of its total power in caches. For this reason the on-chip caches become the most attractive targets for power reduction.

As a result, numerous structural techniques have been proposed to conserve the power dissipated by caches. Sub-banking and bit-line segmentation divides the data into sub-segments. In each cache access, only those sub-segments that contain the desired data can be read out [3][4]. Block buffering [3], multiple line buffer [4], and filter cache [5] place a small cache (i.e., L0 cache) or output latches between the processor and the L1 cache to exploit the spatial locality of reference and reduce L1 cache activities, thereby saving power consumption. Reconfigurable cache [6] and selective cache ways [7] dynamically reorganize the cache architecture (size and associativity et.) for the intended application. A subset of cache components can be disabled to reduce the power consumption when they are not required to achieve good performance. FVC [8][9] and DZC [10] employ compression in cache memory array. Values are stored in an encoded form occupying only a few bits, power can be saved by not

J. Cao, W. Nejdl, and M. Xu (Eds.): APPT 2005, LNCS 3756, pp. 52–60, 2005.
© Springer-Verlag Berlin Heidelberg 2005

driving the other unused bits. There are still some techniques to save power consumption by reducing the way activities, such as way-predicting [11] and sentry-tag [12]. Fortunately, to a significant extent these techniques are orthogonal to each other.

In this paper, we propose an I/D filter scheme to reduce the power consumed by united L2 (ul2) caches shared by instruction and data. By using an I/D indicator bit to classify each cache block into I-block and D-block. For instruction and data accesses, only the corresponding blocks instead of all the blocks in the same set selected are accessed. By this method, we can easy filter the unnecessary way activities and save the power consumption efficiently.

The rest of the paper is organized as follows. Section 2 identifies the problems of the conventional implementation of ul2 caches. Section 3 describes details of our proposed I/D filter scheme and cache architecture. Section 4 gives evaluation results for all benchmarks taken from SPECint95. Section 5 concludes the paper.

2 Conventional Cache Design

In a conventional cache design as shown in Fig. 1, all the cache ways are accessed at the same time to minimize the access delay. In other words, in a W-way set-associative cache (a direct-mapped cache when W = 1), there are always W-way activities per cache access. In other words, in a four-way set-associative cache, there are always four-way activities per cache access, as shown by the gray blocks in Fig. 1. This parallel access scheme is good for the performance, but it is not optimized from the viewpoint of power consumption. This if because that at best only one way would be hit per access, the other (W − 1) ways access would result in a lot of unnecessary way activities, thus large power consumption, especially, when W grows large.

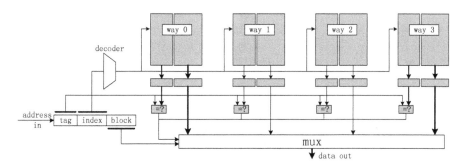

Fig. 1. Conventional four-way set-associative cache architecture (*the gray blocks represent the active ways*)

Form the perspective of united caches shared by instruction and data, cache blocks can be classified into two categories: I-block which holds the instruction values and D-block which holds data values. For instruction and data accesses, only the corresponding blocks instead of all the blocks in the same set selected should be accessed. In other words, for instruction requires resulted from il1 misses, the accesses of D-blocks are unnecessary, it's same for data requires. If we know this result before starting the conventional cache access, we can filter the unnecessary cache activities. For example, in a four-way set-associative cache, the attribute vector of the desired four blocks is "I, D, D, I", then we may only enable way 0 and 3 for an I-access, and way 1 and 2 for a D-access, instead of accessing all the four ways, thus, the cache power consumption can be reduced.

3 I/D Filtered Cache

In this section, we present the I/D filter scheme and the low-power ul2 cache structure in detail. We also present the necessary circuit changes required to implement the I/D filter scheme and discuss the effect of the circuit changes on area and delay.

3.1 Filter Scheme

To distinguish between an I-block and D-block, we add an additional I/D indicator bit for each cache block, an I/D indicator bit of '1' indicates an I-block, and '0' a D-block. By pre-comparing the access type with what's I/D indicator bit contents, we can effectively identify which ways activities are unnecessary, then disable these cache ways in the following cache access, thus, the cache power consumption can be decreased. The I/D indicator bit is stored in a single 8T latch. After the bit value is written into the latch, its content is stable as long as power is supplied. A '1' should be written to the I/D indicator bit if the required block was previously reloaded from the lower level memory for a instruction request miss, otherwise, a '0' should be written for a data request miss.

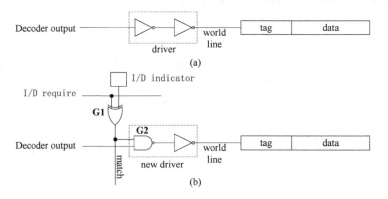

Fig. 2. Control circuit of: (a) Conventional cache and (b) Cache with I/D filter scheme

Unlike in the conventional cache design shown in Fig. 2 (a), the word line is derived from the set decoder directly, additional control circuit was added to enable/disable the cache way access. As shown in Fig. 2 (b), we use a XOR gate (G1) to verify whether the I/D require signal match the content of the I/D indicator or not, then the match signal is use to assert or disable the selected word line though an NADD gate (G2). The world line signal goes active only when the required block's

I/D indicator and I/D require signal matched, otherwise, the cache way is disabled automatically. It is easy to decide what the current I/D require value is, the value is '1' if the access require comes from L1 instruction cache, and '0' from L1 data cache.

3.2   Cache Architecture with I/D Filter Scheme

Fig. 3 depicts a four-way set-associative cache with the proposed I/D filter scheme. Compared to the conventional set-associative caches, the hardware augmentations include only I/D indicator arrays and the control circuit. As shown in Fig. 3, we suppose an instruction access is asserted (the I/D require signal is '1'), the selected I/D indicator vector is "1, 0, 0, 1", then the unmatched way 1 and 2 are disable by the control block shown in Fig. 2 (b). Thus, 50% way activities could be reduced for the current access.

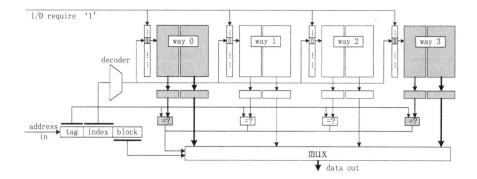

Fig. 3. Four-way set-associative cache architecture with I/D filter scheme (*the gray   blocks represent the active ways*)

3.3   Area and Delay Overhead

The match verification of required blocks is determined in parallel with normal set decoder, the verification time can be completely hidden by the decoder. Consequently, the augmentation of XOR gate will not prolong the cache access latency. But comparing with the origin word line driver, the new driver would increase the driving delay of world line since the NAND gate contains more transistors than an inverter. Fortunately, the total increase to the cache critical path can be avoided if the NAND gate transistor size is tripled, which does not represent a significant overall area increase since those transistors didn't occupy much area to begin with [9]. Thus, we can maintain the same cache access delay with the new cache line driver. Our cache architecture will not increase the cache access latency.

   We use the transistor number as measurement in the area overhead analysis. As described above, the I/D indicator bit was implemented with the 8T latch, in addition, we added a XOR gate and a NADN gate for each cache block to control whether the selected word line should be asserted or not, the area overhead for each cache block is approximately 8 + 6 + 6 = 20 transistors. In this paper, we use the 6T SRAM cell to implement the tag and data array, the area spent in a single cache block is approximately $(B \times 8 \times 6 + T \times 6)$ transistors, in which B is the block size in byte and T is the number bits of tag used. For the base ul2 cache configuration described in the next section, the area overhead is $20 / (64 \times 8 \times 6 + T \times 6) < 1\%$, thus, it is negligible.

4 Experiments

In this section we make experiments to evaluate the performance of our proposed architecture. We first describe the simulation environment and benchmark selection. We then present the results and give some analysis.

4.1 Configurations Studied

We use SimpleScalar [13] to model the ul2 cache with I/D filter scheme in an out-of-order microprocessor. PISA instruction set is adopted in the simulator, and the simulator can simulate the execution of any instruction on each pipeline stage accurately. The necessary modifications to the ul2 cache have been implemented to measure the efficiency of I/D filter scheme. For our base case, the ul2 cache is assumed to be a 256KB, W-way set-associative unified on-chip cache with a block size of 64 bytes. We assume a 16KB, direct-mapped Ll I-cache and a 16KB, 4-way set-associative Ll D-cache. The line sizes for both of these caches are set to 32 bytes. A buddy replacement algorithm, approximating LRU, was used for all the set-associative caches.

Table 1. Benchmarks

Benchmark	Spec dir.	Input data set	Simulation insts
gcc	ref	-o3 genrecog.i	116 million
compress	-	40000 e 2231	124 million
go	-	9 9	133 million
ijpeg	train	vigo.ppm	166 million
li	test	test.lsq (queens 7)	202 million
perl	train	scrable.pl < scrable.in	108 million
m88ksim	train	-c < ctl.in	119 million
vortex	train	vortex.in	101 million

4.2 Benchmarks

All SPECint95 benchmarks were selected for the further analysis and presentation in our studies. The benchmarks used in this paper are compiled with Simplescalar compiler, which is a derivation of *gcc-2.7.2*. All the benchmarks were simulated from the start to the end. The detail input data set is listed in Table 1. Since the train run of benchmark *perl* is too shorter than the other benchmarks, we modified the train input file "*scrabble.in*" and add two words "*abodome*" and "*evilds*", now the file contains {*zed, veil, vanity, abodome, evilds*}.

4.3 Average Filter Rate

The I/D filter scheme reduces the cache power consumption by eliminating the unnecessary cache activities, the amount of power savings by this method depends on the average filter rate (R_{filter}), which can be defined as the ratio of the average unnecessary way activities to the number of cache ways. By definition, the average filter rate is given by

$$R_{filter} = W_{access} / W . \qquad (1)$$

Where W_{access} is the average number of accessible ways in each cache access.

The higher value of R_{filter} means that the I/D filter scheme is more efficient in filtering out the unnecessary way activities. Fig. 4 plots the filter rates of ul2 cache over a range of associativities on all our eight workloads. We found that W play a key rule in the filter efficiency. The filter rate differences between a direct mapped (W=1) and a 8-way (W=8) associative cache are quite significant, especially, we got a very low filter rate (average 2.23%) under direct mapped cache. This is because that decreasing the cache associativities will lower the probability for the co-existence of instruction and data in the same set. As a result, from the viewpoint of filter rate, this technique will be more suitable for highly-associative cache design. For example, the average filter rates are about 24.93%, 33.74% and 38.91% for 2-, 4- and 8-way set-associative caches.

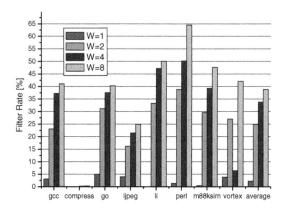

Fig. 4. Filter rates for various numbers of cache ways

Form Fig. 4, we also found that the filter rates vary greatly between different applications, and the filter scheme benefits little for a few applications, especially, the filter rates for *compress* are totally lower than 1%. The reason is that it is out of balance for instruction/data accesses and storage. Fig. 5 gives the access and storage occupations of I-block and D-block in 4-way set-associative ul2 cache. For *compress*, D-block occupies nearly all the ul2 cache accesses (99.93%) and storage (99.54%), the ul2 cache could nearly be seen as a level 2 data cache, obviously, fewer unnecessary way activities can be save. Fortunately, except for quite a few applications, our propose technique get an ideal reduction of way activities for most other applications, since on the whole many applications, such as *gcc, go, li, perl, m88ksim* and *vortex*, are instruction dominated (occupy more than 50%) for cache accesses however data dominated for cache storage. It is means that a lot of D-block activities in these applications can be filtered for instruction accesses that dominate the ul2 cache accesses.

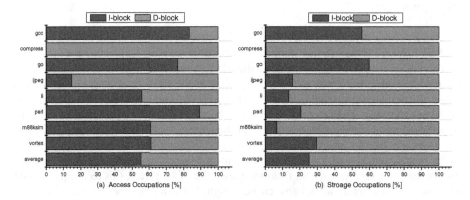

Fig. 5. (a) Access and (b) Storage occupations of I-block and D-block in four-way set-associative ul2 cache

4.4 Energy Savings

Our main goal is to reduce energy consumed by the unnecessary way activities. As mentioned above, the overall energy saving depends on the average filter efficiency. Then the average energy savings per access can be given by

$$E_{save} = R_{filter} \times E_{ways} .$$
(2)

Then the energy saving in percentage is given by

$$R_{save} = R_{filter} \times (E_{ways} / E_{cache}) .$$
(3)

Where (E_{ways} / E_{cache}) is the fraction of the full ways power dissipated in the conventional cache. We used CACTI 3.0 [14] to simulate the energy consumption for the base cache configurations at a 0.18um feature. For all the various numbers of ways used above, the simulation results are illustrated in Table 2.

Table 2. Energy consumption for various numbers of ways

Ways	E_{ways} [nJ]	E_{cache} [nJ]	E_{ways} / E_{cache} [%]
W = 1	1.50600	2.13314	70.60
W = 2	1.52634	2.12812	71.72
W = 4	1.52542	2.16190	70.56
W = 8	1.81505	2.72149	66.69

By using (3), the ul2 cache energy savings of every application are shown in Fig. 6, which is similar to the form of the filter rate shown in Fig. 4, since programs with higher filter rates can reduce more way activities. On average, for direct-mapped cache only 1.57% of cache accessing energy can be saved, but about 17.88%, 23.81% and 25.95% can be saved for 2-, 4- and 8-way set-associative caches. Our proposed filter scheme will be more suitable for set-associative cache design.

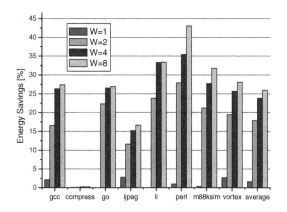

Fig. 6. Energy savings for various numbers of cache ways

5 Conclusion

The on-chip caches usually consume a significant fraction of total power dissipation in modern processors. Form the perspective of united caches shared by instructions and data, cache blocks can be classified into two categories: I-block which holds the instruction values and D-block which holds data values. For instruction or data access we need only read the corresponding blocks instead of all the blocks in the same set selected. Base on this observation, in this paper we propose an I/D filter scheme to reduce the ul2 cache power consumption. By using an I/D indicator bit to classify cache block into I-blocks and D-block, we can easy reduce the unnecessary way activities, thus, the cache power consumption can be reduced.

This technique uses a small amount of additional hardware without increasing the cache access latency, and the area overhead (less than 1%) is negligible. From additional experiments we conclude that the I/D filter scheme will be suitable for the set-associative cache design and reduce the ul2 cache power consumption efficiently. In the baseline cache configurations, on average, about 1.57%, 17.88%, 23.81% and 25.95% of cache accessing energy can be reduced for 1-, 2-, 4- and 8-way set-associative caches.

References

1. Edmondson, J.F. et al.: Internal Organization of the Alpha 21164, A 300-MHz 64-bit Quad-issue CMOS RISC Microprocessor. Digital Tech. J., Vol. 7 (1995)
2. Montenaro, J. et al.: A 160MHz 32b 0.5W CMOS RISC Microprocessor. Int. Solid-State Circuits Conf. (1996)

3. Su, C.L. Despain, A.M.: Cache Design for Energy Efficiency. in Proc. 28th Int. System Sciences Conf. (1995)
4. Ghose, K. Kamble, M.B.: Reducing Power in Superscalar Processor Caches Using Sub-banking, Multiple Line Buffers and Bit-line Segmentation. in Proc. Int. Low Power Electronics and Design Symp. (1999)
5. Kin, J. Gupta, M. Mangione-Smith, W.H.: The Filter Cache: An Energy Efficient Memory Structure. in Proc. 30th Int. Microarchitecture Symp. (1997) 184–193
6. Ranganathan, P. Adve, S. Jouppi, N.: Reconfigurable CACHEs and their Application to Media Processing. International Symposium on Computer Architecture (IACA), (2000) 214–224
7. Albonesi, D.H.: Selective CACHE Ways: On Demand CACHE Resource Allocation. IEEE/ACM International Symposium on Microarchitecture (MICRO-32) (1999) 248–259
8. Yang, J. Gupta, R.: Energy Efficient Frequent Value Data Cache Design. Int. Symp. on Microarchitecture (2002)
9. Zhang, C. Yang, J. Vahid, F.: Low Static-Power Frequent-Value Data Caches. Design, Automation and Test in Europe Conference (DATE '04), Paris, France (2004) 214–219
10. Villa, L. Zhang, M. Asanovic, K.: Dynamic Zero Compression for Cache Energy Reduction. IEEE/ACM International Symposium on Microarchitecture (MICRO-33) (2000) 214–220
11. Inoue, K. Ishihara, T. Murakami, K.: Way-predicting Set-associative Cache for High Performance and Low Energy Consumption. in Proc. Int. Low Power Electronics and Design Symp. (1999) 273–275
12. Chang, Y.J. Lai, F. Ruan, S.J.: An Efficient Two-level Filter Scheme for Low Power Cache. IEEE/ACM 11th Int. Logic and Synthesis Workshop, New Orleans, LA (2002)
13. Burge, D. Austin, T.: The Simplescalar Tool Set, Version 2.0. Technical Report CS-TR-97-1342, Univ. of Wisconsin, Madison (1997)
14. Shivakumar, P. Jouppi, N.: CACTI 3.0: An Integrated Cache Timing, Power, and Area Model. COMPAQ Western Research Lab (2001)

Improving Latency Tolerance of Network Processors Through Simultaneous Multithreading

Bo Liang[1], Hong An[1,2], Fang Lu[1], and Rui Guo[1]

[1] Department of Computer Science and Technology,
University of Science and Technology of China, Hefei 230026, China
{boliang, lufang, timmyguo}@mail.ustc.edu.cn
han@ustc.edu.cn
[2] Computer Architecture Laboratory, Institute of Computing Technology,
Chinese Academy of Sciences, Beijing 100086, China

Abstract. Existing multithreaded network processors architecture with multiple processing engines (PEs), aims at taking advantage of blocked multithreading technique which executes instructions of different user-defined threads in the same PE pipeline, in explicit and interleave way. Multiple PEs, each of which is a multithreaded processor core, process several packets in parallel to hide long memory access latency. Most of them are optimized for throughputs mostly in data-plane. In future network workloads, the boundaries between data-plane and control-plane become blurred, so that PEs are demanded not only wire speed packet forwarding on data-plane, but also highly intelligent and increased complex packet processing function on control-plane. In this paper, we analyze SMT's short latency tolerance potential when used in out-of-order and dynamic scheduling PE cores. We show in this paper that 2~4 issue SMT provides an excellent short memory and branch latency tolerance, which gain higher instructions throughout as well as much simpler structures.

1 Introduction

Network processors (NPs), such as Intel's IXP [1], IBM's PowerNP [2], and Motorola's C-Port [3], are programmable microprocessors designed specifically to build packet switches and optimized for a variety of packet processing functions such as IP forwarding, filtering, network address translation, metering and policing, supporting for virtual private networks, protocol translation, and others. NPs consists of multiple execution engines, each of which is a multithreaded processor core processing several packets in parallel to hide long memory access latency (usually more than 100 cycles) and to increase their overall computing power. In general, processing engines are intended to carry out data-plane functions. Control-plane functions could be implemented in a co-processor or a host processor. At least, above three major NP offering (IXP, Power NP and C-Port) fall in this broad architecture. Multiple NPs may be combined to form a distributed packet switch.

Each of multithreaded processing engines (usually called PE or ME) with above NPs architecture features is connected to link-layer interfaces and to the packet buffers. Additional storage is also present in the form of SRAM and DRAM to store

J. Cao, W. Nejdl, and M. Xu (Eds.): APPT 2005, LNCS 3756, pp. 61–70, 2005.
© Springer-Verlag Berlin Heidelberg 2005

program data. Because handling of one packet is largely independent of another, multithreading can successfully hide various data access latency. However, since the conflicts of storage resource sharing among multiple threads, one of the problems of multithreading is the degradation of the memory system performance, both in terms of miss latency and bandwidth requirements. In addition, NPs keep on growing their capabilities to exploit thread-level parallelism in order to gain higher wire speed packet processing, which makes the negative impact of memory latencies on performance even higher. To alleviate this problem, most current NPs, combined with using multiprocessing and multithreading, devote a high fraction of their transistors to on-chip memory system and sophisticated data access mechanism in order to reduce the average data access time.

Existing multithreaded NPs architecture which PE is usually designed simple (maybe in-order and low issue-width core) in order to reduce issue logic complexity, aims at taking advantage of blocked multithreading (BMT) technique (sometimes also called coarse-grain multithreading) which execute instructions of different user-defined threads in the same pipeline, in explicit and interleave way. It means that a single thread is executed until it reaches a situation that triggers a context switch. Usually such situation arises when the instruction execution reaches a long-latency operation or a situation where a long latency may arise.

At the same time, another key observation is that over the past few years, several vendors have been releasing NPs having a number of different architectures, but most of them are optimized for throughputs mostly in data-plane. Also, existing benchmark suites for network processors primarily contain data-plane workloads, e.g. CommBench[4] and NetBench[5], which perform packet processing for a forwarding function. Although NPs have initially been targeted for data-plane applications, they also play a major role in the control-plane. In fact, future network interfaces not only demand wire speed packet forwarding on data-plane, but also require highly intelligent traffic management and increased complex packet processing function on control-plane. The boundaries between data-plane and control-plane have become blurred [6]. The recent trend is that some control-plane activities, such as TCP and SSL applications, are being considered as a commodity. For example, a new network benchmark NpBench, target towards control-plane (e.g., traffic management, quality of service, etc.) as well as data plane workloads [6].

This work would like to present a simultaneous multithreading (SMT) microarchitecture to tolerate the short latency after the packets have been loaded onto the chip. We believe that SMT include out-of-order and dynamic scheduling techniques, will show to be an effective technique to boost the ILP from future network workloads in which data-plane and control-plane activities have become blurred. In this paper, we analyzed SMT's short latency tolerance potential when used in out-of-order and dynamic scheduling PE cores. We show in this paper that 2~4 issue SMT provides an excellent short memory latency and branch latency tolerance, which gain higher instructions throughout as well as much simpler structures.

The rest of the paper is organized as follows. Section 2 provides the related work on network processors and benchmarks. Section 3 introduces our Simulation Methodology. Section 4 presents our simulation results. And we conclude the paper give out the future work of this study in section 5. Finally, section 6 is our Acknowledgement.

2 Related Works

Tzi-Cker evaluated a series of three progressively more aggressive routing-table cache designs in [9]. He found that the incorporation of hardware caches into network processors, when combined with efficient caching algorithms, can signifi-cantly improve the overall packet forwarding performance due to a sufficiently high degree of temporal locality in the network packet streams.

Intel's IXP [1] provide several types of on-chip memory or buffers which have different capacities and access speed, the data used frequently was kept in the fastest memory while the data used not frequently was kept in the slower memory.

Timothy Sherwood proposed a pipelined memory design that emphasizes worst-case throughput over latency in [10], and he concluded the pipelined memory design is efficient for improving the throughput of NP.

Joan-Manuel Parcerisa and Antonio Gonzalez explored the latency tolerance of SMT architecture through decoupling technology in [12], but the latency referred in [12] is L2 cache access latency, it is not suite for network processors, and the benchmark used in [12] is SPEC which is greatly different from the network processing workload.

Hily examined the behavior of three of the best performing branch prediction strategies (bimod, Gshare, Gselect) while executing several threads of instructions simultaneously Simulation Methodology in [13], and found that in multiprogramming environment if the sizes of the tables (PHT and BTB) are proportional to the number of active threads, there are very few interactions, be they destructive or constructive.

Matt Ramsay evaluate, in [14], the prediction accuracy of four branch predictor configurations: (1) a totally shared predictor, (2) a completely split predictor, (3) a predictor with a shared history and split BHT, and (4) a predictor with a shared BHT and separate history registers, each for two static prediction schemes, a generic 2-bit predictor, a share predictor, and a YAGS predictor. He also concluded that system performance is only marginally affected by branch prediction accuracy in a multithreaded environment because thread-level parallelism allows for the hiding of long latency hazards, such as branch mispredicts.

3 Simulation Methodology

3.1 Simulator

The simulator used in our investigation is called ss-smt[7], a SMT simulator modified from simplescalar [8]. As Figure 1 shows, the instruction queues (i-queues) are per thread distributed, and both RUU and LSQ buffer can be ether shared by all threads or per thread distributed, and they are configured as shared in our simulation. The Fetch stage fetches instructions from the il1-cache, giving priority to the thread that has few instructions in the pipeline. This technique, called ICOUNT in [11], achieved better performance.

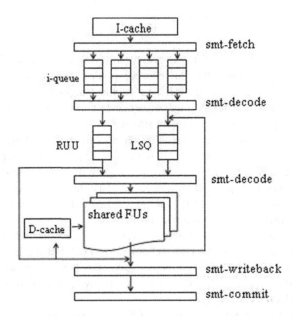

Fig. 1. SMT architecture: (1) i-queue is per thread distributed. (2) Instruction fetch schedule is ICOUNT. (3) RUU and LSQ buffer are shared by all threads.

3.2 Benchmark

NpBench[6] includes 10 applications which can be categorized into three functional groups: traffic-management and quality of service group (TQG), security and media processing group (SMG), and packet processing group (PPG). This categorization is presented in Table 1. We can see that the suite includes several control plane functions as they are missing from the available NP workloads.

3.3 Experimental Framework

The author of NpBench has studied the instruction mix of the 10 applications, and found that the memory access instruction took a portion of 28.2% (19.3% of Load and 8.9% of Store) of the total instructions, while branch instructions took 16.2%. In other words, there is one memory access instruction in each 3.5 instructions, and one branch instruction in each 6 instructions. Additionally the memory access latency is usually longer (tens of cycles) than the branch latency. And all of this told us that NpBench is memory access sensitive.

For a single-issue core, to tolerate a p cycles' latency, we need find p independent instructions, while for a w-issue core, to tolerate the same p cycles' latency, we will have to find $p*w$ independent instructions. Here we can presume that the capability of latency tolerance is an inverse ratio to the issue-width.

We have investigated the memory access latency tolerance and branch latency tolerance. Since the simulator is very slow and the average code size processing a packet is very small (11890 instructions), we run only a portion of 50 M (million) instructions of each thread, after skipping an initial start-up phase and all the threads execute the same application with the same input in a test.

In the experiment of memory access latency, we examined memory access latency tolerance of an idea superscalar (32-issue superscalar, SS-32), a typical superscalar (4-issue superscalar, SS-4), a single-issue SMT (a degraded SMT, SMT-1), a 2-issue SMT (SMT-2) and a 4-issue SMT (SMT-4) respectively. Branch predictor is modeled as always correct (perfect) in this test; therefore, branch pollution effects are not taken into account.

Table 1. Descriptions of the NpBench Suite

Group	Application	Descriptions	Data Plane	Control Plane
TQG	WFQ	Weighted Fair Queuing is a queue scheduling algorithm	X	X
	RED	Random Early Detection is an active queue management algorithm which drops arriving packets probabilistically	X	X
	SSLD	Secure Sockets Layer Dispatcher is an example of content-based switching mechanism	X	X
	MPLS	Multi Protocol Layer Switching is a forwarding technology using short labels	X	X
SMG	MTC	Media Transcoding is the process that a media object in one representation is converted into another representation for wide spectrum of client types	X	X
	AES	Advanced Encryption Standard (RijnDael) is a block cipher that encrypts and decrypts 128, 192 and 256 bits blocks	X	
	MD5	Message Digestion algorithm takes as input a message of arbitrary length and produces as output a 128-bit fingerprint or message digest of the input	X	
	DH	Diffie-Hellman key exchange allows two parties who have not met to exchange keys securely on an insecure communication path	X	
PPG	FRAG	FRAG is a packet fragmentation application	X	
	CRC	Cyclic Redundancy Check is used in Ethernet and ATM Adaptation Layer 5 (AAL-5) checksum calculation	X	

In the experiment of branch latency, we examined the branch tolerance of SMT-1 architecture with a Gshare predictor, with a bimodal predictor and without any predictor respectively, and each thread has an independent predictor in order to evaluate the max performance of each predictor. The PHT (patter history table) size of the Gshare predictor is 1024 and the GHR is 10 bits. The bimodal predictor was configured as having 512 entries. Similar to a previous experiment, the memory access latency was configured as 1 cycle's, and therefore, the memory access pollution effects are not taken into account either.

4 Test Result

4.1 The Memory Latency Hiding Effectiveness of SMT

4.1.1 Superscalar Core
For the purpose of comparison, the memory access latency tolerance of the superscalar architecture is firstly investigated. From the simulated result shown in Figure 2(a), it can be seen that even SS-32 only gets an IPC of 4.67 with 1 cycle's memory access latency, and then drops to an IPC of 1.63 when the latency increases up to 40 cycles. For a typical superscalar, SS-4, the situation follows the same pattern: its IPC gets a max value of 2.45 when the memory access latency is only 1 cycle, and drops rapidly to 1.17 at 10 cycles latency, and 0.72 at 20 cycles, and 0.41 at 40.

It means that the superscalar architecture, though considered an ideal architecture, cannot deal well with the memory access latency in NpBench. And for a multi-issue superscalar architecture, the memory access latency will lead to a large waste in chip resources.

4.1.2 Single Issue SMT
Strictly speaking, "a single issue SMT" is not an accurate term, while SMT is proposed to improved latency tolerance of a superscalar core. When the issue width decreased to 1, the superscalar core degrades to a scalar core, and the SMT degrades

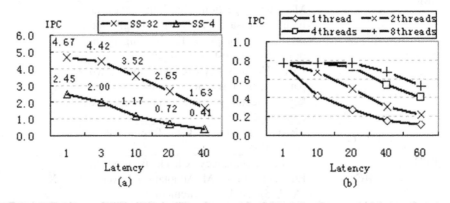

Fig. 2. (a) Memory access latency tolerance of superscalar architecture; (b) Memory access latency tolerance of SMT-1

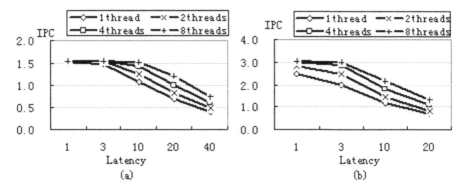

Fig. 3. (a) Memory access latency tolerance of SMT-2; (b) Memory access latency tolerance of SMT-4

to coarse-grain-like multithreading accordingly. But in the following part of this paper we will continue to use the term "single issue SMT" for the sake of consistency.

Figure 2(b) shows the performance of the single-issue SMT core, from which we can see that IPC get it max value of 0.77 at 1-cycle latency, and with the increase of the latency, we can keep IPC almost unchanged by adding threads until the latencies beyond 20 cycles, in other words, by exploiting the thread level parallelism (TLP) we can hide most of the latencies in the range of 1 to 40 cycles. For example, a 4-threads single-issue SMT can keep almost the same IPC when the latency is smaller than 10cycls, or 20 cycles for 8 threads.

4.1.3 2-Issue and 4-Issue SMT

Figure 3 presents the memory access latency tolerance of SMT-2 and SMT-4. As we have discussed in subsection 3.3, SMT-2 have a more strong capability of tolerating latency than SMT-4: for 8 threads, the SMT-2 core can well hide almost all the latency of 10 cycles, SMT-4, in comparison, 3 cycles. Furthermore, the performance,

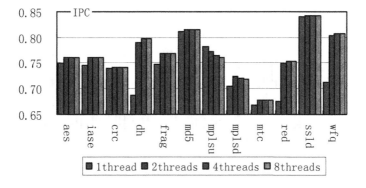

Fig. 4. Branch latency tolerance of Gshare predictor

measured in IPC, of SMT-2 is quite satisfactory, which almost doubles SMT-1's when memory access latency is shorter than 10 cycles. Thus, we can say that SMT-2 is a good choice if the memory access latency is short than 10 cycles.

4.2 The Branch Latency Hiding Effectiveness of SMT

4.2.1 Gshare, a Two Level Branch Predictor

Figure 4 show the IPC of SMT-1 with a complex two-level branch predictor, Gshare. As we can see, IPC of some applications (aes, iaes, dh, frag, mtc, red and wfq) picks up with the increase of the number of threads, while IPC of some applications drop (such as mplsu and mplsd), mainly because of the conflict and disturbance resulting from the sharing use of some on-chip resources between threads. But altogether the average good predict rate is above 98%, which turns out to be a fairly good result.

4.2.2 Bimod-512 Branch Predictor

Figure 5 show the performance of using a bimodal branch predictor with 512 entries. IPC shows in figure 5 are more similar to, even better than, that in figure 4, and which

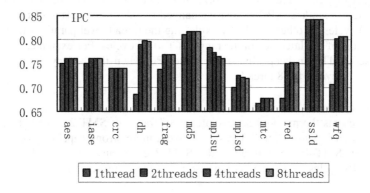

Fig. 5. Branch latency tolerance of bimodal predictor

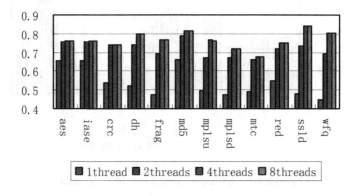

Fig. 6. Tolerate branch latency without any predictor

told us that the bimodal predictor can tolerate the branch latency in NpBench as well as Gshare, although it is simpler.

4.2.3 None Predictor
At last, figure 6 shows the simulation results of without any predictor. In this situation, the 2cycle's branch latency was tolerated only by exploring TLP. And we can see that IPC of all the 10 applications have a remarkable increase (average 33.8%) when the number of threads changed form 1 to 2; IPC of most applications have a increase of average 6.7% when the number of threads changed from 2 to 4; while IPC of most application bears almost no changes when the number of thread changed from 4 to 8. That is to say that 4 threads are enough to hide the 2cycles' branch latency and the performance can be as good as using Gshar or bimodal predictor.

5 Conclusion and Future Work

We have explored the latency tolerance of SMT architecture when used in out-of-order and dynamic scheduling network processors PE cores. From the simulation result we can conclude that: (1) NpBench workload is memory-sensitive in the environment of SMT; (2) the impact of branch latency to system performance is marginal in SMT environment; (3) SMT architecture can well tolerate the short memory access latency (1~20 cycles) in NpBench with the hardware's support for rapid content switching; (4) 2~4 issue SMT provides an excellent short memory and branch latency tolerance, which gain higher instructions throughout as well as much simpler structures.

In our simulation, however, we find that the size of RUU is a systemic bottleneck, especially when the number of thread is over 4. How to solve this problem is the main task in the next steps.

Furthermore, the simulator we used in this investigation is a simulator of GPP, which is not optimized for the network processing. Therefore, in the future, we will adopt a more accurate NP simulator. This work will include the optimization of the instruction set of the simulator, add special function unit for some special application, and some other work.

Acknowledgement

This work has been supported by the National Natural Science Foundation of China under Grant No.60373043; the Natural Science Foundation of Anhui Province of China under Grant No.050420206; the National High Technology Development Program of China under Grant No.2001AA111100; 2002AA110010.

References

1. Intel Corporation: Intel IXP2400 Network Processor Family Hardware Reference Manual.(June 2001)
2. IBM Corporation: The Network Processor: Enabling Technology for High-Performance Networking. IBM Microelectronics (1999)

3. C-Port   Corporation: C-5   Digital   Communications   Processor.   *from*
   http://www.cportcorp.com /solutions/docs/c5brief.pdf (1999)
4. T. Wolf and M. Franklin: CommBench - A Telecommunications Benchmark for Network
   Processors. International Symposium on Performance Analysis of Systems and Software
   (Apr. 2000).
5. G. Memik, W. Mangione-smith and W. Hu: NetBench: A Benchmarking Suite for
   Network Processors. 2001 IEEE/ACM International Conference on Computer-Aided
   Design (2001)
6. Byeong Kil Lee: NpBench: A Benchmark Suite for Control plane and Data plane
   Applications for Network Processors. IEEE International Conference on Computer Design
   (October 2003)
7. Ronaldo Gonçalves, Eduard Ayguadé, Mateo Valero, Philippe Navaux: A Simulator for
   SMT Architectures: Evaluating Instruction Cache Topologies. X II SBAC-PAD, Brazil
   (2000)
8. Simplescalar Simulator, *from* http://www.simplerscalar.com
9. Tzi-Cker Chiueh, Prashant Pradhan: Cache Memory Design for Network Processors. *In
   Proceeding of* the 6th International Symposium. on High Performance Computer
   Architecture, Tolouse, France (January 2000)
10. Timothy Sherwood, George Varghese, Brad Calder: A Pipelined Memory Architecture for
    High Throughput Network Processors, *In Proceedings of* the 30th Annual, ISCA'03
    (2003)
11. Jahangir Hasan, Satish Chandra1, T. N. Vijaykumar: Efficient Use of Memory Bandwidth
    to Improve Network Processor Throughput. *In Proceedings of* the 30th Annual, ISCA'03
    (2003)
12. Joan-Manuel Parcerisa and Antonio Gonzalez: Improving Latency Tolerance of
    Multithreading through Decoupling, IEEE Transactions on Computers, Vol. 50, No. 10
    (October 2001)
13. S. Hily and A. Seznec: Branch Prediction and Simultaneous Multithreading. *In proceeding
    of* International Conference on Parallel Architecture and Compilation Techniques (1996)
14. Matt Ramsay, Chris Feucht, and Mikko H.Lipasti: Exploring Efficient SMT Branch
    Predictor Design. Workshop on Complexity-Effective Design, in conjunction with ISCA
    (June 2003)

RIMP: Runtime Implicit Predication

YuXing Tang, Kun Deng, XiaoDong Wang, Yong Dou, and XingMing Zhou

National Lab for Parallel and distributed Processing, China
{tyx, kundeng, xdwang, xmzhou}@nudt.edu.cn, yongdou@163.com

Abstract. If-conversion and predicated execution are widely adopted to eliminate branch misprediction penalty. Previous predication execution depends on compiler to generate explicit predicated instructions. In this paper, a trace-based predicate mechanism named RIMP (Runtime IMplicit Predication) is discussed. The candidates of if-conversion will be identified during dynamic execution. Conventional trace cache has been modified to store RIMP traces, which include instructions both from fall-through and target block following the conditional branch. Hardware extension will add predication to RIMP trace automatically. With the help of RIMP, legacy applications can benefit from predication mechanism without recompiling source code. Simulation of RIMP implementation under diverse microarchitecture configurations is presented in the paper. Results have shown promising performance improvement. In general, RIMP with 64kB trace storage delivers an average 10.3% IPC improvement while actually speeding up the execution time by over 7%.

Keywords: predication, trace cache, runtime execution, RIMP.

1 Introduction

In popular deep pipeline and wide issue architecture, the misprediction penalty of branch is a significant bottleneck [1][10]. The small basic block and frequent branch prevent the sophisticated pipeline from achieving its performance potential.

Predicated execution is a good way to deal with those conditional branches. Modern processor, such as Itanium [1], MAJC [11], ARM and TI DSP [7], provides the support of predication in various degrees.

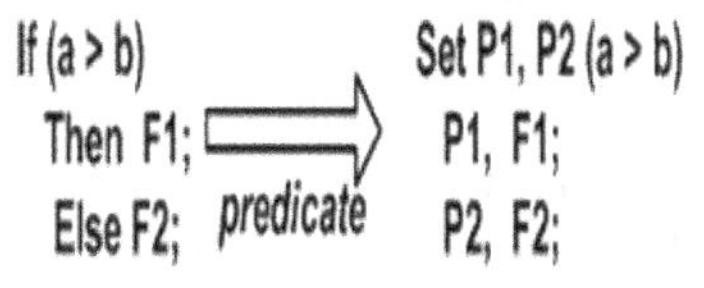

As shown in fig.1, predication eliminates *If-branch*. Control dependent operations in either *Then* or *Else* path will be inverted into data dependent counterpart with proper predication (P1 or P2).

Without if-branch (see right side of fig.1), the straight-line code will improve the instruction fetching performance and remove the penalty of possible misprediction.

Fig. 1. Overview of if-conversion and predication

Generally predication needs the predicated ISA [1][3] or special information from compiler [2] to direct whether the instructions from both paths of conditional branches should commit their result.

J. Cao, W. Nejdl, and M. Xu (Eds.): APPT 2005, LNCS 3756, pp. 71–80, 2005.
© Springer-Verlag Berlin Heidelberg 2005

More ISA support will make the compiler easier to generate code, but will introduce more serious compatibility problem. Previous dynamic mechanisms also need compiler or binary instrumentation tool to identify predicable branch [16]. Those codes, built without the knowledge of predicated hardware, can't benefit from predication execution.

This paper presents a new hardware mechanism for predication, called RIMP (Runtime IMplicit Predication), where a suitable set of instructions will be identified, transferred into predicated form and stored in simplified trace cache. We call it as *implicit predication* because neither compiler nor ISA will be assumed to know the existence of predication. By eliminating the misprediction penalty, RIMP increases the exploited ILP and performance, and releases the urgency for precise and complex branch predictor. Although RIMP uses on-line hardware to realize the whole procedure of predication, all extension are out of the critical datapath of processor and will not increase implement complexity tremendously. Actually besides light-weighted trace cache [4][5] support, just a few of modifications are needed in microarchitecture.

The rest of the paper is organized as follows. Section 2 gives a brief overview of related work. Section 3 describes the RIMP microarchitecture, including RIIMP trace generation, optimization and management in detail. Section 4 presents the simulation and finally, Section 5 concludes with a summary and ideas for future studies.

2 Related Work

Cydra 5 [16] may be the first commercial implementation of predication for high ILP. Mahlke et al. [8] used the hyerblock to support predicated architecures. The guarded execution proposed by Pnevmatikatos and Sohi [13], specifies the predication information in instruction with bit-mask. Previous researches assume the full predicated ISA or predicated-extended ISA.

Tyson [15] and Mahlke [14] studied the potential benefits of predication. They found that almost 30% of the dynamic branch could be removed by full predication. They also pointed out the 5% performance potential of partial predication [17]. RIMP does not need full predication supported ISA, but achieves over 7% raw improvement with tiny partial predicated extension.

Little work can be found by the term of "runtime predication" or "dynamic predication". J.E.Smith and Aramon et al. [19] use dual path hardware to execution both the then-path and else-path at the same time. Chang et al. [18] apply predication in the speculative architecture. Additional source input is redirect false-predication dependence. Klauser [17] proposes dynamic hammock to inject Cmov instruction. A compiler or binary instrumentation tool is used to indicate proper hammock. The predication in RIMP is transparent to all software (application, compiler and OS) and users. No further recompiling or binary instrumentation is needed.

3 RIMP

As illustrated in fig.2, RIMP is out of the microprocessor's critical datapath. RIMP monitors all instructions retired from commit stags. After a corresponding branch has been found, RIMP trace will be compacted, tested, optimized and then stored without stalling regular execution of pipeline.

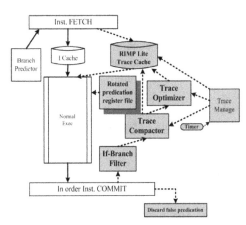

Fig. 2. Overview of RIMP architecture

A special trace cache [5] is used to store and reuse the predicated code. Because RIMP remove all the inside branches, the traditional tag structure and control mechanism will be simplified. We call it as RIMP lite trace cache.

RIMP use a rotated 1-bit register file to support dynamic assignment of predication. If-branch filter and branch predictability are important for RIMP usability. Further information will be found in following section and simulation discussion.

3.1 RIMP Trace Generation

The left side of fig.3 is the normal instruction stream fetched from L1 instruction cache. The right side presents the result after predication. Note the four parts of trace candidate: *if-branch*, *then-path*, *else-path* and *merge-point*. This example comes from the 164.gzip.

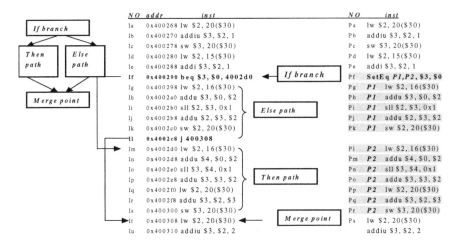

Fig. 3. RIMP trace generation by dynamic if-conversion

In fig.3 the No.If (Beq) instruction is the *if-branch*. At the same time, it indicates that those physical sequential instructions are valid in its *Else-path*. Absolute branch target address is to ensure the exclusive *Then-path*. The distance from if-branch to its target is significant to calculate whether the predicated trace could be saved in a trace cache line.

Comparing these two instruction streams, it is easy to see that RIMP substitutes the *if-branch* instruction into an internal instruction *SetEq*. *SetEq* will set 1-bit predication registers, P1 and P2 (predicates), according to the same if-condition. All instructions in then-path will be under the predicate of P1, and else-path under P2. The unconditional jump at the end of *else-path* (No.Il) has been removed. Finally, there will be no control instruction in RIMP trace.

New PC address will send to both I caches and trace cache. If the RIMP trace cache hits, predicated trace will be fetched into pipeline. Otherwise, processor gets instruction from I cache.

3.2 Eliminate Unnecessary Data Dependences

After beginning to fetch a RIMP trace, the instructions from both else-path and then-path will be mixed up in out-of-order pipeline. The mixed two paths will bring additional data dependences, which will influence the dispatch or issue stage in processor's pipeline.

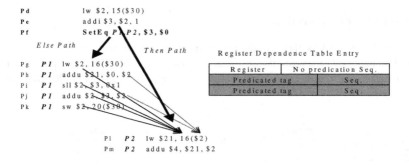

Fig. 4. The dependence in RIMP trace

As shown in fig.4, RAW hazards (Pg->Pl, Pi->Pl, Pj->Pl) over register $2 will prevent instruction No.Pl to be issued before No.Pk has been committed. But the instructions with opposite predication will never be committed at the same time. These data dependences are unnecessary.

RIMP extend the register dependence table (RDT) to prevent illusive hazards. For every architecture register, the field of predicated tag is to record corresponding predication variants, such as P1 or P2; the field of Seq. is to record which instruction will write result to this register. By removing illusive dependences (hazards), RIMP may utilize issuing bandwidth better than traditional predication.

3.3 Trace Manage

Lite trace cache is the kernel component of RIMP. Detail experiments have been done to inspect the effect different trace cache configuration in following section. Conventional trace scheduling and optimizing also can be applied in RIMP.

The basic block is the construct unit for traditional trace cache. But as shown in Fig.3, those instructions, which in front of if-branch or behind the merge point, actually will not do any help for predication. These **impredicative** instructions will occupy precious trace line space. Selective trace-start and end point can be more efficient than just the boundary of basic block, but precise heuristic of trace generation and fetch control are needed.

Predication is a real double-blade sward. Removed branches eliminate misprediction penalty, but almost half of the predicated instructions in RIMP trace are useless. The instruction with false predicates will occupy the execution resources and have their own cost. Only when the execution time for wrong path is less than misprediction penalty, the predication can be valuable. Thanks for today's super-pipeline in microprocessor, it always the case. Experiments have been done to discover RIMP's performance when the number of pipeline stage and penalty is increased.

4 Simulation

RIMP has been simulated in a highly revised simplescalar v3.0d simulator [20]. The baseline architecture parameters are in Table 1.

Table 1. RIMP baseline architecture overview

Instruction	64bits Alpha style instruction, quad word memory access
L1 Icache	32kB, 32 byte lines, 2-way set associative, 1 cycles for hit
L1 Dcache	32kB, 32 byte lines, 4-way set associative, 1 cycles for hit
L2 Cache	256kB, 64 byte lines, 4-way set associative, 6 cycles for hit
RIMP Trace cache	1024 internal operation storage
Predictor	Default bimod in SS with 2048 entries, 90% hit rate in SPEC2k
Fetch width	4 instruction per cycle
Issue width	4 instruction per cycle
Mis-prediction	3 cycle for pipeline flush
Reorder buffer	16 entry and other 8 entry for load/store
Function Units and latency (total/issue)	4 Int ALU (1/1), 1 Int Mult (2/1) / Div (20/19), 4 memory (1/1), 4 FP Add (2/1), 1 FP Mult (4/1) / Div (12/12) / Sqrt (24/24)

The precision of branch predicator is an important factor for RIMP. Default bimod predictor in SS3.0d is used because the compromise of hit rate and implement complexity. Two consecutive fail in predictor will make the branch hard-to-predict. Because the predictability of a branch may change, a timer for periodical flush is used.

The whole SPECint2000 benchmark suit is used, which compiled and linked with peak configuration in alpha-EV6 platform, using SimPoint as input. Three metrics are applied to evaluate and compare different aspects of RIMP.

- *Committed rate* of RIMP instruction. The usability of dynamic predication after static optimization
- *IPC improvement.* Eliminating branch and unnecessary dependence will improve the parallelism.
- *Speedup of execution time.* Because RIMP executes the instructions in false predication, but not commit them, IPC improvement cannot present real performance. Execution time (cycle) can illustrate the raw improvement.

4.1 RIMP Trace Performance

I n fig.5, fig.6 and fig.7, for the same 1024 internal instructions, the RIMP trace cache configuration of 8 sets of 8-way associative, 16 sets of 4-way associative, 32 sets of 2-way associative, and 64 sets of direct mapped are evaluated. All trace lines contain 16 internal predicated instructions. Considering the tradeoff between performance and complexity, 16 sets with 4-way associative trace cache is selected to be default configuration presented at the top of histograms.

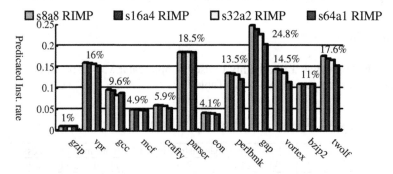

Fig. 5. Committed rate of runtime predicated instruction

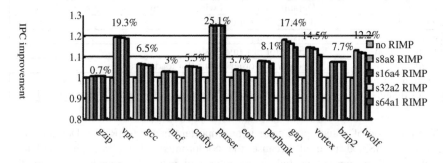

Fig. 6. IPC improvement from RIMP architecture

As shown in fig.5, every benchmark has the opportunity to use runtime predication. Because *163.gzip* contains too many indirect branches, and the default predictor achieves 98% precision in it, only %1 instructions are from RIMP. On average, 11.6% of all committed instructions are fetched from RIMP trace cache.

Fig.6 presents the ILP improvement (IPC) by RIMP, compared with the processor without dynamic predication. The average IPC improvement achieves more than 10.3%. These results come from the elimination of misprediction penalty of difficult branch. Because predication introduces useless instructions in wrong path, speedup of execution time in fig.7 will be the ultimate judge. The average speedup in execution time is above 7.59%.

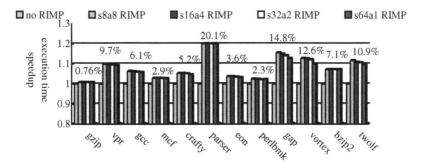

Fig. 7. Speedup of overall execution time

4.2 Enlarge Storage Space of RIMP

Because the replace rate of RIMP trace is much higher than traditional instruction cache, a normal way for improving is to enlarge predicated storage space. There are two ways for larger space, longer RIMP trace line or more trace cache line.

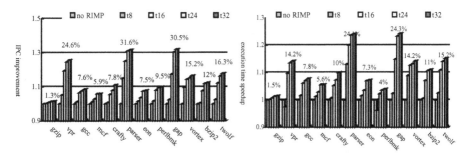

Fig. 8. IPC for different RIMP trace length limits   **Fig. 9.** Speedup for different trace length limits

We use tx to denote that the predicated trace may contain x instructions in single line maximally. In fig.8 and fig.9, tx will be increases from t8 to t32. Then the final trace cache space will be increased from 4KB to 16KB (for 512 to 2K internal 64-bit operation in 16 sets and 4-way association). In fig.10and fig.11, all RIMP traces are in 4-way association and maximum 16 instructions per line, but may have various trace sets, from 16 to 128. It means the whole RIMP trace cache size increases from 8KB to 64KB.

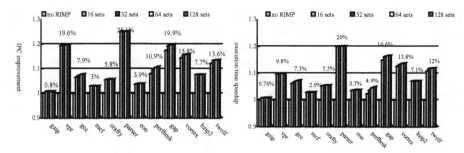

Fig. 10. IPC for increasing RIMP sets

Fig. 11. Time speedup for increasing RIMP sets

Table 2. Two configurations for enlarging RIMP Lite Trace Cache

Name	Assoc	RIMP sets	Line size (insn.)	Insn length	RIMP size
LongLine	4	16	32	64b(8B)	16KB
MoreSets	4	32	16	64b(8B)	16KB

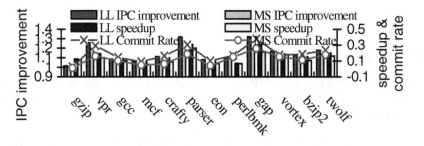

Fig. 12. Compare longer RIMP trace line with more trace sets

The whole performance will be improved when large predicated trace can be generated, stored and reused (fig.9). The simulation results show sustaining increase from t8 to t16, and this increase will be weaken from t24 (see fig.8 and fig.9). More sets will improve the performance, but these IPC improvements are weak. Fig.12 presents the different result for these two enlarging configurations in same RIMP trace cache space (16KB).

As shown in fig.12, in same RIMP trace storage space, enlarging every trace line size should be better than increasing the number of predicated traces. Due to the limit of fetch bandwidth, long trace cache line can be fetched in several sequential cycles. And long RIMP trace line will not increase tag costs or control complexity. More sets means more trace cache line, thus asks for more hardware tag space. When hardware has the ability for more predicated instruction, longer predicated trace will be preferred to more traces.

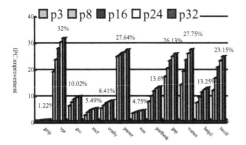
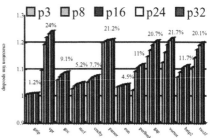

Fig. 13. RIMP's IPC under different mis-prediction penalty

Fig. 14. RIMP's speedup under different mis-prediction penalty

4.3 RIMP in Deep Pipeline

Modern processor may use deep pipeline (superpipelining) to increase frequency. More pipelining stage may have more instructions on-the-fly, thus increase the misprediction penalty to flush those instructions when predictor is fail.

Fig.13 and fig.14 present the improvement of RIMP while the processor architecture has different misprediction penalty. In general, when penalty is increased by deep pipeline, RIMP will be pivotal to eliminate misprediction.

5 Conclusion

RIMP can identify predicable if-branch, generate and store the predicated instructions into a reversed trace cache, without the need of re-compiling or binary instrument. By reusing the RIMP trace in following execution, the whole performance will be improved. As modern processor use deeper pipeline and has higher frequency, RIMP will have more attraction. The impaction of various RIMP configurations has been exposed in this paper. Future work will concentrate on combining RIMP trace cache with traditional trace cache, detecting the influence of interlaced different traces.

References

1. H. Sharangpani and K. Aurora. Itanium processor microarchitecture. IEEE Micro, 20(5):24-43, Sept-Oct 2000.
2. W. Chuang, B. Calder, J. Ferrante. Phi-Predication for Light-Weight If-Conversion. Proceedings of the Intl. Symposium on code generation and optimization, pages 179-190, Mar. 2003
3. J. Sias, H. Hunter, and W. Hwu. Enhancing loop buffering of media and telecommunication applications using low-overhead predication. In Proceedings of the 34th MICRO, Dec. 2001.
4. Quinn Jacobson, James E Smith. Trace preconstruction]. Proceedings of the 27th Annual International Symposium on Computer Architecture (ISCA-00). Vancouver: IEEE Computer Society Press, 2000. 37-46.

5. Eric Rotenberg, Steve Bennett, James E Smith. Trace cache: a low latency approach to high bandwidth instruction fetching [A]. Proceedings of the 29th MICRO: IEEE Computer Society Press, 1996. 24-35.
6. R. L. Sites and R. T. Witek. Alpha AXP Architecture Reference Manual: 2nd Ed. Digital Press, Boston MA, 1995.
7. Oliver Sohm. Variable-Length Decding on the TMS320C6000 DSP platform. Application Report. June 2002. http://www-s.ti.com/sc/psheets/spra805/spra805.pdf
8. S. A. Mahlke, D. C. Lin, W. Y. Chen, R. E. Hank, and R. A. Bringmann, Effective Compiler Support for Predicated Execution Using the Hyperblock. In 25th Intl. Conf. On Microarchitecture, pages 45-54, Dec. 1992.
9. Linlen Gwennap. Intel's P6 uses ducoupled superscalar Design. Microprocessor Report, Vol.9, No.2, Feb. 1995.
10. Hyper-pipelined technology: Intel Pentium 4 Processor – Product Overview. 2004 http://www.intel.com/designPentium4/prodbref/
11. M Tremblay, J Chan, S Chaudhry, A W Conigliaro, S S Tse. The MAJC Architecture: A Synthesis of Parallelism and Scalability. IEEE Micro vol.20 Issue 6, pages 12-25, Nov. 2000.
12. Kevin Krewell. Alhpa ev7 processor: a high-performance tradition continues. Microprocessor Report. In-Stat/MDR, April. 2002.
13. D. N. Pnevmatikatos and G. S. Sohi. Guarded Execution and Branch Prediction in Dynamic ILP processors. In 21st Intl. Symp. on computer architecture, pages 120-129, June 1994.
14. S. A. Mahlke, R. E. Hank, R. A. Bringmann, J. C. Gyllenhaal, D. M. Gallagher, and W. Hwu. Characterizing the Impact of Predicated Execution on Branch Prediction. In 27th Annual Intl. Symp. On Microarchitecture, San Jose, CA, Dec. 1994.
15. G. S. Tyson. The Effects of Predicated Execution on Branch Prediction. In 27th Annual Intl. Symp. On Microarchitecture, pages 196-206, San Jose, CA, Dec. 1994.
16. R. Rau, D. Yen, W. Yen, and R. Towle. The Cydra 5 Departmental Supercomputer. IEEE Computer, 22(1):12-35, Jan. 1989.
17. A. Klauser, T. Austin, D. Grunwald, B. Calder. Dynamic Hammock Predication for Non-predicated Instruction Set Architectures. Proceedings of ICPACT. 1998.
18. P. Y. Chang, E. Hao, Y. Patt, and P. Chang. Using Predicated Execution to Improve the Performance of a Dynamically Scheduled Machine with Speculative Execution. In Intl. Conf. On Parallel Arch. And Compilation Techniques, Limassol, Cyprus, June 1995.
19. J. L. Aramon, J. Gonzalez, A. Gonzalez, J. E. Smith. Dual path instruction processing. Proceeding of the 16th Intl. Conf. On Supercomputing. New York 2002
20. T. Austin, E. Larson, D. Ernst. SimpleScalar: an infrastructure for computer system modeling. IEEE computer. Vol.35, No.2, Pages 59-67 Feb. 2002.

Static Partitioning vs Dynamic Sharing of Resources in Simultaneous MultiThreading Microarchitectures

Chen Liu and Jean-Luc Gaudiot

Department of Electrical Engineering and Computer Science,
University of California, Irvine, CA 92697, USA
{cliu3, gaudiot}@uci.edu
http://pascal.eng.uci.edu

Abstract. Simultaneous MultiThreading (SMT) achieves better system resource utilization and higher performance because it exploits Thread-Level Parallelism (TLP) in addition to "conventional" Instruction-Level Parallelism (ILP). Theoretically, system resources in every pipeline stage of an SMT microarchitecture can be dynamically shared. However, in commercial applications, all the major queues are statically partitioned. From an implementation point of view, static partitioning of resources is easier to implement and has a lower hardware overhead and power consumption. In this paper, we strive to quantitatively determine the trade-off between static partitioning and dynamic sharing. We find that static partitioning of either the instruction fetch queue (IFQ) or the reorder buffer (ROB) is not sufficient if implemented alone (3% and 9% performance decrease respectively in the worst case comparing with dynamic sharing), while statically partitioning both the IFQ and the ROB could achieve an average performance gain of 9% at least, and even reach 148% when running with floating-point benchmarks, when compared with dynamic sharing. We varied the number of functional units in our efforts to isolate the reason for this performance improvement. We found that static partitioning both queues outperformed all the other partitioning mechanisms under the same system configuration. This demonstrates that the performance gain has been achieved by moving from dynamic sharing to static partitioning of the system resources.

1 Introduction

Simultaneous MultiThreading (SMT) has been a hot research area for more than one decade [14,15,16,17,18,19,20,21]. From the embryonic implementation in the CDC 6600 [22], the HEP [9], the TERA [8], the HORIZON [12], and the APRIL [13] architectures, in which there exists some concept of multi-threading or Simultaneous MultiThreading, to the actual commercial implementation of SMT in the latest Pentium 4 [10] and XEON [5] processor families with HyperThreading (HT) technology, all demonstrates the power of SMT (another commercial design of SMT, the COMPAQ EV8 [11], was abandoned

J. Cao, W. Nejdl, and M. Xu (Eds.): APPT 2005, LNCS 3756, pp. 81–90, 2005.
© Springer-Verlag Berlin Heidelberg 2005

even before reaching the manufacturing stage). Because of the limitations of Instruction-Level Parallelism (ILP), the performance gain that traditional superscalar processors could achieve is diminishing even with an increase in the number of execution units. On the other hand, through issuing and executing instructions from multiple threads at every clock cycle, SMT can achieve maximum system resource utilization and higher performance.

However, when it comes to the problem of how to allocate the system resources to the multiple threads, there are different opinions. Sometimes the dynamic sharing method is applied on system resources at every pipeline stage in the SMT microarchitectures [16,17,18] (which means threads can compete for the resources and there is no quota on the resources that one single thread could utilize), could be as low as 0%, or could be 100%. In other cases, all the major queues are statically partitioned [4,5], so that each thread has its own portion of the resources and there is no overlap.

From the implementation point view, static partitioning of resources is easier to implement with lower hardware overhead and less power consumption, which matches exactly with INTEL's implementation goal of hyperthreading – smallest hardware overhead and high enough performance gain [5]. On the other hand, dynamic sharing is normally assumed to be able to maximize the utilization of the system resources and corresponding performance, even though it would come at a higher hardware cost and more power consumption.

The goal of this paper is thus to quantify the impact of static partitioning vs. dynamic sharing on the overall performance of the system. We study the effect of different partitioning mechanisms (static partitioning vs dynamic sharing) on the different system resources (instruction fetch queue and reorder buffer, for example), and their impact on overall system performance.

Prior to our proposed work, we review related work of different partitioning methods on the system resources in Section 2. Section 3 describes our experiment approach. Our simulated work is discussed in more detail in Section 4. Conclusions are presented in Section 5.

2   Related Work

Marr *et al.* [5] presented a commercial implementation of a 2-thread SMT architecture in INTEL's XEON processor family. In their implementation, almost all the queues are statically divided into two, one for each thread. However, the scheduler queues are shared by both threads so that the schedulers can dispatch instructions to the execution engine regardless of which thread they come from, so as to insure timely execution and maintain a high throughput. However, there is still a cap on the number of instructions one thread could have in scheduler queues.

An investigation of the impact of different system resource partitioning mechanisms on SMT processors was performed by Raasch *et al.* in [1]. In this paper, various system resources, like instruction queue, reorder buffer, issue bandwidth, and commit bandwidth are studied under different partitioning mechanisms. The authors conclude that the true power of SMT lies in its ability to issue instructions from different threads in one clock cycle. Hence, the issue bandwidth has to be shared all the time. While different partitioning mechanisms on other system resources like storage queues will result in very little impact on the system performances. However, their work is mainly focused on the back-end of the pipeline, *e.g.*, execution and retirement part, did not affect any of the front-end of the pipeline, *e.g.*, the fetch part. We extended their work by studying the different partitioning techniques on the front-end instruction fetch queue and the back-end reorder buffer, as well as the impact on the overall performance caused by the interaction between them.

3   Our Approach

There are many system resources in a pipeline which could be under different partitioning mechanisms, for example, the instruction fetch queue, the instruction decode queue, the instruction issue queue (sometimes called instruction queue), the reorder buffer, the load/store queue, *etc.* In our proposed work, we selected two resources from above, the front-end instruction fetch queue (IFQ) and the back-end reorder buffer (ROB), and applied different partitioning mechanisms on them separately. Then, we compared the performance of each configuration to find out the impact of different partitioning mechanisms on the overall system performance, which is measured in terms of Instruction per Cycle (IPC). This comparison would lead us to get the optimum configuration. Here we listed all four combinations of architectures to simulate a 2-thread Simultaneous Multi-Threading architecture:

- SMT: Both the front-end instruction fetch queue and the back-end reorder buffer are in the dynamic sharing mode, just like other system resources.
- SIFQ: Only the front-end instruction fetch queue is divided into two, one for each thread, and other system resources are in the dynamic sharing mode.
- SROB: Only the back-end reorder buffer is divided into two, one for each thread, while other system resources are in the dynamic sharing mode.
- STOUS: Both the front-end instruction fetch queue and the back-end reorder buffer are divided into two, one for each thread, and other system resources are in the dynamic sharing mode.

  In each configuration, we perform extensive simulations to obtain the average system performance.

4   Simulation

To properly evaluate the effects of the proposed partitioning mechanism, we designed an execution-driven simulator, based on an SMT simulator developed by Kang *et al.* [7], which is itself derived from SimpleScalar [3], through modifying the *sim-outorder* simulator to implement an SMT processor model. Following the structure of *sim-outorder*, the architectural model contains seven pipeline stages: fetch, decode, dispatch, issue, execute, complete, and commit. Several resources, such as program counter (PC), integer and floating-point register files, and branch predictor, are replicated to allow multiple thread contexts. The simulator uses the 64-bit PISA instruction set.

4.1 Experiment Setup

The major simulator parameters are listed in Table 1. The fetch policy employed is Instruction Count (I-Count). The simulator is configured to issue as many instructions as the total number of functional units at each clock cycle according to the priority set by the I-Count policy.

The simulator has been modified to accommodate the changes of the corresponding sharing policy for IFQ and ROB. In Table 2, we listed the corresponding instruction fetch queue size and the reorder buffer size for each configuration.

The benchmarks used are all from SPEC CPU2000 benchmark suite [6]. The ten benchmarks used (7 integer and 3 floating-point benchmarks) are listed in Table 3.

Since there are 10 sets of benchmark, 4 sets of simulator configuration for the 2-thread input, we run each benchmark with all the benchmarks (including

Table 1. Simulation parameters

Parameter	Value
Instruction Fetch Rate	8
Instruction Decode Rate	8
Instruction Retire Rate	8
L1 Instruction Cache	64Kbytes (256:64:4:LRU)
L1 Data Cache	64Kbytes (512:32:4:LRU)
L2 Cache	1Mbytes (2048:128:4:LRU)
Memory Access Bus Width	32 bytes
Instruction TLB	512Kbytes (32:4096:4:LRU)
Data TLB	1Mbytes (64:4096:4:LRU)
Instruction Issue Queue Size	64
LQ/SQ Size	64/64
INT Units	8
FP Units	4

Table 2. Simulation setup

Configuration Name	Instruction Fetch Queue Size	Reorder Buffer Size
SMT	one 256	one 256
SIFQ	two 128	one 256
SROB	one 256	two 128
STOUS	two 128	two 128

Table 3. SPEC2000 CPU Benchmark used in the simulation

Benchmark	Type	Language	Category
164.gzip	INT	C	Compression
175.vpr	INT	C	FPGA Circuit Placement and Routing
176.gcc	INT	C	C Programming Language Compiler
179.art	FP	C	Image Recognition / Neural Networks
181.mcf	INT	C	Combinatorial Optimization
183.equake	FP	C	Seismic Wave Propagation Simulation
188.ammp	FP	C	Computational Chemistry
197.parser	INT	C	Word Processing
256.bzip2	INT	C	Compression
300.twolf	INT	C	Place and Route Simulator

itself). Hence altogether we run $4 \times 10 \times 10$ iterations of simulation to get all the results. Each iteration of simulation is composed of 1 billion instructions, after fast forwarding through the first 300 million instructions from each thread to skip the initialization part of the benchmark. Then the results (IPC) are averaged to get the average performance for each benchmark under each partitioning configuration.

4.2   Simulation Results

In Fig. 1, we present the average performance in term of IPC for each benchmark under different partitioning architectures. Obviously, the STOUS architecture outperforms other partitioning approaches. We derived the following formulas to compute the performance gain:

$$Gain1 = \frac{IPC_{SIFQ} - IPC_{SMT}}{IPC_{SMT}} \tag{1}$$

$$Gain2 = \frac{IPC_{SROB} - IPC_{SMT}}{IPC_{SMT}} \tag{2}$$

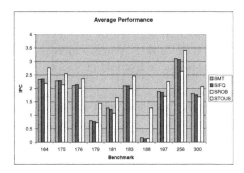

Fig. 1. Average performance gain for different partitioning architectures

Table 4. Performance Gain Comparison

	Gain1(%)	Gain2(%)	Gain3(%)
Overall average performance gain:	19.69	-6.94	148.25
Average performance gain (excluding benchmark 188)	-0.10	-7.66	23.67
Average performance gain (excluding benchmark 179 and 188)	-3.35	-8.66	8.99

$$Gain3 = \frac{IPC_{STOUS} - IPC_{SMT}}{IPC_{SMT}} \qquad (3)$$

When we examine the results more carefully under the light of the above formulas, we observe that when Benchmark 179 and 188 run together with other benchmarks, they could achieve such a huge performance gain (up to 7 or 8 fold), that they may exaggerate the performance gain achieved from other benchmarks. Therefore, in Table 4, we list the average performance gain in three different situations:

- Overall average performance gain, which is computed using the results of running all 10 benchmarks.
- Average performance gain excluding Benchmark 188, which is computing using only the results from running the remaining nine benchmarks.
- Average performance gain excluding Benchmark 179 and 188, which is computed using only the results from running the remaining eight benchmarks.

The reason why we want to compare the performance under these three different situations is because we want to examine the performance gain excluding the interference from those two benchmarks (179 and 188), to see how other benchmarks react to the different system partitioning architectures. From the table, we can see that the STOUS architecture keeps yielding positive performance gain, while other architectures could result in a loss of performance.

4.3   Impact of Functional Units

In order to isolate the reason why Benchmark 188 and 179 could achieve such huge performance gain, we redo the simulation by varying the number of INT and FP functional units as listed in Table 5. We also increased the size of instruction issue queue, load queue/store queue from 64 entries to 128 entries.

Table 5. Functional units configuration

Configuration	I	II	III
INT units	4	8	8
FP units	4	4	8

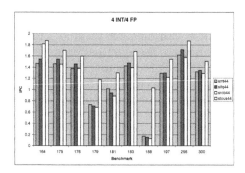

Fig. 2. Average IPC for 4 INT / 4 FP functional units configuration

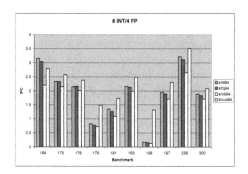

Fig. 3. Average IPC for 8 INT / 4 FP functional units configuration

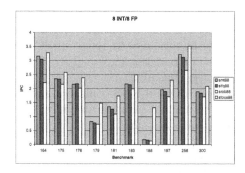

Fig. 4. Average IPC for 8 INT / 8 FP functional units configuration

In Fig. 2, 3, 4, we can see the average IPC for each partitioning mechanism with different functional unit configuration.

From the graph, we can see that with different variations in the number of functional units, the STOUS architecture keeps outperforming the SMT, SIFQ and SROB architectures in term of IPC. This demonstrates that the performance gain is achieved from the difference between static partitioning and dynamic sharing of the system resources, not because of the number of functional units in favor of any of the architectures.

5   Conclusions

From the above tables and graphs, several conclusions can be drawn:

1. Statically partitioning either the IFQ or the ROB solely can only lead to a negative performance gain.
2. Statically partitioning both the IFQ and ROB together could achieve marginal performance gain (even in the worst scenario when running integer benchmarks solely, the STOUS architecture could still achieve 9% performance gain over the SMT architecture).

We feel the reason for this is that static partitioning both the IFQ and ROB is like forcing the input and output of the pipeline to evenly execute the two input threads. Hence we can avoid the situation where one of the threads grabs more resources it could use and clogs the pipeline, while the other thread could not get enough resources and under-executed. Statically partitioning either one of them could not achieve such results because it only controls one end of the pipeline while there is no control over the other end.

The huge performance gain from running Benchmark 188 and 179 results from better system resource utilization with one integer and one floating-point input. Because now in stead of competing with each other for the same type of functional units, the instructions from different threads are running on different types of functional units. Hence the competition for those resources is minimized, the throughput is maximized, which shows the original power of SMT.

Through the static partitioning of the instruction fetch queue and the reorder buffer, we are able to achieve better performance (in terms of IPC) than dynamic sharing. Also, at the same time, static partitioning would require less hardware overhead, and also achieve less power consumption.

Since static partitioning of both IFQ and ROB could bring us this opportunity to achieve better performance with a less complicated mechanism, then the next step is to try different partitioning mechanisms on other system resources, to study the inter-relationship among them, and in the end to find an optimum way to sharing system resources to achieve the best performance with the least hardware overhead and power consumption for the SMT microarchitecture.

Acknowledgment. We would like to thank Dongsoo Kang for his support in our use of his SMT simulator.

References

1. Raasch, Steven E., Reinhardt, Steven K.: The Impact of Resource Partitioning on SMT Processors. Proceedings of the 12th Intenrational Conference on Parallel Architectures and Compilation Techniques (PACT 2003). New Orleans, Louisiana, USA. Sep. 27 - Oct. 01. (2003) 15–26
2. Sazeides, Y., Juan, T.: How to Compare the Performace of Two SMT Microarchitectures. Proceedings of 2001 IEEE International Symposium on Performance Analysis of System and Software (ISPASS-2001). Tucson, Arizona, USA, November 4-6. (2001)
3. Burger, D., Austin, T.: The SimpleScalar Tool Set, Version 2.0. University of Wisconsin-Madison Computer Science Department Technical Report No.1342. June (1997)
4. Koufaty, D., Marr, Deborah T.: Hyperthreading Technology in the Netburst Microarchitecture. IEEE Micro, March-April. (2003)
5. Marr, Deborah T., Binns, F., Hill, David L., Hinton, G., Koufaty, David A., Miller, J.Alan, Upton, M.: Hyper-Threading Technology Architecture and Microarchitecture. Intel Technology Journal Q1. (2002)
6. SPEC CPU 2000 Benchmark Suite: http://www.specbench.org/osg/cpu2000/ (2000)
7. Kang, D., Gaudiot, J-L.: Speculation control for simultaneous Multithreading. Proceedings of the 18th International Parallel and Distributed Processing Symposium (IPDPS 2004). Santa Fe, New Mexico, April 26-30.(2004)
8. Alverson, R., Callahan, D., Cummings, D., Koblenz, B., Porterfield, A., Smith, B.: The TERA Computer System. ACM SIGARCH Computer Architecture News, Vol. 18, No. 3.(1990) 1–6
9. Smith, B.J.: Architecture and Applications of the HEP Multiprocessor Computer System. SPIE Real Time Signal Processing IV. (1981) 241-248
10. Hinton, G., Sager, D., Upton, M., Boggs, D., Carmean, D, Kyker, A., Roussel, P.: The Microarchitecture of the Pentium 4 Processor. Intel Technology Journal Q1. (2001)
11. Preston, Ronald P., Badeau, Roy W., Bailey, Daniel W., Bell, Shane L., et al.: Design of an 8-wide Superscalar RISC Microprocessor with Simultaneous Multithreading. Proceedings of 2002 IEEE International Solid-State Circuits Conference (ISSCC 2002). Vol. 1. (2002)
12. Thistle, Mark R., Smith, Burton J.: A Processor Architecture for HORIZON. Proceedings of the 1988 ACM/IEEE Conference on Supercomputing. Orlando, Florida, USA, Nov. 12-17. (1988) 35–41
13. Agarwal, A., Lim, B-H., Kranz, D., Kubiatowicz, J.:APRIL: A Processor Architecture for Multiprocessing. Proceedings of the 17th Annual International Symposium on Computer Architecture (ISCA 1990). (1990) 104–114
14. Nemirovsky, Mario D., Brewer, F., Wood, Roger C.: DISC: Dynamic Instruction Stream Computer. Proceedings of the 24th annual international symposium on Microarchitecture (Micro-24). Albuquerque, New Mexico, Puerto Rico. (1991) 163–171
15. Yamamoto, W., Nemirovsky, Mario D.: Increasing superscalar performance through multistreaming. Proceedings of the IFIP WG10.3 working conference on Parallel architectures and compilation techniques. Limassol, Cyprus. (1995) 49–58
16. Tullsen, Dean M., Eggers, Susan J., Levy, Henry M.: Simultaneous Multithreading: Maximizing On-chip Parallelism. Procedding of the 22nd Annual International Symposium on Computer Architecture (ISCA 1995). (1995) 392–403

17. Tullsen, Dean M., Eggers, Susan J., Emer, Joel S., Levy, Henry M., Lo, Jack L., Stamm, Rebecca L.: Exploiting choice: Instruction fetch and issue on an implementable simultaneous multithreading processor. Proceedings of the 23rd Annual International Symposium on Computer Architecture (ISCA 1996). (1996) 191–202
18. Eggers, Susan J., Emer, Joel S., Levy, Henry M., Lo, Jack L., Stamm, Rebecca L., Tullsen, Dean M.: Simultaneous Multithreading: A Platform for Next-Generation Processors. IEEE Micro, Vol. 17. No. 5. (1997) 12–19
19. Shin, C-H., Lee, S-W., Gaudiot, J-L.: Dynamic Scheduling Issues in SMT Architectures. Proceedings of the 17th International Parallel and Distributed Processing Symposium (IPDPS'03). Nice, France, April 22-26. (2003) 77-84
20. Burns, J., Gaudiot, J-L.: SMT Layout Overhead and Scalability. IEEE Transactions on Parallel and Distributed Systems, Vol. 13, No. 2. (2002) 142–155
21. Lee, S-W., Gaudiot, J-L.: Clustered Microarchitecture Simultaneous Multithreading. Proceedings of the Euro-Par 2003 International Conference on Parallel and Distributed Computing, Klagenfurt, Austria, August 26-29. (2003)
22. Thornton, J. E.: Design of a computer: the CDC 6600. Scott, Foresman Co., Glenview, Ill. (1970)

Autonomous-Centered Problem Allocation Oriented to Cooperation

Xiping Liu[1,2], Wanchun Dou[1,2], Guihai Chen[1,2], Shijie Cai[1,2], and Jiashan Tang[3]

[1] State Key Laboratory for Novel Software Technology, Nanjing University
[2] Dept. of Computer Science and Technology, Nanjing University, Nanjing, China, 210093
xixi_liu@graphics.nju.edu.cn
[3] Dept. of Applied Mathematics and Physics,
Nanjing University of Posts and Telecommunications, Nanjing, China, 210003

Abstract. By reasonably allocating a cooperative problem which need multiple solvers cope with together, the problem could be performed more effectively and efficiently. A problem could be divided into multiple sub-problems; each has certain ability requirement which is the hinge to relate problem and solver. According to ability requirement, the solver candidate set for each sub-problem could be established. To select suitable solver from candidate set so as to solve a cooperative problem in more autonomous and consistent way, a mathematical allocation model taking the minimization of interaction number as objective function is established. The model solving process is deployed by decreasing two kinds of interactions, i.e. intra-interaction and extra-interaction. Experiment shows this method obtains better performance than general allocation.

1 Introduction

To a complex cooperative problem which need multiple solvers cope with together, it could be divided into multiple sub-problems with more specific and accurate task description, and each solver could handle one or more sub-problems. The solver could be any autonomous entity, such as an intelligent agent or a person. Before the problem solving process, it is necessary to build a reasonable model between problem and problem solver, which provide that what part of cooperative problem should be handled by whom. Nevertheless, such an allocation process is underestimated as most study on cooperation has concentrated on the issue of problem solving process.

In most cases, the cooperation problem could not be split into such a sub-problem set, in which each sub-problem is isolated from others. There exist interactions between sub-problems more or less; accordingly the solvers of those sub-problems need communicate with others for necessary information. To solve a cooperative problem more unified and efficient, the fewer interactions between solvers is high appreciated. The minimization of interaction between solvers of a cooperative problem embodies autonomous-centered idea. With fewer interactions between solvers, each solver could pay less attention to external action and care more about the internal process; consequently, sub-problems assigned to each solver could be solved in more self-governing way and the whole cooperative problem solving system could be more autonomous.

J. Cao, W. Nejdl, and M. Xu (Eds.): APPT 2005, LNCS 3756, pp. 91–100, 2005.
© Springer-Verlag Berlin Heidelberg 2005

There are many scenarios which have autonomous requirement, such as cooperative cognition, cooperative design, and etc. To these cooperative problems, interactions among cooperators are very complex in general and need be put much effort to assure problem could be solved more autonomously and consistently. However, most current researches focus only on issues of how to realize and optimize interaction after interaction between solvers is already determinate [1, 2, 3], while put little effort on how to determine the interaction model for a cooperative system in advance so as to achieve the autonomous.

With these considerations, relative researches about autonomous-centered problem allocation oriented to cooperation are discussed in this paper. The problem handled by multiple solvers usually could be subdivided into multiple sub-problems. Each sub-problem could find many solver candidates by checking if their capability satisfies certain requirement, where a candidate maybe consists of multiple participants. The most suitable candidate should be selected as formal solver for each sub-problem so that the whole cooperative system has fewer interactions and act more autonomously. A mathematical allocation model is established to provide explicit and specific description. To step toward optimal solution of that model, five selection principles are put forward and relevant algorithm is proposed.

This paper is organized as follows: Section 2 presents problem model and ability specification of participant, which underlies the candidate set determining process; the allocation result model is also explored in this section. Section 3 describes the mechanism to determine candidate set according to ability requirement. Section 4 analyzes the criteria of allocation and builds a corresponding mathematical model. An autonomous-centered allocation method to solve the mathematical model is put forward in section5. Section 6 provides the experiment result and relevant analysis. The conclusion and future work is discussed in section 7.

2 Problem Model, Participant Specification and Allocation Result Model

2.1 Problem Model and Ability Specification

Definite 1: CP denotes universe of cooperative problem, which is divided into m sub-problems, denoted as $CP = \{SP_i, i=1, 2,..., m\}$, note any SP_i could not be sub-divided. A complete problem model, could be described as $(CP, IntSP\text{-}SP)$, where $IntSP\text{-}SP$ represent interaction between SPs, which is a relation set [4], the element of set is SP_i pair. For example, a relation set like $\{(SP_1, SP_2), (SP_1, SP_3)\}$, represents there is interaction between SP_1 and SP_2, also between SP_1 and SP_3.

Each sub-problem would have special requirements to solver, which include infrastructure, such as all kinds of devices, machines, etc., and also formless resource, such as professional field knowledge, knowledge background, skill or experience, available work time, and etc. The term *ability* is abstracted to represent such requirements, which is not so easy to represent, partly because knowledge representation is another developing research problem with many open issues [5]. As the effort of this paper is not focus on things like knowledge representation, we just use symbol A to denote certain kind of ability i.e. professional knowledge, skills, devices and etc.

Definition 2: A is ability set of the cooperative problem, representing all the requirements for solving cooperative problem, denoted as $A= \{A_k, k=1, 2,..., l\}$, where A_k has pervasive meaning of requirement to participant.

Definition 3: To SP_i, the ability requirements could be represented as an equivalence class $[A]_{SP_i} = \{A_k \in A$: to handle SP_i, the solver must own $A_k, k \in [1,l]\}$.

2.2 Participant Specification Based on Ability

The solver to certain sub-problems is chosen from all the participants of cooperative problem. Being a participant does not mean that there does be certain sub-problem that could be assigned to this participant, but only represent it is involved in the cooperative problem, as it has the probability to become the formal solver of certain sub-problem.

Definition 4: There are n participants involved in a cooperative problem, denoted as $P= \{P_j, j=1, 2,..., n\}$, the ability that P_j owns is represented as $[A]_{P_j} = \{A_k \in A: P_j$ holds $A_k, k \in [1, l]\}$.

To a sub-problem, if certain participant satisfies the condition $[A]_{P_j} \supseteq [A]_{SP_i}$, it implies P_j have the ability to handle SP_i. Sometimes, the ability set of one single participant does not hold all the ability elements required by SP_i but only a subset of $[A]_{SP_i}$, at this time, the participant need cooperate to handle SP_i with others whose ability sets satisfy the requirement that not satisfied by itself.

2.3 Allocation Result Model

To allocate a cooperative problem, besides determining solver with enough capability for sub-problem, how to deal with the interaction among sub-problems is another issue. As *SPs* are assigned to certain solvers to handle, the interaction among *SPs* is transformed to be the interaction among solvers. If some *SPs* interacting with each other are handled by same solver, as the solver has no need to interact with self, so the interaction among those *SPs* could vanish. Moreover, the interaction among those *SPs* handled by different solvers could be up-transferred to the corresponding solver. In other words, in allocation result, the interaction among sub-problems of problem model would be replaced by interaction among solvers. Besides, If one sub-problem is handled by multiple solvers, all those solvers must interact with each other to negotiate how to solve it.

Definition 5: There are two kinds of interactions between solvers of a cooperative problem. *Intra-interaction*: the interaction among multiple solvers of a sub-problem, which is generated during allocation and used to solve internal process of a sub-problem. *Extra-interaction*: the interaction among solvers of multiple *SPs*, which is transformed from *SPs* interaction existed in problem model and used to solve external process of certain sub-problem that has interactions with other *SPs*.

For example, to $IntSP\text{-}SP = \{(SP_1, SP_2)\}$, solver of SP_1 is P_1 and P_2, solver of SP_2 is P_2, P_3; so there is extra-interaction between P_1 and P_2, P_1 and P_3, also between P_2 and P_3 corresponding to interaction between SP_1 and SP_2, moreover, there is intra-interaction between P_1 and P_2 to SP_1, also between P_2 and P_3 to SP_2. There might be multiple interactions between two solvers. Due to autonomous and intelligent of solver, process to one or more interactions is similar, so the interaction number between two solvers could be supposed as 1 no matter how many interactions exist.

Definition 6: To a cooperative problem, the final result model is $(SP\text{-}Solver, IntP\text{-}P)$, where $SP\text{-}Solver$ consists of $SP_i\text{-}P_j$ pair which represents P_j participate in solving SP_i and $IntP\text{-}P$ consists of $P_{j_1} \text{-} P_{j_2}$ pair which represents there is interaction between P_{j_1} and P_{j_2} (could be intra-interaction or extra-interaction or both).

3   Candidate Set Determining

Allocation has two phases, the first is to determine candidate set for each sub-problem based on ability requirement; and the second is to choose one candidate to solve each sub-problem by certain strategy.

To be a candidate of sub-problem SP_i, P_j should satisfy certain condition. If $[A]_{P_j} \supseteq [A]_{SP_i}$, P_j could be a candidate by single; otherwise, if $[A]_{P_j} \cap [A]_{SP_i} \neq \Phi$

```
For i=1 to m {       //Find all possible candidates for SP_i
  For j=1 to n   //Compute max number of cooperators for P_j
    Compute cn_j=min{|[A]_SP_i|-|[A]_P_j ∩ [A]_SP_i|, n-1}

  For j=1 to n                    //Find candidate encompass P_j
    If cn_j=0                     //P_j can handle SP_i by single
            {Add {P_j} into candidate set CAND_{i,1}}

    Else
      If cn_j<|[A]_SP_i|    //P_j composes candidate with others
        For k=1 to cn_j {       //find other k participants
          Find all composition satisfying
            [A]_P_j ∪[A]_P_{j2} ∪[A]_P_{j3} ∪...∪[A]_P_{j_{k+1}} ⊇[A]_SP_i  j<j_s≤n

          Add {P_j, P_{j2},..., P_{j_{k+1}}} to candidate set CAND_{i,k+1}
        }
  CAND_i=∪CAND_{i,t}
         t
}
```

Fig. 1. Determine candidate sets for each sub-problem

(means P_j has certain ability to participate in solving SP_i), P_j can not be a candidate by single but could composes a candidate with other participants. To find all possible candidate combinations related with P_j for SP_i, the cardinality of $[A]_{SP_i}$ and $[A]_{P_j}$ is provided. The maximum number of cooperators for P_j is computed as $cn=|[A]_{SP_i}|-|[A]_{P_j} \cap [A]_{SP_i}|$ (if $cn>n-1$ then it should be replaced by $n-1$).

Definition 7: To SP_i, the candidate set is represented as $CAND_i = \bigcup_t CAND_{i,t}$ ($t \in [1,|[A]_{SP_i}|]$), where $CAND_{i,t} = \{\{ P_{j_1} .. P_{j_t} \} \subseteq P : [A]_{SP_i} \subseteq \bigcup_s [A]_{P_{j_s}}$, $s=1$, $2,..,t$; to any $s_1, s_2 \in [1, t]$, $[A]_{P_{j_{s1}}} \neq [A]_{P_{j_{s2}}}$ and $[A]_{SP_i} \not\subset \bigcup_{s \setminus s_1} [A]_{P_{j_s}} \}$. Here t means the number of participants in a candidate, especially, when $t=1$, the sub-problem could be handled by single participant.

When determine candidate set for certain sub-problem SP_i, all possible $CAND_{i,t}$ should be obtained in advance so as to provide more freedom to select one most suitable candidate for SP_i. The algorithm is described in figure 1.

4   Selection Criteria and Mathematic Modeling

To find suitable solvers for sub-problems, a precondition is assumed, that is: to $\forall SP_i$ ($i=1, 2,\ldots, m$), $\exists CAND_i \neq \Phi$. Otherwise, SP_i is unsolvable.

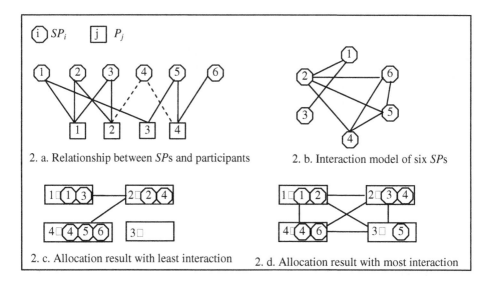

2. a. Relationship between *SP*s and participants   2. b. Interaction model of six *SP*s

2. c. Allocation result with least interaction   2. d. Allocation result with most interaction

Fig. 2. An example of Allocation

4.1 The Criteria of Selection

The criteria of selection mainly refer to the performance requirement, here means number of the interactions between solvers should be minimized as much as possible. Different selection would result in different interaction number. Figure 2 provides an example of different allocation result based on same condition. As figure 2.a showing, there are four participants and six SPs in a cooperative system. The line represents who has the ability to handle which sub-problem; the doted line represents SP_4 need P_2 and P_4 to handle together. The figure 2.b shows the interaction relationship among those SPs. With these precondition, multiple allocation results satisfying function requirement could be obtained, figure 2.c and 2.d denotes two of them. As the figure showing, the interaction number of figure 2.c is four less than 2.d.

4.2 Mathematical Modeling

To a system with not very many participants and SPs, the allocation could be carried out and evaluated at one sight like figure 2; however, to a large scale system with great deal of participants and SPs, it would become very hard to allocate by hand and obtain a reasonable result. So it is quite necessary to construct a method aiming at allocating automatically and decreasing interaction number of cooperative system as much as possible. To make the problem more explicit, a mathematical model to describe the allocation is presented:

1. $T(i_1, i_2)$ $(i_1=1,\ldots,m,\ i_2=1,\ldots,m)$ represents the relativity between two sub-problems, if SP_{i_1} has interaction with SP_{i_2}, $T(i_1, i_2)=1$; otherwise $T(i_1, i_2)=0$. The data derive from problem model $(CP, IntSP\text{-}SP)$.

2. $R(i, j)$ $(i=1,\ldots,m; j=1,\ldots,n)$ is the allocation result matrix, $R(i, j)=1$ represents P_j participate in solving SP_i; otherwise $R(i, j)=0$. $R(i, j)$ reflect the $SP\text{-}Solver$ relation in allocation result model.

3. $Int(j_1, j_2)$ $(j_1=1,\ldots,n; j_2=1,\ldots,n)$ is the interaction relation of allocation result model, $Int(j_1, j_2)=1$ represents P_{j_1} need interact with P_{j_2}, otherwise $Int(j_1, j_2)=0$. $Int(j_1, j_2)$ reflect the $IntP\text{-}P$ relation in allocation result model.

4. The allocation could be described as such an optimization problem: select suitable solver for each sub-problem, makes

$$\min \sum_{j_1=1}^{n-1} \sum_{j_2=j_1+1}^{n} Int(j_1, j_2)$$

Where $Int(j_1, j_2) = 1$ if and only if

$$(R(i_1, j_1)=1 \wedge R(i_2, j_2)=1 \wedge T(i_1, i_2)=1) \vee (R(i_1, j_1)=1 \wedge R(i_1, j_2)=1)$$

$$\text{s. t. } \forall i, \ \bigcup_{R(i,j)=1} \{P_j\} \in CAND_i$$

The model solving is a very complex and hard process, especially for large scale situation. Such a model solving similar to combinatorial optimization is a NP problem in all probability [6]. To try all possible allocation scheme, compute the interaction number and choose one scheme with the fewest as optimal solution is low effective and unrealistic. The best way to solve such a model is to provide a universal algorithm adapt to variable model scale. In next section, an autonomous-centered allocation method is put forward, which is an effective attempt to reach optimal solution; And experiment prove that this strategy bring great performance improvement.

5   Autonomous-Centered Allocation

Before the allocation mechanism is provided, it is necessary to learn the holistic structure of allocation process. The allocation process is performed through m steps, each step take charge of the allocation of one sub-problem based on last step result model. In this way, no sub-problem could know exactly the allocation result of SPs assigned after it, but it could make use of the existed allocation result and anticipate possible allocation result of future to prevent more increase of interactions.

As the allocation is performed step by step, to minimize increased interaction number of each step is a straightforward and effective strategy, which could be realized through reasonable selection principles proposed in following sections.

5.1   Decreasing Intra-interaction

To decrease the intra-interaction, it is necessary to choose as few solvers as possible for each sub-problem. In general, single solver is better than multiple solvers as no unnecessary intra-interaction exists and autonomous process could solve problem more completely and consistently. Therefore, to candidates of SP_i, the fewer members certain candidate has, the higher priority it has. As a result, the candidate with least t is named as *top-candidate* and t' is used to represent the least t.

Selection Principle 1: Solver should belong to certain *top-candidate*. If SP_i need more than one solver to handle together, the interaction among those solvers should be appended to the result model. It seems that all *top-candidates* would generate same number of intra-interaction, i.e. $C_{t'}^2$; nevertheless, increased intra-interaction number of different candidate is variable. In last step result model, if there already exist some interactions among t' participants from certain candidate, then only the left interactions not still existed need be appended. Accordingly, selection principle 2 is proposed to decrease as many as interactions.

Selection Principle 2: Candidate with more existed interactions among its members is preferable.

5.2   Decreasing Extra-interaction

If SP_i has interaction with certain allocated SPs, the interaction between the solvers of those SPs and the solvers of SP_i should be appended based on the last step result model. Here, the participant that need build interaction with solvers of SP_i is called as *pToInt* of SP_i. The candidate encompassing certain *pToInt* is called as *semi-selected candidate*, as by allocating SP_i to this candidate, the solver could melt interaction between SP_i and sub-problem handled by that *pToInt*, so that problem could be solved in more autonomous way.

Selection Principle 3: Solver should belong to certain *semi-selected candidate* if it exists.

Similar with the principle of decreasing intra-interaction, the extra-interaction could be decreased by making full use of existed interactions. Note the extra-interaction is determined after intra-interaction, so it could reuse not only interactions from last step result model but also the new generated intra-interactions to SP_i.

Selection Principle 4: Candidate who has more existed interactions with *pToInt* is preferable.

Supposing certain un-allocated SP_j interacting with SP_i would be assigned to same solver of SP_i in certain later step, there would be no need to append extra-interaction corresponding to SP_i and SP_j because the two sub-problems are handled by same solvers. Consequently, the more un-allocated SPs interacting with SP_i the candidate has capability to handle, the fewer interactions the latter allocation would possibly append. Based on this, selection principle 5 from perspective of expectation to future allocation is proposed as following.

Selection Principle 5: Candidate that has capability to handle more SPs is preferable, here SPs means those have interaction with SP_i and still not be allocated.

5.3 Allocation Order Determining and Specific Algorithm

As allocation is performed in m steps, the allocation order of m SPs is very important to good result. If SP_i with more candidates is allocated at first, then only a few interactions of result model could be reused, so some selection principle could not work well; On the other hand, if SP_i with fewer candidates is allocated at last, then lots of

```
Decide allocation order              //fewer candidates first
For i=1 to m{
          //i denotes the ith sub-problem to be allocated
  Find all top-candidates            //selection principle 1
  Find all semi-selected candidates among top-candidates
                                     //selection principle 3
  For each semi-selected candidate if they exist, other-
  wise for each top-candidate {
    Compute increasedInt      //selection principle 2 and 4
    Find candidates with minimum increasedInt
    If there is more than one
      Selected one candidate        //selection principle 5
    Else
      Take corresponding candidate as solver
  }
  Modify allocation result model
}
```

Fig. 3. Determine solvers for each sub-problem

interactions could be reused, however SP_i could not make good use of selection principle too, because it has little selection space with fewer candidates. Accordingly, the allocation order is achieved by sorting through top-candidate number of each sub-problem with less first strategy, so as to make selection principles work more effectively.

The allocation process consists of m steps in predefined order, each step is to allocate for certain sub-problem, which brings change of interaction number denoted as *increasedInt*, the specific algorithm is provided in figure 3.

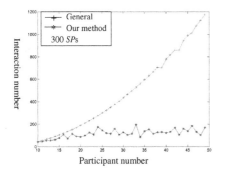

4. a. Variable participant quantity and fixed sub-problem number

4. b. Variable sub-problem quantity and fixed participant number

Fig. 4. Comparison of different allocation method

6   Experiment Result and Discussion

General allocation method is usually to directly choose one candidate as solver without special consideration; the experiment simulates the general allocation method to provide a specific comparison. To check performance of varied possible cooperative system, the candidate set and interaction among sub-problems is determined at random. As the figure 4 showing, to same cooperative problem, the interaction number with autonomous-centered method are much less than that of with general allocation.

7   Conclusion and Future Work

Cooperative problem has been developed greatly along with the requirement of cross-area application. It brings many research issues on allocation to cooperative problem. This paper is a new try to ensure cooperative problem could be solved in more autonomous way so as to improve the consistent and integrity of cooperative system. The problem allocation mechanism presented in this paper could be applied to all kinds of cooperative scenarios with requirement of fewer interactions, such as software development, cooperative cognition, cooperative design, and etc. This mechanism especially facilitates the allocation of large-scale cooperative problem with great deal of participants and sub-problems. With autonomous-centered allocation method, the cooperative problem could be solved by multiple solvers more harmoniously and effectively in much fewer interactions. In future work, more constraint could be taken into account to provide more pervasive allocation modeling, such as precedence among *SP*s, requirement for ability degree, and so on.

Acknowledgement. This paper is based on NSFC (No. 60303025), Jiangsu Provincial NSF research fund (No. BK2004411 and BK2005208).

References

1. Rogers, Erika: Cognitive cooperation through visual interaction. Int. J. Knowledge-Based Systems (1995) 117-125
2. José Antonio Pérez, Rafael Corchuelo, David Ruiz, and Miguel Toro: An Order-Based, Distributed Algorithm for Implementing Multiparty Interactions. In: Farhad Arbab, Carolyn Talcott (eds.): Coordination models and languages. Lecture Notes in Computer Science, Vol. 2315. Berlin, Springer (2002) 250-257
3. David Ruiz, Rafael Corchuelo, José A. Pérez, and Miguel Toro: An Algorithm for Ensuring Fairness and Liveness in Non-deterministic Systems Based on Multiparty Interactions. In: B. Monien and R. Feldmann (Eds.): Euro-Par 2002. Lecture Notes in Computer Science, Vol. 2315. Berlin, Springer (2002) 563-572
4. Luis E. Sanchis: Set theory, an operational approach. Amsterdam, the Netherlands: Gordon and Breach (1996)
5. John F. Sova: Knowledge Representation: Logical, Philosophical, and Computational Foundations. Brooks/Cole (2000)
6. Yue Minyi: Introduction to Combinatorial Optimization. Zhejiang Science & Technology Publishing House (2001)

Contention-Free Communication Scheduling for Irregular Data Redistribution in Parallelizing Compilers[*]

Kun-Ming Yu, Chi-Hsiu Chen, Ching-Hsien Hsu, Chang Wu Yu, and Chiu Kuo Liang

Department of Computer Science and Information Engineering, Chung Hua University, Hsinchu, Taiwan 300, ROC Tel: 886-3-5186412, Fax: 886-3-5329701 yu@chu.edu.tw

Abstract. The data redistribution problems on multi-computers had been extensively studied. Irregular data redistribution has been paid attention recently since it can distribute different size of data segment of each processor to processors according to their own computation capability. *High Performance Fortran Version 2* (HPF-2) provides *GEN_BLOCK* data distribution method for generating irregular data distribution. In this paper, we develop an efficient scheduling algorithm, Smallest Conflict Points Algorithm (SCPA), to schedule HPF2 irregular array redistribution. SCPA is a near optimal scheduling algorithm, which satisfies the minimal number of steps and minimal total messages size of steps for irregular data redistribution.

Keywords: Irregular data redistribution, communication scheduling, GEN_BLOCK, conflict points.

1 Introduction

More and more works had large data or complex computation on run-time in most scientific and engineering application. Those kinds of tasks require parallel programming on distributed system. Appropriate data distribution is critical for efficient execution of a data parallel program on a distributed computing environment. Therefore, an efficient data redistribution communication algorithm is needed to relocate the data among different processors. Data redistribution can be classified into two categories: the regular data redistribution [2, 3, 6] and the irregular data redistribution [1, 4, 10, 11, 12]. The irregular distribution uses user-defined functions to specify unevenly data distribution. High Performance Fortran version 2 (HPF2) provides GEN_BLOCK data distribution instruction which facilitates generalized unequal-size consecutive segments of array mapping onto consecutive processors. This makes it

---

[*] The work is partially supported by National Science Council of Taiwan, under grant number NSC-93-2213-E-216-029.

J. Cao, W. Nejdl, and M. Xu (Eds.): APPT 2005, LNCS 3756, pp. 101–110, 2005.
© Springer-Verlag Berlin Heidelberg 2005

possible to let different processors dealing with appropriate data quantity according to their computation capability. In this scenario, all processors must send and receive message, even if send and receive on the same processor.

In the irregular array redistribution, *Guo et al.* [11] proposed a Divide-and-Conquer algorithm, they utilize Divide and Conquer technique to obtain near optimal scheduling while satisfied minimize the total communication messages size and minimize the number of steps.

In this paper, we present a smallest-conflict-points algorithm (SCPA) to efficiently perform GEN_BLOCK array redistribution. The main idea of the SCPA is to schedule the conflict messages with maximum degree in the first step of data redistribution process. SCPA can effectively reduce communication time in the process of data redistribution. SCPA is not only an optimal algorithm in the term of minimal number of steps, but also a near optimal algorithm satisfied the condition of minimal message size of total steps.

The rest of this paper is organized as follows. In Section 2, a brief survey of related work will be presented. In section 3, we will introduce communication model of irregular data redistribution and give an example of GEN_BLOCK array redistribution as preliminary. Section 4 presents smallest-conflict-points algorithm for irregular redistribution problem. The performance analysis and simulation results will be presented in section 5. Finally, the conclusions will be given in section 6.

2   Related Work

Many data redistribution results have been proposed in the literature. These researches are usually developed for regular or irregular problems [1] in multi-computer compiler techniques or runtime support techniques.

Techniques for communication optimizations category provide different approaches to reduce the communication overheads [5, 7] in a redistribution operation. The communication scheduling approaches [3, 12] avoid node contention and the strip mining approach [9] overlaps communication and computational overheads.

In irregular array redistribution problem, some works have concentrated on the indexing and message generation while some has addressed on the communication efficiency. Guo et al. [10, 11] proposed a divide-and-conquer algorithm for performing irregular array redistribution. In this method, communication messages are first divided into groups using Neighbor Message Set (NMS), messages have the same sender or receiver; the communication steps will be scheduled after those NMSs are merged according to the relationship of contention. Yook and Park [12] presented a relocation algorithm, while their algorithm may lead to high scheduling overheads and degrade the performance of a redistribution algorithm.

3   Preliminaries and Redistribution Communication Models

Data redistribution is a set of routines that transfer all the elements in a set of source processor S to a set of destination processor T. The sizes of the messages are specified by values of user-defined random integer for array mapping from source processor to destination processor. Since node contention considerably influences, a processor can only send messages to other one processor in each communication step. Use the same rule, a processor can only receive messages from other one processor.

To simplify the presentation, notations and terminologies used in this paper are prior defined as follows.

Definition 1 : GEN_BLOCK redistribution on one dimension array A[1:N] over P processors. The source processor is denoted as SP_i, the destination processor is denoted as DP_j, where $0 \leqq i, j \leqq P\text{-}1$.

Definition 2 : The time of redistribution separator the time of startup is denoted as t_s, and the time of communication is denoted as t_{comm}.

Definition 3 : To satisfy the condition of the minimum steps and the processor sends/receives one message at each steps, some messages can not be scheduled in the same communication step are called conflict tuple [11].

Data redistribution implements have two methods: non-blocking scheduling algorithm and blocking scheduling algorithm. The non-blocking scheduling algorithm is faster than the blocking scheduling algorithm. But need more buffer and be better control synchronization. In this paper, we discuss on blocking scheduling algorithm.

Irregular data redistribution is unlike regular has a cyclic message passing pattern. Every message transmission link is not overlapping. Hence, the total number of message links N is $numprocs \leq N \leq 2 \times numprocs - 1$, where *numprocs* is the number of processors. Figure 1 shows an example of redistributing two GEN_BLOCK distributions on an array $A[1:101]$. The communications between source and destination processor sets are depicted in Figure 2. There are totally fifteen communication messages, $m_1, m_2, m_3 \ldots, m_{15}$ among processors involved in the redistribution. In this example, $\{m_2, m_3, m_4\}$ is a conflict tuple since they have common source processor SP_1; $\{m_7, m_8, m_9\}$ is also a conflict point because of the common destination processor DP_4. The maximum degree in the example is equal to 3. Figure 3 shows a simple schedule for this example

Source distribution

Source Processor							
SP	SP_0	SP_1	SP_2	SP_3	SP_4	SP_5	SP_6	SP_7
Size	12	20	15	14	11	9	9	11

Destination distribution

Destination Processor							
DP	DP_0	DP_1	DP_2	DP_3	DP_4	DP_5	DP_6	DP_7
Size	17	10	13	6	17	12	11	15

Fig. 1. An example of distributions

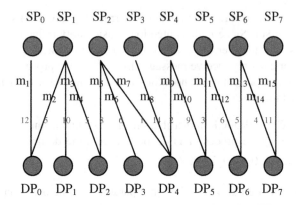

Fig. 2. The communications between source and destination processor sets

Schedule Table	
Step 1	m_2 m_5 m_9 m_{12} m_{14}
Step 2	m_1 m_3 m_6 m_8 m_{11} m_{15}
Step 3	m_4 m_7 m_{10} m_{13}

Fig. 3. A simple schedule

3.1 Explicit Conflict Point and Implicit Conflict Point

The total communication time of a message passing operation using two parameters: the startup time t_s and the unit data transmission time t_m. The startup time is once for each communication event and is independent of the message size to be communicated. The data transmission time is relationship of a message size, size(m). The communication time of one communication step is the maximum of the message in this step. The total communication time of all steps is summary of each the communication time of step. The length of these steps determines the data transmission overheads. The minimum step is equal to maximum degree k, when message can not put into any step of minimum step it must relate to the processor has maximum degree transmission links. Figure 4 shows the maximum degree of figure 1. SP_1, SP_2 and DP_4 had maximum degree (K = 3) from messages m_2~m_9. Because of each one processor can only send/receive at most one message to/from other processor in each communication step. First, we concentrate all processors which have maximum degree transmission links messages. For the sake of simplicity, such messages are referred to as "Maximum Degree Message Set" (MDMS) in the paper, as shown in figure 4. If the messages in MDMSs can put into k steps with no conflict occur, other messages of the processors' degree less than maximum degree will be easier to put into the rest of step without increasing the number of steps.

We say a message to be an explicit conflict point if it belongs to two MDMSs. There exists at most one explicit conflict point between two MDMSs. In figure 4, m_7 is a explicit conflict point since it belongs to two MDMSs $\{m_5, m_6, m_7\}$ and $\{m_7, m_8, m_9\}$. On the other hand, if two MDMSs do not contain the same message, but the neighbor MDMSs each has a message been sent by the same processor, or been received by the same processor. We call this kind of message as an implicit conflict point. As shown by figure 5, m_4 and m_5 are contained by the different MDMSs. DP_2 only receives m_4 and m_5 two messages, so it can not form an MDMS. But m_4 and m_5 are also owned by different MDMSs. Therefore, m_4 is an implicit conflict point. Although, m_5 is also covered by two MDMSs, but it is restricted by m_4. Hence m_5 will not cause conflict. Figure 7 depicts all MDMSs for the example shown in Figure 1.

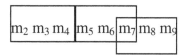

Fig. 4. Maximum Degree Messages Set

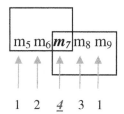

Fig. 5. Example of explicit conflict point

4 Scheduling Algorithm

The main goal of irregular array distribution is to minimize communication step as well as the total message size of steps. We select the smallest conflict points which will really cause conflict to loose the schedule constraint and to minimize the total message size of schedule.

Smallest conflict points algorithm consists of four parts:

(1) Pick out MDMSs from given data redistributed problem.

(2) Find out explicit conflict point and implicit conflict point. And schedule all the conflict point into the same schedule step.

(3) Select messages on MDMSs in non-increasing order of message size. Schedule message into similar message size of that step and keep the relation of each processor send/receive at most one message to/from the processor. Repeat above process until no MDMSs' messages left.

(4) Schedule messages do not belong to MDMSs by non-increasing order of message size. Repeat above process until no messages left.

From Figure 1, we can pick out four MDMSs, $MDMS_1 = \{m_2, m_3, m_4\}$, $MDMS_2 = \{m_4, m_5\}$, $MDMS_3 = \{m_5, m_6, m_7\}$ and $MDMS_4 = \{m_7, m_8, m_9\}$, shown in Figure 8. We schedule m_4 and m_7 into the same step. Then schedule those messages on

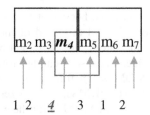

Fig. 6. Example of implicit conflict point

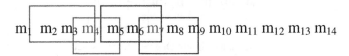

Fig. 7. All MDMSs for the example in Figure 1

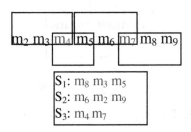

Fig. 8. Results of MDMSs for Figure 1

$$
\begin{array}{l}
S_1: m_8 \ m_3 \ m_5 \ m_1 \ m_{15} \ m_{10} \ m_{12} \\
S_2: m_6 \ m_2 \ m_9 \ m_{13} \ m_{11} \\
S_3: m_4 \ m_7 \ m_{14}
\end{array}
$$

Fig. 9. The schedule obtained form SCPA

MDMSs by non-increasing order of message size as follows: m_8, m_3, m_5, m_6, m_2, m_9. After that, we can schedule the rest messages that are not belong to any MDMSs by non-increasing order of message size as follows: m_1, m_{15}, m_{10}, m_{12}, m_{13}, $m_{14,}$ m_{11}. Figure 9 shows the final schedule obtained form smallest conflict points algorithm.

5   Performance Evaluation and Analysis

To evaluate the performance of the proposed methods, we have implemented the SCPA along with the divide-and-conquer algorithm [11]. The performance simula tion is discussed in two classes, even GEN_BLOCK and uneven GEN_BLOCK distributions. In even GEN_BLOCK distribution, each processor owns similar size of data. Contrast to even distribution, few processors might be allocated grand volume of data in uneven distribution. Since array elements could be centralized to some specific processors, it is also possible for those processors to have the maximum degree of communications.

The simulation program generates a set of random integer number as the size of message. To correctly evaluate the performance of these two algorithms, both programs were written in the single program multiple data (SPMD) programming paradigm with MPI code and executed on an SMP/Linux cluster consisted of 24 SMP nodes. In the figures, "SCPA Better" represents the percentage of the number of

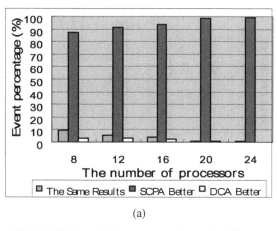

(a)

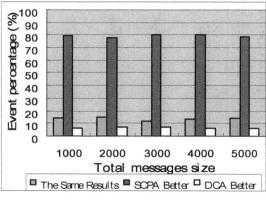

(b)

Fig. 10. The events percentage of computing time is plotted (a) with different number of processors and (b) with different of total messages size in 8 processors, on uneven data set

events that the SCPA has lower total steps of messages size than the divide-and-conquer algorithm (DCA), while "DCA Better" gives the reverse situation. In the uneven distribution, the size of message's up-bound is set to (totalsize/numprocs)*1.5 and low-bound is set to (totalsize/numprocs)*0.3, where totalsize is total size of messages and numprocs is the size of processor. In the even distribution, the size of message's up-bound is set to (totalsize/numprocs)*1.3 and low-bound is set to low-bound is (totalsize/numprocs)*0.7. The total messages size is 1M.

Figure 10 shows the simulation results of both the SCPA and the DCA with different number of processors and total message size. We can observe that SCPA has better performance on uneven data redistribution compared with DCA.

Since the data is concentrated in the even case, from figure 11, we can observe that SCPA have the better performance compared with uneven case. Figure 11 also

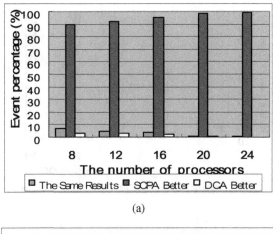

(a)

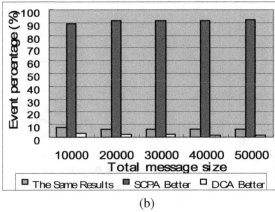

(b)

Fig. 11. The events percentage of computing time is plotted (a) with different number of processors and (b) with different of total messages size in 8 processors, on even data set

illustrates that SCPA has at least 85% supreme than DCA in any size of total messages and any number of processors In both even and uneven case, SCPA performs slightly better than DCA.

6 Conclusion

In this paper, we have presented an efficient scheduling algorithm, smallest conflict points algorithm (SCPA), for irregular data distribution. The algorithm can effectively reduce communication time in the process of data redistribution. Smallest-conflict-points algorithm is not only an optimal algorithm in the term of minimal number of steps, but also a near optimal algorithm satisfied the condition of minimal message size of total steps. Effectiveness of the proposed methods not only avoids node contention but also shortens the overall communication length.

For verifying the performance of our proposed algorithm, we have implemented SCPA as well as the divide-and-conquer redistribution algorithm. The experimental results show improvement of communication costs and high practicability on different processor hierarchy. Also, the experimental results indicate that both of them have good performance on GEN_BLOCK redistribution. But also both have advantages and disadvantages. In many situations, SCPA has better than the divide-and-conquer redistribution algorithm.

References

1. Minyi Guo, "Communication Generation for Irregular Codes," The Journal of Supercomputing, vol. 25, no. 3, pp. 199-214, 2003.
2. Minyi Guo, I. Nakata and Y. Yamashita, "Contention-Free Communication Scheduling for Array Redistribution," Parallel Computing, vol. 26, no.8, pp. 1325-1343, 2000.
3. Minyi Guo, I. Nakata and Y. Yamashita, "An Efficient Data Distribution Technique for Distributed Memory Parallel Computers," JSPP'97, pp.189-196, 1997.
4. Minyi Guo, Yi Pan and Zhen Liu, "Symbolic Communication Set Generation for Irregular Parallel Applications," The Journal of Supercomputing, vol. 25, pp. 199-214, 2003.
5. S. Lee, H. Yook, M. Koo and M. Park, "Processor reordering algorithms toward efficient GEN_BLOCK redistribution," Proceedings of the ACM symposium on Applied computing, pp. 539-543, 2001.
6. Ching-Hsien Hsu, Kun-Ming Yu, Chi-Hsiu Chen, Chang Wu Yu, and Chiu Kuo Liang, "Optimal Processor Replacement for Efficient Communication of Runtime Data Redistribution," Lecture Notes in Computer Science (ISPA'04), Vol. 3358, pp. 268-273, Dec. 2004.
7. C.-H Hsu, Dong-Lin Yang, Yeh-Ching Chung and Chyi-Ren Dow, "A Generalized Processor Mapping Technique for Array Redistribution," IEEE Transactions on Parallel and Distributed Systems, vol. 12, vol. 7, pp. 743-757, July 2001.
8. S. Ramaswamy, B. Simons, and P. Banerjee, "Optimization for Efficient Data redistribution on Distributed Memory Multicomputers," Journal of Parallel and Distributed Computing, vol. 38, pp. 217-228, 1996.

9. Akiyoshi Wakatani and Michael Wolfe, "Optimization of Data redistribution for Distributed Memory Multicomputers," short communication, Parallel Computing, vol. 21, no. 9, pp. 1485-1490, September 1995.

10. Hui Wang, Minyi Guo and Wenxi Chen, "An Efficient Algorithm for Irregular Redistribution in Parallelizing Compilers," Proceedings of 2003 International Symposium on Parallel and Distributed Processing with Applications, LNCS 2745, 2003.

11. Hui Wang, Minyi Guo and Daming Wei, "Divide-and-conquer Algorithm for Irregular Redistributions in Parallelizing Compilers", The Journal of Supercomputing, vol. 29, no. 2, pp. 157-170, 2004.

12. H.-G. Yook and Myung-Soon Park, "Scheduling GEN_BLOCK Array Redistribution," Proceedings of the IASTED International Conference Parallel and Distributed Computing and Systems, November, 1999.

Experiments on Asynchronous Partial Gauss-Seidel Method

Hiroshi Nishida and Hairong Kuang

Computer Science Department, California State Polytechnic University,
Pomona, 3801 West Temple Avenue, CA 91768, USA
{hnishida,hkuang}@csupomona.edu

Abstract. This paper presents design and experimental results of a parallel linear equation solver by asynchronous partial Gauss-Seidel method. The basic idea of this method is derived from the asynchronous iterative method; newly computed values of unknowns are broadcast to all other processors and are incorporated into computing the next value immediately after they are received. However, since the asynchronous iterative method requires frequent data passing, it is difficult to achieve high performance on practical cluster computing systems due to its enormous communication overhead. To avoid it, the asynchronous partial Gauss-Seidel method reduces frequency of broadcasting new values of unknowns by passing multiple values in a chunk. The experimental results show the advantage of the asynchronous partial Gauss-Seidel method.

1 Introduction

The most representative sequential algorithms for solving systems of linear equations are the Jacobi method and the Gauss-Seidel method, while the parallel Jacobi method and the asynchronous iterative method are the parallel algorithms cited most frequently [1, 2, 3].

The sequential Gauss-Seidel method generally converges in less number of iterations than the sequential Jacobi method by incorporating newly computed values of unknowns into the computation of the next value of the unknown. However, the Gauss-Seidel method cannot be parallelized because of its nature of dependency. On the other hand, the Jacobi method is easily parallelizable; partitioning the input matrix into blocks so that one processor is responsible for computing one of the blocks, and exchanging the values of unknowns at the end of each iteration. Although the design of the parallel Jacobi method is simple, it requires barrier synchronization at the end of each iteration, which causes a significant degradation of performance. The asynchronous iterative method, which is based on chaotic relaxation introduced by Chazan and Miranker in 1969, was proposed by Baudet in 1978 [1, 2]. It performs fast parallel computation by using older data received earlier in time and by removing the barrier of synchronization inherent in the parallel Jacobi method. One of sub-methods of the asynchronous iterative method which passes newly computed values of unknowns one by one is called the *purely asynchronous method* [2]. Baudet's experimental results show that the purely asynchronous iterative method converges in fewer iterations than the parallel Jacobi method [2]. However, as far as the elapsed

J. Cao, W. Nejdl, and M. Xu (Eds.): APPT 2005, LNCS 3756, pp. 111–120, 2005.
© Springer-Verlag Berlin Heidelberg 2005

time is concerned, it could be disadvantageous on practical cluster computing systems due to its huge communication overhead.

In this paper, we introduce the *asynchronous partial Gauss-Seidel method* which passes multiple values of unknowns in a chunk and reduces communication overhead. The most important parameters which decide its performance are frequency of data sending and frequency of data receiving. Reduction of the frequency of data sending and the frequency of data receiving decreases communication overhead. However it may increase the number of iterations to converge solving systems of linear equations, since the algorithm becomes closer to that of *asynchronous Gauss-Seidel's method* [2].

In section 2, we explain the detail of the asynchronous partial Gauss-Seidel method. Section 3 presents and analyzes experimental results. A summary and a discussion of future work are described in section 4.

2   Asynchronous Partial Gauss-Seidel Method

2.1 Basic Concept

A system of linear equations with vector of unknown x whose size is n can be represented in a matrix form as follows:

$$Ax = b$$

where A is an n by n matrix, and x and b are vectors with n elements.

The asynchronous iterative method is a parallel method for solving sparse systems of linear equations. The simplest way of allocating tasks to processors is partitioning A and b equally by rows, as well as the parallel Jacobi method. Each processor is responsible for solving a portion of unknown x. When p processors exist, the matrix A and the vector b are divided into p tasks, each of which consists of n/p rows of A and b. Each processor is allocated one of the tasks and is in charge of computing x within the range of the given task. For example, processor k computes $x_{nk/p}, ..., x_{n(k+1)/p-1}$, by using the partition k of matrix A consisting of rows $nk/p, ..., n(k+1)/p$ and the partition k of b consisting of elements $b_{nk/p}, ..., b_{n(k+1)/p-1}$.

Baudet classifies the asynchronous iterative method into three different sub-methods - asynchronous Jacobi's method, asynchronous Gauss-Seidel's method and purely asynchronous method - according to timing of exchanging new values of unknowns, or choice of the values [2]. The purely asynchronous method releases each new value immediately after its computation, while the asynchronous Jacobi's method and the asynchronous Gauss-Seidel's method exchange new values only at the end of each iteration. The only difference between the asynchronous Jacobi's method and the asynchronous Gauss-Seidel's method is the choice of the values of unknowns within each iteration. The asynchronous Gauss-Seidel's method uses new values of unknowns in its subset as soon as they are computed for further computation in the same iteration, while the asynchronous Jacobi's method uses only values of unknowns known at the beginning of an iteration.

Baudet's experimental results show that the purely asynchronous method converges in less iterations than the asynchronous Jacobi's method and the asynchronous Gauss-Seidel's method [2]. The results also show that the asynchronous Gauss-Seidel's method increases the number of iterations with the increase of processors.

A drawback of the purely asynchronous method is that the communication overhead by exchanging new values one by one is huge on practical cluster computing systems. As shown by experimental results in section 3, it is obviously difficult to achieve desirable performance on modern cluster computing systems.

The asynchronous partial Gauss-Seidel method, introduced in this paper, lessens the frequency of data passing and improves the drawback of the purely asynchronous method. It sends multiple new values of unknowns in a chunk and reduces the communication overhead. The choice of the values of unknowns used in the computation is the same as that of the purely asynchronous method; the most recent and available values are used. However, the asynchronous partial Gauss-Seidel method differs in the timing of releasing new values. It releases the new values right after the number of unsent values reaches a certain fixed number. For instance, suppose we define the number of values of unknowns passed in a chunk as 50. Each processor computes 50 new values of unknowns using available values including the most recent values computed on the processor. After the computation of the 50 new values, the processor broadcasts them simultaneously to all other processors. Chunks of new values from other processors are received asynchronously. As soon as the values are received, each processor incorporates them into its buffered x and makes them available to the next computation.

The most important parameter in the asynchronous partial Gauss-Seidel method is the frequency of sending new values of unknowns. A decrease in the frequency of data sending reduces communication overhead. However, at the same time, it may cause an increase in the number of iterations to converge. In Baudet's experiments, the asynchronous Gauss-Seidel's method increases the number of iterations to converge with the increase of processors [2]. The asynchronous partial Gauss-Seidel method becomes closer to the asynchronous Gauss-Seidel's method with decrease in frequency of data passing; the same phenomenon may occur in the case of asynchronous partial Gauss-Seidel method. Hence the tradeoff between the reduction of communication and the increase of iterations becomes a significant issue of this method. In section 3, we discuss it with the practical experimental results.

Another important parameter is the frequency of receiving new values from other processors. In order to avoid blocking at receiving new values, our programs periodically check whether new packets from other processors arrive and are stored in the operating system's buffer. A processor calls select() system call on UNIX or equivalent system calls on other operating systems at each time it checks network data buffered in the operating system. Calling a system call and waiting for its return requires a certain period of time. Therefore, frequent receiving, or checking new values from other processors causes the increase of runtime overhead. However, by the immediate incorporation of new values into the processors' buffered x, the asynchronous partial Gauss-Seidel method may finish its computation faster because the new values can be used to evaluate the next value in earlier time. It is not easy to guess the relationship between the frequency of data receiving and the practical speedup. In section 3, we show the experimental results with different frequencies of data receiving.

2.2 Design and Implementation

The basic algorithm to compute new values of unknowns is the same as those of the other asynchronous sub-methods and is expressed as follows:

$$x_i = \frac{1}{a_{i,i}} (b_i - \sum_{j=1}^{i-1} a_{i,j} x_j - \sum_{j=i+1}^{n} a_{i,j} x_j)$$

The three asynchronous sub-methods - the asynchronous Jacobi's method, the asynchronous Gauss-Seidel's method and the purely asynchronous method - only differ by the choices of the values used in computation. The asynchronous partial Gauss-Seidel method always uses available x to compute a new value, as well as the purely asynchronous method. The difference between the purely asynchronous method and the asynchronous Gauss-Seidel method is the frequency of exchanging new values of unknowns.

Suppose we have 24 unknowns: x_0, x_1, x_2, ..., x_{23}. And suppose 2 processors P_0, P_1 are used for solving the system of linear equations, each of which is in charge of computing 12 unknowns; P_0 computes $\{x_0, ..., x_{11}\}$, P_1 computes $\{x_{12}, ..., x_{23}\}$ respectively. In the purely asynchronous method, each new value of x is broadcast immediately after its computation. In the asynchronous practical Gauss-Seidel method, multiple values of x are broadcast in a chunk. For example, suppose 4 values of unknowns x_k, x_{k+1}, x_{k+2}, x_{k+3} are broadcast together, they are bundled into a chunk and are broadcast after the computation of these 4 values. Another parameter we must define is the frequency of data receiving. Here we assume that the new values are received after every computation of 4 values. The execution and data exchange of this model is expressed in Figure 1 and 2.

Figure 1 shows a sequence of computation and data exchanges. On processor 0, after computing x_0 through x_3, the new values are broadcast to other processors – in this case they are sent only to processor 1. Afterwards processor 0 checks the values of unknowns

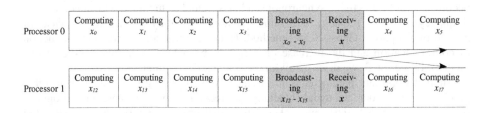

Fig. 1. Execution of the asynchronous partial Gauss-Seidel method 1

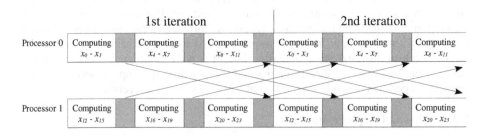

Fig. 2. Execution of the asynchronous partial Gauss-Seidel method 2. Shaded areas represent time spent for communication consisting of broadcasting and receiving x.

sent from other processors. If any values are stored in the operating system's buffer, processor 0 incorporates them into its x buffer. Figure 2 illustrates a phase of iterations.

As described in 2.1, there are two important parameters in the asynchronous partial Gauss-Seidel method: the frequency of data sending and the frequency of data receiving. In the example described above, we define that after every computation of 4 values, the new values are broadcast and, at the same time, values from other processors are checked. Figure 3 shows another model in which values from other processors are checked after every computation of 2 values.

Processor 0	Computing x_{01}	Computing x_1	Receiving x	Computing x_2	Computing x_3	Broadcasting $x_0 - x_3$	Receiving x	Computing x_4	

Fig. 3. Execution of the asynchronous partial Gauss-Seidel method with a different data receiving frequency

If new values are broadcast and checked at the same time one by one, this performs the same algorithm as the purely asynchronous method. And if new values are broadcast at the end of each iteration and values from other processors are checked after every computation of each value, then it becomes the asynchronous Gauss-Seidel's method. If only one processor is used, the algorithms of the purely asynchronous method and the asynchronous partial Gauss-Seidel method equal that of the sequential Gauss-Seidel method.

2.3 Convergence Detection

One of the biggest issues in the asynchronous iterative method is a methodology of convergence detection. *Chaotic relaxation* [1] states the convergence conditions as follows;

> there must be a fixed positive integer s such that, in carrying out the evaluation of the ith iterate, a process cannot make use of any value of the components of the jth iterate if $j - s$ [2, 3].

Though we feel the necessity of further research on the convergence detection methodology, we use a simple detection technique in our experiments. First, while computing new values of unknowns, processors compute a difference between a new value and an old value of each unknown. If the difference is within a given error tolerance, the processors set their convergence flags true, otherwise they are set false. These flags are cleared at the beginning of each iteration. Processor 0 collects the values of these flags from all processors, and it terminates the computation if all flags are true. Theoretically, it is ideal to detect convergence using time stamps or periodical synchronization. However, we focus only on practical usage of the asynchronous iterative method, and we assume the conditions in which the asynchronous iterative method is used as follows;

> Numbers of unknowns in systems of linear equations are large enough, and time spent for data transmission among processors is much shorter than time spent for an iteration of computation. In other words, no values older than the previous iteration are used for evaluation on any processor.

Our experiments show that in our system, delivering a new value of an unknown takes approximately the same time spent for computing 30 values of unknowns. This can be considered small enough compared to the time taken for an iteration of computation in big systems of linear equations.

3 Experimental Results

3.1 Experiments

The experiments have been carried out on 8 machines with 34 different systems of linear equations. Each measurement is repeated 10 times. The following 11 algorithms have been used in the experiments: the asynchronous partial Gauss-Seidel method with 9 different combinations of data sending-receiving frequencies, the purely asynchronous method and the parallel Jacobi method.

The specification of a machine is as follows:

> Model: Sun Blade 2500, CPU: Ultra SPARC IIIi 1.6GHz, LAN: 100Mbps, Memory: 2GB, OS: Solaris 9

Input matrices A and vectors b are generated by a random generator with different random seeds [8]. The approximate density of a matrix A is 38%. The generated matrices are compressed into a *zeros skipped* format. The compression rate is approximately 40%. This compression technique helps reduce not only initial task assignment time but also computation time. The input matrices A and vectors b are equally partitioned and are statically assigned to all machines. The size of unknowns is 3360.

The 9 combinations of data sending-receiving frequencies in the asynchronous Gauss-Seidel method are as follows:

Table 1. Frequencies of data sending and receiving

Frequency of sending	Frequency of receiving
10	1
10	5
10	10
50	10
50	25
50	50
100	10
100	50
100	100

A combination of the frequency of sending '50' and the frequency of receiving '25' means that after every computation of 50 values, the new 50 values are broadcast and values from other machines are checked after every computation of 25 values. This is expressed as "APGS 50-25" in 3.2.

Table 2 (a). Elapsed time compared to the parallel Jacobi method (%)

# of proc.	PA	APGS 10-1	APGS 10-5	APGS 10-10	APGS 50-10	APGS 50-25	APGS 50-50	APGS 100-10	APGS 100-50	APGS 100-100
1	60.1	60.3	60.1	60.3	60.4	60.3	60.4	60.2	60.1	60.2
2	63.4	55.1	51.1	53.2	49.5	49.1	54.1	52.4	53.4	54.3
3	65.1	64.4	54.2	58.4	52.2	52.5	52.2	60.6	64.1	59.4
4	79.2	55.1	52.5	54.3	61.9	57.9	61.1	56.7	57.8	67.0
5	83.1	62.4	58.4	57.6	62.2	61.4	59.4	64.2	61.0	63.3
6	97.2	59.1	58.3	58.1	56.6	59.9	59.2	63.0	63.8	66.3
7	100.9	65.7	60.3	61.9	58.7	59.3	63.1	62.0	65.8	69.7
8	111.2	62.3	58.6	62.5	61.2	61.9	64.0	62.9	65.4	73.0

Table 2 (b). Number of iterations compared to the parallel Jacobi method (%)

# of proc.	PA	APGS 10-1	APGS 10-5	APGS 10-10	APGS 50-10	APGS 50-25	APGS 50-50	APGS 100-10	APGS 100-50	APGS 100-100
1	59.5	59.5	59.5	59.5	59.5	59.5	59.5	59.5	59.5	59.5
2	52.2	52.5	49.0	51.0	48.0	47.7	52.9	51.5	52.5	53.5
3	45.9	59.6	50.4	54.5	49.4	49.9	49.7	59.3	63.0	58.0
4	50.0	48.3	46.9	48.6	58.9	55.0	58.4	54.3	55.4	65.4
5	46.7	54.0	51.3	50.2	58.5	58.0	56.0	61.9	58.5	61.2
6	48.5	49.1	50.0	49.9	51.6	55.4	55.0	60.0	61.1	64.3
7	46.2	53.9	50.6	52.1	53.4	54.3	57.6	58.6	63.1	68.0
8	46.7	48.9	47.4	51.5	55.5	56.8	58.9	59.1	62.2	71.7

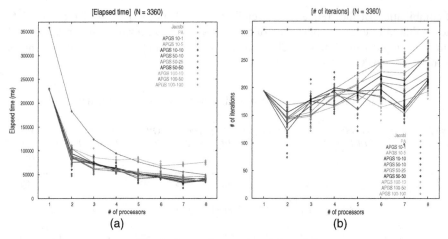

Fig. 4. A sample experimental result 1

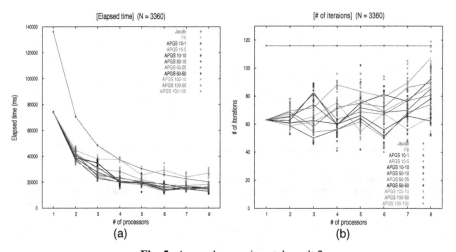

Fig. 5. A sample experimental result 2

3.2 Results

The parallel Jacobi method converges with 26 systems of linear equations out of 34 systems. On the other hand, the purely asynchronous method and all the asynchronous partial Gauss-Seidel methods converge with 32 systems. The sequential Gauss-Seidel method converges with all the systems.

The comparisons on the elapsed time and the number of iterations between the parallel Jacobi method and other methods are shown in Table 2. The results are calculated using 26 systems with which the parallel Jacobi method converges. PA stands for the purely asynchronous method. APGS represents the asynchronous partial

Gauss-Seidel method and the following numbers mean the combinations of data sending-receiving frequencies (see 3.1).

Table 2 (a) shows the average ratios of the elapsed time taken by the PA method or the APGS methods to the elapsed time taken by the parallel Jacobi method with the same number of processors. The measured elapsed time includes the time spent for network communication: the initial task assignment and exchanges of values of unknowns. In all algorithms, the APGS 10-5 method constantly records short elapsed time. On the whole, the elapsed time taken by the APGS 10-5 method is 40-50% shorter than the elapsed time taken by the parallel Jacobi method. The other APGS methods also show fairly good results. The APGS 100-100 method, whose results are probably the worst in the APGS methods', still result in being 28-46% faster than the parallel Jacobi method. On the other hand, the purely asynchronous method performs slow computation in this experiment. It becomes less efficient with the increase of processors. With more than 6 processors, the purely asynchronous method becomes slower than the parallel Jacobi method. As a while, the APGS methods tend to become slower with the decrease of the data sending-receiving frequencies. Table 2 (b) shows the average ratios of the number of iterations. Obviously, both the decrease of the frequency of data sending and the decrease of the frequency of data receiving cause the increase of the number of iterations. The exception is the APGS 10-1 method. In most cases, the APGS 10-1 method takes more iterations than the APGS 10-5 method. Further investigation on this phenomenon is needed.

Sample experimental results are shown in Figure 4 and Figure 5. The horizontal axes in the figures represent the number of processors (machines). The vertical axes in Figure 4 (a) and Figure 5 (a) represent the elapsed time. The vertical axes in Figure 4 (b) and Figure 5 (b) represent the number of iterations.

4 Conclusions and Future Work

In the practical usage of parallel iterative algorithms for solving systems of linear equations, the reduction of the communication overhead and the reduction of the number of iterations are the most important factors which decide the computation speed. In this paper, we describe the design and experimental results of the asynchronous partial Gauss-Seidel method, which requires less communication overhead than the purely asynchronous method and, at the same time, requires less iterations than the parallel Jacobi method. The experimental results show the advantage of the asynchronous partial Gauss-Seidel method. However, the asynchronous partial Gauss-Seidel method has a disadvantage that finding the best combination of the data sending-receiving frequencies is difficult. Further research is needed on this issue.

In our experiments, we use the broadcast to send values of unknowns. It will be interesting to try other message passing methods in order to reduce more communication overhead. Also, our experiments are limited to the computation on small cluster computing systems. Examination on bigger cluster computing systems is needed.

References

1. D. Chazan and W. Miranker, *Chaotic Relaxation*. Linear Algebra and its Applications, Vol 2, pp. 199-222, 1969.
2. G. M. Baudet, *Asynchronous Iterative Methods for Multiprocessors*, Journal of the Association for Computing Machinery, Vol 25, No 2, pp 226-244, 1978
3. B. Wilkinson and M. Allen, PARALLEL PROGRAMMING, *Techniques and Applications Using Networked Workstations and Parallel Computers, Second Edition,* Ch. 6 and Ch.11, 2004
4. K. Blathras, D. B. Szyld and Y. Shi, *Timing Models and Local Stopping Criteria for Asynchronous Iterative Algorithms,* Journal of Parallel and Distributed Computing, vol. 58, pages 446-465, 1999.
5. E. J. Lu, M. G. Hilgers and B. McMillin, *Asynchronous Parallel Schemes: A Survey*, Technical Report, Computer Science Department, University of Missouri - Rolla, 1993
6. J. C. Strikwerda, *A Convergence Theorem for Chaotic Asynchronous Relaxation*, Linear Algorithms and Applications, 253 (1997) pp.15-24.
7. P. Christen, *A parallel iterative linear system solver with dynamic load balancing*, Proceedings of the 12th international conference on Supercomputing, Melbourne, Australia, pages: 7 – 12, 1998
8. M. Matsumoto and T. Nishimura, *Mersenne Twister: A 623-dimensionally equidistributed uniform pseudorandom number generator*, in ACM Transactions on Modeling and Computer Simulation (TOMACS), Special issue on uniform random number generation, Volume 8 , Issue 1 pages 3-30, January 1998

Improved Program Dependence Graph and Algorithm for Static Slicing Concurrent Programs

Jianyu Xiao[1,2], Deyun Zhang[1], Haiquan Chen[1], and Hao Dong[1]

[1] School of Electronics and Information Engineering, Xi'an Jiaotong University,
Xi'an 710049, China
{Xjy, dzhang, chq, dongh}@xanet.edu.cn
[2] Department of Computer Science, Shaoyang University, Shaoyang 422000, China

Abstract. Based on the comparison among existing slicing algorithms and analysis of the fact that Krinke's algorithm [9] produces imprecise program slice for the program structure which has loops nested with one or more threads, a conclusion is drawn that the reason for the impreciseness is that Krinke's data structure—threaded program dependence graph—had over coarse definitions of data dependence relations between threads, and the constraint put on the execution path in concurrent program is unduly loose. An improved threaded program dependence graph is proposed which adds a new dependence relation of loop-carried data dependence crossing thread boundaries. An improved slicing algorithm is also proposed which introduces a new concept of regioned execution witness to further constrain the execution path. The pseudo code of the algorithm adding loop-carried data dependence relations crossing thread boundaries is given. The pseudo code of the new slicing algorithm is also given whose complexity has been analyzed. Examples show that the improved slicing algorithm designed on the improved data structure can restrain the impreciseness of Krinke's.

1 Introduction

Program slicing is based on the deletion of statements that preserve the original behavior of the program with respect to a slicing criterion which is a pair $<p,x>$, where p is a program point and x is a program variable. Program slicing is the important basis of static analysis of programs and is extensively used in program understanding and software maintenance [1]. Program slicing includes static slicing and dynamic slicing [1]. Static slicing technique for sequential program has been mature with over 20 years' development and the mainstream method is based on graphic reachability algorithm with program dependence graph (PDG) as program's internal representation data structure. PDG is constructed based on Control Flow Graph (CFG) with control flow edges deleted and data dependence and control dependence edges added. Data dependence includes flow dependence and def-order dependence, and flow dependence includes loop independent dependence and loop-carried dependence. According to [4], these dependence relations are adequate to express relations between statements of sequential program.

J. Cao, W. Nejdl, and M. Xu (Eds.): APPT 2005, LNCS 3756, pp. 121–130, 2005.
© Springer-Verlag Berlin Heidelberg 2005

Concurrent program slicing now attracts more and more attention as concurrent systems were increasingly adopted. The execution order of statements in concurrent program is undetermined and dependence relations between statements include choice dependence, synchronization dependence, communication dependence and interference dependence etc. besides the conventional control and data dependence. So, static slicing concurrent program is very complex, and the conventional slicing algorithm of sequential program cannot be adopted which is based on the assumption that the execution order of statements is determined. The first method for static slicing concurrent program was proposed by Cheng [6], which was followed by Zhao [7] and Chen [8]. The principle of their method was to construct an extended PDG data structure as concurrent program's internal representation (such as Cheng's Process Dependence Net [6], Zhao's Multi-threaded Dependence Graph [7] and Chen's Concurrent Program Dependence Graph [8]) and use the graphic reachability algorithm. The problem of this method is that many dependence relations of concurrent program have no nature of transitivity that is the precondition of graphic reachability algorithm. The aftereffect of the problem is impreciseness of program slice. Krinke [9][10] pointed out Cheng's problem and proposed a new static slicing algorithm to solve it which was based on a new data structure −threaded program dependence graph (tPDG) and a new concept −threaded execution witness to constrain the execution path of program. However, we find that Krinke's algorithm cannot properly handle the data dependence in program structure with one or more threads nested in a loop and will produce imprecise program slice also. We conclude that the reason for the impreciseness is that tPDG has over coarse definitions of data dependence between statements and the slicing algorithm puts unduly loose constraint on the execution path in concurrent program.

This article is an extension of Krinke's work which proposes a new strategy of static slicing concurrent program with improvement of program's internal representation data structure and slicing algorithm. The improvement of data structure is to add a new dependence of loop-carried data dependence crossing thread's boundary. The improvement of slicing algorithm is to introduce a new concept of regioned execution witness to further constrain the execution path in concurrent program. Examples shows that the improved slicing algorithm designed on the improved program dependence graph can restrain the impreciseness of Krinke's. Definitions of terms related to program slicing are given in section 2. Impreciseness of Krinke's algorithm is analyzed in section 3. In section 4, the strategy of improvement of data structure and slicing algorithm are described with examples being analyzed to show its effectiveness; the construction algorithm of the improved data structure is given with emphasis on the addition of loop-carried data dependence crossing thread's boundary; the pseudo code of the improved slicing algorithm is given and its complexity is analyzed. In section 5, we give an ending remark and a direction of further study.

2   Terms Related to Static Slicing Concurrent Programs

In this section, terms related to static slicing concurrent program will be formally defined for easy description of the proposed data structure and slicing algorithm. $\Theta=\{\theta_0,\theta_1,...,\theta_n\}$ is assumed to be a set of threads in program with θ_0 being the main thread. Function $\theta(p)$ is assumed to return identity of the innermost thread including statement p. Function $para(\theta_i,\theta_j)$ is defined as {true, if θ_i and θ_j may execute concurrently | false, else}.

Def.1. Execution witness: A node sequence $<n_1,...,n_k>$ is said to be an execution witness on CFG G iff n_{i+1} is reachable from n_i (written $n_i \xrightarrow{*} n_{i+1}(1\leq i<k)$).

Def.2. Threaded Control Flow Graph (tCFG): tCFG $<N,E,s,e,cobegin,coend>$ is an extension of CFG, where $cobegin,coend\in N$ represent $cobegin$ and $coend$ statements separately.

Def.3. Threaded execution witness: A sequence of nodes $l=<n_1,...,n_k>$ is a threaded execution witness on tCFG iff $\forall_{t\in\Theta}:ll_t=<m_1,...,m_j>\Rightarrow\forall_{i=1}^{j-1}:m_i \xrightarrow{cf\cdot pf}{}^{*} m_{i+1}$, where ll_t is a sub-sequence satisfying $\theta(m_i)=t$ ($1\leq i\leq j$); $\xrightarrow{cf}{}^{*}$ says reachable through sequential control flow edges and $\xrightarrow{pf}{}^{*}$ says reachable through concurrent control flow edges.

Def.4. Program Dependence Graph (PDG): PDG is a variant of CFG with two kinds of edges—control dependence and data dependence added.

Def.5. Loop-carried dependence: In PDG G, node j is said to be loop-carried dependent on node i (written $i\xrightarrow{lc(L)}j$) if: ① i is data dependent on j , (i.e. $i\xrightarrow{dd}j$); ② i, j are nested in loop L ; ③ in the corresponding CFG, there exists a execution path P from i to j which includes a back edge pointing to L's condition predicate.

Def.6. Interference Dependence (id): In PDG G, node j is said to be interference dependent on node i (written $i\xrightarrow{id}j$) if: ① $\theta(i)\neq\theta(j)$ and $\theta(i)$ may execute concurrently with $\theta(j)$; ② there exists a variable v satisfying $v\in def(i)\wedge v\in ref(j)$.

Def.7. Threaded Program Dependence Graph (tPDG)[9] is an extension of tCFG with control dependence, data dependence and interference dependence edges added.

3 Analysis of Krinke's Slicing Algorithm

3.1 Principle of Krinke's Algorithm

Krinke's method is based on the data structure tPDG. Its principle is: In tPDG G, node p is assumed to be the slicing criterion, $S_\theta(p)$ is assumed to be the program slice with respect to p ,

$$S_\theta(p)=\{q|P=<n_1,...,n_k>,q=n_1 \xrightarrow{d_1}...\xrightarrow{d_{k-1}} n_k=p,d_{1\le i<k}\in\{cd,dd,id\} \qquad (1)$$

where P is a threaded execution witness.}

3.2 The Fact of Krinke's Impreciseness

We found that Krinke's algorithm produced imprecise program slice for the program structure which has loops nested with one or more threads. The tCFG showed in fig.1 (a) has a loop nested with two threads. S_2 is assumed to be slicing criterion. According to Krinke's algorithm, $<S_4,S_6,S_2>$ is a valid threaded execution witness on tCFG as S_4 can reach S_2 through loop predicate S_0. As shown in the figure, there exists $S_4 \xrightarrow{id} S_6$ and $S_6 \xrightarrow{id} S_2$, so S_4 should be included in the program slice $S_\theta(S_2)$ according to Krinke's algorithm. But in fact, in program's behavior, if S_4 reaches S_2 through S_0, then the definition to e by S_4 should have been redefined by S_5 and the definition to d by S_6 is redefined by S_1. This means that S_4's definition to e would not affect c or d in S_2 and S_4 should not be in $S_\theta(S_2)$. S_8 is just the same as S_4. That is to say, the program slice computed by Krinke's algorithm includes unrelated statements.

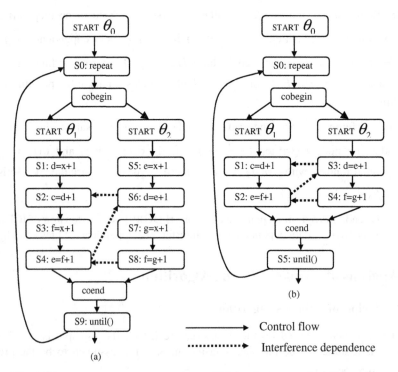

Fig. 1. The concurrent program structure which has loops nested with threads

3.3 The Reason for Krinke's Impreciseness

We conclude that Krinke's impreciseness stems from the fact that its algorithm improperly handles the loop-carried data dependence relation crossing thread's boundary. Krinke didn't realize that during the execution path constructed the loop-carried dependence, the included loop back edge's execution meant the old s-instance of the thread has finished and the new created instance's execution has overlaid some statements' behavior of the old instance. So, the data dependence relation in the old instance of thread should be re-computed. But Krinke still used the invalid data dependence which produced impreciseness.

4 The Strategy of Improved Data Structure and Slicing Algorithm

4.1 The Improvement of Data Structure

Based on Krinke's tPDG, we further refine the data dependence relation between threads. We introduce a new kind of relation of loop-carried data dependence crossing thread's boundary. The improved concurrent program's internal representation is named tPGD'.

Def.8. Loop-carried (data) dependence between threads: In tPDG G, a node S_i in thread θ_i is said to be loop-carried (data) dependent between threads on a node S_j in θ_j if the corresponding tCFG satisfies: ① θ_i and θ_j were nested in the same loop; ② S_i interference dependent on S_j; ③ the execution path constructing the interference dependence includes a loop back edge.

Def.9. Loop-carried (data) dependence between instances of the same thread: In tPDG G, a node S_i in thread θ is said to be loop-carried (data) dependent on a node S_j in θ between instances of the same thread if the corresponding tCFG satisfies: ① θ is nested in a loop; ② S_i data dependent on S_j; ③the path from S_j to S_i includes a loop back edge.

Def.10. Loop-carried (data) dependence crossing thread's boundary (ddl): Loop-carried (data) dependence between threads and Loop-carried (data) dependence between instances of the same thread are called by a joint name -- loop-carried (data) dependence crossing thread's boundary (written $\xrightarrow{ddl}$).

Def.11. Improved threaded program dependence graph tPDG': tPDG' is based on tPDG with loop-carried data dependence edges being changed to be ordinary data dependence edges and loop-carried (data) dependence crossing thread's boundary being added.

The pseudo code of the algorithm for adding loop-carried data dependence edges on tPDG cannot be given here due to the limit of space. The principle of this algorithm is: given tPDG of a concurrent program structure with loops nested with threads, for each loop body from the outermost one to the innermost one, all variable-defining nodes which can reach loop head node are first computed and recorded; then all variable-referencing nodes in the current loop which can be reached from loop-head node are computed; lastly, the corresponding loop-carried data dependence edges crossing thread's boundary can be found. The complexity of the algorithm is $O(n)$, where n is the number of program's statements.

4.2 The Improvement of Slicing Algorithm

The root cause of the impreciseness of Cheng's algorithm [6] is that some sequences of nodes in paths which reach slicing criterion through all kinds of dependence edges do not obey the constraint upon concurrent program's execution behavior. Krinke [9] improved Cheng's algorithm by introducing a concept of threaded execution witness to constrain the qualification of execution paths. We considered Krinke's algorithm still put unduly loose constraint on execution paths. Our improved slicing algorithm built on the new data structure tPDG' introduced another new concept of regioned execution witness to further constrain the valid execution paths in TCFG.

Def.12. Region: Region R is a sub-graph of tCFG G. A node m in R is said to be *In* node if there exists an edge (v,m) in G where v is not in R; A node n in R is said to be *out* node if there exists as edge (n,v) in G where v is not in R. R is said to be a Single-In-Single-Out (SISO) region if there is only a pair of *In* and *out* nodes in R.

Def.13. Regioned execution path: Regioned execution path in tCFG G is a sequence of nodes $<n_1,...,n_k>$ satisfying one of the following: ①In G, R is assumed to be a SISO region representing a basic thread which has no nodes of *cobegin/coend*. n_s is assumed the *In* node of R, and n_e is assumed the *out* node of R. A regioned execution path in R is a path from n_s to n_e; ②In G, R is assumed to be nested with one or more SISO regions. R_i' is assumed to be a SISO region between *cobegin* and the corresponding *coend*. A regioned execution path in R is constructed as a regioned execution path according to ① with R_i' being looked as a single node N_i', and then N_i' is replaced by a regioned execution path in R_i'; ③In G, R is assumed to be an SISO region and is nested with a set of SISO regions, each of whose element R_i represents a basic thread. A regioned execution path in R is an arbitrary interleaving of regioned execution paths in R_i.

Def.14. Regioned execution witness: In tCFG G, a regioned execution witness is a sub-sequence of a regioned execution path in G.

The principle of our new slicing algorithm designed on tPDG' is: In tPDG' G, node p is assumed the slicing criterion, $S'_\theta(p)$ is the program slice with respect to p,

$$S'_\theta(p)=\{q|P=<n_1,...,n_k>,q=n_1 \xrightarrow{d_1} ... \xrightarrow{d_{k-1}} n_k=p, d_{1\leq i<k}\in\{cd,dd,id,ddl\}\} \qquad (2)$$

where P is a regioned execution witness in the corresponding tCFG. }

Theorem 1: A sequence of nodes $<n_1,n_2,...,n_k>$ in region R is a regioned execution witness iff the sequence's two arbitrary nodes n_i and n_j ($i<j$) satisfy that $para(\theta(n_i),\theta(n_j))$ returns true or $path(n_i,n_j,R)$ returns true, where Function $path(n,m,R)$ is defined as { $true$, if there exists a regioned execution path in R from node n to node m | $false$, else}.

Proof: It can be directly drawn from Def.14 and Def.13.

4.3 Description and Analysis of the New Slicing Algorithm

The improved slicing algorithm is an iterative one based on a work-list, which starts from the sling criterion and adds slice nodes successively. The algorithm takes an existed slice node as work node and traverses backwards along the data, loop-carried data, and control and interference dependence edges. All nodes reached via only data or control dependence edges can be directly added to the slice nodes set as regioned execution witnesses exist for these nodes according to definitions of data and control dependence and their transitivity. Our algorithm is different from Krinke's in handling interference dependence edges and loop-carried data dependence edges crossing thread's boundary.

When a node is reached via an interference edge, it is possible that a valid path on tPDG' could have no valid regioned execution witness on the corresponding tCFG. The strategy to handle interference edge is:

$t = T[\theta(y)] \wedge R = Region(x, y)$
IF $t ==\perp$ || $path(y,t,R) == true$ **THEN**

FORALL $i \in (0..N-1) \wedge para(\theta_i,\theta(y)) == false$ **DO** $T[i] = y$ **OD** $c' = (y,T)$

IF c' has not been handled
THEN mark c' as already handled and set w and S as $w=w\cup\{c'\};S=S\cup\{y\}$

Where x is the work node, y is the currently evaluated node. $T[N]$ is the array for recording trace of nodes of threads which the algorithm accesses. In $T=[t_0,t_1,...,t_n]$, $t_{0\leq i\leq n}$ corresponds to the thread $\theta(i)$, which records the algorithm's last reached node in $\theta(i)$ or in threads executing non-concurrently with $\theta(i)$. $\perp$ means the position has not been defined. This information is used to decide if there exists regioned execution witness. S is the set of slice nodes, c' is the node triple $(x,T[N])$, and w is the set of node triples. $coregion(i,j)$ returns the smallest SISO region which includes i and j . $path(i,j,R)$ decides whether there exists a regioned execution path in R from i to j . x is assumed to be the current work node which is already a slice node. As for a node q where exists $q \xrightarrow{id} x$, there is a valid path in tPDG' from q to slicing criterion through x . But there may be no corresponding regioned execution witness in the TCFG and we may draw different conclusions for different scenes. x being a slice node and according to Theorem 1, there is a regioned execution witness from x to slicing criterion which is a sequence nodes named s whose direction is assumed to be from left to right. ①If there is no nodes in s which executes sequentially with q, that is, the position of $\theta(q)$ in the trace record array associated with x is $\perp$. Then in the new sequence s' which is formed by adding q to the tail of s, q and any other node v satisfy that $para(\theta(q),\theta(v))$ returns true. According to Theorem 1, s' is still a valid regioned execution witness and q should be considered as a slice node. ②If there are nodes in s which execute sequentially with q, the tail node of the sub-sequence s'' formed by the nodes which execute sequentially with q is recorded on the position (assumed to be t) of $\theta(q)$ in the trace record array associated with x. If there is a path from q to t, (i.e. $path(q,t,coregion(x,q))$ returns true), then in the new sequence s''' formed by adding q to the tail of s, q and any other node v in s'' satisfy $path(q,v,coregion(q,v))$ being true, and q and any node w not in s'' satisfy $para(\theta(q),\theta(w))$ being true. According to Theorem 1, s''' is still a valid regioned execution witness and q should be considered as a slice node; ③Under any other conditions, q is not a slice node.

For the nodes which reach through the loop-carried data dependence edge crossing thread's boundary, they can be added as slice nodes, but the reverse edge which involves in the construction of the dependence should be handled specially as follows.

FORALL $i \in (0..N-1) \wedge i \in L$ **DO** $T[i]$= the source node of the back-edge of L **OD**
    FORALL $i \in (0..N-1) \wedge para(\theta_i,\theta(y))$==$false$ **DO** $T[i]=y$ **OD**
      $c'=(y,T)$
IF c' has not been handled
THEN mark c' as already handled and set w and S as $w=w\cup\{c'\};S=S\cup\{y\}$
OD

x is assumed to be the current work node. As for a node q where exists $q \xrightarrow{ddl} x$, there is a valid regioned execution witness from q to x in the region of $coregion(q,x)$ according to Def.8, 9, 10. According to Theorem 1, q is a slice node. The execution path s from q to x which builds the loop-carried data dependence edge crossing thread's boundary includes a loop back edge, meaning that s first traverses the source of the loop back edge and then successively reaches the loop head and the thread-creating statement nodes. As the source of the loop back edge executes sequentially with threads in the loop body, the position of the loop's nested thread in the trace record array associated with x should be set as the source of the loop back edge.

The main complexity of the slicing algorithm comes from interference dependence. A program is assumed to have totally N statements nodes and t threads. In the worst condition, these threads execute concurrently and every node in a thread is interference dependent on a node in another thread. The number of nodes directly adjoining slicing criterion through interference dependence edge is approximately $(\frac{N}{t})$, that is, in the first iteration of the algorithm there are nearly $(\frac{N}{t})$ nodes to be evaluated; In the second iteration, there are nearly $(\frac{N}{t} \times \frac{N}{t})$ nodes to be evaluated; ... ; In the t-th iteration, there are nearly $(\frac{N}{t})^t$ nodes to be evaluated. The algorithm complexity is approximately $O(N^t)$, which is exponential to the number of threads.

4.4 Examples

Example 1. In fig.1 (a), S_2 is assumed to be the slicing criterion and there exist $S_2 \xrightarrow{id} S_6$ and $S_6 \xrightarrow{id} S_4$. But there does not exist regioned execution witness from S_4 to S_2 because the loop back is not included in the co-region of S_4 and S_2. According to our algorithm, S_4 cannot be added to program slice, which conforms to the analysis in 2.2 and has conquered the impreciseness of Krinke's.

Example 2. In fig.1 (b), S_1 is assumed to be the slicing criterion and there exist $S_1 \xrightarrow{id} S_3$ and $S_3 \xrightarrow{id} S_2$. But there doesn't exist regioned execution witness from S_2 to S_1 in the co-region of S_3 and S_2. According to our slicing algorithm, S_2 cannot be added to program slice. But there exists $S_2 \xrightarrow{ddl} S_3$ in the figure, and S_2 will be included in program slice, which conforms to intuitive analysis. This means that the improved algorithm would not lose any related statements in program slice.

Examples analysis shows that the improved slicing algorithm can restrain the impreciseness of Krinke's for the program structure with loops nested with threads.

5 Conclusion

According to [11], slicing algorithm of program written in a concurrent language which has procedure and synchronization primitives is always un-optimal, because the reachability of statements in this kind of concurrent program is undecidable. So, there is always room for optimization for the existing slicing algorithm of concurrent program. In this article, based on the analysis of the impreciseness of Krinke's slicing algorithm, an improved threaded program dependence graph is proposed and the algorithm constructing the new data structure is given in pseudo code. Upon the new data structure, the improved slicing algorithm is also given. Examples show that the improved slicing algorithm designed on the improved data structure can restrain the impreciseness of Krinke's. Due to the fact that the worst complexity of the slicing algorithm is exponential to the number of program, which is the same for Krinke's, the future work is to optimize the slicing algorithm.

References

1. D. W. Binkley, M. Harman, A survey of empirical results on program slicing [J], Advances in Computers 2004. 62:105-178.
2. E. M. Clarke, M. Fujita, S. P. Rajan, T. Reps, S. Shankar, and T. Teitelbaum. Program slicing for VHDL [A]. In Charme'99, Bad Herrenalb, Germany, September 1999.
3. L. Millett and T. Teitelbaum. Issues in slicing PROMELA and its applications to model checking, protocol understanding, and simulation [J]. STTT, 2000. 2(4): 343~349.

4. S. Horwitz, J. Prins, and T. Reps. On the adequacy of program dependence graphs for representing programs [A]. Proceedings of Conference Record of the Fifteenth Annual ACM Symposium on Principles of Programming Languages, 1988. p.146–157.

5. Zhenqiang Chen, Baowen Xu, Jianjun Zhao. An overview of methods for dependence analysis of concurrent programs [J]. SIGPLAN Notices, 2002. 37(8): 45-52.

6. J. Cheng. Slicing Concurrent Programs - A Graph-Theoretical Approach [A]. 1st International Workshop on Automated and Algorithmic Debugging, 1993.

7. Jianjun Zhao. Slicing concurrent Java programs [A]. Proceedings of the 7th IEEE International Workshop on Program Comprehension, 1999. p.126-133.

8. Zhenqiang Chen, Baowen Xu, Hongji Yang etc. An Approach to Analyzing Dependency of Concurrent Programs [A]. Proceedings of the The First Asia-Pacific Conference on Quality Software, 2000. p.39-43.

9. J. Krinke. Context-sensitive slicing of concurrent programs [A]. Proceedings ESEC/FSE, 2003. p. 178–187.

10. J. Krinke. Static slicing of threaded programs [A]. Proc. ACM SIGPLAN/SIGFSOFT Workshop on Program Analysis for Software Tools and Engineering (PASTE'98), 1998. p.35--42.

11. M. Muller-Olm and H. Seidl. On optimal slicing of parallel programs [A]. 33th ACM Symposium on Theory of Computing, 2001. p.647–656.

Parallelisation of Sequential Programs by Invasive Composition and Aspect Weaving

Mikhail Chalabine and Christoph Kessler

Programming Environments Laboratory,
Dept. of Computer and Information Science,
Linköping University, Linköping, Sweden

Abstract. We propose a new method of interactively parallelising programs that is based on aspect weaving and invasive software composition. This can be seen as an alternative to skeleton programming. We give motivating examples for how our method could be applied.

1 Introduction

The compositional approach to parallel programming is thoroughly elaborated in the research community [5,17]. It postulates compositionality of algorithmic *skeletons* focusing on their homomorphic transformations. In general, nevertheless, the design of efficient parallel code is still a black art as it is not possible to derive it universally from given sequential fragments. Likewise, a systematic reuse of efficient hand-crafted parallel parts (not expressible as skeletons) remains onerous. Parallel library routines are only a first advancing step and further research is needed to enable true composition of parallel programs.

In this position paper we consider interactive parallelisation of given sequential code written in a high-performance computing language such as Fortran, C, C++ or (with restrictions) Java. Currently, automatic parallelisation by compilers works only for very few special cases such as regular loop nests scanning arrays with statically analysable data dependencies, but fails in more general situations. To overcome this, the programmer (or the parallelisation expert as an advisor) should give hints to the compiler to guide the load-balancing, data layout, and the parallelisation process in general. This can either be done in the form of directives, as in HPF and OpenMP, or in the form of explicitly parallel language extensions such as skeletons. In both cases, the sequential source program must be modified. In the former directives must be added, and in the latter, larger parts must be reorganised or rewritten. All this may require reengineering at the conceptual level with the following disadvantages:

- reengineering is a complex, resource-critical operation;
- with contemporary techniques the original application program is obscured by the extra directives, code, or reorganisations that handle parallelism. This makes maintenance and future extensions of the application more difficult;
- simultaneous editing by the program author (minor modifications in the application code) and the parallelisation expert is generally not possible;

J. Cao, W. Nejdl, and M. Xu (Eds.): APPT 2005, LNCS 3756, pp. 131–140, 2005.
© Springer-Verlag Berlin Heidelberg 2005

- the use of a modern high-level approach (see Section 2.3) requires a shift in the programming paradigm where the programmer operates on a limited set of expressible patterns.

We see a key problem in the fact that different aspects, such as parallelisation, data layout, synchronisation, and other platform-specific refinements, are hard-coded into the application core. Instead we advocate a compositional approach where such aspects are separated. To do this we apply concepts from aspect-oriented programming (AOP) [11] and invasive software composition (ISC) [1].

Parallelising program modifications, let us call them *recipes*, are formulated as rewrite rules. A rewrite rule consists of a *pattern* (also called a *joinpoint*) and a *modification* (also called an *advice*). Patterns are specified in terms of a high-level intermediate representation (IR) of the (application) program, where special program constructs, such as loop headers, variable definitions or function calls, are identified by symbolic references to the IR, so-called *hooks*. Recipes that concern different aspects are defined in separate recipe files. The recipe files are woven together with the application program by a weaver tool to produce a parallelised program. The weaver matches recipe patterns to the given program and executes rewrite rules where applicable. We see the following advantages:

- the parallelisation process is simplified to specifying rewrite rules;
- the level of abstraction leaves the original application intact, which simplifies reasoning about large programs;
- the programming paradigm is not changed while the expressiveness of structural programming is addressed in a new systematic way using rewrite rules instead of highorder functions (see Section 2.3);
- programmability is improved by allowing for interactive parallelisation and stepwise refinement [16] where the woven program is displayed in real-time;
- reusability is also improved as advices can be reused;
- further aspects, beyond parallelism, can be woven into the code;
- contemporary approaches (including high-order functions) can be imitated by generic advices;
- porting applications to an alternative parallel environment becomes easier, as it only requires (ex)changing one or a few advice files;
- testing is also simplified as it is possible to step back and forth in the parallelisation process by replaying desired transformations.

Note that the suggested method is more powerful than AOP, allowing arbitrary patterns and declared hooks that are not limited to object-oriented (OO) constructs such as class definitions, method calls or field access. In particular, the application needs not be given in an OO language at all.

In the following we elaborate on these points with a number of motivating examples. In Section 2 we give a short overview of contemporary approaches to parallelisation and metaprogramming - the technique we use in our approach. In Section 3 we present our ideas supported by intuitive examples. In Section 4 we further consider a simple parallelisation case and in Section 5 we conclude.

2   Background

There is a number of methods of assembling parallel programs. These range from low-level manual coding to elaborate frameworks based on skeletons and complex loop transformations. In this section we give a short comparative survey.

2.1   Automatic Parallelisation by Dependence Analysis

This class of parallelisers represents the classical approach in parallelising compilers. It mainly focuses on loop nests with statically analysable data dependencies and suggests application of complex transformations (*e.g., skewing, interchange, etc.*) to identify parallelisable loop levels. Despite the moderate progress in the area and substantial difficulties with static analysis of structures other than loops, significance of the dependence theory can hardly be overestimated. We refer to [7,10,20] for more details.

2.2   Domain-Specific Automatic Parallelisation

In this class of automatic parallelisation we make certain assumptions about the target domain. We then prune the solution space of possible parallelisation strategies and arrive at heuristics capable of transforming the original sequential code. The transformation engines are often graph- or logic-based [6]. For example, PARAMAT targets two application domains in numerical computations, namely, linear algebra and partial differential equations. The approach relies on a hierarchical concept recognition system where complex entities are defined via their less complex compounds. The pattern matching consists in executing a matching automaton that represents the concept hierarchy. This is realised through a bottom-up traversal of the source program's syntax tree. As for the logic, the parallelisable algorithmic pattern recogniser (PAP) is based on the deductive inference engine of Prolog. The system does a structural analysis of the input code, *i.e.*, a hierarchical parsing of the program dependence graph, driven by concept-recognition rules; the latter are based on attribute grammars. Each concept is recursively specified by its compositional hierarchy and by control and data-dependence relationships. See [6] for more details.

2.3   Skeleton Approach

Today's structural epitome is the skeleton approach, introduced by Cole in 1989 [5] and since then developed in several projects such as P3L [15], SkiL [12], HSM

```
D&C(isTrivial, Solve, Divide, Conquer, Pr) {
  IF (isTrivial(Pr))
  THEN Solve(Pr)
  ELSE Conquer( Map D&C (Divide (Pr))) }
```

Fig. 1. Pseudocode for the *Divide-and-conquer* skeleton

[13], HDC [8] and others, see also Rabhi and Gorlatch [17]. These, along with the general success, discovered substantial engineering difficulties and limitations in skeletons; would it be a tight connection to the compiler [8] or the reiterated elaboration on obstructive confines within the set of expressible patterns. In general, the technique consists of defining a small number of parallel algorithmic building blocks such as *task-parallel farming, pipelining, parallel divide-and-conquer* and other *data-parallel* and *control-parallel* patterns (see Figure 1). These are usually given in the form of (compiler-known) higher-order functions and a mechanism for composing and instantiating them for the given problem. The latter is usually done by parametrisation in the problem-specific user code where the skeleton is adopted as a generic parallel subroutine, *ipso facto*, a black-box component. The user code is plugged into skeleton instances by passing function names. Special tools can analyse and optimise skeleton programs by applying static transformations. For pragmatic reasons, however, and to embody a larger set of coverable patterns, the skeleton literature recommends to admit *ad-hoc* parallelism provided by the underlying programming environments such as OpenMP [4], threads or MPI (see, *e.g.*, [14]). Unfortunately, this may cause unexpected side effects and thus compromise the compositionality of skeletons. Furthermore, the composition of two optimal parallel algorithms does not necessarily yield an optimal parallel algorithm. Thus parallelisation of compositions of skeletons is a significant problem in skeleton-based approaches.

In order to reach parallelism with skeletons the sequential program must be explicitly refactored into an appropriate form that matches the interfaces of the skeletons. Such a paradigm shift is costly. We advocate to leave the user code intact and, instead, inject parallelisation code invasively. We also encourage to relax the black-box property allowing weaving of other aspects into the existing skeleton implementations, *i.e.*, make the interfaces adaptable. Another problem we see is that coordination of multiple processors by a skeleton is limited to a source code subtree rooted at the skeleton instance. This is an inherent property of structured parallel programming, as control and data flow are tied to the code structure; it is a *single-entry-single-exit* region in the program. In contrast, our approach allows to express coordination that is spread over the entire program.

2.4 Metaprogramming

Metaprogramming goes beyond the scope of homomorphic transformations and allows for more flexibility in approaching parallelism. Transformations and adaptations in metaprogramming are directed by the procedural specifications at a higher level of abstraction. The technique is widely used in contemporary sequential programming and is an important tool for composition of components [1]. In our work we use an abstraction to metaarchitecture, where we develop an intelligent preprocessor capable of advanced code transformations at compile time. It is common to look at metalevel architectures as introducing an intermediate layer between the application and the system. Such a layer is configurable to a certain degree based on the internal and external requirements.

For us metaprogramming is a way of supporting static change in component interconnection topologies, *i.e.*, adaptations in context and behaviour, necessary for introduction of parallelism.

2.5   Aspects in Parallel Programming

One of the first attempts to introduce aspects explicitly is to consider distributed data structures, so-called *covers* [19]. This work originates from the fact that optimal data distribution cannot be determined automatically for most non-trivial algorithms. It is therefore necessary to represent the knowledge about the problem domain and either perform data distribution appropriately or introduce a mechanism to make data accesses transparent. The work adopts the second approach and tackles spatial aspects in functional setting by parameterising skeletons with covers.

There are further projects that study aspects in concurrent systems [2,9,18]. Our ideas differ from these in the following way. We mainly concentrate on parallelisation of existing sequential software. We adapt and extend components at hooks by destructive weaving. We aim at typed transformations that can be checked statically and, in contrast to aspect oriented systems that work directly on the abstract syntax representation, we rely on a component model. We consider dependencies between the parallelisation concerns and integrate this knowledge into the concern model. The ultimate goal is to do reasoning on the concern model instead of primitive code patching found in many aspect-oriented systems. We mainly operate on the declared hook level where we can rewrite parts of the code while others (AOP) mainly add code.

3   Suggested Parallelisation Method

In this section we discuss the technique which relies on the recently developed invasive software composition approach (ISC) [1]. The method evolves from the field of modular composition systems focusing on component reuse. It allows for invasive component formations via code transformations, where the programmer operates on a hierarchical composition system. For integration of parallelism we suggest a system consisting of a *component model, concern model, composition technique* and a *composition language*. The component model outlines a *composition unit*, which is a software item subject to composition. We call it a *component* or a *fragment box* in the context of ISC. The concern model classifies domain-specific properties crosscutting an application and formalises interdependencies among them. Concerns are hallmarks that can not be embraced by a single fragment box, such as a procedure, class or a package; concerns appear scattered over the code. The concern model structures these properties and serves as a foundation for reasoning on the mutual effects any consequent application of a number of concerns might have. It also frames the coalescing of concerns with components. The actual technique capable of type-safe fragment-based parameterisation of components, subject to concern-model constraints, is an invasive composition technique. We refer to it as *weaving*. The principal difference of the ISC weaving from the AOP weaving is that ISC modifies (renames) code fragments comprising the composition interface through *composers*. These comprise an algebra of operators, uniform for all programming languages. It contains, for example, *bind*, *rename*, and *extend* composers. In AOP, on the other hand, a separate weaving tool (a compiler) injecting pieces of code at pre-defined join points must be re-implemented for every given language. ISC results in a two-stage compilation which unifies a number of software engineering paradigms, including, generic programming, architecture systems, inheritance, view-based programming and aspect systems [1]. In our work we aim to apply and extend the approach for parallel programming, allowing to weave parallelism invasively into sequential cores. So our approach is dual to that taken by the skeleton community which plugs sequential user code into well-defined skeletons – parallel components. We see parallelism as consisting of a number of dependent parallelisation concerns crosscutting applications.

As a simple example, consider an MPI-based parallelisation of a large sequential core [3]. We require an extra parameter, namely, an MPI communicator object, to be added to a wide range of functions. A function in an imperative language has a well-defined parameter declaration section and we use this fact to automatise the process by targeting these specific positions and binding them to a new parameter. Thus, for every method M we advice:

```
findPosition("M.Parameters").bind("MPI::Intracomm COMM").
```

In this way we transform void M(type1: *parameter1*) into void M(type1: *parameter1*, MPI::Intracomm *COMM*). With such flexible parameterisations of fragment boxes, we can even weave components defined for clashing paradigms. For instance, we can weave a D&C skeleton-based implementation of the parallel Quicksort (an instance of the *divide-and-conquer* skeleton shown in Figure 1) into sequential code; this corresponds to a manual infusion according to the skeleton-based programming model. But we can also extend it with irregularity handling or introduce fault tolerance into the existing skeleton either after or before it is actually woven into a core (see Figure 3 for the result and Figure 2 for a recipe sketch). Thus we allow a transitive composition procedure where the sequential core is incrementally enhanced with parallelism. We suggest to write parallelisation recipes for programs; recipes that catch the structure of a program and specialise components. Such compositions are specified as scripts in the composition language.

```
D&C-FaultTolerance-Recipe() {
    IF (FindInstance("D&C"))
    THEN EXTEND "D&C.Solve()" :: FAULT_TOLERANCE_BASIC }
```

Fig. 2. *Divide-and-conquer* programmable transformation recipe extending all instances of D&C with a basic fault tolerance handling. FAULT_TOLERANCE_BASIC is a macro.

3.1   Parallelisation Concerns

In parallelisation we refer to parallelism-specific concerns. We see them as highly interdependent and distinguish the following set: Data distribution, Parallelism, Synchronisation, Communication, Cross-processor Data Flow, Data and Control Dependencies, Load balancing. Parallelism is a concern that creates the necessary conditions for running programs in parallel. For example, in Java this requires inheriting from the Thread superclass. In OpenMP, on the other hand, this corresponds to application of the omp_parallel pragma that marks a parallel region. Data distribution maps data sets to processor sets. A natural example will be the HPF's distribute and align directives. Synchronisation appears in many languages and can be further partitioned into synchronisation concerns for mutual exclusion, preventing race conditions, and conditional blocking [9]. In Java synchronisation also appears in terms of the synchronize, sleep, notify, and notifyAll methods. Communication can be seen as parameter passing via shared variables or via communication in a distributed-memory system.

3.2   Parallelisation Process

The integral system view is as follows. The user is given sequential code and a distribution of the composition system. The job of the user is to write parallelisation recipes for the parallelisation concerns described above. The user enters recipes into the system and the weaver seeks to coalesce the described concerns with the sequential core. The join points are defined in terms of explicitly and implicitly declared *hooks*. An *implicit hook* is a point in the abstract syntax tree constrained by the language structure (recall the parameter declaration example). Declared hooks are named and have a type in the component model. The explicit declaration of hooks, patterns, and parameters can be alleviated by drag-and-drop operations on a graphical user interface displaying the content status by the application of the already defined recipes.

Instead of unordered weaving *before*, *after*, and *around* code fragments we advocate the following procedure. As the first step, a subset of join points is considered that are triggered by automated reasoning on the control and data flow information as in (domain-specific) automatic parallelisation. This corresponds to direct integration of parallelism at the lowest composition level operating directly on the IR (*e.g.*, loop parallelisation). As the next step concepts are woven at the default composition interfaces prescribed by the language structure. Reasoning on the concept model and the set of available implicit hooks is performed

```
D&C(isTrivial, Solve, Divide, Conquer, Pr) {
1  IF (isTrivial(Pr))
2  THEN
3     WHILE (Solve(Pr) != TRUE)
4  ELSE Conquer( Map D&C (Divide (Pr) ...)) }
```

Fig. 3. *Divide-and-conquer* skeleton with a primitive completion assurance

here. The remaining part of parallelism is then pinpointed explicitly by the parallelisation expert via interactive system dialogues and parallelisation recipes.

4   An Example

As an illustration consider parallelisation of a simple Java recursion for computing Fibonacci numbers. There are two independent recursive calls that can be executed in parallel (line 5, Listing 1.1) and we weave in four concerns: parallelism, synchronisation, data dependencies and communication. Parallelisation is started by creating an instance of the composition system:

```
CompositionSystem cs = new CompositionSystem(outputPath);
ClassBox fibGenClass = cs.createClassBox("fibGenerator.java");
```

Now we can create parallel tasks and set up data flows wrapping all read/write accesses to shared data in a `synchronized` section. To do that we first augment our `fibGenerator` class with parallelism by referring to the Java `Thread` super class and adding the `run()` method in the core; this is done at the default composition level and is triggered automatically or by advising:

```
cs.findHook("fibGenClass.SuperClass").bind("Thread");
cs.findHook("fibGenClass.Members").bind("void run() { }");
```

We now make a copy of the original recursion for further transformations (compare Line 19 in Listing 1.2 with Listing 1.1). We then extend the run method with the ability to compute a given Fibonacci number by advising:

```
cs.findMethod("fibGenClass.run").
  findHook("MethodEntry").bind("addend2=figRecSeq(THIS.probSize)");
cs.findHook("fibGenClass.ClassEntry").
  bind("static addend1; static addend2");
```

The last statement introduces temporary storage used by parallel workers. In general it should allow the system to conclude that `addend2` is a global variable. From this fact the system should trigger the synchronisation concern assuring a safe write access as shown in line 24, Listing 1.2. The expert can now introduce parallel workers by referring to a proper constructor and the `run` method in the following way:

```
cs.findMethod("fibGenClass.fibRec").findPositionAfter("ELSE").
  bind("fibGenerator fib1 = NEW fibGenerator(n − 1)");
```

As the next step we rebind communication hooks, *i.e.*, substitute recursive calls in line 5 with functionality to retrieve data from parallel workers:

```
cs.findFunctionCallHook("fibGenClass.fibRec-1").bind("addend1");
cs.findFunctionCallHook("fibGenClass.fibRec-2").bind("addend2");
```

Note that `fibRec-1` and `fibRec-2` hooks are uniquely identified in the IR. We introduce synchronisation by executing the following advice:

```
cs.findHook(writeAccess('addend1')).insertAfter(SIMPLE_SYNC);
```

`SIMPLE_SYNC` is a macro expanded in lines 13 to 15, Listing 1.2.

```
1    PUBLIC CLASS fibGenerator {
2        INT PROB_SIZE = 10;
3        PUBLIC STATIC INT fibRec (INT n) {
4                IF (n ≤2) RETURN 1;
5                ELSE RETURN fibRec(n - 1) + fibRec(n - 2);
6        }
7        PUBLIC STATIC VOID main(String[] args) {
8                fibonacci(PROB_SIZE);
9    }}
```

Listing 1.1. Computing the nth Fibonacci number

```
1    PUBLIC CLASS fibGenerator EXTENDS Thread {
2    ...
3    STATIC INT addend2 = -1; // temporary storage
4    STATIC INT addend1 = 0;
5    fibGenerator(INT n) {THIS.probSize = n; }
6
7    PUBLIC STATIC INT fibRec (INT n) {
8    IF (n ≤2) { RETURN 1;}
9    ELSE {
10          fibGenerator fib1 = NEW fibGenerator(n-1);
11          fib1.start();
12          addend1 = fibRecSeq(n-2);
13          WHILE (addend2 == -1) {
14                  TRY { fibGenerator.currentThread().wait();
15                  } CATCH (InterruptedException e) { e.printStackTrace(); }}
16          RETURN (addend1 + addend2);
17    } }
18
19   PUBLIC STATIC INT fibRecSeq (INT n) {
20   IF (n ≤2) RETURN 1;
21   ELSE RETURN (fibRecSeq(n - 1) + fibRecSeq(n - 2)); }
22
23   PUBLIC VOID run() {
24   SYNCHRONIZED(THIS) {
25          addend2 = fibRecSeq(THIS.probSize);
26          THIS.notifyAll();
27   }}
```

Listing 1.2. Transformed Fibonacci

5   Conclusions and Future Work

We have presented a new method of parallelising sequential programs that is based on invasive software composition. It allows for incremental parallelisation of sequential cores written in high-level programming languages. The process is based on a levelled weaving on abstract syntax tree representations, implicit hooks, and declared hooks. We argued that such a composition hierarchy simplifies and structures parallelisation of programs. The method prescribes to write parallelisation recipes describing integration of parallelism. Such recipes are interpreted statically by the composition system that weaves parallelism-specific concerns into sequential cores. The main purpose of this position paper was to present the technique we are working on. As future work we intend to introduce coherence into concern/recipe structures in terms of a concern model as to avoid primitive patching of code. We also aim to elaborate on the reasoning mechanisms as to improve guidance in selecting transformations targeted at parallelism. Among other things, this will question the circumstances under which the parallelisation concerns apply.

References

1. Uwe Aßmann. *Invasive Software Composition*. Springer-Verlag, 2003.
2. Mariano Ceccato and Paolo Tonella. Adding distribution to existing applications by means of aspect oriented programming. In *4th IEEE International Workshop on Source Code Analysis and Manipulation*, 2004.
3. Mikhail Chalabine, Christoph Kessler, and Staffan Wiklund. Optimising intensive interprocess communication in a parallelised telecommunication traffic simulator. In *Proc. Int. High-Performance Computing Symposium (part of the Advanced Simulation Technology Conference)*, Orlando, Florida, USA, 2003.
4. Rohit Chandra, Leonardo Dagum, Dave Kohr, Dror Maydan, Jeff McDonald, and Ramesh Menon. *Parallel Programming in OpenMP*. 2001.
5. Murray Cole. *Algorithmic Skeletons: A Structured Approach to the Management of Parallel Computation*. MIT Press, 1989.
6. Beniamino di Martino and Christoph W. Keßler. Two program comprehension tools for automatic parallelization. *IEEE Concurrency*, 8(1 (Spring)):37–47, 2000.
7. Martin Griebl. *Automatic Parallelization of Loop Programs for Distributed Memory Architectures*. Habilitation thesis, University of Passau, Germany, 2004.
8. Christoph A. Herrmann and Christian Lengauer. HDC: A higher-order language for divide-and-conquer. *Parallel Processing Letters*, 10(2/3):239–250, 2000.
9. David Holmes, James Noble, and John Potter. Aspects of Synchronisation. In *Proceedings of TOOLS-25'97*. IEEE, 1997.
10. Ken Kennedy and John R. Allen. *Optimizing compilers for modern architectures: a dependence-based approach*. Morgan Kaufmann Publishers Inc., San Francisco, CA, USA, 2002.
11. G. Kiczales, J. Lamping, A. Mendhekar, C. Maeda, C. Videira Lopes, J.-M. Loingtier, and J. Irwin. Aspect-oriented programming. In *Proc. of ECOOP*. Springer-Verlag, 1997.
12. Herbert Kuchen. A skeleton library. In *Euro-Par '02: Proceedings of the 8th International Euro-Par Conference on Parallel Processing*, pages 620–629, London, UK, 2002. Springer-Verlag.
13. M. I. Marr and M. Cole. Hierarchical skeletons and "ad hoc" parallelism. In *Parallel Computing: State-of-the-Art and Perspectives*, volume 11. Elsevier.
14. Marcus I Marr. *PhD dissertation: Descriptive Simplicity in Parallel Computing*. University of Edinburgh, 1997.
15. S. Pelagatti. *Structured development of parallel programs*. Taylor & Francis, 1997.
16. Roger Pressman. *Software Engineering: A Practitioner's Approach*. McGraw Hill, 1992.
17. F. A. Rabhi and S. Gorlatch. *Patterns and Skeletons for Parallel and Distributed Computing*. Springer, 2002.
18. Rafael Ramirez and Andrew E. Santosa. An aspect-oriented framework for concurrent applications. In *Proc. of the 3rd German Workshop on Aspect Oriented Software Development*, Essen, Germany, 2003. German Informatics Society.
19. Mario Südholt. *The Transformational Derivation of Parallel Programs using Data Distribution Algebras and Skeletons*. PhD thesis.
20. Hans Zima and Barbara Chapman. *Supercompilers for parallel and vector computers*. ACM Press, New York, NY, USA, 1991.

Revisiting the Election Problem
in Asynchronous Distributed Systems

SungUoon Bauk

School of Electrical and Computer Engineering,
Chungbuk National Unvi. Cheongju,
ChungBuk 361-763, Korea
spark@chungbuk.ac.kr

Abstract. This paper is about the relationship between the Election Problem and Failure Detectors in asynchronous distributed systems. It is stated in [7] that a Perfect Failure Detector P is needed to solve the Election problem. But in contrast to the result, there is a failure detector that solves Election weaker than the Perfect Failure Detector. We introduce the Confirmatory failure detector C. We show that to solve Election, C is necessary while P is not, whereas $C+\diamondsuit S$ is sufficient when a majority of the processes are correct.

1 Introduction

Electing a *Leader* (or Coordinator) in a distributed system is widely recognized as one of the central paradigms in the theory of distributed computing. The first formations of the problem appeared in a variety of papers [2,3,4]. Since then the problem has been studied extensively in various models and network topologies, like rings, complete network topology, unidirectional networks, and grids. One reason for this wide interest is that many distributed protocols need an election protocol.

To elect a leader in a distributed system, an *agreement* problem must be solved among a set of participating processes. This problem, called the *Election* problem, requires the participants to agree on only one leader in the system [1].

Consensus and Election are similar problems in that they are both agreement problems. The so-called FLP impossibility result, which states that it is impossible to solve any non-trivial agreement in an asynchronous system even with a single crash failure, applies to both problems [5]. Many interesting theoretical results have been stated for Consensus [2,3,4] and it is not clear whether those results apply to Election because of the differences between Consensus and Election. The motivation of this work is to explore the applicability of those results to Election.

The rest of the paper is organized as follows. Section 2 describes motivations and the related works. In Section 3 we describe our system model. In Section 4 we introduce the *Confirmatory* Failure Detector C and show that to solve Election, C is necessary while P is not, whereas $C+\diamondsuit S$ is sufficient when a majority of the processes are correct. Finally, Section 5 summarizes the main contributions of this paper and discusses related and future work.

J. Cao, W. Nejdl, and M. Xu (Eds.): APPT 2005, LNCS 3756, pp. 141–150, 2005.
© Springer-Verlag Berlin Heidelberg 2005

2 Motivations and Related Works

Actually, the main difficulty in solving the election problem in presence of process crashes lies in the detection of crashes. As a way of getting around the impossibility of Consensus, Chandra and Toug extended the asynchronous model of computation with unreliable *failure detectors* and showed in [8] that the FLP impossibility can be circumvented using failure detectors.

Can we also circumvent the impossibility of solving Election using some failure detector? The answer is of course "yes". The bully algorithm of Garcia-Molina [14] solves the election problem with the failure detector P (*Perfect*) in asynchronous distributed systems. It is stated in [7] that Failure detector $\Diamond S$ cannot solve Election, even if only one process may crash. This means that Election is strictly harder than Consensus, i.e., Election requires more knowledge about failures than Consensus.

An interesting question is then "What is the weakest failure detector for solving the Election problem in asynchronous systems with unreliable failure detectors?" It is stated in [7] that a *Perfect Failure Detector* is needed to solve the Election problem; hence, a *Perfect Failure Detector* is the weakest failure detector that is sufficient to solve the election problem.

But in contrast to the result, we show that there is a failure detector that solves Election weaker than the Perfect Failure Detector. This means that the weakest failure detector for election is not a Perfect Failure Detector P.

In this paper, we introduce the *Confirmatory* Failure Detector C and show that to solve Election, C is necessary while P is not, whereas $C+\Diamond S$ is sufficient when a majority of the processes are correct.

3 Model and Definitions

Our model of asynchronous computation with failure detection is the one described in [5]. In the following, we only recall some informal definitions and results that are needed in this paper.

3.1 Processes

We consider a distributed system composed of a finite set of processes $\Omega=\{1,2,..,n\}$ completely connected through a set of channels. Each process has a unique id and its priority is decided based on the id, i.e., a process with the lowest id has the highest priority. Communication is by *message passing*, *asynchronous* and *reliable*. Processes fail by crashing and the crashed process does not recover. Byzantine failures are not considered.

To simplify the presentation of the model, it is convenient to assume the existence of a discrete global clock. A history of a process $i \in \Omega$ is a sequence of events $h_i = e_i^0 \cdot e_i^1 \cdot e_i^2 \cdots e_i^k$, where e_i^k denotes an event of process i occurred at time k. Histories of correct processes are infinite. If not infinite, the process history of i terminates with the event $crash_i^k$ (process i crashes at time k). Processes can fail at any time, and we use f to denote the number of processes that may crash. We consider systems where at least one process is correct (i.e. $f < |\Omega|$).

A failure detector is a distributed oracle which gives hints on failed processes. We consider algorithms that use failure detectors. An algorithm defines a set of runs, and a run of algorithm A using a failure detector D is a tuple $R = < F, H, I, S, T>$: I is an initial configuration of A; S is an infinite sequence of events of A (made of process histories); T is a list of increasing time values indicating when each event in S occurred; F is a failure pattern that denotes the set $F(t)$ of processes that have crashed through any time t. A *failure pattern* is a function F from T to 2^{Ω}. The set of correct processes in a failure pattern F is noted *correct(F)* and the set of incorrect processes in a failure pattern F is noted *crashed(F)*; H is a failure detector history, which gives each process p and at any time t, a (possibly false) view $H(p,t)$ of the failure pattern. $H(p,t)$ denotes a set of processes, and $q \in H(p,t)$ means that process p *suspects* process q at time t.

3.2 Failure Detector Classes

Failure detectors are abstractly characterized by *completeness* and *accuracy* properties [8]. Completeness characterizes the degree to which crashed processes are permanently suspected by correct processes. Accuracy restricts the false suspicions that a process can make.

Two completeness properties have been identified. *Strong Completeness*, i.e. there is a time after which every process that crashes is permanently suspected by every correct process, and *Weak Completeness*, i.e. there is a time after which every process that crashes is permanently suspected by some correct process. Four accuracy properties have been identified. *Strong Accuracy*, i.e. no process is never suspected before it crashes. *Weak Accuracy*, i.e. some correct process is never suspected. *Eventual Strong Accuracy* ($\Diamond$Strong), i.e. there is a time after which correct processes are not suspected by any correct process; and *Eventual Weak Accuracy* ($\Diamond$Weak), i.e. there is a time after which some correct process is never suspected by any correct process. A failure detector class is a set of failure detectors characterized by the same completeness and the same accuracy properties (Fig. 1). For example, the failure detector class P, called *Perfect Failure Detector*, is the set of failure detectors characterized by Strong Completeness and Strong Accuracy. Failure detectors characterized by Strong Accuracy are reliable: no false suspicions are made. Otherwise, they are unreliable.

For example, failure detectors of S, called Strong Failure Detector, are *unreliable*, whereas the failure detectors of P are *reliable*.

Completeness	Accuracy			
	Strong	Weak	$\Diamond$Strong	$\Diamond$Weak
Strong	P	S	$\Diamond P$	$\Diamond S$
Weak	Q	W	$\Diamond Q$	$\Diamond W$

Fig. 1. Failure detector classes

3.3 Reducibility and Transformation

The notation of *problem reduction* first has been introduced in the problem complexity theory [10], and in the formal language theory [9]. It has been also used in the distributed computing [11,12]. We consider the following definition of problem reduction. An algorithm A *solves* a problem B if every run of A satisfies the specification of B. A problem B is said to be *solvable with* a class C if there is an algorithm which solves B using any failure detector of C. A problem B_1 is said to be reducible to a problem B_2 with class C, if any algorithm that solves B_2 with C can be transformed to solve B_1 with C. If B_1 is not reducible to B_2, we say that B_1 is *harder than* B_2.

A failure detector class C_1 is said to be *stronger than* a class C_2, (written $C_1 \succeq C_2$), if there is an algorithm which, using any failure detector of C_1, can emulate a failure detector of C_2. Hence if C_1 is stronger than C_2 and a problem B is solvable with C_2, then B is solvable with C_1. The following relations are obvious: $P \succeq Q$, $P \succeq S$, $\Diamond P \succeq \Diamond Q$, $\Diamond P \succeq \Diamond S$, $S \succeq W$, $\Diamond S \succeq \Diamond W$, $Q \succeq W$, and $\Diamond Q \succeq \Diamond W$. As it has been shown that any failure detector with *Weak Completeness* can be transformed into a failure detector with *Strong Completeness* [8], we also have the following relations: $Q \succeq P$, $\Diamond Q \succeq \Diamond P$, $W \succeq S$ and $\Diamond W \succeq \Diamond S$. Classes S and $\Diamond P$ are incomparable.

3.4 Election Problem

The Election problem is described as follows: At any time, at most one process considers itself the leader, and at any time, if there is no leader, a leader is eventually elected. More formally, the Election Problem is specified by the following two properties:

- *Safety*: All processes never disagree on a *leader*.
- *Liveness*: At any time, if there is no *leader*, a *leader* is eventually elected.

3.5 Consensus Problem

In the *Consensus* problem (or simply Consensus), every participant *proposes* an input value, and correct participant must eventually *decide* on some common output value [7,13]. Consensus is specified by the following conditions.

- *Agreement*: no two correct participant decide different values;
- *Uniform-Validity*: if a participant decides v, then v must have been proposed by some participant;
- *Termination*: every correct participant eventually decides.

Chandra and Toueg have stated the following two fundamental results [6]:

- If $f < |\Omega|$, Consensus is solvable with either S or W.
- If $f < \lceil |\Omega|/2 \rceil$, Consensus is solvable with either $\Diamond S$ or $\Diamond W$.

4 Failure Detector to Solve Election

As we pointed out in the motivation and related works, Election can be solved with the *Perfect* failure detector P, which can indeed be implemented in a synchronous system. One might naturally wonder whether P is indeed the weakest failure detector for Election among failure detectors that are implement-able only in a synchronous system. We show in the following that the answer is "no" and we derive an interesting observation on the practical solvability of Election.

We define the *Confirmatory* failure detector C, which is weaker than P. We show that, to solve Election, (1) C is necessary (for any environment); (2) $C+\Diamond S$ is sufficient for any environment with a majority of correct processes. We then show that (3) P is strictly stronger than $C+\Diamond S$ for any environment where at least one processes can crash in a system of at least three processes.

4.1 Confirmatory Failure Detector

Each module of failure detector C outputs a subset of the range 2^Ω. Internally every failure detector module C uses two lists, i.e. a failure detection list, FL and a confirm detection list, CL. Initially the FL and the CL are empty; i.e. there is none which the failure detector C suspected or confirmed in Ω. If any process is once confirmed to be correct by any correct process, then the confirmed process id is inserted into the conforming process CL list. But if the confirmed process is suspected to be crash, the confirmed process id is removed from the CL list and it is inserted into the FL list. The module of failure detector C in process i has the variables, FL_i and CL_i respectively. Let H_C be any history of such a C. Then $H_C(i,t)$ represents the set of processes that process i suspects at time t. For each failure pattern F, $C(F)$ is defined by the set of all failure detector histories H_C that satisfy the following properties:

- *Confirmatory Completeness:* If a process that was confirmed to be correct by any correct process crashes, then there is a time after which some confirming process permanently detects the process crash. More precisely:

$$\forall i,j \in \Omega, \exists i \in correct(F): j \in CL_i \wedge j \in crashed(F) \Rightarrow \exists t \in T, \forall t' \geq t,$$
$$\exists i \in \Omega, j \in H(i, t')$$

- *Confirmatory Accuracy:* A process that has been confirmed to be correct is not suspected again before crash by the conforming processes. More precisely:

$$\forall t \in T, \forall i,j \in \Omega, \forall i,j \in correct(F): j \in CL_i \wedge j \notin crashed(F) \Rightarrow j \notin H(i, t)$$

Note that *Confirmatory Completeness* does not require the every crashed process to be eventually suspected, but only requires that if a process that was at least once confirmed to be correct by any process crashes, then eventually, the failure detector module C of the correct process i who has already confirmed its living keeps permanently outputting the process id.

If the failure detector module C of a process i outputs some crashed process ids, then the process i does accurately know that they have crashed since they had already been confirmed to be correct. But about those processes that had never been confirmed, the failure detector module C of i does not necessarily know whether they crash (or which processes crash).

Note also that *Confirmatory Accuracy* does preclude the possibility for a confirmed process crash to be detected before a confirmed process has actually crashed; i.e., the process that is confirmed to be correct by a correct process is never suspected before crash.

4.2 The Necessary Condition for Election

We show here that if a failure detector D solves Election then D can be transformed into C. We give an algorithm in Fig. 2 that uses Election to emulate, within a distributed variable $output(C)$, the behavior of failure detector C. We assume the existence of a function $election()$ which elects a high priority process as a leader. Different instances of this function are distinguished with an integer k. Each process i has a local copy of $output(C)$, denoted by $output(C)_i$, which provides the information that should be given by the local failure detector module of C at process i.

The basic idea of our algorithm is the following. The value of $output(C)_i$ is initially set to Φ. Every process i performs a sequence of rounds $1,..k,...$ Within each round k, process i invokes $election()$ and waits until a result of election is returned. If the newly elected leader is identical with the current leader, then i directly move to the next round. Otherwise, i puts the id of current leader into $output(C)_i$ since the current leader has crashed.

```
      /* Algorithm executed by every process i */
1 k := 1;
2 FLᵢ : = Φ; CLᵢ := Φ;
3 current_leaderᵢ := election();
4 CLᵢ := current_leaderᵢ;
5 while (election() = current_leaderᵢ ) do
6    k := k + 1;
7 FLᵢ := current_leaderᵢ;
8 output (C)ᵢ := FLᵢ ;
```

Fig. 2. Emulating C using Election

Lemma 4.1. The algorithm of Fig.2 uses Election to implement C.

Proof. We show below that output(C) satisfies Confirmatory Completeness and Confirmatory Accuracy properties of C.

- Consider Completeness. Let j be the leader that crashes at time t and let i be any correct process. By Liveness property of Election, the j was confirmed to be correct by all correct processes. Therefore j had been inserted in CL_i of i after t. Assume by contradiction that i does never put the process j into $output(C)_i$. This means that process i remains in the while loop of the algorithm. There are two cases of this scenario; (1) blocked forever waiting for the election() function to return or (2) keeps infinitely incrementing k, invoking election() and only deciding the process j as a current leader. But by the *liveness* property of Election, i does never remain blocked forever waiting for the election() function to return. Since the leader process j crashes at time t, there is an integer k such that for every $k' \geq k$, the process i does not invoke instance k' of Election. By the *safety* and the *liveness* properties of Election, the instance k' of election() returns the new leader process that is different from the current leader, i.e., not the crashed leader j:

a contradiction. By the algorithm, once i puts the j into FL_i and output$(C)_i$, i does never change the value of output$(C)_i$. Hence, there is a time after which output$(C)_i$ permanently contains the crashed processes which have already been confirmed to be correct. So it means that output(C) satisfies Confirmatory Completeness.

- Consider now Accuracy. Let i be a correct process and assume that output$(C)_i$ contains j after a time t. By the algorithm of Fig. 2, this can only be possible if some instance k of election() returns a newly elected leader that is different from the current one at process i. Given that all correct processes agree with one leader by the *safety* property of Election, the new leader can be returned only if the current leader crashes. That means that a leader, the conformed process, is not suspected before crash. Hence output(C) satisfies Confirmatory Accuracy.   □

The following proposition follows directly from Lemma 4.1.

Proposition 4.2. If any failure detector D solves election, then $C \preceq D$.

4.3   The Sufficient Condition for Election

Figure 3 describes a simple algorithm that transforms Consensus into Election using the failure detector C. The algorithm uses Consensus as a black-box, represented by a function *consensus*(): a process calls the function with new candidate for leader as a parameter and the function eventually returns *a new leader*. So the function satisfies the *Safety* and *Liveness* properties of Election.

The basic idea of our Election algorithm of Fig.3 is the following. When the *election*() is invoked, the process waits for an output from C to ensure the leader crash. If the process received from C the information that current leader has crashed, the process invokes *consensus*() with a new candidate for leader and decides the new leader returned by *consensus*(). Otherwise the process decides the current leader. We assume that every process i, either crashes, or invokes *election*() in Fig.3. The new leader candidate of participant i is denoted *new_candidate* that is decided by the *Next* function. The Function *election*() terminates by the execution of a "**return** *election_result*" statement, where *election_result* is a leader that is current one or newly elected one: when i executes **return** *election_result*, we consider that i decides its a leader.

---

function election( )
/* Algorithm executed by every process i */
1 **wait until** received from *output*(C_i);
2 **if** (*current_leader* $\in C_i$) **then**
3   *new_candidate_i* := *Next* (*current_leader_i*);
4   *election_result_i* := *consensus*(*new_candidate_i*) ;
5 **else**
6   *election_result_i* := *current_leader_i* ;
7 **return** *election_result_i* ;

Fig. 3. Transforming Consensus into Election with C

Lemma 4.3. The algorithm of Fig.3 uses C to transform Consensus into Election.

Proof. We consider the properties of Election separately.

- Safety. By the *Confirmatory Accuracy* property of C (*current_leader* $\in C_i$), a leader crash is detected only if the current leader has crashed. Any participant that decides an *election_result* must have decided *election_result* through *consensus()*. By the *Agreement* property of Consensus, no two processes can decide differently, which implies the *Safety* property of Election.
- Liveness. If a leader process crashes, then by the *Confirmatory Completeness* property of C, some process eventually detects a leader crash. By the *Validity* property of Consensus, a process decides its leader only if some process i has invoked *consensus()* with a prospected leader as a parameter. By the *Termination* property of Consensus, every correct process eventually decides a *leader* which ensures the *Liveness* property of Election.

We define here the failure detector $C+\diamond S$. Each module of $C+\diamond S$ outputs a subset of Ω. Failure detector $C+\diamond S$ satisfies the *Confirmatory Completeness* and *Confirmatory Accuracy* properties of C, together with the *Strong Completeness* and *Eventual Weak Accuracy* properties of $\diamond S$. Since Consensus is solvable with $\diamond S$ for any environment with a majority of correct processes [8], then the following proposition follows from Lemma 4.3:

Proposition 4.4. $C+\diamond S$ solves Election for any environment where a majority of processes are correct.

4.4 Confirmatory Failure Perfection Is Not Perfection

Obviously, failure detector P can be used to emulate $C+\diamond S$ for any environment, i.e., $C+\diamond S \preceq P$. We state in the following that the converse is not true for any environment where at least one processes can crash in a system of at least three processes.

Proposition 4.5. $P \npreceq C+\diamond S$ for any environment where at least one process can crash in a system of at least three processes.

Proof. *(By contradiction).* We assume that there is an algorithm $A_{C+\diamond S \to P}$ that transforms $C+\diamond S$ into failure detector P. Then we show the fact that P transformed by above algorithm satisfies *Strong Completeness* but it does not satisfy *Strong Accuracy*: So it is a contradiction. We denote by *output(P)* the variable used by $A_{C+\diamond S \to P}$ to emulate failure detector P, i.e., *output(P)*$_i$ denotes the value of that variable at a given process i. Consider three different processes i, j and k in Ω. Let F_1 be the failure pattern where the process i, which is not the conformed process, crashes at time t_0 and other process crashes. Let H be the failure detector history where all processes output $\{i\}$ at time t_1 where $t_0 \leq t_1$. Clearly, H belongs to $C+\diamond S(F_1)$. Since variable *output(P)* satisfies *Strong Completeness*, then there is a partial run of $A_{C+\diamond S \to P}$, $R_1 = < F_1, H, I, S, T >$, such that, at t_2 where $t_2 \in T$ and $t_1 \leq t_2$, $\{i\} \subset$ *output(P)*$_k$. Now consider F_2 the failure pattern where no process crashes. Clearly, H also belongs to $C+\diamond S(F_2)$. Since $C+\diamond S$

outputs exactly the same values in F_1 and F_2 (History H), then $R_2 = < F_2, H, I, S, T >$ is also a partial run of $A_{C+\diamond S \to P}$. But in R_2, at t_2, $i \in output(P)_k$ and $i \in correct(F_2)$, which means that P violates *Strong Accuracy*: a contradiction.   □

5 Concluding Remarks

The importance of this paper is in extending the applicability field of the results, which Chandra and Toueg have studied on solving problems, into the Election problem in asynchronous system (with crash failures and reliable channels) augmented with unreliable failure detectors.

So far the applicability of these results to problems other than Consensus has been discussed in [6,13,14,15]. In [8], it is shown that Consensus is sometimes solvable where Election is not. In [7], it was shown that the weakest failure detector for Election is the *Perfect* failure detector P, if we consider Election to be defined among every pair of processes. If we consider however Election to be defined among a set of at least three processes and at least one can crash, this paper shows that P is not necessary for Election. An interesting consequence of this result is that there exists a failure detector that is weaker than *Perfect* failure detector P.

This paper introduces the *Confirmatory* failure detector C, and shows that: (1) C is necessary to solve Election, and (2) $C+\diamond S$ is sufficient to solve Election when a majority of the processes are correct. A corollary of our result above is that we can construct a failure detector that is strictly weaker than P, and yet that solves Election.

Is this only theoretically interested? We believe not, as we will discuss below. Interestingly, failure detector $C+\diamond S$ helps deconstruct Election: intuitively, $\diamond S$ conveys the pure agreement part of Election whereas C conveys the specific nature of detecting a leader crash. Besides better understanding the problem, this deconstruction provides some practical insights about how to adjust failure detector values in election protocols.

In terms of the practical distributed applications, we can induce some interesting results from the very structure of $C+\diamond S$ on the solvability of Election. In real distributed systems, failure detectors are typically approximated using time-outs. To implement a failure detector C, one needs to choose a large time-out value in order to reduce false leader failure suspicions. But to implement $\diamond S$, a time-out value that is not larger than the one for C is needed. Therefore an election algorithm based on such a $C+\diamond S$ might reduce a possibility to violate the safety condition but speed up the consensus of electing new leader in the case of a leader crash.

References

1. G. LeLann: Distributed Systems–towards a Formal Approach. Information Processing 77, B. Gilchrist, Ed. North–Holland, 1977.
2. H. Garcia-Molina: Elections in a Distributed Computing System. IEEE Transactions on Computers, C-31 (1982) 49-59
3. H. Abu-Amara and J. Lokre: Election in Asynchronous Complete Networks with Intermittent Link Failures. IEEE Transactions on Computers, 43 (1994) 778-788

4. G. Singh: Leader Election in the Presence of Link Failures. IEEE Transactions on Parallel and Distributed Systems, 7 (1996) 231-236
5. M. Fischer, N. Lynch, and M. Paterson: Impossibility of Distributed Consensus with One Faulty Process. Journal of ACM, (32) 1985 374-382
6. T. Chandra and S.Toueg: Unreliable Failure Detectors for Reliable Distributed Systems. Journal of ACM, 43 (1996) 225-267
7. L. Sabel and K. Marzullo. Election Vs. Consensus in Asynchronous Distributed Systems. In Technical Report Cornell Univ., Oct. 1995
8. T. Chandra, V. Hadzilacos and S. Toueg: The Weakest Failure Detector for Solving Consensus. Journal of ACM, 43 (1996) 685-722
9. J. E. Hopcroft and J. D. Ullman: Introduction to Automata Theory, Languages and Computation. Addison Wesley, Reading, Mass., 1979
10. Garey M.R. and Johnson D.S: Computers and Intractability: A Guide to the Theory of NP-Completeness. Freeman W.H & Co, New York, 1979
11. Eddy Fromentin, Michel RAY and Frederic TRONEL: On Classes of Problems in Asynchronous Distributed Systems. In Proceedings of Distributed Computing Conference. IEEE, June 1999
12. Hadzilacos V. and Toueg S: Reliable Broadcast and Related Problems. Distributed Systems (Second Edition), ACM Press, New York, pp.97-145, 1993
13. R. Guerraoui: Indulgent Algorithms. In: Proceedings of the ACM Symposium on Principles of Distributed Computing, New York: ACM Press 2000
14. Schiper and A. Sandoz: Primary Partition: Virtually-Synchronous Communication harder than Consensus. In Proceedings of the 8th Workshop on Distributed Algorithms, 1994
15. R. Guerraoui and A. Schiper: Transaction model vs. Virtual Synchrony model: bridging the gap. In: K. Birman, F. Mattern and A. Schiper (eds.): Distributed Systems: From Theory to Practice. Lecture Notes in Computer Science, Vol. 938. Springer- Verlag, Berlin Heidelberg New York (1995) 121-132.

Scheduling Scheme with Fairness and Adaptation in the Joint Allocation of Heterogeneous Resources*

Yu Hua[1], Chanle Wu[1], and Mengxiao Wu[2]

[1] School of Computer, Wuhan University, Wuhan, China
[2] Department of Computer Science, Vrije University, Amsterdam, Netherlands
yhuastarmpls@hotmail.com

Abstract. Generally speaking, packets transmitted from source to destination need not only bandwidth and buffer but also computing and processing capacity in each node, available wavelength in optical networks and power in wireless sensor networks. So, joint and fair allocation of heterogeneous resources (such as, bandwidth, buffer and processing) is rather pivotal and necessary for realizing differentiated end-to-end QoS guarantee. In this paper, we proposed the framework and algorithm, realized fair allocation of heterogeneous resources, and designed the Heterogeneous-Deficit Round Robin (H-DRR) algorithm, which was the improvement to DRR. The comparison of performance with and without unfriendly packets stated that H-DRR could realize the fair allocation of heterogeneous resources and provide better QoS guarantee, especially in the differentiated situation.

1 Introduction

Packets transmitted from source to destination need not only bandwidth and buffer but also computing and processing capacity in each node, available wavelength in optical networks and power in wireless sensor networks. In particular, high-speed optical networks have been employed as major technology in the backbone internet. In consequence, the rates of transmission become faster and faster and bandwidth may not be bottleneck in some conditions. On the other hand, various novel techniques and new service need processor check optional headers of packets and deal with all kinds of network events according to the control information. The capacity of processing can be the bottleneck in real network environments [1].

However, traditional best-effort service models can't provide fair allocation of heterogeneous resources. In order to realize QoS provision, the scheduling algorithm can be considered as a good approach and key component in the QoS-enabled networks. So, the ideal implementation is the scheduling algorithm supporting the allocation of heterogeneous resources in the unified and fair manner. This is the motivation of the

---

* Supported by the National High Technology Development 863 Program of China under Grant No. 2003AA001032; National Key Laboratory in Software Engineering under Grant No.SKLSE03-14; the Natural Science Foundation of Hubei Province of China under Grant No. 2001B057.

J. Cao, W. Nejdl, and M. Xu (Eds.): APPT 2005, LNCS 3756, pp. 151–163, 2005.
© Springer-Verlag Berlin Heidelberg 2005

paper. As far as scheduling algorithm is concerned, its main function is to select the next packet to transmit, decide when it can be transmitted and which port it will be sent, based on pre-defined transmission metrics. Lots of works in this research area has been done during last decade. In addition, the current scheduling algorithms, can be classified as sorted-priority and frame-based methods. In the frame-based algorithms, time is divided into several frames and packets can be transmitted on the per-frame basis. Deficit Round Robin (DRR) [3] is a credit-based extension of classical Weighted Round Robin (WRR) with lower complexity than WFQ. DRR has O (1) worst-case per packet complexity if the operation number selecting the next packet is constant with respect to the number of active flows [4]. Corresponding improvements include other round-robin algorithms, such as Pre-order Deficit Round-Robin (PDRR) [5] and Smoothed Round-Robin (SRR) [6]. As stated in [3] and [4], flows in DRR can share the bandwidth fairly with variable packet lengths.

2 Related Works

To the best of our knowledge, many previous research works are based on fair allocation of single resource. Recently some researchers begin to pay more attention to the allocation of multiple resources. Yunkai Zhou and Harish Sethu [1], [7] proposed the methods of fair allocation among multiple resources. Vijay Ramachandran, Raju Pandey and S-H. Gary Chan [8] also presented some ideas on fair allocation of CPU and bandwidth resources. The method utilized feedback mechanism to improve the performance.

The main disadvantages of current ideas and methods can be summarized as follows: Firstly, most works focus on the fair allocation of the single resource, such as bandwidth or buffer. Therefore, the end-to-end QoS guarantee can't be achieved in real environments, due to lack of fairness among multiple heterogeneous resources. Secondly, lots of algorithms may be implemented in the static or partly dynamic environments, but once faced with burst traffic, they are hard to work well owing to lack of the adjustable feedback mechanisms and information exchange. Thus, high complexity is increased naturally. Thirdly, although several papers have presented some concrete algorithms for fair allocation of heterogeneous resources, their core ideas were to append other kinds of resources as new parts to existing scheduling algorithms. Characters of resources and relationships among them could seldom be considered and analyzed comprehensively.

3 Contributions in This Paper

Our contributions can be stated in brief as follows: Firstly, a kind of scheduling architecture was introduced and explained in detail with guarantee bounds based on fairness measure. In this paper, fair mechanisms for different resources were separate in data plane and interactive each other in the control plane. Secondly, we look on the real-time utilization of buffer space as indication of congestions happening. According to the information, proposed methods may be used in the dynamic conditions and adjustment with feedback could be obtained in simple way. Thirdly, in order to enhance the robustness and security, the system also presented techniques to detect and drop the unfriendly packets and ill-behaved flows. Fourthly, we bring forward improved DRR algorithm (i.e., Heterogeneous-DRR), which was the prototype of fair allocation of heterogeneous resources and could be implemented in current network environments.

4 Scheduling Architecture and Analysis

4.1 Fairness Measure

The definition of fairness is rather important for achieving the fair allocation of heterogeneous resources. In classical Generalized Processor Sharing (GPS) theory [9], the fairness of GPS server can be defined as:

$$\frac{W_i(t_1,t_2)}{W_j(t_1,t_2)} \geq \frac{\phi_i}{\phi_j}, i, j = 1,2,\cdots, N \tag{1}$$

which confines transmitted amounts of flow i, $W_i(t_1,t_2)$, during the time interval $[t_1,t_2]$ (GPS server serving N flows is characterized by N positive real number, $\phi_1,\phi_2,\cdots,\phi_N$). The GPS model is rather idealized and need partition packets into infinitesimal parts in order to realize the policy of fair allocation. In fact, a kind of scheduling algorithm is fair if and only if the fair measure is bounded. In the series of WRR scheduling algorithms, the basic idea is based on the defined weights, which may determine the fair bounds. For example, in the classical WRR, the form can be described as:

$$\left|\frac{S_i(t_1,t_2)}{\omega_i} - \frac{S_j(t_1,t_2)}{\omega_j}\right| \leq 1 \tag{2}$$

in which, $S_i[t_1,t_2]$ is the amount of flow i transmitted during the time interval $[t_1,t_2]$, and ω_i is the integer weight of flow i. Furthermore, in the design of round robin approach [10], the service of different flows i and j can be presented as:

$$\left|\frac{S_i(t_1,t_2)}{\omega_i} - \frac{S_j(t_1,t_2)}{\omega_j}\right| \leq \frac{\omega_j - \omega_i + 1}{\omega_j} = \varepsilon \tag{3}$$

Through analyzing the definition and methods above, we present a generalized fairness measure. Usually, the fairness measure should manifest different contexts. Thus, the fairness bound should be made up of two parts, which include the common service discrepancy [4] and adjustable factor based on the feedback information. In this paper, the representation of fairness measure can be given as:

$$\frac{\sum_N R}{C}\left|\frac{S_i(t_1,t_2)}{R_i} - \frac{S_j(t_1,t_2)}{R_j}\right| \leq \varepsilon + \sigma_{Feedback} \tag{4}$$

C is the capacity of the system and R_i is the number of resources required by flow i. ε is the same expression as common service discrepancy bound [4]. $\sigma_{Feedback}$ is the adjustable factor based on the feedback information, which may reflect current network state. Without loss of generality, the meaning of $S_i(t_1,t_2)$ is extended in order to support fair allocation of heterogeneous resources and can be defined to represent the service provided by the corresponding single or compound resources during the time interval $[t_1,t_2]$. Here, the service is the combination of computing and bandwidth abilities with adapted weights, which have the interval between zero and one. Of course, the allocation of weights between two resources should be according to the actual environments and current state information of resources available.

4.2 Proposed Architecture

The proposed architecture of fair scheduling for heterogeneous resources is presented in Fig.1. The architecture is composed of two parts: control plane and data plane. When the classified packets arrive, they may enter different queues according to their priorities. The different DC (i.e., Deficit Counter) values represent different amounts

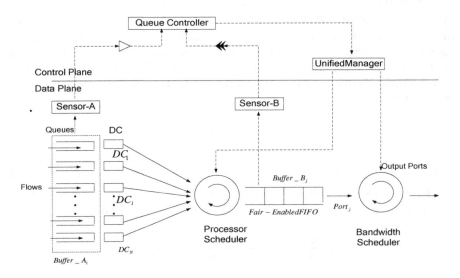

Fig. 1. Architecture of Fair Scheduling for Heterogeneous Resources in Dynamic Environments

of resources, which determine corresponding levels of service. Without loss of generality, different DC values are sorted in order and each DC value is defined for each queue before processor scheduler. DC_1 is the biggest number and DC_N is the smallest number. In the data plane, two queue sensors are used to collect information of queue lengths in order to make dynamic adjustment. The sensors check the changes of queue length, gather state information from different queues and then send them to the queue controller in the control plane.

After the queue information is received and analyzed in the queue controller, the corresponding adjustable strategy based on the pre-defined SLA will be transmitted to the *UnifiedManager*. Then, it may send out modified DC values to processor scheduler or revised rates to bandwidth scheduler in coordinated manner. In this way, two schedulers may adjust their rates for managing arriving packets according to the dynamic feedback information. This is a simple description of the proposed architecture.

The strategy in this architecture is differentiated based on the packets' priorities and attributes. The main aim of the processor scheduler is to fairly allocate time slices of processing data to the prioritized queues in the round-robin manner. During this process, packets in high-prioritized queue have better guarantee in throughout and delay performance, but corresponding guarantee should be limited and bounded in order to avoid other packets starving in transmission time. Once the scenario happens, packets in lower priority queues would be dropped largely and nearly not sent normally. It is acceptable that packets in higher priority queues may achieve better QoS guarantee and nevertheless should not be permitted to occupy resources all the time. At the same time, among different lower priority queues, the system may drop packets in lower priority queues, but the policy of dropping must be made fairly.

When packets are permitted to enter the fair-enabled FIFO queue (i.e., the queue before bandwidth scheduler), they should be served and usually can't be dropped anyway before they are allocated into corresponding output ports. An example of analogy is helpful to explain and understand this viewpoint. The process described here is similar to that of checking tickets in airport waiting room and boarding the plane. That is to say, when passengers arrive at the airport waiting room, they are classified and enter different passages (i.e., queues before processor scheduler) according to their destination and plane ticket grade (i.e., common or VIP). Once checkers look over tickets and baggage, passengers can be permitted to enter the boarding queues (i.e., fair-enabled FIFO queue) waiting for boarding the planes and of course they could not be dropped in the general way. Therefore, fair-enabled FIFO queue should be considered to need better guarantee than queues before processor scheduler.

The queue before bandwidth scheduler is called fair-enabled FIFO queue. Because the scheduling in processor scheduler is in DRR way for time slices of processing, the allocation has been fair largely among different queues. The packets in queues with higher priority have *already* contained longer processing time represented by the bigger DC values. Basic and essential opportunity of transmission has also been provided to the packets in queues with lower priority. Thus, differentiated DRR manner can provide corresponding QoS guarantee to packets in queues with different priorities. In addition, this manner also avoids some packets occupying the resources all the time and may provide valid transmission time to packets in queues with lower priority. This is the essence of fairness. This queue is based on the fair allocation and can be called fair-enabled. We emphasize that packets in higher priority queue should not occupy the bandwidth at all times. If many resources are occupied all the while, it is unfair. Therefore, the fair-enabled queue should be FIFO, not parallel classified queues. Thus, packets of different levels would be connected with each other in one-by-one way and only difference among them is that they have different sizes, which can be proportional to allocated time slices in previous processing stage.

Table 1. H-DRR scheduling algorithm

Initialization:
ActiveList=NULL;
Buffer_A_i = Buffer_B_j = ready ; /* buffers before processor and bandwidth schedulers.*/
Put initial values to parameters, *P_DC, B_DC, PQ, BQ*, respectively.
Enqueue Module; /*Invoked by packet *p* from flow *i* arrives*/
p : =ArrivingPacket; *i* : =Flow (*p*);
If (*Length* (*p*) > *Threshold $_i$*) /* If the length of packet *p* exceeds the threshold of flow *i* */
Then *Drop the packet p ;* /* the packet *p* is considered as the ill behavior according to SLA. */
End If;
If (ExistsInActiveList $_p$ (*i*) = = FALSE)
Then Append flow *i* to ActiveList $_p$; /* ActiveList $_p$ is the list in processor scheduler.*/
    $P\_DC_i = 0$;
End If;
If (ExistsInActiveList $_B$ (*i*) = = FALSE)
Then Append port $_j$ to ActiveList $_B$; /* ActiveList $_B$ is the list in bandwidth scheduler.*/
    $B\_DC_j = 0$;
End If;
Dequeue Module: /* Always running */
While (TRUE) **do**
If (ActiveList $_p$ ≠ NULL) **then**
i : =HeadOf ActiveList $_p$;
Remove flow *i* from ActiveList $_p$;
  $P\_DC_i = P\_DC_i + PQ_i$;
End If;
If (ActiveList $_B$ ≠ NULL) **then**
 j : =HeadOf ActiveList $_B$;
Remove port $_j$ from ActiveList $_B$;
 $B\_DC_j = B\_DC_j + BQ_j$;
End If;
If *Buffer_B = Exceeded* /* the length of *Buffer_B* exceeds pre-defined constant.*/
  Then If there exists the excess bandwidth available
        Then $B\_DC_j = B\_DC_j + \beta_j$

Else $P\_DC_i = P\_DC_i - \alpha_i$;

      If $Buffer\_A_i = Exceeded$

          Then Carry out dropping strategy according to pre-defined SLA;

          End If;

    End If;

  Else If $Buffer\_A_i = Exceeded$

      Then If $Excess\_process\sin g = Avaiable$

          Then $P\_DC_i = P\_DC_i + \alpha_i$;

          Else Carry out dropping strategy according to pre-defined SLA;

      End If;

    End If;

  End If;

While (QueueIsEmpty(i) = = FALSE) **do**

  p =HeadOfLinePacketInQueue(i);

  If ($PacketSize(p) > B\_DC_j$) or ($Process\_Cost(p) > P\_DC_i$)

  /* $PacketSize$ (p) is the size of current packet p ; $P\_Cost(p)$ is processing cost of packet p */

    Then break;

  End if;

  $P\_DC_i = P\_DC_i - Process\_Cost(p)$;

  $B\_DC_j = B\_DC_j - PacketSize(p)$;

  Execute Scheduling p ; /* this includes the concrete operations on packets.*/

End While;

If $Empty(Buffer\_A_i)$

  Then $P\_DC_i = 0$;

  Else $InsertActiveList_p(Buffer\_A_i)$;

End If;

If $Empty(Buffer\_B_j)$

  Then $B\_DC_j = 0$;

  Else $InsertActiveList_B(Buffer\_B_j)$;

End If;

End While

4.3 Practical Scheduling Algorithm: H-DRR

H-DRR inherits the advantages of PPLS [1] and other improvements for DRR to some extent. The pseudo-code of H-DRR is shown in table 1. We defined two kinds of variables, Processor Deficit Counter (P_DC) and Bandwidth Deficit Counter (B_DC), which represent the deviation of ideally fair share by amounts of respective resource served actually. At the same time, *PQ* and *BQ* are the Processor Quantum and Bandwidth Quantum, respectively. Heterogeneous resources can be divided into units of resources and resource allocation may be realized according to the number of units. When packets arrive, *Enqueue* model will be executed at once. If packets are from a new flow, the flow will be appended to *ActiveList*. Then, security checking based on the pre-defined SLA is made in order to avoid the unfriendly packets. The main function of *Dequeue* module is to serve the arriving flows in round-robin way. When a flow is visited, deficit counters, *P_DC* and *B_DC*, are added by corresponding quantum. The packets with the same destination arrive at fair-enabled FIFO queue in $port_j$ and the bandwidth may be allocated among different ports with round-robin way. In this paper, fair-enabled FIFO queue in $port_j$ is also called *Buffer_B_j* for simplicity. Furthermore, since the packets in *Buffer_B_j* need better QoS guarantee than packets in *Buffer_A_i* queues (The viewpoint has been stated and analyzed in previous sections), the state of *Buffer_B_j* is checked firstly. If the state of *Buffer_B_j* is *Exceeded* (i.e., the length of the queue reaches or exceeds pre-defined threshold) and system has excess resources, more units of bandwidth may be allocated. But if there is no excess resource, the processing rate should be reduced in order to decrease the number of packets arriving at *Buffer_B_j*. Effective methods are to reduce P_DC_i or use dropping strategy if the state in *Buffer_A_i* is also *Exceeded*. In addition, if only *Buffer_A_i* is *Exceeded*, more units of processing capacity should be allocated. At the same time, packets from flow i can be scheduled if the length of packets is smaller than bandwidth deficit counter and processing cost is smaller than processing deficit counter. After scheduling, both deficit counters are reduced by different costs. At last, if the current flow is still backlogged, it is inserted into the end of corresponding *ActiveList*. Otherwise, deficit counters are set to zero again.

α_i and β_j are adjustable factors in processor and bandwidth scheduler, respectively. They can be used to increase or decrease loads in current router in order to enhance the efficiency of resource utilization and avoid network congestion in advance. The whole scheduling process is step-by-step. Concretely speaking, if there are the idle resources in networks, values of α_i and β_j can be gradually added to the different DC values and more resources may be used in order to improved network performance. On the contrary, if there are risks of network congestion, their negative values can be gradually added to related DC values in similar ways. Then, traffic and load would decrease and avoid network congestion. Of course, values of α_i and β_j are rather important and critical, but they are difficult to determine accurately. Because this is a kind of fuzzy problem, to the best of our knowledge, analysis methods based on statistical history data and fuzzy mathematical techniques could be utilized for efficiently determining their values. In addition, their concrete values may be defined as several units of corresponding resource capacity in actual environments. There exists the mapping relationship between service levels and values of α_i and β_j. Values of α_i and β_j can be determined according to the percentage of corresponding DC values. Naturally, queues with higher priority have larger percentage.

5 Performance Study

5.1 Simulation Settings

The simulation process was divided into two parts.

Firstly, in order to verify the performance of different models in single node, the network topology is designed in fig. 2 (left). Eight hosts were set as input traffic sources and each host had 10 flows, which were corresponding to flow ID. Then, 80 flows can be classified into four levels from the highest priority to the lowest one in sequence. In addition, each host generated packets according to ON (20 packets/s) and OFF (no packets) periods independently and identically distributed with Pareto distribution.

Furthermore, the sizes of packets were randomly selected with a uniform distribution varying from 300 bits to 3000 bits. Flows with higher priority require capacities of processor varying in the rage of 30 M to 50 M and bandwidth in the range of 10Kbps to 30Kbps. Otherwise, flows with lower priority require capacities of processor varying in the rage of 1M to 10M and bandwidth in the range of 1Kbps to 6Kbps. All output link bandwidth were set to 800Kbps, and power of processor was 80 MHz. The time of simulation could last for 400 seconds. In addition, the sources 4 and 5 that would send higher rates of packets (i.e., 50 packets/s) from 200 second on were set as unfriendly flows in order to verify the adaptation of different scheduling algorithms.

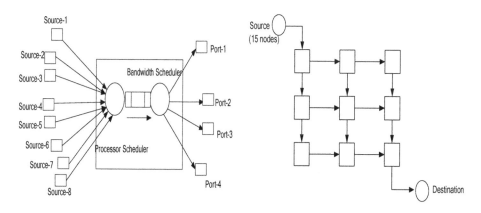

Fig. 2. Scenarios of single node with eight sources and four output ports and 3×3 mesh network

Secondly, the scenario of multiple paths and nodes was shown in fig.2 (right). The traffic sources from15 nodes produced UDP packets with sizes of uniform distribution varying from 200 to 2000 bytes. The routing tables in intermediate nodes were produced in advance according to the distance from destination. In each node, H-DRR algorithm was executed and scheduled for heterogeneous resources. 15 nodes were classified as five levels with different quanta. At the same time, the requirements from different priorities could be described as the vector $(\text{Processing}, \text{Bandwidth}, \text{Quanta})_k$, which denoted that flows with priority-k required the capacities of processing and bandwidth and pre-defined quanta. Thus, the requirements from five levels were defined as $(45cycles,35kbps,2000b)$, $(30cycles,25kbps,1500b)$, $(20cycles,17kbps,1200b)$, $(12cycles,10kbps,700b)$ and $(6cycles,4kbps,300b)$, respectively from the highest priority to the lowest one . At the same time, $\alpha = \beta = [600,500,400,300,200]^T$ .The power of processor was 140 MHz and the output link was 1000Kbps for different flows shared. Packets from two nodes, node 6 and 10 with double sending rates, were unfriendly from 250 second on. Other settings were the same as the first simulation. All results were average values of simulation for 30 times.

5.2   Comparison and Analysis

In the simulation, the comparison of performance was made among FCFS (First Come First Serve), WF2Q [2], DDRR (Double Deficit Round Robin, two independent DRR schedulers utilized), PPLS [1] and H-DRR in the aspects of delay, throughput and fairness guarantees with and without unfriendly packets according to different intervals. Before unfriendly packets arrived, the comparisons of performance were in usual conditions. The parameters, delay and throughput, was defined like other common principles. The fairness guarantee could be defined according to the *deviation* from ideal GPS scheduling. At the same time, except PPLS, two schedulers (i.e., processor and bandwidth schedulers) were set for FCFS, WF2Q, and DDRR in order to realize the allocation of heterogeneous resources.

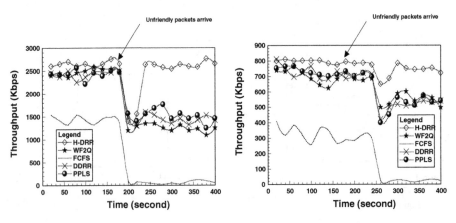

Fig. 3. Throughput in scenario of single node   **Fig. 4.** Throughput in scenario of multiple nodes

Throughput:

The chats of comparison in throughput are presented in Fig.3 and 4. In Fig.3, the packets in first 10 seconds were ignored in order to make traffic reach steady state. Before the unfriendly packets arrived, the worst performance was FCFS among the presented algorithms. The average throughputs for H-DRR, WF2Q, DDRR and PPLS were 2654.32Kbps, 2487.56 Kbps, 2458.64 Kbps, 2486.24 Kbps, respectively. H-DRR has some advantages of throughput in single node because processor and bandwidth were scheduled cooperatively. However, when the unfriendly packets arrived, the whole performance was affected and the throughput of FCFS was close to zero because of no fair scheduling and management for preempted situations. At the same time, adaptation of other algorithms worked and negative effects may be reduced to some extent. Nevertheless, the throughput of only H-DRR was approximate to former data because the pre-checked mechanisms were utilized. In this point, the advantage of H-DRR was rather clear.

In Fig.4, in the topology with multiple paths and nodes, H-DRR had the ability to adapt and adjust when the wicked flows arrived. Furthermore, the average throughputs of H-DRR, WF2Q, FCFS, DDRR and PPLS in the whole process were 795.24 Kbps, 602.64 Kbps, 175.39 Kbps, 548.67 Kbps and 624.68 Kbps, respectively.

Delay:

It is understood that fairness could not be guaranteed in FCFS for lack of fair scheduling mechanism. Therefore, H-DRR, DDRR, and PPLS were selected for the comparison of delay guarantee. The results of simulation were shown in Fig.5-6 with delay versus flow ID. However, H-DRR, PPLS and DDRR may have better differentiated QoS guarantees with fair round-robin way. The average values in delay of H-DRR, PPLS and DDRR were 0.017s, 0.067 and 0.084s, respectively, in fig.5. The approximate conclusion could also be made from the results in Fig.6. The average delay in H-DRR was close to 0.027 and those in PPLS and DDRR were 0.094s and 0.245s respectively among all flows.

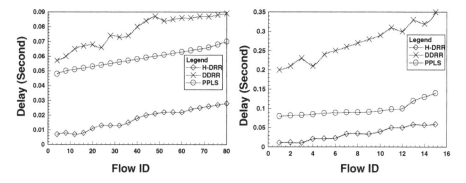

Fig. 5. Delay in scenario of single node   **Fig. 6.** Delay in scenario of multiple nodes

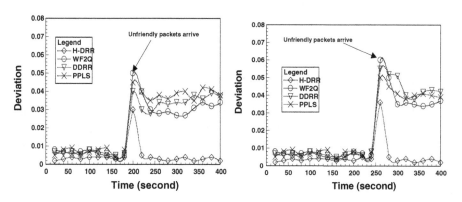

Fig. 7. Deviation in scenario of single node   **Fig. 8.** Deviation in scenario of multiple nodes

General Fairness:

The performance in fairness could be reflected by the deviation from ideal GPS scheduling. Because FCFS has no fair scheduling mechanism, FCFS was excluded in this comparison again. In Fig.7, before 200 second, the results of fairness among algorithms may be approximate overall. However, when the unfriendly packets arrived, it may be observed that only H-DRR could provide efficient and robust fairness guarantee. Although other algorithms had adaptation and made some adjustments, corresponding mechanisms were derived from fair round-robin way without efficient mechanisms to protect legal packets. Thus, the overall performance in fairness was on the decline. In Fig.8, the corresponding conclusion could also be made according to the simulation results. The average deviation values of algorithms other than H-DRR were about 0.021 in Fig.7 and 0.026 in Fig.8.

Therefore, from the results of the different simulation environments, we can state that H-DRR may achieve better performance in throughput, delay and fairness guarantee with and without unfriendly packets.

Instance Fairness:

Now, a concrete example is shown in order to verify the fairness performance of H-DRR. One video application, one voice application and three data applications were looked on as the traffic sources. Hereafter, the voice application has the higher priority and the data applications have the lower priority. Then, G.726 Encoder was utilized for producing voice packets with bit rate (32kbps) and packet length (80bytes). MPEG-4 Encoder produced the packets of video application with bit rate (128kbps), frame rate (50) and frame number (4000). At the same time, data applications were based on the arrival rates with Poisson distribution. Their mean packet length was 400 bytes and average packet rate was 50 packets/s. In addition, the output bandwidth was 800kbps and initial vector $PQ=[900, 700, 400, 350, 300]^T$ for five applications and $BQ=[1500, 800]^T$ for two output ports. Then, other settings were the same as the previous simulation.

Table 2. Delay results based on H-DRR algorithm

Scenario/Application	Video	Voice	Data-1	Data-2	Data-3
Single node (ms)	96.5	135.7	250.6	269.5	347.5
multiple nodes (ms)	122.3	204.6	347.8	401.2	456.3

The delay results can be found in Table.2. Differentiated performance may be realized based on the different pre-defined priorities. In contrast, packets with higher priority has better delay guarantee than ones with lower priority. At the same time, the differentiated percentage of the delay data is corresponding to the pre-defined PQ values approximately. In two scenarios, the proportion of three types (video, voice and data, and the average PQ value for data application is 350) is $2.5:2:1$ according to the defined PQ vector above. The simulation results show that the average proportion of transmission delay is $2.94:1.62:1$. The result is close to the pre-defined proportion and fairness for different types may be guaranteed efficiently.

6 Conclusion

In this paper, we proposed an idea of improved fairness allocation. The idea emphasized not only fair allocation of heterogeneous resources but also the separate scheduling mechanisms according to different adaptation requirements. This method could reduce waste of resources and improve the utilization. Corresponding architecture and H-DRR algorithm have also been presented and analyzed in detail. Extensive simulation results proved the methods feasible and efficient. Thus, the methods proposed may efficiently solve the problem of network heterogeneous resources allocation in cooperative manner. In addition, it may also contain other types of resources, such as wavelength resource in optical network, capacity of power in wireless network, ability of sensor in sensor network. Because the solutions in different environments hold the same essence of fair scheduling, the method may be used for other concrete applications and can also be improved and expanded easily.

References

1. Yunkai Zhou, Harish Sethu, On Achieving Fairness in the Joint Allocation of Processing and Bandwidth Resources, IWQoS 2003, LNCS 2707, (2003) 97–114.
2. Bennett, J.C.R.; Hui Zhang, WF2Q: Worst-case fair weighted fair queuing, in Proc. IEEE INFOCOM, vol.1, (1996) 120–128.
3. M. Shreedhar, G. Varghese, Efficient fair queuing using deficit round-robin, IEEE/ACM Trans. Networking, vol. 4, (1996). 375–385
4. Luciano Lenzini, Enzo Mingozzi, and Giovanni Stea Tradeoffs Between Low Complexity, Low Latency, and Fairness with Deficit Round-Robin Schedulers IEEE/ACM Transaction on networking, vol. 12,no. 4, (2004) 681 - 693
5. Shih-Chiang Tsao and Ying-Dar Lin   Pre-order deficit round robin: a new scheduling algorithm for packet-switched networks Computer Networks, Volume 35, Issue 2-3 (2001) 287 - 305
6. Chuanxiong Guo SRR: an O(1) time-complexity packet scheduler for flows in multiservice packet networks IEEE/ACM Transactions on Networking Volume 12, Issue 6 (2004) 1144 - 1155
7. Yunkai Zhou and Harish Sethu Towards end-to-end fairness: a framework for the allocation of multiple prioritized resources, IEEE International Performance, Computing, and Communications Conference, (2003) 495 - 504
8. Vijay Ramachandran, Raju Pandey and S-H. Gary Chan Fair resource allocation in active networks Proceedings of Ninth International Conference on Computer Communications and Networks, (2000).468-475
9. Parekh, A.K. Gallager, R.G. A generalized processor sharing approach to flow control in integrated services networks: the single-node case IEEE/ACM Transactions on Networking Volume: 1, Issue: 3 (1993) 344 – 357
10. Hemant M. Chaskar and Upamanyu Madhow Fair scheduling with tunable latency: a round-robin approach IEEE/ACM Transactions on Networking (2003) 592 - 601

Solving the Symmetric Tridiagonal Eigenproblem Using MPI/OpenMP Hybrid Parallelization[*]

Yonghua Zhao[1,2,3], Jiang Chen[1], and Xuebin Chi[1]

[1] Supercomputing Center, Computer Network Information Center,
Chinese Academy of Sciences, 100080, Beijing, China
{yhzhao, tschj, chi}@sccas.cn
[2] Graduate School, Chinese Academy of Sciences, 100080, Beijing, China
yhzhao@sc.cnic.cn
[3] Department of Computer Science, Dezhou University, 253000, Shandong, China

Abstract. We present a hybrid MPI/OpenMP parallel implementation for the eigenvalues of symmetric tridiagonal matrices on cluster of SMP's environments. The algorithm is based on a divide-and-conquer method which uses the split-merge technique and Laguerre's iteration. We study two different implementations of the algorithm: one based on MPI and the other based on a hybrid parallel paradigm with MPI/OpenMP. We take a coarse grain OpenMP approach to parallel implementation for solving the eigenvalues of symmetric tridiagonal submatrices within a SMP node. And dynamic work sharing is used in Laguerre's iterations. This has two effects: first, the amount of synchronization has been reduced; secondly, this could have an effect on the load balance. In addition, we analyze the communication overhead on two different implementations. An experimental analysis on the DeepComp 6800 shows the hybrid algorithm performs good scalability.

1 Introduction

The symmetric tridiagonal eigenvalue problem has been intensively studied on parallel computers and hence forms a group of excellent algorithms developed for this problem. There are quite a few reliable algorithms with various features; these include parallel QR method [1,2], bisection and multisection method (B/M) [3], the Cuppon's Divide-and-Conquer method (D&C) [4] and homotopy method [5]. Recently, Dhillon&Parlett proposed algorithm MRRR [7,8] that gives the first stable $O(nk)$ algorithm to compute k eigenvalues and eigenvalues of a symmetric tridiagonal matrix. Not only does the Dhillon/Parlett's approach have better serial complexity than other methods, but it also has a natural parallel structure.

[*] This work is supported by the Chinese Hitech Program (863) "Supercomputing Grid Node Construction" (2002aa104540), and the Informatization Construction of Knowledge Innovation Project of the Chinese Academy of Sciences "Supercomputing Environment construction and Applications" (INF105-SCE).

J. Cao, W. Nejdl, and M. Xu (Eds.): APPT 2005, LNCS 3756, pp. 164–173, 2005.
© Springer-Verlag Berlin Heidelberg 2005

The divide and conquer algorithm has natural parallelism as the initial problem is partitioned into several subproblems that can be solved independently. Basic strategy of the algorithm is to express the tridiagonal matrix as a row-rank modification of a direct sum of two smaller tridiagonal matrices. The entire eigenproblem can then be approximated by some methods in terms of the eigenproblems of the smaller tridiagonal matrices, and this process can be repeated recursively. Parallelizing tridiagonal eigensolvers based on the Divide-and-Conquer algorithm have been researched by many individuals [6,10,11,12,14].

We chose to study the Divide-and-Conquer algorithm based on the rank-two modification to divide the matrix and applying the Laguerre iteration to approximate the eigenvalues. This algorithm is designed originally for shared-memory parallel architectures by Li and Zeng [11]. Afterward, the algorithm is designed and implemented on distributed-memory parallel architectures by Treffz and Huang, et al [14]. The reason for choosing the algorithm is that it contains most of the advantages of above mentioned algorithms:

1. The same parallel structure as Cuppen's method but less memory contention.
2. The same accuracy as B/M.
3. The same flexibility as B/M in evaluating partial spectrum.
4. The same advantage as QR and B/M in separating the evaluation of eigenvalues and eigenvectors.

Largescale highly parallel systems based on cluster of SMP architecture are today's dominant computing. They can be thought of as a hierarchical two-level parallel architecture, since they combine features of shared and distributed memory machines. As a consequence, the hybrid programming paradigm combining message passing and shared-memory parallelism has become popular in the past few years.

The hybrid model has already been applied in many applications, ranging from [9,13] to atmospheric research [10], to molecular dynamics analysis [16]. Usually, programmers resort to MPI for the message passing communication, using OpenMP as an interface for writing multi-threaded application. The adoption of this model is facilitated by both the architectural developments of modern supercomputers and the characteristics of a wide range of applications [15,18].

In this paper we focus on how to accommodate and exploit particular features of Cluster of SMP's environments in order to improve performance of eigensolvers of symmetric tridiagonal matrices. We present a hybrid MPI/OpenMP parallel implementation for the eigenvalues of symmetric tridiagonal matrices on cluster of SMP architecture. We take into account two different implementations of the algorithm: one based on MPI and the other one based on a hybrid parallel paradigm with MPI/OpenMP. We take a coarse grain OpenMP approach to parallel implementation for the eigenvalues of symmetric tridiagonal submatrices within SMP node. And dynamic work sharing is used in Laguerre's iterations for load balance. An experimental analysis on the DeepComp 6800 shows the hybrid algorithm shows good scalability.

This paper is organized as follows: the second section gives the basic structure of the split-merge algorithm. The third section discusses the parallel algorithm in MPI and MPI/OpenMP hybrid. The fourth section analyzes the performances of these two paradigms. We give some conclusions in the last section.

2   The Basic Structure of the Split-Merge Algorithm

Let T be an $n \times n$ tridiagonal matrix, in order to find some or all the zeros of $f(\lambda) \equiv \det[T - \lambda I]$, let us introduce the split matrix $\hat{T}$ by replacing some b_k in T with zero, and write

$$T = \hat{T} + \begin{pmatrix} & \vdots & \\ \cdots & 0 \ b_k & \cdots \\ & b_k \ 0 & \\ & \vdots & \end{pmatrix}. \tag{1}$$

The eigenvalues of $\hat{T}$,

$$\hat{T} = \begin{pmatrix} T_0 & \\ & T_1 \end{pmatrix}, \tag{2}$$

consists of eigenvalues of T_0 and T_1.

Theorem 1. [11] *Let*

$$\lambda_1 < \lambda_2 < \cdots < \lambda_n \quad and \quad \hat{\lambda}_1 < \hat{\lambda}_2 < \cdots < \hat{\lambda}_n$$

be the eigenvalues of T and $\hat{T}$ respectively. Then

$$\begin{cases} \hat{\lambda}_1 \in (\lambda_1, \lambda_2) \\ \hat{\lambda}_i \in (\lambda_{i-1}, \lambda_{i+1}), & 2 \leq i \leq n-1 \\ \hat{\lambda}_n \in (\lambda_{n-1}, \lambda_n) \end{cases} \tag{3}$$

$$\begin{cases} \lambda_1 \in (\hat{\lambda}_1 - |b_k|, \hat{\lambda}_1) \\ \lambda_i \in (\hat{\lambda}_{i-1}, \hat{\lambda}_{i+1}), & 2 \leq i \leq n-1 \\ \lambda_n \in (\hat{\lambda}_n, \hat{\lambda}_n + |b_k|) \end{cases} \tag{4}$$

It follows immediately from the above theorem the following corollary.

Corollary 1. *For each $i = 1, \cdots, n$, in the open interval $(\hat{\lambda}_i, \lambda_i)$ (or $(\lambda_i, \hat{\lambda}_i)$ if $\lambda_i < \hat{\lambda}_i$), there is no λ_j or $\hat{\lambda}_j$ for $j = 1, \cdots, n$. In the other words, for $i = 1, \cdots, n$, let L_i the open internal with end points λ_i and $\hat{\lambda}_i$, then $\{L_i\}_{i=1}^n$ is the collection of disjoint intervals.*

According to this corollary, for each $i = 1, \cdots, n$ the Laguerre iteration starting from $\hat{\lambda}_i$ can reach λ_i without any obstacles. Thus, to evaluate the i-th smallest eigenvalue λ_i of T, the corresponding eigenvalue $\hat{\lambda}_i$ of $\hat{T}$ is always used as a starting point in our algorithm.

The eigenvalues of $\hat{T}$ consist of the eigenvalues of T_0 and T_1. To find the eigenvalues of T_0 and T_1, the splitting process may be applied recursively until it become a 2×2 or 1×1 eigenvalue problem or it can be terminated with an $m \times m$ problem and we can use the QR algorithm or some other method to solve the tridiagonal problem. The sequential split-merge algorithm has the structure which can be viewed as a "tree" of tasks, as shown in Fig. 1. Task TL(m) is to solve a small matrix of order m that can be found by some algorithm such as QR decomposition. TL($2^i m$) is to approximate the eigenvalues for the submatrix of order $2^i m$ by Laguerre's iteration.

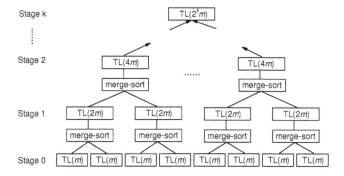

Fig. 1. Task tree of the algorithm

3   Parallel Algorithm

3.1   MPI Parallelization Algorithm

The algorithm is well-suited to efficient parallelization. First, when the matrix T is split in a tree as in Fig. 1, solving the eigenproblem of each submatrix is independent of the others and thus can be done by one processor or a group of processors. Next, from down to up, we can regard the root in each subtree of the graph as a master processor and all other processors are leaves. Each processor at leaf sends its approximated eigenvalue to the master processor of subtree. That processor then gathers and sorts the eigenvalues, and scatters them back to all the other processors in the subtree. Eigenvalues of symmetric tridiagonal submatrix at stages can be evaluated separately by performing independent Laguerre's iteration on all processors in the subtree.

For convenience, let us assume that the number of processors and the order of input matrix are both a power of two, 2^k and 2^t, respectively. Fig. 2 shows MPI parallel algorithm. The algorithm first divides the original matrix into 2^k submatirces of order 2^s ($s = t - k$), then computes the eigenvalues of these submatrices in parallel on different processors. At stage 0 those computations that are carried out on each processor are completely independent and they do not involve performing any kind of communications. At the next stage, the master processor gathers eigenvalues from other processors, merges and sorts the eigenvalues, then scatters them among the processors; use Laguerre's iteration to find 2^s eigenvalues corresponding to task TL(2^{s+j}).

Algorithm 1. MPI parallel split-merge method

```
execute task T(2^s)
do j=1 to k
    master_proc = int(myid/2^j)*2^j
    if (master_proc .eq. myid)
        gather eigenvalues from other proc in the same group
        merge and sort 2^{s+k} received eigenvalues
        distributes them evenly among myid, ..., myid+2^j - 1
    else
        Send 2^s approximated eigenvalues to master_proc
        Receive 2^s from process master_proc
    endif
    use Laguerre's iteration to find 2^s eigenvalues corresponding to
task TL(2^{s+j})
enddo
```

Fig. 2. MPI parallel split-merge algorithm

This algorithm has been successfully implemented on distributed memory machines [14]. The inherent variance in the number of Laguerre's iterations justified more sophisticated approaches to load balancing.

3.2   Hybrid MPI/OpenMP Parallelization

In the MPI/OpenMP hybrid paradigm, each SMP node runs a multi-thread MPI process. In order to create an overall high efficient algorithm, we have to stick the algorithms for hierarchical architecture. In the MPI algorithm of eigenvalue problem of symmetric tridiagonal matrix, there are three different operations: (I) each process computes the eigenvalues of a divided submatrix; (II) approximated eigenvalues are gathered, sorted and broadcast among processes; (III) each process solves the approximated eigenvalues of combined submatrices by Laguerre's iterations. However in the MPI/OpenMP hybrid parallelization, the algorithm divides the matrix onto the processes of node, so task (I) and task (III) can be finished by the threads within node.

Multi-thread parallelization in task (I) is similar to the multi-process parallelization in MPI. It includes the submatrix division and solving among threads on the same node, approximated eigenvalues exchange among threads and solving approximated eigenvalues of combined submatrices by Laguerre's iterations. This can employ coarse grain SPMD OpenMP model. Due to shared memory among threads within a node, when the matrix is divided onto threads, the array region of each thread needs to be computed to map the array onto the threads. The eigenvalues gathering and broadcasting can be operated via data duplication rather than message passing.

OpenMP parallelism of task (III) solves the approximated eigenvaules of combined submatrices on nodes. In a process, solving the eigenvalues of the submatrix by Laguerre's iteration is a loop operation. It can be simply performed by OpenMP *PARALLEL DO* directive. Because the numbers of Laguerre's iterations on different approximated threads may be different, this can lead to load imbalance. To improve the load balance and reduce the amount of synchronization involved, *PARALLEL DO* directive with *SCHEDULE* clause should be adopted to distributed iterations among threads dynamically.

In the MPI/OpenMP hybrid paradigm, a process is performed by multiple threads in parallel on a node. Comparing to the pure MPI paradigm, the MPI/OpenMP hybrid paradigm can decrease the number of processes on the same number of processors. In Fig. 2, because the number of iteration j equals to $\log_2$[number of processes], the number of iterations in the MPI/OpenMP hybrid paradigm is less than that in the pure MPI program, which would diminish communication overhead between processors. Moreover, because each iteration needs a synchronization to interchange computed eigenvalues, the MPI/OpenMP hybrid paradigm decreases the synchronization overhead as well.

In addition, all MPI communication in the MPI/OpenMP hybrid paradigm occur within the OpenMP parallel region, but are handled by only one thread within each SMP node. This configuration assembles the nodal messages into a single large message and thus reduces the network latency overhead.

4   Results and Discussion

In the analysis of our algorithms we make 4 test programs that compute the eigenvalues of a fixed size tridiagonal symmetric matrix. We utilize a group of test matrices similar to that used in [12]. This group consists of 12 types of matrices that produce different behaviors. In this section we give out the test result on a Wilkinson matrix that is type 6 in that group, and the order of test matrix is 16384.

The test programs are run on DeepComp 6800 at Supercomputing Center, Chinese Academy of Sciences. DeepComp 6800 consists of 265 nodes; each node owns 4-way Intel Itanium 2 1.3 GHz processors, 8 GB or 16 GB memory; all the nodes are inter-connected by QsNet with bandwidth of 300 MB/s and latency of 7 μs. The operating system is Redhat Linux Advanced Server 64-bit 2.1, the MPI library is QsNet optimized MPICH 1.2.4. All the test programs are compiled by Intel C compiler 64-bit 7.1 with the option "-O3".

The number of processors is the product of the number of MPI processes and the number of OpenMP threads per process, the number of MPI processes is varied from 1 to 64, and the number of OpenMP threads per process is varied between 2 and 4. While using 4 OpenMP threads, we adopt 2 different schedules in work sharing among threads: static and dynamic, and only static work sharing is adopted when using 2 OpenMP threads. Therefore there are 4 programs in total: the pure MPI program, in which the number of processors equals to the number of MPI processes that is varied form 4 to 128 (4, 8, 16, 32, 64 and 128); and the MPI/OpenMP hybrid program with 2 threads, in which the

Table 1. Running times of the pure MPI and the MPI/OpenMP hybrid programs

Num. of Processors	Time (s)			
	Pure MPI	MPI/OpenMP 2 threads	MPI/OpenMP 4 threads static	MPI/OpenMP 4 threads dynamic
4	71.06	70.98	70.10	70.10
8	42.06	41.38	40.82	35.56
16	22.81	22.56	20.98	19.92
32	11.95	11.85	10.32	9.51
64	6.21	6.15	5.03	4.77
128	3.30	3.24	2.42	2.22

Table 2. Speedups of the pure MPI and the MPI/OpenMP hybrid programs

Num. of Processors	Speedup			
	Pure MPI	MPI/OpenMP 2 threads	MPI/OpenMP 4 threads static	MPI/OpenMP 4 threads dynamic
4	1	1	1	1
8	1.690	1.715	1.717	1.971
16	3.115	3.146	3.341	3.519
32	5.948	5.991	6.793	7.371
64	11.452	11.546	13.936	14.696
128	21.533	21.891	28.967	31.577

number of MPI processes is varied from 2 to 64 (2, 4, 8, 16, 32 and 64); and the MPI/OpenMP program with 4 threads, in which the number of MPI processes is varied from 1 to 32 (1, 2, 4, 8, 16 and 32).

The running time and parallel speedup of these 4 programs are listed in table 1 and table 2, and their figures are shown in Fig. 3, respectively. The parallel speedup is computed by the following formula defined by Liu et al [20] that is different to the common definition:

$$S_{n\_node} = \frac{T_p(1\_node)}{T_p(n\_node)}, \tag{5}$$

where $T_p(1\_node)$ is the parallel running time on 1 node, $T_p(n\_node)$ is the parallel running time on n nodes.

These two figures illustrate the wall clock running time and parallel speedup versus the total number of processors for two programming paradigms, including the pure MPI and the MPI/OpenMP hybrid. In the 4 programs, The pure MPI and the MPI/OpenMP hybrid with 2 threads are the slowest ones that their performance are almost same while the latter one is slightly faster, the MPI/OpenMP hybrid with 4 threads using static schedule is faster than these two and the MPI/OpenMP hybrid with 4 threads using dynamic schedule achieves the best performance. The speed gaps between them are getting larger with the increase of the number of processors. The MPI/OpenMP hybrid with 4

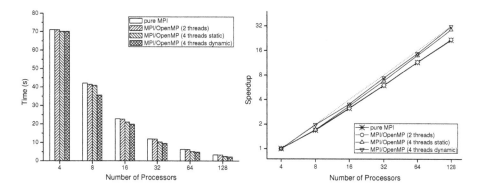

Fig. 3. wall clock running time (left) and parallel speedup (right) of pure MPI program, MPI/OpenMP hybrid program with 2 threads and MPI/OpenMP hybrid program with 4 threads

threads using dynamic schedule is about 13%–30% faster than the pure MPI and the MPI/OpenMP Hybrid with 2 threads, and it is 5%–12% faster than the program using static schedule; the gap between the pure MPI and the MPI/OpenMP hybrid with 2 threads is less than 2%.

The parallel speedups of all the programs demonstrate good scalability in the parallel speedup. Apparently, the pure MPI and the MPI/OpenMP hybrid with 2 threads have the worst speedup, the curves of them are almost overlapped; and the MPI/OpenMP hybrid with 4 threads using dynamic schedule also performs the best in speedup that is very close to the ideal speedup; the speedup of the MPI/OpenMP hybrid with 4 threads using static schedule is still in the middle. When the number of processors is more than 8, the two MPI/OpenMP hybrid with 4 threads programs get much better speedups.

OpenMP reduces the communication overhead within the same node which is one of the reasons that make the MPI/OpenMP hybrid programs achieve better performance. It also diminishes the number of iterations by $\log_2 N_{th}$, where N_{th} is the number of OpenMP threads per process. Moreover, dynamic work sharing schedule in OpenMP can help getting better load balance, which is another reason for the performance increase.

5   Conclusions

On a cluster of SMP's, the MPI/OpenMP hybrid paradigm outperforms the MPI paradigm in solving the eigenvalues of a tridiagonal matrix. With the increase of the number of OpenMP threads within a MPI process and using the same number of processors, the performance of the MPI/OpenMP hybrid increases correspondingly. Generally, dynamic work sharing schedule can get better performance than static schedule because it promotes load balance.

The relative performance of the MPI/OpenMP hybrid and MPI is determined by several factors. Therefore, not all problems can get the performance benefit from using the MPI/OpenMP hybrid paradigms as that is presented in this article. Programmers should make the tradeoff according to their problem.

In practice, serious consideration must be given to the nature of codes before embarking on a hybrid parallelization implementation. In some situations however significant benefit may be obtained from a hybrid parallelization implementation. For example, an MPI code has the following problems:

- Load imbalance or fine-grain granularity may cause poor scalability.
- Fast processors make the communication performance significant and the level of parallelization is sufficient.
- MPI code suffers from memory limitations due to the use of a replicated data strategy.
- A hybrid MPI/OpenMP implementation could be more efficient for a larger problems giving poor scalability for a high number of processors.

References

1. Sameh, A., Kuck, D.: A parallel QR algorithm for symmetric tridiagonal matrices. IEEE tans. Comput. **C-26** (1977) 81–91
2. Arbenz, P., Gates, D., Sprenger, C.: A parallel implementation of the symmetric tridiagonal QR algorithm. In Proc. Fourth Symp. On the Frontiers of Massively Parallel Computation (IEEE CS Press, 1992)
3. Cuppen, J.J.M.: A divide and conquer method for symmetric tridiagonal eigenproblem. Numer. Mathematik **2** (36) (1981) 177–195
4. Wilkinson, J.H.: The Algebraic Eigenvalue Problem. Oxford, Clarendon Press, 1965
5. Li, T.Y., Zhang, H., Sun, X.H.: Parallel homotopy algorithm for the symmetric tridiagonal eigenvalue problem. SIAM J. Scientific and Statistical Comput. **12** (May, 1991) 469–487
6. Dongarra, J.J., Soorenson, D.C.: A Fully Parallel Algorithm for the symmetric Eigenvalue Problem. SIAM J.Sci.Stat. Comput., vol. 8, no. 2. pp.s139–s154, 1987
7. Dhillon, I.S., Fannm, G., Parlett, B.N.: Application of new algorithm for the symmetric eigenproblem to computational quantum chemistry. In processings of the Eigen SIAM Conference on Parallel processing for Scientific Computing, Minneapolis, MN, March 1997. SIAM
8. Dhillon, I.S., Parlett, B.N.: Multiple representations compute orthogonal eigenvertors of symmetric tridiagonal matrices. Lin. Alg. Appl., **387** (2004) 1–28
9. Luong, P., Breshears, C.P., Ly, L.N.: Coastal ocean modeling of the U.S. west coast with multiblock grid and dual-level parallelism. In Supercomputing 2001: High Performance Networking and Computing (SC2001)
10. Pavani, R., De Ros, U.: Solving the tridiagonal symmetric eigenvalue problem on a transputer network. n. 146/p, Dipartimento di mathematica, Politecino di Milano, 1994
11. LI, T.Y., Zeng, Z.: Lagurre's iteration in solving the symmetric tridiagonal eigenproblem. SIAM J. Scientific Comput. **15** (1994) 1145–1173
12. Pavani, R., De Ros, U.: A Distributed divide-and-conquer approach to the parallel symmetric eigenvalue problem. The International Conference on High-Performance Computing and Networking, Milano, 1995

13. Bova, S.W., Breshears, C., Cuicchi, C., Demirbilek, Z., Gabb, H.A.: Dual-level parallel analysis of harbor wave response using MPI and OpenMP. Int. J. High Perform Comput. Appl. **14** (2000) 49–64

14. Trefftz, C., Huang, C.C., Li, T.Y., Zeng, Z.: A scalable eigenvalue solver for symmetric tridiagonal matrices. Parallel Computing **21** (1995) 1213–1240

15. Cappello, F., Etiemble, D.: MPI versus MPI+OpenMP on the IBM SP for the NAS Benchmarks. In Supercomputing 2000: High Performance Networking and Computing (SC2000)

16. Loft, R.D., Thomas, S.J., Dennis, J.M.: Terascale spectral element dynamical core for atmospheric general circulation models. Supercomputing 2001: High Performance Networking and Computing (SC2001)

17. Crawford, C.H., Evangelinos, C., Newman, D., Karniadakis, G.E.: Parallel benchmarks of turbulence in complex geometries. Comput. Fluids **25** (1996) 677–698

18. Henty, D.S.: Performance of hybrid message-passing and shared-memory parallelism for discrete element modeling. In Supercomputing 2000: High Performance Networking and Computing (SC2000)

19. Dong, S.H., Em Karniadakis, G.: Dual-level parallelism for high-order CFD methods. Parallel Computing **30** (2004) 1–20

20. Liu, W., Zheng, W.M., Zheng, X.W.: The concept of node-oriented speedup on SMP cluster. Computer engineering and design, Vol. 21, No. 5, Oct. 2000

Trust Management with Safe Privilege Propagation[*]

Gang Yin[1], Huai-min Wang[1], Tao Liu[2], Ming-feng Chen[3], and Dian-xi Shi[1]

[1] School of Computer Science,
National University of Defense Technology, China
jack_nudt@yahoo.com.cn
[2] School of Electronic Science and Engineering,
National University of Defense Technology, China
bravewendy@163.com
[3] China Xi'an Satellite Control Center,
cmf1968@sina.com

Abstract. Trust management uses delegation to enable decentralized authorization across administrative domains. Delegation passes one's authority over resources to trusted entities and thus enables more flexible and scalable authorization. However, unrestricted delegation may result in privilege proliferation and breach the privacy of information systems. The delegation models of existing trust management systems do not provide effective control on delegation propagation, and the correctness of constraint enforcement mechanisms is not formally analyzed, which may lead to privilege proliferation. In this paper, we propose a role-based constrained delegation model (RCDM), which restricts the propagation scope of delegation trees by a novel delegation constraint mechanism named *spacial constraint*. This paper also introduces a rule-based language to specify the policies and the deduction algorithm for constrained delegation defined in RCDM. The soundness and completeness properties of the deduction algorithm ensure the safety and availability of our delegation model.

1 Introduction

Trust management (TM) is an attractive approach to authorization in decentralized environments. Several TM systems have been proposed in recent years, e.g., SPKI/SDSI [2], PolicyMaker [4], KeyNote [5], Delegation Logic [7], RT [8] and Cassandra [6]. One important characteristic of TM systems is using delegation to transfer authority between entities, and the delegation chains enable the authorization from resource owner to users through multiple delegation steps. However, delegation chains may lead to "privilege proliferation" and breach the privacy of information systems.

---

[*] This work is supported by Grand Fundamental Research 973 Program of China (No.2005CB321804), National Natural Science Foundation under Grant No.90412011; the National High Technology Development 863 Program of China (No.2003AA115210; No.2004AA112020).

J. Cao, W. Nejdl, and M. Xu (Eds.): APPT 2005, LNCS 3756, pp. 174–183, 2005.
© Springer-Verlag Berlin Heidelberg 2005

The essential reason for privilege proliferation in delegation-based systems is inefficient control over delegation chains. The typical constraints on delegation are boolean control and integer control over the depth of delegation chains. Boolean control prohibits further delegation or allows unrestricted re-delegation. Keynote [5], SPKI [2] and RT [8] support this kind of constraint. DL [7] supports integer control which provides more flexibility than boolean control, but it supposes that the trust relationships are transitive within the upper-bound of delegation depth, which is too optimistic and may lead to undesired propagation of privileges. DL also supports constraints on delegation width, but to enforce such constraints, it uses a temp key to sign an assistant policy, which may introduce risks and inconvenience. Furthermore, existing TM systems have not analyzed the correctness of the delegation deduction algorithm to ensure that the constraints are really enforced when make authorization decision.

In this paper, we propose a delegation model with safer privilege propagation named RCDM (Role-based Constrained Delegation Model). RCDM restricts the propagation scope of delegation trees by a novel delegation constraint structure named spacial constraint. A rule-based policy language is also introduced to specify the core policies and deduction algorithm for constrained delegation in RCDM. The rest of this paper is organized as follows. Section 2 defines the main components of RCDM including a basic model and spacial constraint model. In section 3, we describe the syntax and semantics of a rule-based specification language designed for RCDM. The correctness of the deduction algorithm is formally analyzed. Finally we discuss the related work and conclude this paper.

2 Role-Based Constrained Delegation

In this section, the main components of the role-based constrained delegation model (RCDM) are defined.

2.1 Basic Model

We first define a role-based authorization system, based on which a constrained delegation system will be defined as a general extension.

Definition 1. An *authorization system* (AS) is a 7-tuple (E, R, AR, A, S, $\ni$, $\propto$), where E, R, AR, A and S are sets of entities, roles, administrative roles, actions and statements; $\ni$ and $\propto$ are relations:

- A role $r \in R \subseteq E \times N$ is denoted as e.rn, where N is the set of role names.
- A administrative role $ar \in AR \subseteq E \times 2^N$ is denoted as e.ar(rn_1, rn_2,…,rn_s), represents the privileges to manage the authorization of roles. {$e.rn_1$,…,$e.rn_s$} are *dominated roles* of ar; DE is a function: AR→E, DE(ar)=e is the defining entity.
- An action $a \in A \subseteq E \times R$ is denoted as (e, r). ATE is a function: A→E, ATE(a)=e is the target entity of a; ATR is a function: A→R, ATR(a)=r is the target role of a.
- A statement $s \in S \subseteq E \times A$ is denoted as (e, a), which means e asserts a.
- $\ni \subseteq E \times AR$ is a *control* relation. Given $e \in E$, $ar \in AR$, $e \ni ar$ means e *controls* ar, and e is also called the source of the authority [4] over ar. DE(ar)$\ni$ar always holds.

- $\propto\subseteq AR\times R$ is a *dominate* relation. Given $ar\in AR$, $r\in R$, $ar\propto r$ means ar *dominate* r. Suppose $ar=e.ar(rn_1, rn_2,...,rn_s)\in AR$, $ar\propto e.rn_i$, $i\in[1..s]$.

We now define a constrained delegation system based on AS.

Definition 2. A *constrained delegation system* (CDS) is a 5-tuple (AS, D, C, DP, $\triangleright$), where D, C and DP are sets of delegation, delegation constraint and delegation path; $\triangleright$ is a relation:

- A delegation $d\in D\subseteq E\times E\times AR\times C$ is denoted as a 4-tuple (x, y, ar, c), which means that x delegate ar to y, and uses c to restrict the further propagation of ar.
- A delegation path $\xi\in DP\subseteq 2^D\times S$ is denoted as $(x_0, c_0),...,(x_{n-1}, c_n)(x_n, a)_{ar}$, which means x_i delegates ar to x_{i+1}, using c_i to restrict the propagation of ar, $i\in[0..n-1]$, and finally x_n performs the action a. The natural number *n* is called the delegation depth of ξ. If $n=0$, $\xi=(x_0, a)_{ar}\in DP$ is called a *dummy* delegation path.
- $\triangleright\subseteq DP\times C$ is a *satisfy* relation. Given $\xi\in DP$, $c\in C$, $\xi\triangleright c$ means that ξ *satisfies* c.

Definition 3. Given $\xi=(x_0, c_0)(x_1, c_1),...,(x_{n-1}, c_{n-1})(x_n, a)_{ar}\in DP$, ξ is a *sound delegation path* if and only if $\xi_i\triangleright c_i$ for all $i\in[0..n-1]$, where $\xi_i=(x_i, c_i),...,(x_{n-1}, c_{n-1})(x_n, a)_{ar}$. A dummy delegation path is always sound. Given $s=(x, a)\in S$, s is a *sound statement* if and only if there exists a sound delegation path $\xi=(x_0, c_0)(x_1, c_1),...,(x_{n-1}, c_{n-1})(x, a)_{ar}\in DP$, and $x_0\ni ar$, $ar\propto ATR(a)$.

When an entity makes authorization decision, the action in a sound statement will be allowed by the authorization entity.

2.2 Spacial Constraint on Delegation

CDS allows an entity to pass privilege along delegation path; the entity may delegate privileges to more than one entity and thus the delegation process forms a delegation tree, which is formally defined as follows.

Definition 4. A *delegation tree* $t\in T$ starting from a delegation $d\in D$ is denoted as a 4-tuple (d, Nodes, Leafs, Arcs), $d=(e_0, e_1, ar, c_0)$ is the root delegation of *t*; Nodes$\subseteq E$ and Arcs$\subseteq E\times E$ are sets of nodes and edges of *t*. Given $\xi=(e_0, c_0)(e_1, c_1)...(e_{n-1}, c_{n-1})(e_n, a)_{ar}\in DP$, $e_i\in t.$Nodes, $i\in[1..n]$; $\langle e_i, e_{i+1}\rangle\in t.$Arcs, $i\in[1..n-1]$; ATE(a)$\in t.$Leafs.

The delegation tree *t* clearly defines a propagation structure of privilege starting from d, where *t*.Nodes is the set of *delegation-agencies* who may be delegated with ar; *t*.Leafs is the set of *delegation-targets* who are targets of actions performed by delegation-agencies. The propagation of privileges will be controlled if we can control the shape of delegation tree. Base on this observation, we propose a constraint model based on delegation tree, named *spacial constraint*.

Definition 5. A *spacial constraint structure* is a 2-tuple (SC, $\triangleright$), where SC$\subseteq TS\times Nat\times TS$ is a set of spacial constraints, TS is a set of trust scopes [3], Nat is a set of natural numbers. A spacial constraint $c\in SC$ is written sc(ac, dc, tc), where c.ac$\in$TS is the constraint on the scope of delegation-agencies; c.dc$\in$Nat is the constraint on the upper-bound of delegation depth, c.tc$\in$TS is the constraint on delegation-targets. More detailed description follows:

- A trust scope ts$\in$TS$\subseteq 2^E$ is a set of entities with some trusted attributes. Currently we use roles to define the trust scopes, for example, the trust scope $\{e\,|\,e$ is a member of $e.n$, $e.n\in$R$\}\in$TS can be denoted as $e.n$.
- The constant ε=sc(E, ∞, E)$\in$SC defines a null constraint, where $\infty\in$Nat is a constant upper-bound of all the constraints of delegation depth.
- Given ξ_i=$(e_i,c_i)\ldots(e_{n-1},c_{n-1})(e_n,a)_{ar}\in$DP, $i<n$, $\xi_i\triangleright c_i$ if and only if $(e_j\in c_i.ac)\wedge(n-i\leq c_i.dc)\wedge(ATE(a)\in c_i.tc)$, $j\in[i+1..n]$.

Note spacial constraint also supports numerical control on delegation depth, which is discussed in the introduction section. Here we use depth control to avoid delegation loops and unnecessary inference beyond the upper-bound of delegation depth. According to the definition of CDS, SC can be a realization of C, which will provide CDS with practical constraint mechanism on delegation.

3 Policy Specification and Semantics

The authors have proposed REAL05 [3], a rule-based extensible authorization language for controllable authorization in decentralized environments. In this section, we present a revised subset of REAL05 for management-level authorization, which is denoted by REAL^M.

3.1 Syntax

The syntax of REAL^M contains terms, predicates and rules. REAL^M also introduces two new predicates, PRA and hasActivated, which make the language more practical.

Definition 5 (Predicate). A predicate x.P($\vec{e}$) is used to define security assertions, where x is the *principle* of the predicate, P is the *predicate name*, $\vec{e}$ is a list of terms and also called the *parameter list* of the predicate. x.P($\vec{e}$) can be read as: "x says P($\vec{e}$)". The following RCDM components are expressed with predicates:

- A statement s=(x, a)$\in$S$\subseteq$E$\times$A where a=(y, z.rn)$\in$A, is expressed with *UAS* predicate: x.UAS(y, z.rn), which means that x authorizes y with the role z.rn.
- A delegation d=(x, y, ar, c)$\in$D$\subseteq$E$\times$E$\times$AR$\times$SC is expressed with *DRA* predicate: x.DRA(y, ar, c), which means that x delegates ar to y, using c to restrict the further propagation of ar.
- The predicate *PRA* is used for binding permissions to roles: x.PRA(pm, x.r), which means entity x binds a permission pm to role x.r. A permission identifies a protected resource or function.
- The predicate *hasActivated* is used for role activation: x.hasActivated(y, x.rn) means that entity y has activated the role x.rn under the grant of entity x.

REAL^M extend the expressive power of RCDM with rules, which add conditions to predicates and thus can express more complex and dynamic policies.

Definition 6 (Rule). A rule has the following form,

$$H \leftarrow B_1, B_2, ..., B_n.$$

where H, $B_i(i \in [1..n])$ are predicates. H is the *head* of the rule, the principle of H is the *issuer* of the rule. $B_1, B_2, ..., B_n$ are *body* of the rule. The parameters in the predicates can be variables which start with "?". If $n=0$ then the rule is called a *fact*. The meaning of the rule can be read as: to deduce H, we must deduce $B_1, B_2, ..., B_n$.

Given a rule set P, if a predicate p is deducible from P, we say $P \vdash p$. Then the meaning of a rule can be described as: given a rule r, if $P \vdash p$ for each p in $B(r)$, then $P \vdash H(r)$. Given a rule r, and r is a UAS rule and DRA rule, if $P \vdash H(r)$, then $H(r)$ uniquely maps to an element in S or D. For example, given a UAS rule "x.UAS(y, z.r) $\leftarrow B_1, B_2, ..., B_n$", if $P \vdash x.UAS(y, z.r)$, then $(x, y, z.r) \in S$.

3.2 Semantics

$REAL^M$ uses rules to define the policies based on RCDM. But to enforce the semantics of RCDM, such as the sound delegation and the sound statement, we need to define semantic rules for RCDM to facilitate delegation deduction.

Definition (Delegation Trace Structure). The delegation trace structure is a triple $(TR, SC, \succ_R)$, where $t \in TR \subseteq 2^E \times Nat \times 2^E$ is called a delegation trace or trace for short, written as tr(as, dd, ts), which traces a delegation path. t.as is the set of delegation-agencies; t.dd is the delegation depth and t.ts is the set of delegation-targets. $\succ_R$ $\subseteq TR \times SC$ is a relation, given $t \in TR$, $c \in SC$, $(t, c) \in \succ_R$ if and only if $(t.as \subseteq c.ac) \wedge (t.dd \leq c.dc) \wedge (t.ts \subseteq c.tc)$.

The delegation trace will be used to verify the soundness of sub-delegation chains during inference (see m2 in tab.1). For example, the trace of the delegation path $\xi = (e_0,c_0)(e_1, c_1)...(e_{n-1},c_{n-1})(e_n,a)_{ar} \in DP$ is tr($\{e_1,...,e_n\}$, n, $\{ATE(a)\}$).

Definition (Semantic Predicate). The semantic predicates are used to construct the algorithms for delegation reduction, which are defined as follows:

- $x.doa(y, ar, t)$ means entity x passes the admin-role ar to entity y through one or more delegation steps, and t is the trace of the delegation chain.
- $x \succ y$ means $(x, y) \in \succ_R$, where x is a trace and y is a spacial constraint.
- $x \subseteq y$ means every entity in entity set x belong to the trust scope defined by y.
- $x \leq y$ means x is no more than y, where x and y are natural numbers.

The language engine supports two kinds of queries, which are used to make decision on role activation requests and resource access request.

Definition (Query). There are two classes of queries that $REAL^M$ can answer:

- Role activation query: ?$\leftarrow x.canActivate(y, x.rn)$, which means whether entity x will allow y to activate the role $x.rn$.
- Resource access query: ?$\leftarrow x.canAccess(y, p)$, which means whether entity x will allow y to access the resource identified by the permission p.

The role activation query is raised by authorization entity when a user wants to activate some roles to use in current session. If the query is deduced, then the authorization entity will create a hasActivated predicate, indicating that the user can use the role to access resources. When the user makes request to access resources controlled by the authorization entity, the authorization entity will create a resource access query to decide whether to allow the current request. The semantics of the two queries are defined in tab.1.

The meta-rules in tab. 1 define an algorithm for proof of compliance with $REAL^M$. We use M to denote these rules. Given a set of $REAL^M$ rules P and a query Q, Q is deducible if $(P \cup M) \vdash Q$. $m1$ and $m2$ define recursive process for multi-step delegation reduction. $m3$ and $m4$ define the semantics of role-activation-query, while $m5$ defines the semantics of resource-access-query. $m6$ defines the semantics of predicate ">". The semantics of "$\subseteq$" is not listed in table 1 because it needs to deal with some special cases, please see [3] for a detailed implementation.

Table 1. The Deduction Algorithm for Constrained Delegation

[Meta Rules for Delegation of Authority]

(m1)   ?x.doa(?y, ?ar, ?t) ← ?x.DRA(?y, ?ar, ?c), tr({?y}∪?t.as, ?t.dd, ?t.ts) ≻ ?c.

(m2)   ?x.doa(?y, ?ar, ?t) ← ?z.DRA(?y, ?ar, ?c), tr({?y}∪?t.as, ?t.dd, ?t.ts) ≻ ?c,
       ?x.doa(?z, ?ar, tr({?y}∪?t.as, ?t.dd+1, ?t.ts)).

(m3)   ?x.canActivate(?y, ?x.?rn) ← ?x.UAS(?y, ?x.?rn).

(m4)   ?x.canActivate(?y, ?x.?rn) ← ?z.UAS(?y, ?x.?rn)), ?x.doa(?z, ?x.ar(?rn), tr([], 1, [?y])).

[Meta Rules for Access Control]

(m5)   ?x.canAccess(?y, ?p) ←?x.hasActivated(?y, ?x.?rn), ?x.PRA(?p, ?x.?rn).

[Meta Rules for Delegation Constraint Computation]

(m6)   ?t ≻ ?c ← ?t.as ⊆ ?c.ac, ?t.dd ≤ ?c.dc, ?t.ts ⊆ ?c.tc.

Now we need to demonstrate the correctness of delegation reduction algorithm defined in tab.1, i.e., soundness and completeness. The soundness in this context means that if a query like "?←x.canActivate(y, x.rn)" is deduced, then there exists a *sound statement* with the form "(x, y, x.rn)". The completeness here means that if a statement like "(x, y, x.rn)" is proved to be a *sound statement*, then the query like "?←x.canActivate(y, x.rn)" can be deduced based on $REAL^M$ semantics. First we give a lemma to facilitate the proof of the soundness and completeness results.

Lemma. Given a set of $REAL^M$ rules P, let $PS = P \cup M$, $PS \vdash x.doa(y, ar, t)$ if and only if there is a chain of rules $\sigma = r_0, r_1, ..., r_m$ in P, where $H(r_i) = x_i.DRA(x_{i+1}, ar, c_i)$, $t_i = tr(\{x_{i+1}, ..., x_{m+1}\}, t.dd + (m-i), \varnothing)$, $x_0 = x$, $x_{m+1} = y$, $i \in [0..m]$, and,

- $t \succ c_i$ and $t_i \succ c_i$ for each $i \in [0..m]$;
- $P \vdash H(r_i)$ for each $i \in [0..m]$.

Proof. We first prove the if part. Do induction on the length l of σ. If $l=1$, $P \vdash x_0.DRA(x_1, ar, c_0)$. According to $m1$ and $t \succ c_0$, $PS \vdash x_0.doa(x_1, ar, t)$ holds, i.e., $PS \vdash x.doa(y, ar, t)$. Suppose the if part holds for $l=k>1$, now consider the case when

$l=k+1$. There exists $\sigma_{k+1}=r_0,r_1,\ldots,r_k$ in P, where $H(r_i)=x_i.DRA(x_{i+1}, ar, c_i)$, $i\in[0..k]$, and (a, b, c) hold for each $i\in[0..k]$: (a) $P\vdash H(r_i)$; (b) there exists $t\in TR$ and $t\succ c_i$; (c) $t_i=tr(\{x_{i+1},\ldots,x_{k+1}\}, t.dd+(k-i), \varnothing)\succ c_i$. Now we need to prove $PS\vdash x_0.doa(x_{k+1}, ar, t)$.

Consider $\sigma_k=r_0,r_1,\ldots,r_{k-1}$, $t'=tr(\{x_{k+1}\}\cup t.as, t.dd+1, t.ts)$. From (c), $x_{k+1}\in c_i.ac$ and $t.dd+1\leq t.dd+(k-i)\leq c_i.dc$ for each $i\in[0..k-1]$. So it is clear that $t'\succ c_i$ for each $i\in[0..k-1]$. From (c), $tr(\{x_{i+1},\ldots,x_k\}, t.dd+(k-1-i), \varnothing)\succ c_i$ for each $i\in[0..k-1]$. Then by induction assumption, $PS\vdash x_0.doa(x_k, ar, t')$. From (b) and (c), $tr(\{x_{K+1}\}\cup t.as, t.dd, t.ts)\succ c_k$. Consider the rule r_k in P, $H(r_k)=x_k.DRA(x_{k+1}, ar, c_k)$ and $P\vdash H(r_k)$. Therefore by using m2, $PS\vdash x_0.doa(x_{k+1}, ar, t)$ follows.

Now we prove the only if part. Clearly, there exists a sequence of proof steps that ends with x.doa(y, ar, t). Do induction on the steps s of the sequence. If $s=1$, m1 will be used to deduce the result. There must be a rule $r_0\in P$, $H(r_0)=x_0.DRA(x_1, ar, c_0)$ and $P\vdash H(r_0)$, $x_0=x$, $x_1=y$, $t\succ c_0$, $tr(\varnothing, t.dd, \varnothing)\succ c_0$, so the only if part holds. Suppose the only if part holds for $s=k>1$, consider the case when $s=k+1$. The proof must start from m2, so (d, e, f) must hold: (d) there is a rule $r\in P$, $H(r)=z.DRA(y, ar, c)$ and $P\vdash H(r)$; (e) $tr(\{y\}\cup t.as, t.dd, t.ts)\succ c$; (f) $PS\vdash x.doa(z, ar, t')$, where $t'=tr(\{y\}\cup t.as, t.dd+1, t.ts)$. Consider (f), it is clear that the steps of the proof sequence that ends with x.doa(z, ar, t') is less than k. By induction assumption, there exists a chain $\sigma_t=r_0,r_1,\ldots,r_v$ in P, where $H(r_i)=x_i.DRA(x_{i+1}, ar, c_i)$, $x_0=x$, $x_{v+1}=z$, $i\in[0..v]$, and (g, h, i) hold for each $i\in[0..v]$: (g) $P\vdash H(r_i)$; (h) $t'\succ c_i$; (i) $tr(\{x_{i+1},\ldots,x_{v+1}\}, t'.dd+(v-i), \varnothing)\succ c_i$.

Therefore, the chain $\sigma_w=r_0,r_1,\ldots,r_w$ is in P, where $H(r_i)=x_i.DRA(x_{i+1}, ar, c_i)$, $P\vdash H(r_i)$, $w=v+1$, $r_w=r$ (i.e., $x_{w+1}=y$, $c_w=c$), $i\in[0..w]$. From (e), $t\succ c_w$; from (f) and (h), $t\succ c_i$ for each $i\in[0..w-1]$. Then $t\succ c_i$ for $i\in[0..w]$. From (e), $t_w=tr(\{x_{w+1}\}, t.dd, \varnothing)\succ c_w$. From (f), (h) and (i), $t_i=tr(\{x_{i+1},\ldots,x_{w+1}\}, t.dd+1+(v-i), \varnothing)\succ c_i$, $i\in[0..w-1]$. Therefore $t_i\succ c_i$, where $t_i=tr(\{x_{i+1},\ldots,x_{w+1}\}, t.dd+(w-i), \varnothing)$, $i\in[0..w]$. ∎

The lemma shows some important properties of delegation traces during inference process defined in table 1. Now we prove the soundness and completeness result.

Theorem (Soundness and Completeness of Semantics). Given a set of REAL^M rules P, let $PS=P\cup M$, $PS\vdash x.canActivate(y, x.rn)$ if and only if there is a rule $r\in P$, $H(r)=z.UAS(y, x.rn)$ and $P\vdash H(r)$, (z, y, x.rn) is a *sound statement*.

Proof. We first prove the only if part (soundness). According to the rules in M, the result x.canActivate(y, x.rn) must be deduced from m3 or m4. In the first case: there must be a rule $r\in P$, $H(r)=x.UAS(y, x.rn)$ and $P\vdash H(r)$. It is clear that $x\ni x.ar(rn)$ and $ar\propto ATR(a)$, where $a=(y, x.rn)\in A$. So there exists a "dummy" delegation path $\xi=(x, a)_{x.ar(rn)}\in DP$, and (x, a) is a *sound statement*.

In the second case, i.e., the deduction starts from m4, then following results hold: (a) there is a rule $r\in P$, $H(r)=z.UAS(y, x.rn)$ and $P\vdash H(r)$; (b) $PS\vdash x.doa(z, ar, t)$, where $ar=x.ar(rn)$, $t=tr(\varnothing, 1, \{y\})$. From (b) and lemma, there is a chain of rules $\sigma =$

$r_0, r_1, \ldots, r_m$ in P, where $H(r_i) = x_i.DRA(x_{i+1}, ar, c_i)$, $t_i = tr(\{x_{i+1}, \ldots, x_{m+1}\}$, $t.dd+(m-i)$, $\varnothing)$, $x_0 = x$, $x_{m+1} = z$, $i \in [0..m]$, and the following result hold for each $i \in [0..m]$: (c) $t \succ c_i$ and $t_i \succ c_i$; (d) $P \vdash H(r_i)$. Then there is a delegation path $\xi = (x_0, c_0)(x_1, c_1), \ldots, (x_m, c_m)(x_{m+1}, a)_{ar} \in DP$, where $a = (y, x.rn)$. From (b) and (c), it is clear that $(x_j \in c_i.ac) \wedge (m+1-i \leq c_i.dc) \wedge (ATE(a) \in c_i.tc)$, $i \in [0..m]$, $j \in [i+1..m+1]$. According to def. 5, it is clear that $\xi_i = (x_i, c_i), \ldots, (x_{m-1}, c_m)(x_{m+1}, a)_{ar} \rhd c_i$, $i \in [0..m]$. Therefore ξ is sound, and because $ar \propto ATR(a)$ and $x \ni ar$, (x_m, a) is also sound, i.e., $(z, y, x.rn)$ is a sound statement.

Now we prove the if part (completeness). Given a rule $r \in P$, $H(r) = z.UAS(y, x.rn)$ and $P \vdash H(r)$, $(z, y, x.rn) \in S$ is a *sound statement*. According to def. 3, there exists a sound delegation path $\xi = (x_0, c_0)(x_1, c_1), \ldots, (x_{m-1}, c_{m-1})(x_m, a)_{ar} \in DP$, where $x_0 = x$, $x_m = z$, $ATE(a) = y$, $ar = x.ar(rn)$ and $x_0 \ni ar$, $ar \propto ATR(a)$. Now consider two cases: $m = 0$ and $m \geq 1$. If $m = 0$, then there exists a rule r in P, $H(r) = x.UAS(y, x.rn)$ and $P \vdash H(r)$. By using *m3*, $PS \vdash x.canActivate(y, x.rn)$ follows.

If $m \geq 1$, then there exists a rule r in P, $H(r) = x_m.UAS(y, x.rn)$ and $P \vdash H(r)$; and a chain $\sigma = r_0, r_1, \ldots, r_{m-1}$ in P, where $H(r_i) = x_i.DRA(x_{i+1}, ar, c_i)$ and $P \vdash (r_i)$, $i \in [0..m-1]$. According to def. 3 and def. 5, $(x_j \in c_i.ac) \wedge (m-i \leq c_i.dc) \wedge (ATE(a) \in c_i.tc)$ holds for each $i \in [0..m-1]$ and $j \in [i+1..m]$. Let $t = tr(\{\}, 1, \{y\})$, clearly, $t \succ c_i$ and $t_i \succ c_i$, where $t_i = tr(\{x_{i+1}, \ldots, x_m\}$, $t.dd+(m-1-i)$, $\varnothing)$, $i \in [0..m-1]$. Then according to lemma, $PS \vdash x_0.doa(x_m, ar, t)$. Therefore by using *m4*, $PS \vdash x.canActivate(y, x.rn)$ follows.   ■

The soundness result ensures the safe propagation of privileges when we use delegation to realize decentralized authorization. The completeness result ensures the availability of an authorization system, i.e., the sound statements should be authorized.

4   Related Work

We have briefly reviewed some of the related work. Now we give further comparison of our work with some highly related work. PolicyMaker [4] and KeyNote [5] are systems where the concept of trust management was motivated and evaluated. PolicyMaker allows arbitrary programs to be used in credentials and policies. KeyNote uses a special assertion language to define delegation policies. Both PolicyMaker and KeyNote do not provide mechanisms to control the privilege proliferation during delegation. RT [8] is a family of role-based trust management languages whose semantics are built upon Datalog rules. RT supports boolean control over delegation of role authorities. The role intersections in RT can be viewed as a kind of constraint on the scope of delegation targets.

Cassandra [6] expresses policies in a language based on Datalog^C [9], which bears some similarities to our system. The expressiveness of Cassandra (and its computational complexity) can be tuned by choosing an appropriate constraint domain. The rules in Cassandra can refer to remote policies (for automatic credential retrieval and trust negotiation). However, Cassandra does not embed any delegation control mechanism in its reserved semantics. For example, the integer control on delegation is totally managed by security administrators in Cassandra, which will easily lead to mistakes in security management.

PeerTrust [10] is a TM language that provides trust negotiation capabilities for servers and clients, with facilities to import and reason about access control policies, digital credentials, and metadata about local resources requiring protection. The authors demonstrate how to use PeerTrust to avoid an explicit registration step on the Semantic Web. An implementation of implicit registration and authentication that runs under the Java-based MINERVA Prolog engine was introduced. PeerTrust is closely related to SD3 [11], which is an extension of Datalog, security policies are a set of assumptions and inference rules in SD3. However, both PeerTrust and SD3 lack control on delegation.

B. C. Neumann uses restricted proxy model [1] to support a variety of restrictions on authorization and delegation, including grantee, for-use-by-group, issued-for, quota, authorized, group-membership, accept-once. But the restricted proxy model does not provide restriction specification and semantics computation. Some of these restrictions can be expressed by $REAL^M$. For example, the authorized restriction can be viewed as an access-level constraint on delegation targets [3]. RCDM model can be easily modulated to adapt the delegation at access-level.

The authors have presented REAL05 [3], which can be used to define constrained delegation policies both for management-level and access-level. But the correctness of the deduction algorithm in [3] is not analyzed. In this paper, the soundness and completeness of the semantics are proved. During the analysis work, we find that some of the components in GCDM model [3] could be redefined to be more intelligible and reasonable. We add the component R and S into authorization system AS, and redefine the main components such as delegation path, delegation tree, and the semantics of relation $\triangleright$ (which is denoted by $\Rightarrow$ in REAL05). These work results in a well-defined management-level delegation model named RCDM. The main components in RCDM can be clearly expressed with $REAL^M$. Based on the semantics of REAL05, the semantics of $REAL^M$ defines a deduction algorithm for delegation of role authorities, whose correctness is formally analyzed in this paper.

5 Conclusion

Trust management is an attractive alternative for authorization in future distributed systems. The constraints on delegation will enable TM systems to adapt scenarios where authorization can not totally depend on trust relationships between entities. Existing TM systems support boolean or integer to control re-delegation, some supports width control. However, these systems did not prove that their constraint mechanisms do work correctly, which we think it is necessary for the whole security of decentralized systems. This paper is trying to do such a work, and we find it is very helpful to find the inconsistency between the constrained delegation model and its deduction semantics.

Acknowledgements

The authors would like to thank Hai-ya Gu and Yan-qing Chen for their helpful discussions and the anonymous reviewers for their valuable comments.

References

1. B. C. Neumann, "Proxy-Based Authorization and Accounting for Distributed Systems," in Proceedings of the 13th International Conference on Distributed Computing Systems, Pittsburgh, PA, May 1993.
2. C. M. Ellison, B. Frantz, B. Lampson, R. Rivest, B. M. Thomas, and T. Ylonen. SPKI Certificate Theory. IETF RFC 2693, 1998.
3. Gang Yin, Huaimin Wang, Dianxi Shi, Haiya Gu, "Towards more Controllable and Practical Delegation", Mathematical Methods, Models and Architectures for Computer Networks Security Workshop (MMM-ACNS'05), St. Petersburg, Russia, LNCS 3685, Springer Verlag, 2005.
4. M. Blaze, J. Feigenbaum, and J. Lacy. Decentralized trust management. In Proceedings of 17th Symposium on Security and Privacy, pages 164-173, Oakland, 1996. IEEE.
5. M. Blaze, J. Feigenbaum, John Ioannidis, and Angelos D. Keromytis. The KeyNote trust-management system, version 2. IETF RFC 2704, September 1999.
6. Moritz Y. Becker, Peter Sewell, Cassandra: Flexible Trust Management, Applied to Electronic Health Records Proceedings of the 17th IEEE Computer Security Foundations Workshop (CSFW'04).
7. Ninghui Li, Benjamin N. Grosof, and Joan Feigenbaum. Delegation logic: A logic-based approach to distributed authorization. ACM Transaction on Information and System Security (TISSEC), February 2003.
8. Ninghui Li, John C. Mitchell, and William H. Winsborough. Design of a role-based trust management frame-work. In Proceedings of the 2002 IEEE Symposium on Security and Privacy, pages 114-130. IEEE Computer Society Press, May 2002.
9. Paris C. Kanellakis, Gabriel M. Kuper, and Peter Z. Revesz. Constraint query languages. Journal of Computer and System Sciences, 51(1):26-52, August 1995.
10. Rita Gavriloaie, Wolfgang Nejdl, Daniel Olmedilla, Kent E. Seamons, and MarianneWinslett. No Registration Needed: How to Use Declarative Policies and Negotiation to Access Sensitive Resources on the Semantic Web. The 1st European Semantic Web Symposium, May. 2004, Heraklion, Greece.
11. T. Jim. SD3: A Trust Management System With Certified Evaluation. In IEEE Symposium on Security and Privacy, Oakland, CA, May 2001.
12. Vijay Varadharajan, Philip Allen, Stewart Black. An Analysis of the Proxy Problem in Distributed systems. IEEE Symposium on Research in Security and Privacy. Oakland, CA 1991.

Vector Space Based on Hierarchical Weighting: A Component Ranking Approach to Component Retrieval

Gui Gui and Paul D. Scott

Department of Computer Science, University of Essex, Colchester ,UK
{gqui, scotp}@essex.ac.uk

Abstract. In this paper, we present an approach to software component ranking intended for use in searching for such components on the internet. The method used introduces a novel method of weighting keywords that takes account of where within the structure of a component the keyword is found. This hierarchical weighting scheme is used in two ranking algorithms: one using summed weights, the other using a vector space model. Experimental comparisons with algorithms using TF-IDF weighting that ignore component structure are described. The results demonstrate consistent superiority of the hierarchical weighting approach.

1 Introduction

The internet is rapidly becoming a major repository of software components. In order to realise the potential of this resource, software developers need effective search engines that can retrieve those components that match their requirements. Conventional search engines are unsuitable for this task because they are oriented towards retrieving documents comprised largely of natural language text. Software components, viewed as text objects, differ from natural language documents in that they have a much more rigid structure but a greatly impoverished vocabulary.

The simplest possible approach is to retrieve all components that contain the keywords that appear in the query. Unfortunately, this typically returns a large number of components of which only a small percentage are actually relevant; that is, only a few of the many components retrieved match the user's requirements. A more sophisticated approach, in which the retrieved components are ranked by relevance, is thus essential.

In this paper, we propose a new ranking algorithm that exploits component structure by combining text information ranking techniques with the hierarchical structure analysis of components. In our proposal, keywords are retrieved from naming information at different hierarchical levels within components and used as index words. A hierarchy weighting (HW) scheme is developed to define the relative importance of keywords found at different levels in the structural hierarchy of components. The similarity of components and queries can then be defined using a vector space representation [3], [4] of the set of index keywords. In our work, we focus on Java

J. Cao, W. Nejdl, and M. Xu (Eds.): APPT 2005, LNCS 3756, pp. 184–193, 2005.
© Springer-Verlag Berlin Heidelberg 2005

components. The main contribution of our proposal is that it demonstrates the importance of component structure to deriving an accurate ranking of the relevance of components to queries.

The paper is organized as follows: Section 2 reviews related work and introduces the novel features of our approach; Section 3 describes the proposed hierarchy weighting scheme and the vector space based on hierarchy weighting for component ranking; Section 4 presents some experiments and analysis. The paper concludes with a discussion of the implications of our findings.

2   Related Work

Much of the existing literature on component retrieval techniques is based on their application to component repositories that take the form of well organized, documented and maintained libraries of software components. Such repositories are undoubtedly a valuable resource to a software engineer but, by their very nature, they are highly labour intensive. In contrast, our concern in the present paper is with retrieving components from the internet. Typically little or no documentation is available for such components and any indexing must be done automatically because of the huge numbers of components to be considered.

There are two main approaches to ranking in existing component retrieval systems: ranking by usability and ranking by relevance (Table 1). Ranking by usability, which attempts to identify well engineered components that can readily be incorporated as part of a complex piece of software, is outside the scope of the present paper. Here we are concerned only with ranking by relevance which attempts to identify components that meet the users' functional requirements.

Table 1. Categories of component ranking

Ranking by usability	Ranking by relevance
SourceForge [12]	Domain, facet classification [15], [11]
Based on formal specification [10], [7]	TF-IDF [14]
Based on usage relationship (Spars [8])	Natural language processing [13]
	Concept lattice [9]
	Ontology processing [5], [6]

The various approaches to ranking components by relevance differ chiefly in the complexity of the representations used to describe both components and queries. Semantic based component retrieval [13] uses a natural language representation. Components are described by specific domain models in natural language. Relevant components are retrieved using a closeness measure based on semantic and syntactic analysis. The use of natural language provides a natural and flexible way to express users' queries and describe the context of a component. Unfortunately, this approach presupposes that natural language descriptions of components either exist or can be constructed so it is not really a practical method for retrieving components from the internet where documentation may be sparse or non-existent. Furthermore, the natural language processing involved is likely to make any such system very slow if large numbers of components must be examined.

Some approaches simplify natural language processing by using concept lattices [9] and ontology processing [5], [6]. Ranking by adopting an ontology builds a relationship between domain models and reuse repository. The limitations of this proposal are that it assumes the existence of an ontology and search quality depends on the availability and accuracy of ontology and domain model. Alternatively concept lattices can be used to represent the relationship between keywords and components. As with the natural language approach, these methods presuppose that a great deal of work has been done in constructing an appropriate ontology and categorizing the components appropriately. Hence again, it appears unsuitable for retrieving components from the internet.

Components can be classified into different categories according to domain and facet information [11], [15]. Ranking of components is based on the relationship of domains. However, the granularity of ranking based on domain classification is too big to rank each individual component properly. Some components that can not be classified into any pre-existing categories or the intersections of categories could result in loss of some components. The flexibility of component retrieval can be lost by forced classification of components.

All of the above approaches seem more suited to aid retrieval from well maintained and systematically organized software repositories than for searching the internet. An alternative is to make use of the standard text information retrieval techniques that have proved so effective in searching for ordinary text documents on the internet. Such approaches typically index a document by the keywords it contains and match queries to documents on the basis of the keywords they have in common. The majority of such systems associate a weight with each occurrence of a keyword that provides an estimate of how much evidence the keyword provides about the content of the document.

Washizaki and Fukazawa [14] have ranked components using TF-IDF weighting, in which the weight is proportional to frequency of the keyword in the component and inversely proportional to the number of components in the repository that include that keyword. Although this approach can achieve some success, it is limited because it does not exploit all the information that is readily available from a component. In particular, it only considers keyword frequency and takes no account of the structure of a component; the significance of a keyword may well depend on where it appears within the component. Naming information has proved to be effective in component clustering [1], [2].

Consequently it appears worthwhile to develop a weighting scheme that reflects not only the frequency of a keyword but also its location within the structure of a software component.

3 Hierarchical Weighting and Vector Space Representation

The work reported in this paper is concerned with retrieving Java components. The decision to concentrate on Java was taken for two main reasons. First, very large numbers of Java components, often taking the form of Java Beans, are available in the public domain on the internet. Second, the Java facilities of reflection and introspection permit the exploration of the structure of such components.

Programmers often provide clues to the function of a component in the names they give to the various entities of which it is composed. These include the names given to the classes and methods and the name of the file containing the component. Hence index keywords can be obtained by searching for legitimate words within the collection of names used by the programmer. Some of these keywords will provide more information than others about the function of the component. Thus, in addition to extracting the keywords, it is desirable to associate a weight with a each keyword, in a each component, that reflects how much evidence it provides about component functionality. The main hypothesis of this paper is that the location of a keyword within the structure of a component provides a strong indication of how much functional evidence it provides.

Let $C = \{c_1, c_2, \ldots c_i, \ldots c_n\}$ be the set of all the components in the entire collection where n is the total number of components. $K = \{k_1, k_2, \ldots k_i, \ldots k_m\}$ is the set of index keywords extracted from the entire collection of components. A weight W_{ij} is assigned to each index keyword k_i in the component c_j in order to measure the importance of k_i in expressing the content of component c_j. If keyword k_i appears in the component c_j, $W_{ij} > 0$; if not, $W_{ij} = 0$.

Java components are normally compressed as jar files that can include several classes, each of which contains several methods. Keywords are extracted from the file names, class names and method names. Set $k^{jar} = (k_1^{jar}, k_2^{jar}, \ldots, k_N^{jar})$ comprises the keywords retrieved from the jar file name, set $k^{c_i} = (k_1^{c_i}, k_2^{c_i}, \ldots k_r^{c_i})$ comprises the keywords retrieved from the name of class c_i, and set $k^{f_i} = (k_1^{f_i}, k_2^{f_i}, \ldots k_s^{f_i})$ contains the keywords retrieved from name of method f_i. The relationship of keywords from different levels is shown in Fig. 1.

Thus, the complete set of keywords for a given component is:

$$K = (k_1^{jar}, k_2^{jar}, \ldots k_N^{jar}) \cup (\bigcup_{i=1}^{n} (k_1^{c_i}, k_2^{c_i}, \ldots k_{r_i}^{c_i})) \cup (\bigcup_{j=1}^{m} (k_1^{f_j}, k_2^{f_j}, \ldots k_{s_j}^{f_j})) \tag{1}$$

Keywords from higher levels of the hierarchy are more likely to convey information about the overall function of the component. Hence it is appropriate to give greater weight to keywords from higher levels in the hierarchy. Thus, $W_{ij}^{jar} > W_{ij}^{class} > W_{ij}^{method}$

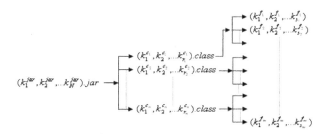

Fig. 1. Keywords hierarchical level breakdown

where W_{ij}^{lev} denotes the weight assigned to the ith keyword at level lev in the jth component. The number of retrieved keywords usually increases as one moves down the hierarchy. Typically there is only one jar file which will contain several classes, each of which contains several methods. Hence a simple way of assigning greater weight to keywords retrieved from higher levels is:

$$W_{ij}^{lev} = \frac{f_{ij}}{\psi^{lev}} \qquad (2)$$

where f_{ij} is the frequency of occurrences of keyword k_i and ψ^{lev} is the number of entities at hierarchy level lev in the jth component. It is clear that, provided a jar file contains more than one class and each class contains more than one method, then if a given keyword occurs with the same frequency at all levels of the hierarchy $W_{ij}^{jar} >$ $W_{ij}^{class} > W_{ij}^{method}$. The total weight assigned to a give keyword i in a particular component j is thus:

$$W_{ij} = W_{ij}^{jar} + W_{ij}^{class} + W_{ij}^{method} = f_{ij}^{jar} + \frac{f_{ij}^{class}}{\psi^{class}} + \frac{f_{ij}^{method}}{\psi^{method}} \qquad (3)$$

Hence the frequency of keywords and their location within the structure of a component are combined to estimate their importance in describing the function of that component.

These hierarchical keyword weights can be used in two distinct ways to match queries to components. The first, termed HW, is simply to give each component a similarity score formed as the sum of the hierarchical weights of all words appearing in the query. A more sophisticated alternative, termed VS-HW, is to adopt a vector space representation [3]. Each component is represented as a vector as $\vec{c}_j = (W_{1j}, W_{2j}, ..., W_{mj})$ in the space of all index terms. A query is represented as a vector $\vec{q} = (W_{1q}, W_{2q}, ..., W_{mq})$ where W_{iq} denotes the importance of the ith keyword to the query q (always either 0 or 1 in the present study). The similarity of component and the query, $Sim(q, c_j)$, is then defined as:

$$Sim(q, c_j) = \frac{\sum_{i=1}^{t} W_{ij} \cdot W_{iq}}{\sqrt{\sum_{i=1}^{m}(W_{ij})^2} \sqrt{\sum_{i=1}^{t}(W_{iq})^2}} = \frac{\sum_{i=1}^{t}(f_{ij}^{jar} + \frac{f_{ij}^{class}}{\psi^{class}} + \frac{f_{ij}^{method}}{\psi^{method}})}{\sqrt{\sum_{i=1}^{m}(f_{ij}^{jar} + \frac{f_{ij}^{class}}{\psi^{class}} + \frac{f_{ij}^{method}}{\psi^{method}})^2} \sqrt{\sum_{i=1}^{t}(W_{iq})^2}} \qquad (4)$$

4 Experimental Results

In order to investigate the efficacy of this hierarchical weighting scheme for reliably ranking components, approximately 10,000 components were retrieved from the internet using a randomly seeded spider program. Keywords were then identified as fragments of filenames, class names and method names using a dictionary based approach. A total of 21,778 such keywords were found. In addition to the two ranking schemes based on hierarchical weightings, comparative experiments were also conducted using the TF-IDF weighting scheme [3] which is widely used in conventional text retrieval. It differs from our hierarchical weighting scheme chiefly in that in that it assigns the same weight to a keyword wherever it appears within the structure of a component. Both direct (TF-IDF) and vector space (VS-TF-IDF) variants of TF-IDF ranking were implemented.

Searches were carried out using 50 different queries; 20 were single word queries, the remainder comprised either two or three keywords. The total number of components retrieved (i.e. achieving a score greater than zero) for each query varied but fell in the range 140-180 with an average of 158. The components returned were categorized as relevant or irrelevant to the query by inspection; the average number of relevant results was 67.

Three methods were used to compare the performance of the four ranking schemes: average cumulative relevant results distribution, precision distribution and precision histograms [3].

4.1 Average Cumulative Relevant Results Distribution

The cumulative relevant results distribution graph is a plot of the number of relevant results retrieved in the highest ranked x components as x is increased from 1 to T, the total number of components retrieved. An ideal ranking scheme would rank all the relevant components before all those that are irrelevant. Hence the ideal distribution takes the form of a straight line from 0,0 to R,R, where R is the number of relevant components, followed by a second (flat) straight line from R,R to T,R. The quality of a ranking scheme is indicated by how closely it approaches this ideal. Completely random ranking would result in a single straight line from 0,0 to T,R.

Fig. 2 shows the average cumulative relevant results distribution for all four ranking schemes over the 50 queries. It can be seen that all four algorithms generally perform much better than random ranking, although the TF-IDF results are indistinguishable from random ranking in the 20 highest ranked components. None achieves ideal

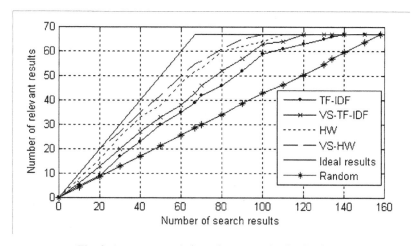

Fig. 2. Average cumulative relevant results distributions

performance, but it is clear that VS-HW performs better than the other three procedures. The performance of HW is only slightly worse. The rankings achieved by the two TF-IDF procedures are markedly worse.

4.2 Distribution of Mean Precision

The precision of a retrieval procedure is simply defined as the proportion of all retrieved items that are relevant. The rth-precision is the precision achieved for the r highest ranked retrieved items. The mean value of rth-precision over all values of r provides another way of comparing the rankings achieved by different ranking procedures, since a higher mean value indicates that more of the highly ranked items were relevant.

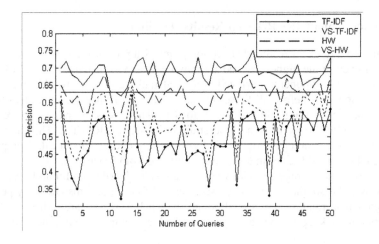

Fig. 3. Mean rth-precision distributions

Fig. 3 shows the mean rth-precision values achieved by the four ranking procedures for each of the 50 queries. To provide a baseline indication of performance across all the queries, the horizontal lines indicate the average mean rth-precision achieved by each procedure over all queries. That is, each line indicates the value of:

$$\overline{MeanRP}(A) = \frac{\sum_{i=0}^{N} \overline{RP}(A \mid i)}{N} \tag{5}$$

where $\overline{RP}(A \mid i)$ is the mean rth-precision achieved by ranking algorithm A on the ith query and N is the total number of queries.

It is clear that, as in the cumulative relevant results distributions, the two algorithms using hierarchical weighting perform considerably better than the two based on TF-IDF. Not only are their overall performances higher, as indicated by the baselines, but they are also more consistent since the deviations from the baseline are much smaller. Once again, the best results are obtained using VS-HW and the worst using TF-IDF.

4.3 Precision Histograms

Precision histograms [3] allow comparisons between two ranking algorithms by considering the differences between their rth-precision values at a specific value of r: the total number of relevant results for the current query. Consider two ranking algorithms, A and B, and let M be the total number of relevant results. Let $MP_A(i)$ and $MP_B(i)$ be the M-th precision values achieved by algorithm A and B for query i. Hence the difference between these two value $MP_{A/B}(i)$, is:

$$MP_{A/B}(i) = MP_A(i) - MP_B(i) \qquad (6)$$

If $MP_{A/B}(i)$ is zero, the two algorithms have equivalent performance for that query. Positive values of $MP_{A/B}(i)$ imply that algorithm A performs better than algorithm B while a negative values of imply B has better performance.

Figure 4 shows the histograms achieved when each of the hierarchical weighting procedures are compared with each of the TF-IDF procedures on all 50 queries.   The results are consistently positive demonstrating that the hierarchical weighting procedures perform better on the entire test set.

A quantitative measure of the difference between two algorithms across a set of queries can be obtained by considering the mean value of $MP_{A/B}(i)$:

$$\overline{MP}_{A/B} = \frac{\sum_{i=1}^{N}(MP_A(i) - MP_B(i))}{N} \qquad (7)$$

Computing these mean differences for the four histograms demonstrates again that the performance of the two procedures using hierarchical weighting is superior to both of those using TF-IDF weighting (See Table 2).

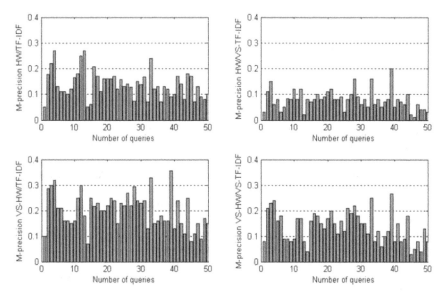

Fig. 4. M-th Precision Histograms

Table 2. Mean Mth-precision differences

Procedures Compared	Mean Mth-precision difference
HW/TF-IDF	0.138
HW/VS-TF-IDF	0.078
VS-HW/TF-IDF	0.20
VS-HW/VS-TF-IDF	0.136

5 Conclusion

In this paper, we have proposed a new approach to software component ranking that exploits the hierarchical structure of components to estimate how much information a keyword, extracted from an entity's name, provides about the overall function of the component. A simple scheme has been proposed, in which the weight associated with the occurrence of a keyword within an entity's name is inversely proportional to the number of entities of the same type in the component. This hierarchical weighting scheme has been incorporated into two ranking algorithms: one which simply uses the summed weight of all keyword and one which uses a vector space representation. The performance of these algorithms has been assessed by comparing them to their counterparts that use TF-IDF weighting and take no account of the component structure. Our results demonstrate the consistent superiority of the hierarchical weighting algorithms. The vector space algorithm performed slightly better than the simple summed weight method.

The work reported here is part of a larger project to develop a complete system for retrieving software components from the internet and it is clearly to this field that it makes the most direct contribution. However, we suggest that our findings may be of wider interest. In particular, the use of hierarchical weighting may be of similar value in information retrieval applications that involve any type of semi-structured text documents.

References

1. Andritsos, P., Tzerpos, V.: Information-Theoretic Software Clustering. IEEE Transactions on Software Engineering. Vol. 31, Issue 2 (2005) 150 - 165
2. Anquetil, N., Lethbridge, T.C.: "Recovering Software Architecture from the Names of Source Files," J. Software Maintenance: Research and Practice, vol. 11, pp. 201-221, May 1999.
3. Baeza, R., Neto, B.: Modern Information Retrieval. ACM Press, Addison Wesley, New York (1999)
4. Belew, R. K.: Finding Out About A Cognitive Perspective on Search Engine Technology and the WWW. Cambridge University Press (2000)
5. Bernstein, A., Klein, M.: Towards High-Precision Service Retrieval. Proc. International Semantic Web Conference (ISWC-02), Sardinia, Italy (2002)
6. Braga R.M.M., Mattoso, M., Werner, C.M.L.: The use of mediation and ontology technologies for software component information retrieval. Symposium on Software Reuse (SSR'01), Toronto, Canada, (2001)

7. Fischer, B.: Specification-Based Browsing of Software Component Libraries. Journal of Automated Software Engineering, Vol. 7, No. 2, (2000) 179-200

8. Inoue, K., Yokomori, R., Fujiwara, H., Yamamoto, T., Matsushita, M., Kusumoto, S.: Component Rank: Relative Significance Rank for Software Component Search. In: Proc. International Conf. on Software Engineering (ICSE2003), Portland, OR, (2003) 14–24

9. Lindig, C.: Concept-based component retrieval. In: Working Notes of the ZJCAI-95 Workshop: Formal Approaches to the Reuse of Plans, Proofs, and Programs. Kohler, J., Giunchiglia, F., Green, C., Walther, C. (eds) (1995) 21-25

10. Penix, J., Alexander, P.: Efficient Specification-Based Component Retrieval. Automated Software Engineering, Vol. 6. Kluwer Academic Publishers (1999) 139-170

11. Seacord, R., Hissan, S., Wallnau, K.: Agora: A Search Engine for Software Components. IEEE Internet Computing, Vol.2, No.6 (1998)

12. Sourceforge.: Sourceforge.net. http://sourceforge.net/. Accessed August 4th (2005)

13. Sugumaran, V., Storey, V.C.: A Semantic-Based Approach to Component Retrieval. The DATA BASE for Advances in Information Systems. Vol. 34, No. 3. ACM SIG Management Information Systems (2003) 8-24

14. Washizaki, H., Fukazawa, Y.: Component-Extraction-based Search System for Object Oriented Programs. Proc. 8th International Conference on Software Reuse, Lecture Notes in Computer Science, Vol. 3107, Springer-Verlag, Berlin Heidelberg New York (2004)

15. Zhang, Z., Svensson, L., Snis, U., Srensen, C., Fgerlind, H., Lindroth, T., Magnusson, M., Stlund, C.: Enhancing Component Reuse Using Search Techniques. Proceedings of IRIS 23. Laboratorium for Interaction Technology, University of Trollhttan Uddevalla (2000)

A High Availability Mechanism
for Parallel File System[*]

Hu Zhang[1], Weiguo Wu[1], Xiaoshe Dong[1], and Depei Qian[1,2]

[1] Department of Computer Science, Xi'an Jiaotong Univ.,
Xi'an, Shaanxi, China 710049
zhanghu@mailst.xjtu.edu.cn
[2] School of Computer Science, Beihang Univ., Beijing, China 100083

Abstract. Parallel file systems achieve a high I/O throughput by dividing a file into multiple blocks and storing them on multiple I/O nodes. However, the reliability and availability of the parallel file systems are sacrificed for the stripping of file data over multi I/O nodes. A new mechanism named Logic Mirror Ring (LMR), has been developed to improve the reliability and availability of the parallel file systems in this study. A logic mirror ring is built over all I/O nodes to indicate the mirror relationship among the nodes, i.e., each node maintains not only its own data but also the mirror data of other nodes. The fault tolerant capability of the system is improved because the node maintaining the mirror data of the failed node will take over the requests to the failed node. The mirror depth can be adjusted to different levels based on the requirements of the reliability and availability. A model is developed to evaluate the reliability and availability of the parallel file systems. The effects of LMR on the reliability and availability of the parallel file system is studied. The results show that LMR can be used to improve the reliability and availability of the parallel file systems effectively.

1   Introduction

Parallel file system is widely used in clusters dedicating to I/O-intensive parallel applications. As a common way, the independent storage devices attached to I/O nodes are connected together as a whole single storage space via parallel file system. The file data is divided into stripes or blocks and stored in multi I/O nodes, and the meta data server is used to provide a single name space and directory hierarchy. Consequently, the cluster that employs parallel file system can get high I/O performance and scalability if the network of cluster can provides enough bandwidth.However, the parallel file system's reliability is sacrificed for stripping of file data over multi I/O nodes. For example, there is a parallel file system with N identical I/O nodes, assuming that the Mean Time To Failure ($MTTF$) of the I/O nodes is H hours and all other components of the cluster, such as network, are fault-free, then, the $MTTF$ of whole parallel

---

[*] This research is supported by National 863 Plan under grant No.2004AA111110 and 2002AA104550.

J. Cao, W. Nejdl, and M. Xu (Eds.): APPT 2005, LNCS 3756, pp. 194–203, 2005.
© Springer-Verlag Berlin Heidelberg 2005

file system is H/N hours. It means that if we want to achieve higher I/O performance by adding more I/O nodes, the whole file system subjects to failure more easily. Therefore, the way to improve the reliability and availability of parallel file system becomes an important issue.

In this paper, a new mechanism, named Logic Mirror Ring (LMR), is developed. The nature of LMR is to mirror data to enhance the reliability of the parallel file system. Its feature is to construct the mirroring and failover relationship between current I/O nodes, but not to require any special nodes to store backup data, and not to change the topology of the original parallel file system. Each piece of data is replicated and stored in multi nodes, and the different copies of data would be synchronized to be identical. As another important feature of LMR, the mirror depth of LMR can be adjusted to meet different requirements of reliability of parallel file system according to the applications. A model is also developed to evaluate the reliability and availability of the parallel file system employing LMR. Through the model, Effects of LMRs are studied herein at various nodes number, mirror depth, and other factors. The result shows that: LMR can improve the reliability of parallel file system with a factor of 32 when the mirror depth is 2 with a tradeoff of write performance degrading to 50% and double storage space. Furthermore, LMR is able to make parallel file system tolerate any single point failure of I/O nodes when the mirror depth is larger than 1.

2   The Mechanism Based on *LMR*

2.1   Model of Parallel File System

Modern distributed storage systems have a feature that the metadata and data is divided into two individual parts to be stored and managed [8]. For a parallel file system, there must have one or more metadata servers to manage and store the file's metadata, and multi data servers to maintain the file data [3,5,6,7]. Clients are the applications that access the parallel file system via the APIs, which are provided by the parallel file system. These three entities are connected through networks. Fig. 1 is a typical architecture of parallel file system.The research of this paper are based on such parallel file system model.

2.2   Logic Mirror Ring (*LMR*)

Definition 1. Logic Mirror Ring (LMR) *is a virtual directed circle that indicates the mirror relationship of the data servers in the parallel file system.* LMR *is constructed with multi nodes connected as a loop, and each node of the loop represents a data server of the parallel file system.*

For example, there is a parallel file system with n nodes (data servers), and these data servers can be marked with $S_0, S_1, \cdots, S_n$. A LMR can be constructed in any orders with all of these nodes, and each node must appear in the LMR exactly once. In this paper, the LMR $R_0 = \{S_0, S_1, \cdots, S_n\}$ is selected to facilitate description of mechnism.

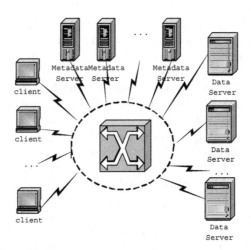

Fig. 1. Architecture of Parallel File System

Definition 2. Mirror depth *is a numeric number m that indicates how many replicas of each piece of data in the parallel file system. It implies that each piece of data have m copies residing in different nodes. For a parallel file system with n nodes, mirror depth will be in the range of $0 < m \leq n$.*

Definition 3. *The concept of* **Adjacent Distance** *relate to the concept of LMR. Given a LMR R and two nodes S_k and S_l, defined the positive direction as the direction of LMR, if there are $i - 1$ nodes between S_k and S_l, we define the adjacent distance from S_k to S_l as i,therefore, the adjacent distance from S_l to S_k is $-i$ for opposite direction.*

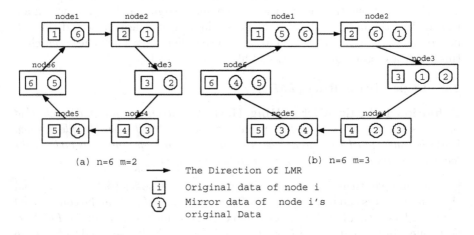

Fig. 2. Data layout of a parallel file system employing *LMR*

In a parallel file system with LMR R_0 , when the mirror depth is set to m, then for any node S_i in the R_0, The data that is originally designated to store in node S_i should be mirrored to the nodes which adjacent distance to S_i are less than m. For example, if the mirror depth is 3, the original data of each node should be mirrored to the next node and next of the next node, in other words, for any node S_k in R_0, there are three share of data stored in S_k , one is the original data of S_k, one is the original data of $S_{(n+k-1) \ mod \ n}$, and the other one is the original data of $S_{(n+k-2) \ mod \ n}$. Fig. 2(a) and Fig. 2(b) illustrate the data layout of a parallel file system which employs LMR. The LMR have 6 nodes, and its mirror depth is set to 2 and 3 respectively. As shown in Fig. 2, a data server of parallel file system employing LMR maintains not only its original data but also the replicas of other data servers.

2.3   Adjacent Replication

Definition 4. *The* **Adjacent Replication** *is a mechanism used to synchronize the original data and its replicas between original node and its backup nodes. During the runtime of a parallel file system, every modification of original data should be synchronized to the backup nodes instantly through adjacent replication. /bf Adjacent Replication is a mechanism which used to synchronize the original data and its replica between original node and its backup nodes.*

2.4   Theory of Fault Tolerant

For a parallel file system with LMR R_0 and mirror depth of m, every node will response the requests of accessing original data of that node, and all replicas located in the other nodes will be synchronized with the original data through adjacent replication when all nodes are in normal state. If a node fails, assuming node S_k, its next node S_{k+1} (its adjacent distance from is 1) will takeover all the request to the original data of S_k, and the other backup data of S_k will be synchronized with the backup data of S_k which reside in S_{k+1}. As a result, node deals with the original requests both to S_k and S_{k+1}. If the node S_{k+1} also fails at that moment, then node S_{k+2} will takeover the requests both to S_k and S_{k+1}.

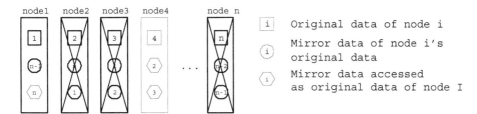

Note: *   The Mirror Depth of the illustrated system is 3.
         *   The big cross over node means that the node is failed.

Fig. 3. The fault-tolerant theory of LMR

In this way, the node that is next to a failed node will takeover the requests to all consecutive failure nodes just in front of it. And the whole parallel file system works until the number of failed node reaches the mirror depth (Fig. 3 shows how the mechanism works). Therefore, in the extreme situation that the parallel file system still work when some nodes fail. There must have a node which responds all requests to m consecutive failure nodes. For example, when the mirror depth is m, if nodes S_k to S_{k+m-2} are all down, then node S_{k+m-1} will respond the requests sent to node $S_k, S_{k+1}, \cdots, S_{k+m-1}$ (to facilitate describe, $k+m-1 < n$).

The following theorem can be concluded from the description above.

Theorem 1. *For a parallel file system with* LMR *and mirror depth of* m, *it would not loss any data when the number of consecutive failed nodes is less than* $(m-1)$.

3   Analysis of Reliability and Availability

In this section, a Markov-Chain model is developed to analyze the reliability and availability of the parallel system with *LMR*. The following assumptions are made to make the model easy to be analyzed:

a. The node changes its status instantly, which imply that no different nodes change status simultaneously. This assumption eliminates some middle statuses. These middle statues have a negligible impact on the results of reliability analysis;
b. The probabilities to failure and to recovery of different nodes are independent, and all the nodes have same failure rate and recovery rate;
c. The replicas in different nodes are consistent in the moment of a node fails.

3.1   Reliability Evaluation

The reliability can be evaluated by Mean Time To Data Loss ($MTTDL$) of the system. $MTTDL$ can be derived from a Markov Chain model. Fig. 4 shows the

Table 1. Notations used in this section

Symbol	Description	Symbol	Description
$MTTDL$	Mean Time To Data Loss	n	Numbers of I/O nodes.
μ_m	Failure rate of node when Mirror Depth is m, and $\mu_m = 1/MTTF_m$.	m	Mirror Depth
γ_m	Repair rate of node when Mirror Depth is m, and $\gamma_m = 1/MTTR_m$.	s	Number of status in Markov model
$MTTF_m$	Mean Time To Failure when Mirror Depth is m	A	Availability of whole system
$MTTR_m$	Mean Time To Repair when Mirror Depth is m		

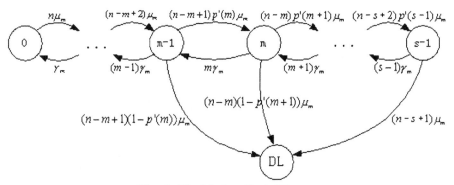

Fig. 4. The Markov State Diagram

Markov state diagram of the parallel file system [4,9,11], which have n nodes forming a LMR and the mirror depth is set to m.

In Fig. 4, states marked with the number i represent the situation that there is no data loss while i nodes fail. According to Theorem. 1, one node can takeover all the data services of m consecutive nodes. Therefore, the maximal state note $(s-1)$ satisfies (1).

$$s - 1 = n - \lceil \frac{n}{m} \rceil \tag{1}$$

$p'(i)$ is defined as a function which presents the probability of the system being active when one more node fails in state $(i-1)$. For the states $i < m-1$, failure of any one more node will not be result in data loss. But when system is in the states of $i \le m-1$, one more node failure may lead to data loss. Therefore, the $p'(i)$ can be expressed as:

$$p'(i) = p(i|i-1) = \frac{p(i \cap i-1)}{p(i-1)} = \begin{cases} 1 & i < m \\ \frac{p(i)}{p(i-1)} & i \le m \end{cases} \tag{2}$$

The $p(i)$ in (2) represents the probability of being active while the system has i failure nodes. It seems that $p(i)$ can't be calculate from given n, m and i with formula methods. Therefore, we developed a program to calculate $p(i)$, and then get every $p'(i)$. Finally, with the given μ_m and γ_m, all transfer probability between states of Markov-Chain diagram can be figured out, and then the $MTTDL$ of system can be calculated from the Markov-Chain model [9].

Fig. 5 illustrates the reliability of the system with different n and m. As shown in Fig. 5, this mechanism improves the system reliability dramatically. According to the calculated result of $MTTDL$, e.g. the mechanism improves the reliability by a factor of around 32 when the Mirror Depth is 2. Fig. 5(a) and Fig. 5(b) is a comparison of the system with different value of and $MTTF_1$ and $MTTR_1$ of its node. It can be seen that $MTTF_1$ and $MTTR_1$ have a strong impact on the reliability of the whole system.

The failure rate μ_m and the repair rate γ_m of a node should have different value at different Mirror Depth m. If μ_1 and γ_1 represent the failure rate and repair rate of a single node at $m = 1$, and current system has a mirror depth

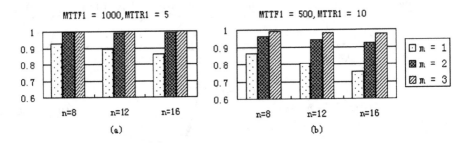

Fig. 5. Compare of MTTDL of system with different setup

of m, then, in general, one node responds the same quantity of read requests and m times write requests as that of a node in the original system (A parallel file system is called original system when its mirror depth is 1), Assuming the failure rate of a single node depends on the hardware and software which build the node instead of the number of the requests arrived it. Therefore, the failure rate of a single node is(The mirror depth is m):

$$\mu_m = \mu_1 \tag{3}$$

If a node fails, the time to re-synchronize all m portions of mirror data in the node is the most time-consuming process in the whole repair process. It should invoke m times synchronization operations than the original system. Therefore, the repair time of one node is:

$$MTTR_m = m \cdot MTTR_1 \tag{4}$$

According to the definition of repair rate and (4):

$$\gamma_m = \frac{1}{MTTR_m} = \frac{1}{m \cdot MTTR_1} = \frac{\gamma_1}{m} \tag{5}$$

3.2 Availability Analysis

A parallel file system with LMR still work until meets a data loss, the Mean Time To Failure ($MTTF$) of it can be considered identical with its $MTTDL$. Furthermore, when sysytem down, there must have m consecutive nodes fail, and the m consecutive nodes must be repaired sequentially, then the $MTTR$ of whole system is:

$$MTTR = m \cdot MTTR_m \tag{6}$$

The following equation can be obtained from (4) and (6):

$$MTTR = m^2 \cdot MTTR_1 \tag{7}$$

From the definition of availability [13] and (7), the availability of the parallel file system can be expressed as:

$$A = \frac{MTTF}{MTTF + MTTR} = \frac{MTTDL}{MTTDL + m^2 MTTR_1} \tag{8}$$

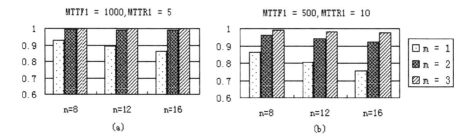

Fig. 6. Availability of system with different setup

Fig. 6 illustrates the availability of the system when it is in different configurations which are shown in Fig. 5. It is obvious that the new mechanism improves the system availability significantly when the Mirror Depth set to 2 and 3.

When constructing a parallel file system with node number n, given the $MTTF_1$ and $MTTR_1$ of single node and the requirement of availability (A) of who file system, the $MTTDL$ can be derived from (8). Therefore, an appropriate Mirror Depth which meets the requirement of availability can be obtained in a numerical method. And the method is derived from the process of calculating the $MTTDL$.

4   Related Works

Many companies and research organizations dedicate to improve the reliability and availability of parallel file system. And many parallel file systems with different availability mechanism are developed, such as GPFS, Lustre, CEFT-PVFS, etc.

GPFS [3,4] is a parallel shared-disk file system that designed by IBM's Almaden Research Center, and it has been wildly used as a mature product. In GPFS, disks and computer nodes are connected through a dedicated storage network to achieve high I/O throughput, the file data is divided into stripe and stored in multi disks to make it possible to be accessed concurrently. The GPFS emploies dual-attached RAID controller and file level duplication to tolerate disk failures. Furthermore, GPFS can tolerate the partial failure of network, the data related to the network will be insolated from other parts of the file system, and the data relating to the network would not be accessed, but the whole file system still work. In contrast with GPFS, *LMR* doesn't need dedicated devices to guarantee high reliability and availability of parallel file system.

Lustre [5,6] is a SAN file system built with three components: clients, Cluster control system, and Storage Target. SAN connects the three components together. The Cluster control system maintain the name space and file system meta-data coherence, cluster recovery etc. the Storage Target are OSDs (Objected Storage Device), which is programmable, the OSDs can be configured as RAID0, RAID1 or RAID5 to enhance the system's ability of fault-tolerate. Just like GPFS, Lustre isolate the failure parts of system, and maintain the other parts available when some nodes fail. And the recovery operation will cure the system soon. Comparing with *LMR*, Luster need dedicated device like SAN and OSD.

PVFS [7] is an open source parallel file system developed by Clemson University. PVFS has three components: clients, I/O server, meta-data server. File data are declustered into stripes and stored in I/O servers in a Round-Robin style. The meta-data server manages the name space and file meta-data. PVFS can provide high aggregate I/O bandwidth and high scalability. But there is no special mechanism employed to improve the reliability of PVFS. Several research projects have been done to enhance PVFS's reliability. Such as, CEFT-PVFS project. CEFT-PVFS [4] (Cost-Effective and Fault-Tolerate PVFS) implement a soft RAID10 over PVFS, it combines the origin system and another identical system together, and the two system mirror mutually, when one node fail, client can redirect the requests of that node to its mirror node. Comparing with *LMR*, CEFT-PVFS have a little higher reliability than *LMR* when Mirror Depth set to 2, but, CEFT-PVFS change the topology of original system and has a worse performance in read operations.

5   Conclusion

In this article, a new mechanism named *LMR* has been developed to improve the reliability and availability of parallel file system. *LMR* constructs the mirroring and failover relationship between current I/O nodes, and doesn't change the topology of the original system. The effects of *LMR* on the reliability and availability of systems with different node number n and Mirror depth m are studied. The result shows that *LMR* can satisfy the different requirements of availability and reliability by adjusting system mirror depth. Furthermore, the write performance is reduced to $1/m$ of original system, but the read performance is not decreased. The statistics of file operation shows that the read operations takes much larger portion than the write operations. [14,15], which indicates *LMR* is effective in the parallel file system.

References

1. J. Wu, P. Wyckoff, and D. Panda, "PVFS over InfiniBand: Design and Performance Evaluation", The International Conference on Parallel Processing (ICPP-03) , Taiwan 2003
2. IA64 Cluster Document
   (www.hlrs.de/hw-access/platforms/zx6000/user_oc.pdf) 2003
3. F. Schmuck and R. Haskin, "GPFS: A Shared-Disk File System," in Proceedings of the Conference on File and Storage Technologies (FAST'02) Monterey, CA 2002
4. Y. Zhu, H. Hong, X. Xin, D. Feng and D. R. Swanson, "Design, Implementation and Performance Evaluation of A Cost-Effective Fault-Tolerant Parallel Virtual File System", The International Workshop on Storage Network Architecture and Parallel I/O, New Orleans, LA, 2003

5. O. Rodeh, A. Teperman. "zFs - A Scalable Distributed File System Using Object Disks", in Proceedings of the 20th IEEE/11th NASA Goddard Conference on Mass Storage Systems and Technologies(MSS'03) San Diego, California, 2003
6. P. J. Braam, "The Lustre Storage Architecture", (http://www.clusterfs.com) 2004
7. P. H. Carns, W. B. Ligon III, R. B. Ross, and R. Thakur, "PVFS: A Parallel File System For Linux Clusters", (http://www.parl.clemson.edu/pvfs/papers.html) 2000.
8. S. A. Brandt, E. L. Miller, D. E. Long, L. Xue, "Efficient Metadata Management in Large Distributed Storage System", in Proceedings of the 20th IEEE/11th NASA Goddard Conference on Mass Storage Systems and Technologies(MSS'03) San Diego, California, 2003
9. Sung Hoon Baek, Bong Wan Kim, Eui Joung Joung, and Chong Won Park, "Reliability and performance of hierarchical RAID with multiple controllers", in Proceedings of the 20th annual ACM symposium on Principles of Distributed Computing, 2001
10. Q. Xin E. L. Miller, T. Schwarz, D. E. Long, "Reliability Mechanisms for Very Large Storage System", in Proceedings of the 20th IEEE/11th NASA Goddard Conference on Mass Storage Systems and Technologies(MSS'03) San Diego, California, 2003
11. David McDysan, "QoS & Traffic Management in IP & ATM Networks", TsingHua University Press, 2000, pages 153-164.
12. Li Yuya, "The mathematic of Reliability", Huazhong Univ of Science and Technology Press,1990
13. Chris Oggerino, "The Fundamental of High Availability", China Electrical Power Press, 2002
14. Mary Baker, Ohn Hartman, Michael Kupfer, Ken Shirriff, and John Ousterhout, "Measurements of a Distributed File System", in Proceedings of the 13th SOSP, October 1991 15.
15. Nils Nieuwejaar, David Kotz, Apratim Purakayastha, Carla Schlatter Ellis, and Michael L. Best, "File-access characteristics of parallel scientific workloads" IEEE Transaction on Parallel and Distributed Systems, Vol. 7. No. 10. 1996

A User-Guided Semi-automatic Parallelization Method and Its Implementation

Chuliang Weng, Zhongguo Chen, Xinda Lu, Minglu Li, and Yong Yin

Department of Computer Science and Engineering,
Shanghai Jiao Tong University, Shanghai 200030, China
weng-cl@cs.sjtu.edu.cn

Abstract. In this paper, we propose a user-guided semi-automatic parallelization method, which is based on code templates corresponding to parallel programming paradigms and the concept of meta-task independent with each other. As an implementation of this method, we develop the system *Metaparallel*, which is based on Java language and MPICH, and the framework of *Metaparallel* is discussed. At last, the parallelization flow is studied with a case. In addition, we test the usability of *Metaparallel* by the practical engineering problem.

1 Introduction

Parallel programming could be classified into two kinds: implicit parallelism and explicit parallelism. Implicit parallelism means the programmer does not explicitly specify parallelism, but lets the compiler and the run-time support system automatically exploit it [1]. Usually the automatic parallelization of sequential programs is the approach of implicit parallelism. This is undertaken by compiler which analyzes the dependence of the sequential program and converts the sequential program into the corresponding parallel program. Whereas, when the programmer explicitly specifies parallelism in programs, it is called as explicit parallel programming. Explicit parallelism includes three parallel programming models: shared-variable, data-parallel and message-passing.

Implicit parallelism or automatic parallelism is expected to parallelize the sequential codes into the parallel program codes without additional efforts of the sequential programmers. However, the parallel performance of automatic parallelism is poor and much more parallelism of sequential codes can not be exploited automatically [2][3]. On the other hand, with the explicit parallelism model, programmers should be versed in the parallel programming knowledge such as parallel architectures and parallel programming models in order to develop efficient parallel programs.

In this paper, we present and implement a user-guided semi-automatic parallelization mechanism, in which sequential codes can be parallelized automatically with the user's guidance. This method integrates the accessibility of implicit parallelism and the efficiency of explicit parallelism, and is helpful for programmers in the engineering domain to develop parallel programs without the redundant burden.

J. Cao, W. Nejdl, and M. Xu (Eds.): APPT 2005, LNCS 3756, pp. 204–213, 2005.
© Springer-Verlag Berlin Heidelberg 2005

2   Related Works

The message-passing model is one kind of explicit parallelism, and PVM (Parallel Virtual Machine) and MPI (Message Passing Interface) are two important libraries of the message-passing model. Currently, MPI is the standard of the message-passing program, and the message-passing model and MPI had been adopted widely in scientific computing research and industry domains.

There are many research efforts on automatic parallelization of programs for a long time. One kind of approaches is using parallel compilers for automatic parallelization [4][5][6][7]. The automatic parallelizing approach cannot generate parallel codes as good as hand coding for many applications, because the complexities of practical applications cannot be manipulated effectively. Another important approach is the interactive parallelization, which includes works [8][9][10]. With this interactive approach, programmers can input their knowledge to the parallel compiler for improving the capability and efficiency of parallel compilers. In addition, the source code transformation is the other important method for parallelization, which includes works [11][12][13].

In this paper, we focus on the semi-automatic parallelization. The proposed method and the implemented system are based on meta-task and template, and an interactive graphic interface is provided for users to guide the procedure of parallelization, for achieving the goal of accessibility, structuredness and template-based scalability.

3   Semi-automatic Parallelization Method

In this paper, our parallelization goal is to transform sequential codes into parallel codes, which conforms with the message-passing model and can run in the MPICH environment. The architecture is shown as Fig. 1.

As Fig. 1 shows that the sequential application is transformed into the parallel program by the *Metaparallel* system, which had been implemented with Java language. Behind the *Metaparallel* system, the following key technologies are adopted.

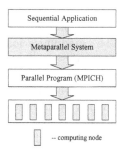

Fig. 1. The architecture of parallelization

3.1   Programming Template

In the message-passing model, there are some parallel programming *paradigms*: master-slave, pipeline, workpool, divide-and-conquer, and phase-parallel [1]. Usually, parallel programs include two kinds of executing parts, that is, computation parts and communication parts. Different parallel programs have the different combination means of computation parts and communication parts. Through analyzing a large number of the pratical parallel codes of different applications, it can be found out that there are some common characteristics of these parallel codes, which can be reflected as programming paradigms. So we implemented the programming templates based on the above five programming paradigms. Users can select one appropriate template for generating their parallel applications. In addition, users can make a new programming template to satisfy the practical requirement in order to implement the *scalability* of this method.

In the template, the main characteristic of multiple processes in the parallel program is determined. For example, in the master-slave paradigm, it is the master process that is responsible for assigning data to the other slave processes and gathering the computing results. However, in the pipeline paradigm, processes are organized as a pipeline, and one process receives data from its front process and sends data to the process behind it.

Moreover, the *specific* computing codes is needed to be inserted to the template code for generating executable parallel programs. The inserting procedure is accomplished with the guidance of users. The user should provide sequential codes, determine the parts that can be executed in parallel, and indicate the data association of parts that have to be executed sequentially.

There are also the other two operations for users to generate parallel programs. One operation is the modification of some existing codes in the template, which is not suitable for specific applications. And the other operation is the deletion of some existing codes in the template, which is no use for users' applications.

With the guidance of users involving the selection of programming paradigms, the inserting of specific computing codes and the modification of existing codes in the template, the parallel program corresponding to a sequential application can be generated by the *Metaparallel* system.

3.2   Meta-task

We adopt the term "meta-task" to represent the segment of codes, which has to be executed in sequence. The communication between one meta-task and others in the parallel program is the data exchange such as receiving data from other meta-tasks or sending data to other meta-tasks in the parallel program. And meta-tasks are the practical computing parts of the specific application. The codes of meta-tasks are provided by users and are inserted to the template codes in order to generate the finial parallel program of the application.

There are two ways for inserting meta-tasks into the template codes. One way is inserting the codes of meta-tasks directly into the template codes, which is very simple. However, there are some problems with this means, one is the name conflict of variables and functions, and the other is that the meta-task can not be debugged and executed independently for it is just a segment of program codes, which will increase the difficulty of debugging the problem of a large scientific computing program. The other way is to adopt the object-oriented mechanism to encapsulate the meta-task. The interface is defined for the corresponding meta-task, and is called in the template codes. Users implement the functions in the interface according to the specific application, and these functions are the practical computing parts of the specific application. In this paper, the second means of inserting meta-tasks into the template codes is adopted.

3.3   Transparency of Heterogeneity

Usually, a parallel computing environment consists of heterogeneous machines and different operating systems. An effective automatic parallelization tool should shield users from the heterogeneity for lightening the burden of users. So Java language is adopted to implement the template by us, and users are also expected to implement or encapsulate the legacy codes into the meta-task with Java language in response to their specific scientific computing applications. As Java language is a crossing-platform programming language, which can transparent the heterogeneous characteristic of different kinds of computers. The performance of Java language had been discussed in many literatures and it is validated that Java language is also suitable for high performance computing.

Through *Metaparallel*, a large-scale computing application will be transformed into parallel programs executed in the MPICH environment. As MPICH is implemented for C language and Fortran language, so the mpiJava toolkit is adopted to bridge the difference in the programming language between the java-implemented parallelization environment and the C-implemented MPICH. mpiJava is an object-oriented Java interface to the standard Message Passing Interface, and provides a Java-based wrapper of MPI, through which MPICH functions can be invoked in the Java programs.

3.4   Structuredness

Based on template and meta-task, a method of the user-guided semi-automatic parallelization is proposed. The main idea is that a meta-task is an individual sequential "atom", which maybe need input data from other meta-tasks or send

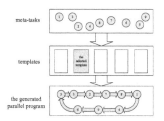

Fig. 2. The generation procedure of parallel programs

output data to other meta-tasks. It is the template that will describe the data association of these meta-tasks and organize these meta-tasks so that meta-tasks can run harmoniously in parallel. This is illustrated as Fig. 2.

As Fig. 2 shows that with the selected template, the nine meta-tasks are organized as a logic circle, and there is a continuous data stream flowed in the looped pipeline, and the meta-tasks execute at different stages simultaneously in an overlapped fashion. This scenario existed in the solving of the large-scale equation system derived from the structure dynamics analysis [14].

4   System Design and Implementation

In this section, *Metaparallel*, a system of the user-guided semi-automatic parallelization, is discussed in terms of the system design, and the implementation of the system is refereed.

4.1   The Framework of the *Metaparallel* System

The proposed user-guided semi-automatic parallelization tool is an integrated development environment, illustrated as Fig. 3. The system includes the general integrated development environment, and parallelizing modules.

Firstly, according to the specific application, the user should code or encapsulate the *meta-task* as a *SingleTaskSubmit* object, which is the sequential program in Java and should implement the interface *Metatask*. Then the user should debug and execute the sequential codes as an individual program, ensuring the correctness of meta-tasks. From the perspective of software engineering, it is helpful to develop and debug a large-scale application, also helpful to maintain the existed application, because it has good structuredness.

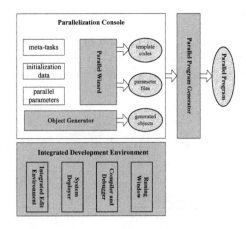

Fig. 3. The Framework of *Metaparallel*

One of the parallelizing modules is *Parallel Wizard*, which is a graphic interactive interface. With *Parallel Wizard*, Users input the parallelization parameters (including *initialization data, parallel parameters*, etc.), which define parallelization strategies and parallelization rules for a specific application, and then select the appropriate program template from available templates, and the template codes are generated. Another aspect to be considered is to handle the data. Users should specify data types and covert them into the types that could be recognized by the generated program, and these data will be stored in a *parameter file*.

The other parallelizing module is Object Generator. It is Object Generator that generates some relative objects from the meta-tasks, which will be called by the final parallel program. The Object Generator is an important part of the system, which not only creates source codes of objects but also compiles them into Java *class* file. The most important object to be created is *DataManger* for handling data. It is used to fetch data from data sources and parse the initial data according to the configuration parameters. The object can also convert data into user-defined types when the user-defined type is fit for the communication between computing nodes. It also has member functions to store results back to relative storages. The instance of *DataManger* will be employed in the final parallel program.

Parallel Program Generator is responsible for generating the final parallel program, which will be executed in the MPICH environment. During the course of creating the codes of the parallel program, the determined template will be the skeleton of the parallel program, and the related objects are instanced as one part of the parallel program. Also a *parameter file* will be created that records configuration parameters input by user during *Parallel Wizard*, which is also used

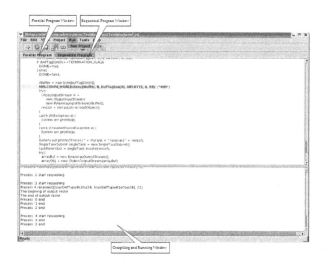

Fig. 4. The *Metaparallel* environment

to organize the multiple processes when the parallel program is executed. One more important operation is to insert the instance of the meat-task implemented as a Java object to the program skeleton with the necessary modification.

Besides above components, the system also includes the general integrated development environment, which consists of *Integrated Edit Environment, System Deployer, Compiler and Debugger*, and *Running Window. Integrated Edit Environment* is used to edit the program. *Compiler and Debugger* is for compiling and debugging sequential programs and parallel programs based on the Java toolkit, *System Deployer* is responsible for deploying the parallel program in the cluster of machines. Finally, the sequential program and the parallel program can be executed in *Running Window*.

4.2   Implementation

The *Metaparallel* system is implemented with Java language based on the following software toolkits: MPICH1.2.5, mpiJava1.24, JDK1.4.2, Fortran and C/C++. The whole environment of *Metaparallel* is shown as Fig. 4. We have implemented the functions discussed in above sections.

5   Case Study

In this section, A case of parallelization of fractal computing by *Metaparallel* is given as follows.

1. Application Analysis. In this phase, users should analyze the application to determine the parallel programming paradigm, and then select an appropriate template from available templates. As fractal computing is the embarrassing parallel computation, where each iteration of fractal computing is independent from the other iterations, so the master-slave paradigm is adopted. Therefore, the corresponding master-slave template is chosen as the skeleton of the parallel program.
2. Parallel Wizard. Through this phase the user guides the procedure of parallelization. The user should input parameters required by the system, and import the source codes of fractal computing as the meta-task.
3. Generating Relative Objects. After the system has got the necessary information, the system should generate the code of relative objects (in this example, they are DataManagement object and ViewFrame object) and the system implicitly compiles the generated codes into Java classes.
4. Generating Parallel Codes. With necessary objects and the code template, the parallel code can be generated now. The result of the generation is the parallel source code shown as Fig. 5. The brief comparison between the meta-task source code and the generated parallel source code is also displayed in Fig. 5. We can find out that the parallel source code is generated with many parallel instructions and some invocations of the class of the meta-task.
5. Compiling and Debugging. Users can compile and debug the generated parallel code to make sure that the parallel program is correct.

Fig. 5. A case of *Metaparallel*

Fig. 6. The result of execution

6. Deploying and Running. The compiled parallel program is deployed in the parallel computing environment based on MPICH in this phase. Finally, the user can run the program in *Running Window* of *Metaparallel*, although the program can also be executed manually in the command mode.

In this case, a master process is created for assigning and reclaiming the computing data, and displaying the computing result, and three slave processes

Table 1. Execution time and speedup

node number	execution time(s)	speedup	parallel efficience(%)
1	5813	1.000	100.0
2	2915	1.994	99.7
3	1952	2.978	99.2
4	1460	3.982	99.5
5	1182	4.918	98.4

are created to calculate the fractal fern leaf. The execution result of the generated parallel program is shown as Fig. 6.

In addition to this example, we have parallelized the practical engineering computing application, the dynamic analysis for the track structure [14]. In this engineering case, the pipeline parallel programming paradigm is adopted to reduce the execution time of solving a large-scale equation system derived from the structure dynamics problem. The execution time and speedup are listed as Table 1 with increasing the homogeneous computing nodes.

6   Conclusion

In this paper, we challenge the issue of parallelizing the program. Based on meta-task and template, a user-guided semi-automatic parallelization method is proposed and a parallelizing system *Metaparallel* is implemented.

The accessibility of *Metaparallel* is achieved with the graphic interactive mode, and the sturcturedness of *Metaparallel* is implemented based on the concept of meta-task, and the customized template can be used for the scalability. Through the case study, the parallelizing procedure is discussed, and the parallelizing result of the practical engineering application indicates that the proposed approach is feasible and the system is effective for parallelization.

Acknowledgements

This research was supported by the project "General High Performance Engineering Numerical Simulation Computing Platform" of Shanghai Municipal Informatization Commission, and the National Natural Science Foundation of China (No. 60173031 and No. 60473092).

References

1. Hwang, K., Xu, Z.: Scalable Parallel Computing - Technology, Architecture, Programming. McGraw-Hill, New York (1998)
2. Boulet, P., Brandes, T.: Evaluation of automatic parallelization strategies for HPF compilers. In: Proceedings of HPCN Europe 1996. Volume 1067 of Lecture Notes in Computer Science. (1996) 778–783

3. Dion, M., Robert, Y., Philippe, J.L.: Parallelizing compilers: what can be achieved? In: Proceedings of HPCN Europe 1994. Volume 797 of Lecture Notes in Computer Science. (1994) 447–456

4. Eigenmann, R., Hoeflinger, J., Padua, D.: On the automatic parallelization of the perfect benchmarks. IEEE Transactions on Parallel and Distributed Systems **9** (1998) 5–23

5. Lim, A.W., Lam, M.S.: Maximizing parallelism and minimizing synchronization with affine partitions. Parallel Computing **24** (1998) 445–475

6. Wilson, R.P., French, R.S., Wilson, C.S., Amarasinghe, S.P., Anderson, J.M., Tjiang, S.W.K., Liao, S.W., Tseng, C.W., Hall, M.W., Lam, M.S., Hennessy, J.L.: SUIF: an infrastructure for research on parallelizing and optimizing compilers. ACM SIGPLAN Notices **29** (1994) 31–37

7. Blume, W., Doallo, R., Eigenmann, R.: Parallel programming with Polaris. Computer **29** (1996) 78–82

8. Ierotheou, C.S., Johnson, S.P., Cross, M., Leggett, P.F.: Computer aided parallelisation tools (CAPTools) - conceptual overview and performance on the parallelisation of structured mesh codes. Parallel Computing **22** (1996) 163–195

9. Hiranandani, S., Kennedy, K., Tseng, C.W., Warren, S.K.: The D editor : a new interactive parallel programming tool. In: Proceedings of the Supercomputing'94, IEEE Computer Society Press (1994) 733–742

10. Yang, B., Wang, D., Zheng, W.: Several critical techniques in constructing interactive environment of parallelizing compiler. Journal of Software (Chinese) **12** (2001) 698–705

11. Kuck, Inc., A.: Parallel performance of standard codes on the compaq professional workstation 8000: Experiences with visual KAP and the KAP/Pro toolset under windows NT. (1997) Champaign, IL.

12. Mitra, S., Kothari, S.C., Cho, J., Krishnaswarmy, A.: ParAgent: A domain-specific semi-automatic parallelization tool. In: Proceedings of the 7th International Conference on High Performance Computing. Volume 1970 of Lecture Notes In Computer Science. (2000) 141–148

13. Felber, P.: Semi-automatic parallelization of java applications. In: Proceedings of the International Symposium on Distributed Objects and Applications (DOA'03). (2003) http://www.eurecom.fr/~felber/publications/DOA-03.pdf.

14. Weng, C., Lu, X.: Application of network-based parallel computing to dynamic analysis for track structure. Journal of Shanghai Jiaotong University (Chinese) **38** (2004) 497–500

CAPU: Enhancing P2P File Sharing System with Capacity Aware Topology[*]

Hongliang Yu, Weimin Zheng, Dongsheng Wang, Haitao Dong, and Lu Li

Department of Computer Science and Technology, Tsinghua University,
Beijing, 100084, China
{hlyu, zwm-dcs}@tsinghua.edu.cn

Abstract. Measurement works show that the unstructured P2P file sharing systems such as Gnutella face the problem of poor scalability and inefficiency search for unpopular items. In this paper, we propose new mechanisms that greatly enhance the performance of file sharing system. Our work exploits the prevalent heterogeneity of the nodes in existing unstructured networks in terms of capacity to construct a quasi-hierarchical topology-aware topology which achieves approximately optimal system throughput. Based on this overlay topology, we propose proactive file index propagation scheme to facilitate search. We also introduce a two-stage search algorithm integrate probabilistic biased random walk that search for popular items and low-redundant multicast (MPR) searching for rare items, achieving approximately O(1) search efficiency for popular items and receivable search latency for rare items respectively. We evaluate our design through simulations and the results show 3 to 5 orders of magnitude improvement in total system capacity compared to other Gnutella-like system.

1 Introduction

Among various usages of P2P architecture, file sharing is one of the dominant applications. From Napster [1] (the earliest P2P system, emerging in 1999), Gnutella [2], Freenet [3], KaZaA [4], to BitTorrent [5], P2P file sharing systems attract millions of users and become the biggest bandwidth consumer of the internet.

Based on their overlay topology and data placement strategy, P2P file sharing system can be put into two categories: unstructured and structured. Because of the lawsuit of RIAA, Napster and its various centralized-search mocker had to face their destination. Decentralized system, such as Gnutella and KaZaa that decentralized-download as well as decentralized-search, emerged as the most prevailing P2P file sharing system. Peers joining into such overlay systems connect with each other in an ad-hoc fashion, which form random overlay topology. And the placement of the files in unstructured systems is also extremely unrestrained. When a node queries with some key, it sends the query message to all its neighbors in the system. On receiving such a query request, if the

---

[*] This work was supported by the National Natural Science Foundation of China under Grant No. 60433040.

J. Cao, W. Nejdl, and M. Xu (Eds.): APPT 2005, LNCS 3756, pp. 214–225, 2005.
© Springer-Verlag Berlin Heidelberg 2005

node has files satisfied this query, it returns a list of content according with the query to the originating node, else it forwards the request recursively within a given TTL scope. Clearly, this approach is not scalable, as load on each node grows linearly with the total number of query, which grows with system size. Besides, it can only get partial answers, that is to say unless the flooded query request traverses all the nodes in the overlay, it can not find distinct copies of desired file, which is presented in [6] as the problem of searching 'needle and hay'. [6] indicates that Gnutella-like unstructured file sharing system is so called mass-market system, it can easily find popular files (hay) but it suffers long response time and low recall rate to find rare files (needle).

To solve the searching scalable problem, several approaches have been simultaneously but independently proposed, all of which support a distributed hash table (DHT) functionality. In these systems[6][9][10], which we call DHTs, peers are connected according to some strict rules, and files are associated with a key, which is produced, for instance, by hashing the file name, and each node in the system is responsible for storing a certain range of keys. By these means, DHTs provide scalable query responsible time and acceptable load balance mechanism. Though for several years there have been durative mania in research community for DHT, such systems still have no sign to be put into use for file sharing. This phenomenon is due to two factors, the first is that DHTs are great vulnerable to peer churn rate, the second is that DHTs are not easy provide partial query techniques, such as keyword searching. Based on the facts mentioned above, [6][11] address that unstructured systems are more suitable for file sharing application than structured networks.

Our work in this paper is based on the following unfathomed problems in unstructured networks:

1. Scalability: When faced with a high aggregate query rate and a large system size, nodes become overloaded and the throughput of the system degrade significantly.
2. Search Efficiency: Previous works have observed that the flooding-based approach is an efficient, simple solution for finding copies of popular files, but they also identify the poor latency and result quality for queries that focus on rare items.

The Gnutella measurements presented in [7] prove our points in that Gnutella is less effective for locating rare items: 41% of all queries receive 10 or fewer results, and 18% of queries receive no results though there are matching content existing in the system. Furthermore, the results have poor response times. For queries that return a single result, the first result arrives after 73 seconds on average. For queries that return 10 or fewer results, 50 seconds elapsed on average before receiving the first result.

There are several previous works to improve the scalability of Gnutella-like system, such as use biased random walk to substitute flood-based search and active algorithm to limit the query load into each node to avoid overloading[6][11]. We admit these works are in the right direction to solve problems mentioned above. But their approach still can not provide an all-purpose effective scheme both for the scalability of system load and for the efficiency of searching for rare items. The design of CAPU are based on the prevalent heterogeneity of the nodes in existing unstructured networks in terms of capacity, including processing power, disk latencies, access bandwidth, and etc[12].

The first feature of CAPU is that we construct a quasi-hierarchical topology-aware topology that balance query, and maintenance cost among nodes, thus achieve the approximately optimal system throughput. The second feature is based on this quasi-hierarchical overlay topology, every node report pointers of the files stored in it to their high-level neighbors, and the high-level neighbors combine pointers of its files and the pointers reported by other node together, and report to its higher level neighbor. Simulation results show that this recursive proactive file index propagation great expedite the search process. And finally we scheme out a query algorithm, which integrate biased random walk and low-redundant multicast (MPR), achieving approximately O(1) search efficiency for popular items while receivable search latency for rare items.

We evaluate CAPU through simulation-based experiments. The results show that our approach can provide 3 to 5 times improvements in terms of length of search path.

The rest of this paper is organized as following. In section B, we present the detailed design of CAPU, including the overlay protocol, the file index update strategy, and search algorithm. Simulation results are given in Section C. Section D and Section E are related works and final conclusion.

2   CAPU Design

The key components of our work include:

1. *A quasi-hierarchical capacity- aware topology protocol* that ensure a approximately optimal overlay topology in terms of high system throughput, load balance, low diameter, and high resilience to failure.
2. *Proactive file index propagation* exploit the capacity-aware topology, every node recursive propagate local file pointers and the file pointers reported by its lower-level neighbors to the its upper level neighbors, thus make full use of the quasi-hierarchical topology to provide cache for query in advance, and shorten length of query path.
3. *Two-stage search algorithm* take advantage of the capacity-topology and file index strategy, the first stage bias random search message to high capacity node dynamic-adaptive to query load, if not getting query result until reaching the highest level node, then the second stage search algorithm is triggered. The second stage introducing Multi-Point Relays (MPRs)[13] to multicast query request among highest level nodes, reducing the number of re-transmission packet compared with flooding. So far as we know, this is the first work to introduce MPR into multicast in P2P system.

The concept of capacity or heterogeneity level is essential to our protocol. Unlike the design in[6][11], which consider nodes distributed continuously in term of capacity, we classify different heterogeneity into finite discrete levels, which make the construction of quasi-hierarchical topology realistic. Though there are several means to mark off capacity level, in general, one could model the capability of a node as a vector consisting of several components including access network bandwidth, CPU frequency, memory size, disk access latency, etc. These parameters are highly relevant to the amount of resources a node is willing to contribute to the P2P system. In our model, we consider bandwidth is the most critical resource for file sharing application. It is observed in Gnutella [14] that nodes show several orders of magnitude difference in their access network bandwidth (slow dial-up connection, DSL connections, Cable connections, to high speed LAN connections). So, it is easy to classify node capacity according to the bandwidth. In the following of this paper, we assume that the set of node classes are given to us based on access network bandwidth.

Algorithm 1: Quasi-hierarchical capacity aware topology protocol

$L(I)$: capacity level

$Th(I)$: degree upper-bound

$B(I)$: bootstrap node set

$N(I),D(I)$: neighbor node set and its size

$N_p(I),D_p(I)$: neighbors with higher capacity level than X and its size

$N_c(I),D_c(I)$: neighbors with lower capacity level than X and its size

$N_s(I),D_s(I)$: neighbors with same capacity level with X and its size

{new node I join}

   if I is new node into system **then**

     Get $B(I)$ from web cache;

     For each node $J \in B(I)$, such that $L(J) \geq L(I)$

        Send PING msg to J

   end if

   {when node I receive PING message for node J }

   if $D(I)<Th(I)$ **then**

     Send PONG back msg to I ;

   else {node I has no free degree }

      if $\exists M \in N_c(I)$, such that $L(M)<L(J)$ **then**

       Send PONG back msg to I ;

      end if

   end if

   {when node I receive PING message from at least $Th(I)$ nodes}

   Denote the set of all the nodes returning PONG msg as S ;

   Select the top $Th(I)$ nodes with max level and free degrees;

   Send CONNECT request message to all these nodes;

   {when node I receive CONNECT message for node J }

   if $D(I)<Th(I)$ **then**

     Add J into $N_c(I)$, accept the connect request;

  else

    if

      Select node $M \in N_c(I)$, such that $L(M)$ is minimum;

      Delete M from $N_c(I)$; Redirect M 's link to J ;

      Add J into $N_c(I)$, accept the connect request;

    end if

  end if

2.1 Quasi-Hierarchical Capacity-Aware Topology Protocol

The capacity adaptive topology protocol is the key algorithm for node to select, connect, drop, and maintain neighbors. The goals of this algorithm include:

1. Ensure that the degree of node is directly proportional to the nodes' capacity, nodes of the same capacity level have balanced degree.
2. The capacity level of all the nodes in the directional shortest path from node A to node B (suppose A's capacity level is lower than B) are intervenient between capacity level of them.
3. A node, whose capacity level is not maximum, has at least one neighbor with higher capacity level than it.

The first goal is to fully exploit the heterogeneous capacity of participating nodes as well as to balance topology maintenance and query load among nodes in term of node capacity. The second goal is to ensure that weak nodes are prevented from becoming hotspots or bottlenecks that throttle the performance of the system, thus avoiding bottle-necks to decrease the average query response time. Hence, a powerful node would maintain more connections and more file pointers, receive more queries on average compared with a weak node. The third goal ensure that weak node is within O(1) reach of higher capacity ones, which at most equals the number of capacity levels, thus shorten the file indices propagating and query hops. Besides, since the high capacity-level nodes are likely to stay connected to the system for a longer period of time, the probability that a node would loose a connection with a higher-level node is lower than that of a lower-level node, thereby increase fault tolerance of the system. Consequently, from a comprehensive perspective of the system design, this quasi-hierarchical load balanced architectures obtain high overall system throughput.

The *neighbor list* of a node X, denoted as $N(X)$, is divided into two groups, which are *parent neighbor list($N_p(X)$)*, and *child neighbor list($N_c(X)$)*. *Parent neighbor list* of node X include all the neighbors with higher capacity-level nodes than X or the same level. Similarly, *child neighbor list* consists of neighbors with lower capacity-level. We define node X's degree $(D(X))$ as the sum of size of X's neighbor list.

$$D(X) = D_p(X) + D_c(X)$$

For every node, it has a degree threshold $(Th(X))$ exponentially proportional to its capacity-level.

When a node X join the network, it obtains a list of node addresses using a rendezvous mechanism by either contacting a host cache server as Gnutella[2] or consulting its own cache from a previous session in a fashion similar to an initial connection. We denote the capacity level of node X as $L(X)$, and the set of bootstrap address returned from host cache as $B(X)$. Node X select all the on-line nodes whose capacity-level are higher than or the same as $L(X)$, and send PING message to these nodes, the PING message carrying information of X, such as $L(X)$, and $N(X)$(in case of rejoining). On receiving this PING message, node N check if $D(N)$ is lower than $Th(N)$, or if N finds that its has neighbors with lower capacity level than X, it reply a PONG message to the source node X, piggybacking the information of N, such as $L(N)$, and $D_c(N)$.

X prefers connecting to nodes with higher capacity level, and if two nodes are of the same level, X prefers selecting to the one with smaller *children neighbor list*. With this priority constraint, X sorts the nodes, which send back PONG message and have higher or at least the same capacity-level, into a sorted list. Then X send the CONNECT request message to the top $Th(X)$ nodes in the sorted list. If the size of the sorted list is less than $Th(X)$, X fetch bootstrap nodes from host cache again, and continue PING process. If a node N receive this CONNECT message from X, it checks if its $D(N)$ is smaller than $Th(N)$, if $D(N) < Th(N)$, X accept this CONNECT request, and add X into its child neighbor list($Nc(X)$), and accept to connect with X. If $D(N)$ is equal to $Th(N)$, node N select the neighbor with the lowest level, say M, and send message to M to redirect M's link from N to X, add X into its child neighbor list ($Nc(X)$), and establish connection with X.

The node in CAPU system periodically send *keep alive* message to nodes in its *neighbor list* to keep the connection. If the node has not received this heartbeat message for a node in the for several period, the node is consider dead and will be flushed. In next section, we will show that this *keep alive* message is attached with other data, such as file index, neighbor list, and weight of link. A detailed pseudo-code of this protocol is shown in Algorithm 1. Results from experiments measuring the topology adaptation process are discussed later in Section C.

2.2 Proactive File Index Propagation

To fully exploit the heterogeneous capacity of nodes in the peer-to-peer networks, every node in CAPU system proactively propagate the file index of the content in it and the file indices propagated by its neighbor to it to its higher capacity-level neighbors. This scheme is different from the approach in other works, such as[6][15], in which node only maintain node information of one-hop neighbors. In CAPU, the quasi-hierarchical capacity-aware topology enable node report their file information recursively from down to top, consequently from weak nodes to powerful nodes. In the *keep alive* message from node to its higher capacity level neighbors, it reports the change of its previously reported file pointers. If the file index has not been updated in a certain time interval, the file index will be prune. This ensures that all index information remains mostly up-to-date and consistent throughout the lifetime of the node. When a node receives a query, it lookups locally to see if there is some contents matching this query. If only the nodes storing the matching content has a directed path in the overlay topology to this node, the node will respond via a simple local query, thus greatly shorten the query path length, and the response time.

The rationality of this approach is related on the work of [14] they prove that the proactive cache mechanism can improve search efficiency significantly in peer-to-peer networks. But because their work is based on Path Caching with Expiration (PCX), which can reduce the search path length only after successfully found the content. In the setting of unstructured file sharing system, because of lacking scalable search algorithm, the first time for a user to look for the rare content is very difficult, that is to say, CUP can only benefit search for popular items. So, unlike CUP, we place the file cache in the possible search path (refer to CAPU's search algorithm in next section) beforehand. And because we only propagate indices from low-level node to high-level node, nodes with large file index are all powerful nodes and have little chance to be over loaded. This approach shorten the search path length significantly, which reduce

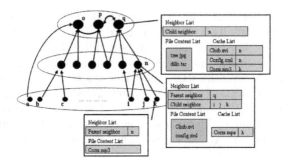

Fig. 1. Proactive file index propagation

the query message in the system and reduce the query load of low-level nodes. Besides, due to the propagation scheme, upper level nodes are aware of large amount of file information in the system, which makes it possible to design scalable search algorithm to find rare items.

Figure 1 show an example of CAPU approach.

2.3   Two-Stage Search Algorithm

The two-stage search algorithm consists of two step, probabilistic biased random walk and MPR-based multicast. Proactive file index propagation protocol provide efficient illumination for query process, in that unlike other cache mechanism who place cache along the succeeded search path, CAPU prepare file index in the possible search path transcendentally. And because of the quasi-hierarchical topology, the query request is relayed from originating node to the higher capacity-level gradually so it is highly possible that the query is answered in the middle path, which avoids the high level nodes to be over loaded, and the possibility of a query message revisit certain node is low. To avoid some node become the hotspot of query and thus overloaded, we adopt load balance bias random walk mechanism. If the query has reached a node of highest capacity-level while still finding no content matching it, the key of this query is considered as rare key, and the second stage of CAPU search algorithm is triggered. We multicast this query request among the highest capacity-level nodes. Instead of using the flood-based multicast like Gnutella, we introduce MPR-based multicast to reduce the number of duplicate re-transmissions while forwarding a broadcast query packet, thus to cut down the consumption of bandwidth caused by message collision.

In the following part, we present the search algorithm in detail. When a node originate or receive a query message, if there are content matching the query in its local content list or in its file index list provided by its lower-level neighbor, it respond to the originator with the corresponding node address storing the content. If it can not find locally, it then forward the query to a neighbor in its parent neighbor list.

To ensure the load balance between nodes we design probabilistic biased random walk query forwarding algorithm. When node I is to forward the query request, the probability to send the query to node J in $N_p(I)$ is $w_{i,j}$.

We specify $w_{i,j}$ as following:

$$\sum_{J \in N_p(I)} w_{i,j} = 1 \tag{1}$$

$$\sum_{I \in N_c(J)} w_{i,j} = 1 \tag{2}$$

A matrix of nonnegative weights $w_{i,j}$ satisfying (1) is called stochastic; if it satisfies (2) as well, it is called doubly stochastic. If a matrix is irreducible (the associated graph is connected) and doubly stochastic, then a Markov chain with this transition matrix has the uniform distribution as its unique steady state distribution.

Algorithm 2: Updating link weights

$w_{i,j}$: the weight of node I associated with parent neighbor J

$w_{j,i}$: the weight of node I associated with child neighbor J

$$w_{out} = \sum_{J \in N_p(I)} w_{i,j} \; ; w_{in} = \sum_{J \in N_c(I)} w_{j,i} \; ;$$

for all $J \in N_p(I)$ {update parent link weights}

$$w_{i,j} = \frac{w_{i,j}}{w_{out}} \; ; \text{send } w_{i,j} \text{ with } keep \; alive \text{ to } J$$

end for
for all $J \in N_c(I)$ {update child link weights}

$$w_{j,i} = \frac{w_{j,i}}{w_{in}} \; ; \text{send } w_{i,j} \text{ with } keep \; alive \text{ to } J$$

end for

The weights of links are periodically updated by node I attached with *keep alive* message as follows:

$$w_{out}(i) \leftarrow \sum_{J \in N_p(I)} w_{i,j}, \forall J \in N_p(I) : w_{i,j} \leftarrow \frac{w_{i,j}}{w_{out}(i)} \tag{3}$$

$$w_{in}(i) \leftarrow \sum_{j \in N_c(I)} w_{j,i}, \forall J \in N_c(I) : w_{j,i} \leftarrow \frac{w_{j,i}}{w_{in}(i)} \tag{4}$$

After an update of the, node I communicates the new weights $w_{i,j}$ in equation (3) with all the neighbors in its *parent neighbor list*, and communicates updated in $w_{j,i}$ equation (4) with neighbors in its *child neighbor list*. This update process can be piggybacked in the keep live message in the topology maintenance process. When a new node is integrated either in the parent neighbor list or in the child neighbor list of a node, its weight is initialized to the mean weight of the nodes already in the neighbor list. This update algorithm is a special case of iterative scaling, for details please refer [15].

If until the query request reach the highest capacity-level, it still can not find the matching content, then the MPR-based multicast algorithm is triggered. This technique restricts the number of re-transmitters to a small set of neighbor nodes, instead of all neighbor, like in pure flooding. This set is kept small as much as possible by efficiently selecting the neighbors which covers (in terms of one-hop radio range) the same network region as the complete set of neighbors does. This small subset of neighbors is called *multipoint relays* of a given network node. Multipoint relaying technique works in a distributed manner, each node calculates its own set of multi point relays, which is completely independent of other nodes' selection of their MPRs. Each node reacts when its neighborhood nodes change and accordingly modifies its MPR set to continue covering its two-hop neighbors. The information required to calculate the multipoint relays is the set of one-hop neighbors and the two-hop neighbors, i.e. the neighbors of the one-hop neighbors. To obtain the information of two-hop neighbors, in CAPU protocol, the highest capacity-level node attaches the list of its own neighbors, while sending its *keep alive* message to other highest capacity-level nodes in its sibling neighbor list. With these information, a node can heuristically calculate its approximately optimal MPR set (for finding a multipoint relay set with minimal size is NP-hard). Algorithm 3 is the pseudo-code of this heuristic algorithm. The multicast process is illustrated in Figure 2. The detailed two-stage search algorithm is shown in Algorithm 4.

Algorithm 3: Calculate MPR

$N_s^2(X)$: the union of sibling neighbor list of all the nodes in X's sibling neighbor list
$MPR(X)$: Multipoint relaying set of X
$MPR(X) \leftarrow \phi$

$S \leftarrow N_s^1(X)$

for all node $M \in N_s(X)$, such that $|N_s(M) \cap N_s^2(X)| = 1$

    Put M into $MPR(X)$

    $S \leftarrow S - N_s(M)$

end for
while ($S \neq \phi$)

        For node $M \in N_s(X) - MPR(X)$, such that $|N_s(M) \cap S|$ is the maximum

        Put M   in $MPR(X)$

        $S \leftarrow S - N_s(M)$

end while

Algorithm 4: Two-stage search algorithm

$K(I)$: file content stored in node I ;
$F(I)$: file index propagated to I from its child neighbors
if X is the initial node **then**

    Choose node J from $N_p(I)$ proportional to $w_{i,j}$

    Send query to J
else

if $K(I)$ has matching file for the query **then**
   send the matching file back to initial node; return;
else
   if $\exists J \in N_p(I)$, such that $L(J) > L(I)$

      Choose node J from $N_p(I)$ proportional to $w_{i,j}$

      Send query to J
   else
  MPR multicast the query
  end if
  end if
end if

3   Experimental Result

In this section, we present experimental results obtained with a simulator for CAPU system. The simulator was implemented on ONSP, an overlay networks simulation

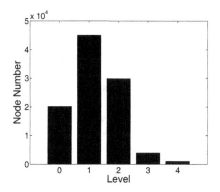

Fig. 2. Node number distribution over node capacity level

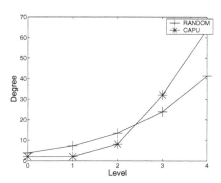

Fig. 3. Degree distribution over node capacity level

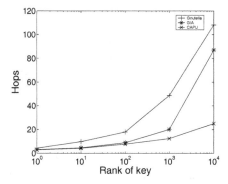

Fig. 4. Average length of query path distribution over popularity of key

platform, which can parallel simulating the function of most off-the-shelf peer-to-peer protocols. By implementing the event logic according to the protocol's definition, the user can easily simulate various protocols. Our experiments were performed on a 32 processors cluster (Pentium IV CPU and 2G memory), running Linux Redhat 7.0. We use a set of real query trace obtained from Gnutella.

Our experiments focus on the aggregate system behavior in terms of its capacity to handle queries under a variety of conditions. We show how the individual components of our system and the synergies between them affect the total system capacity. In all the experiment showed below, number of nodes is 10,000. The key distribution is according to Zipf distribution, with α equal 0.95. Figure 2 shows the node distribution over capacity level. This data is based on [12]. Figure 3 shows degree distribution over node's capacity level. The RANDOM represent the random connected overlay, we can see that our capacity-aware effectively ensure high capacity nodes have high degree. Figure 4 is the experimental result of search hops of Gnutella, Gia[6], and CAPU, In this experiment, we use Zipf key distribution, rank of key refers to the rank of the key popular in the system. We can see that CAPU is especially efficient in searching for rare items.

4 Related Works

For the topology construction strategy, Chawathe et al [6] suggest Gia and use dynamic topology adaptation that puts most nodes within short reach of high capacity nodes. Their topology adaptation scheme defines a level of satisfaction and ensures that high capacity nodes are indeed the ones with high degree and that low capacity nodes are within short reach of higher capacity ones. This scheme is different from our protocol, as Gia ensures that high capacity nodes are indeed the ones with high degree and that low capacity nodes are within short reach of higher capacity ones, while CAPU ensure that the construction of a quasi-hierarchical topology as the foundation for cache propagation and search algorithm. Mudhakar et al [2] present a similar topology as CAPU. But in their paper, they did not provide any effective protocol to construct this topology. There are also proposals to solve the scalability problem caused by flood-based query scheme, Lv. et al [11] propose to replace flooding-based query-forwarding with random walks. Researchers provide different cache scheme to enhance search efficiency. Recent searching results are cached for quick searching in the next time[14]. Considering peers' heterogeneity, Kazaa[4] utilizes powerful peers as supernodes that they hold much larger file lookup tables and provide query resolving for many other peers like a searching hub. Our work can be considered as an expansion of this simple super-node based solution. Boon T. L. et al [7] propose a hybrid solution, which flood query for popular items but use DHT for rare items. But the authors point out that the publishing load for rare item is a heavy burden for the super nodes.

5 Conclusion

We have proposed CAPU, a enhanced unstructured P2P file sharing system, which include quasi-hierarchical capacity-aware topology, proactive file index propagation, and two-stage search algorithm integrating probabilistic biased random searching for popular items and MPR-based multicast algorithm searching for rare items. Our simulation results suggest that these approaches provide 3 to 5 of magnitude improvement in the total capacity of the system while retaining significant robustness to failures. We also compare the impacts of different components of our system, which show that the reciprocation of these algorithms benefit the system greatly.

References

[1] Napster. http://www.napster.com
[2] Gnutella. http://gnutella.wego.com
[3] Clarke, I., Sandberg, O., Wiley, B., and Hong, T.W. Freenet: A distributed anonymous information storage and retrieval system. http://freenet.sourceforge.net.
[4] KaZaA. http://kazaa.com.
[5] BitTorrent. http://bitconjurer.org/BitTorrent/
[6] Y. Chawathe, S. Ratnasamy, L. Breslau, N. Lanham, and S. Shenker. Making Gnutella-like P2P Systems Scalable. In Proceedings of ACM SIGCOMM 2003, Germany, August 2003 .
[7] B. T. Loo, R. Huebsch, I. Stoica, and J. Hellerstein. The Case for a Hyrid P2P Search Infrastructure. In IPTPS 2004.
[8] Ben Zhao, John Kubiatowicz, and Anthony Joseph. Tapestry: An infrastructure for fault-tolerant wide-area location and routing. Technical Report UCB/CSD-01-1141, Computer Science Division, U. C. Berkeley. April 2001.
[9] A. Rowstron and P. Druschel. Pastry: Scalable, distributed object location and routing for large-scale peer-to-peer systems. (Middleware 2001). November 2001.
[10] I. Stoica, R. Morris, D. Karger, M. F. Kaashoek, and H. Balakrishnan. Chord: A scalable peer-to-peer lookup service for Internet applications. In Proceedings of ACM SIGCOMM 2001 . August 2001.
[11] Lv, Q., Ratnasamy, S., and Shenker, S. Can Heterogeneity Make Gnutella Scalable. In Proceedings of IPTPS '02. Cambridge, MA, Mar. 2002
[12] Saroiu, S., Gummadi, P. K., and Gribble, S. D. A Measurement Study of Peer-to-Peer File Sharing Systems. In Proceedings of Multimedia Computing and Networking 2002 (MMCN'02) (San Jose, CA, Jan. 2002).
[13] A. Qayyum, L. Viennot, A. Laouiti. "Multipoint relaying: An efficient technique for flooding in mobile wireless networks". INRIA research report RR-3898, 2000
[14] M. Russopoulos, and M. Baker. "CUP: Controlled Update Propagation in Peer-to-Peer Networks." USENIX 2003 Annual Technical Conference, San Antonio TX, Jun 2003
[15] M. Naor, U. Wieder, Know thy Neighbor's Neighbor: Better Routing for Skip-Graphs and Small Worlds, In Proceedings of IPTPS'04, San Diego, USA, Feb 2004
[16] I. Csisza´r, "Information Theoretic Methods in Probability and Statistics," Information Theory Soc. Rev. articles
[17] Mudhakar S., Bugra G. and Ling L. "Scaling Unstructured Peer-to-Peer Networks With Multi-Tier Capacity-Aware Overlay Topologies" Proceeding of ICPADS 2004.

Implementing Component Persistence in CCM Based on StarPSS

Jingbin An, Yan Jia, and Zhiying Wang

School of Computer Science, National University of Defence Technology,
Changsha, Hunan, China 410073
ajb@nudt.edu.cn

Abstract. In distributed computing environment, we always need to store the objects' state. CORBA Persistent State Service (PSS) provide a high-level approach to realize the persistence of CORBA object. In this paper, we present StarPSS, a design and implementation of PSS in C++ language, and based on the StarPSS, we propose a design of mechanism to implement component persistence in CORBA Component Model (CCM) environment.

1 Introduction

Nowadays, most distributed applications need support of data entities with various lifetimes, and such a support includes objects whose lifetime equals the execution time of a certain block of code, or objects whose lifetime is dynamically controlled by the application itself.

Generally, the support of lifetime is viewed as consisting of two groups with the separation criterion being the ability to exceed lifetime of a single application execution. In accordance with this criterion, the ability to outlive a single application execution is referred to as *persistence*, and the data entities that have this ability are referred to as *persistent*. Conversely, the data entities that lack persistence are denoted as *transient*.

In order to make objects persistent, programmers is obliged to detract from there focus to database programming. To relieve the programmer of this burden, it is helpful to provide persistence as a seamless extension of the existing types of data lifetimes supported by the environment. If the persistence property of data does not interfere with other features of the environment, e.g. the type system, the data access mechanisms, etc, the environment is said to provide *orthogonal persistence*[1]. Orthogonal persistence minimizes cost of building persistent applications by separating the details of persistence support from the rest of the application design.

In distributed computing environment, especially the OMG's CORBA platform, a service to supports persistence, eventually transparent orthogonal persistence, of server-side objects is a direction of research all along. In middle of 1990's, OMG proposed the first CORBA persistence service, Persistent Object Service (POS). In middle of the 1990's, OMG proposed the POS (Persistent Object Service) which is the first CORBA service for CORBA object persistence. POS provides a group of

J. Cao, W. Nejdl, and M. Xu (Eds.): APPT 2005, LNCS 3756, pp. 226–233, 2005.
© Springer-Verlag Berlin Heidelberg 2005

interfaces and a suit of constructs to hold and manage the persistent state of objects[2]. But it is a pity that there have not any real implementation of POS after it is put out owing to some reasons [3][4].Therefore a brand new service, PSS(Persistent State Service), is established to substitute POS in year 2000.

At the same time, the limitations of traditional CORBA computing model is more and more obvious, and rapidly increasing large-scale enterprise applications brought forward more advanced demands, such as rapid development, lower cost, high reliability, scalability, easy deployment. To meet these requirements, OMG adopted the CORBA Component Model (CCM) to extend and subsume the CORBA Object Model.

With the dramatic evolution of distributed computing model, the persistence technology has also got a great progress. Now there are various persistence solution to distributed application, such as CMP/SMP of CCM, CMP of Sun's EJB [9], JDO[10][11], the open source project Hibernate, etc.

In this paper, we propose our design and implementation of CORBA Persistent State Service, which is named StarPSS, and present a solution to implement the component persistence in CCM based on the service.

The rest of this paper is organized as follows. Section 2 describes the CORBA PSS, and puts out the details of design and implementation of StarPSS. Section 3 shows how to implement CCM persistence via our persistent service. Section 4 summarizes this paper.

2 StarPSS

2.1 CORBA Persistent State Service

The CORBA Persistent State Service provides a service to programmers who develop CORBA object implementations. A client has no way to tell if the implementation of an object uses this service. The task of PSS is just to define the interface between the CORBA servant domain and persistent datastore domain, and by this internal interface servants in CORBA server can access one or several datastores to save or retrieve server objects' state information from persistent storage[5].

2.2 Basic Conceptions

PSS presents persistent information as storage objects stored in storage homes which themselves are stored in datastores.

A *datastore* is an entity that manages data, for example a database, a set of files or a schema in relational database.

A datastore is a set of *storage homes*, and each storage home has a type. Within a datastore, there is at most one storage home of a given type.

A storage home contains *storage objects.* Each storage object has an ID unique within its storage home, which is called *short-pid*, and a global ID, called a *pid*. A storage home can only contain storage objects of a given type.

Within a datastore, a storage home manages its own storage objects and the storage objects of all derived storage homes. A storage home and all its derived storage homes are called a storage home family.

In a storage home, there certainly is a list of state members of its storage type which is called a *key,* and it identifies a storage object managed by the home uniquely. A storage home can have any number of keys.

2.3 StarPSS

StarPSS is our implementation of CORBA PSS in C++ language which consists of:

1. PSDL Compiler
2. Generated Persistent Code
3. PSS Runtime Library

PSDL Compiler: In order to make a server object to be a persistent one, eveloper should use PSDL to describe the storage object's type information, and then compile it via the PSDL compiler. The compiler generates the corresponding persistent code for each type of storage object respectively.

The PSDL compiler is constituted by two parts: a fore-end used to scan and parse the PSDL source file, and a back-end which is up to generated target persistent code based on the grammar tree constructed by the fore-end.

Generated Persistent Code: The codes generated by the PSDL compiler includes two main parts, one is interfaces via which PSS service users(CORBA server developers) can access storage objects, and the other is the corresponding implementations of these interfaces.

PSS Runtime Library: The implementation code of PSS is the application independent part of the whole PSS service, and it is carried out as a common runtime library. Each CORBA application using PSS must link this library. PSS runtime library includes four parts: initializer, connector, catalogs, and database connections.

Initializer is responsible for the setup of the service, and it creates a local object *ConnectorRegistry* by which service users can get access to PSS. In StarPSS, Initializer is implemented as a local object named *Initializer* which implements the ORBIntializer interface.

Connector corresponds to a type of real datastore. It manages storage home factories, storage object factories, and catalogs as well. Furthermore, connector also provides a group of operations, by which users can register factory objects or create all kinds of catalogs.

Catalog in StarPSS represents two fold of meanings. Firstly, catalog can be looked as a manager of storage home incarnations, and on the other side, catalog represents a connection or a group of connections to datastore.

Since StarPSS is a service for users to access persistent datastore, we need to establish connections to database to store and retrieve state information of CORBA objects, and apparently, the features, such as creating, destroying, and management of database connections make up a basic part of PSS runtime. We will provide the details in a section later dedicatedly.

2.4 Store Object State in RDBMS

StarPSS uses relational database as backend datastore to save persistent objects' state information. Figure 1 shows the mechanism how the StarPSS store the storage objects in RDBMS. A storage home may manage several storage objects with the same type. Each storage object has several state members m_0, m_1, .., m_n, which are defined in the PSDL file, and a unique pid created by PSS automatically. StarPSS maps a specific type of storage home to a table in relational database, and relevantly maps each one of the storage objects managed by the storage home to a single record in the table. Each field of the record is corresponding to the state members and pig respectively.

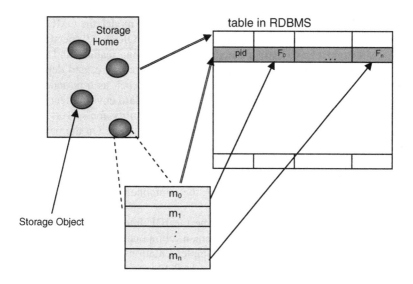

Fig. 1. Store object state in RDBMS

2.5 Session and Session Management

In PSS, we call a connection to datastore a *session*, and correspondingly call the manager of connections as session manager. Session management of StarPSS has three levels. The first level, namely the lowest one, is the basic session, the second one is session pool, and the third is session manager.

Session/Transactional Session: StarPSS supports two kinds of basic sessions, one has not transactional support and the other has. Users must create and destroy basic sessions explicitly.

Session Pool: On top of basic session, session pools provide the user a more sophisticated mechanism for implicit session management. Users only need to create an instance of session pool, all things about establishment, destroying of database connections will be taken over by the pool itself.

Session Manager: Session pool provides a service to share database connections, but the connections can only be shared within the same catalog, in another words, the share of connections is limited to catalog level. In order to share database connections across different PSS application, StarPSS supports the third level of session management, Session manager, which coordinates the usage of connections among multiple catalogs.

3   Implementation of CCM Persistence Based on StarPSS

3.1   CORBA Component Model

To address the limitations with the earlier CORBA object model, the OMG adopted the CORBA Component Model (CCM) as a part of the new CORBA 3.0 specification [6].

The CCM extends the CORBA object model by defining features and services that enable application developers to implement, manage, configure, and deploy components that integrate commonly used CORBA services, such as transaction, security, persistent state, and event notification services, in a standard environment.

The CCM standard provides greater software reusability for servers and greater flexibility for dynamic configuration of CORBA applications. On the base of widely used CORBA, CCM is well positioned for use in scalable, mission-critical client/server applications [7].

CCM components are the basic building blocks in a CCM system. A major contribution of CCM derives from standardizing the component development cycle using CORBA as its middleware infrastructure.

Component developers using CCM define the IDL interfaces that component implementations will support, and then implement components using tools supplied by CCM providers. The resulting component implementations can then be packaged into an assembly file, such as a shared library, a JAR file, or a DLL, and linked dynamically. Finally, a deployment mechanism supplied by a CCM provider is used to deploy the component in a component server that hosts component implementations by loading their assembly files. Thus, components execute in component servers and are available to process client requests [8].

The CCM container provides the runtime environment for components. A container's runtime environment provides services, such as transaction, notification, persistence and security, to the managed component. Each container manages components and is responsible for initializing the managed component and connecting it to other components and ORB services.

A component server is indeed an application server which can provides environment for multiple containers to manage the components. It stands between the clients and DBMS in the 3-tiers architecture and upon which the business logic is executed as components managed by specific containers.

3.2   Component Persistence

CCM supports the use of persistence mechanism for making component state durable, e.g. storing it in a persistent store like a database. The CCM entity container API type defines two forms of persistence support:

Container-Managed Persistence (CMP): With the CMP, the persistence of components is managed by the container, and the component developer simply defines the state which is to be made persistent. The container (in conjunction with generated code) automatically saves and restores state as required.

Self-Managed Persistence (SMP): With the SMP, the component developer assumes the responsibility for saving and restoring state when requested to do so by the container. Self-managed persistence is triggered by the container invoking the callback interfaces, which the component must implement.

Since there is no any compelling restriction in CCM specification, developers and CCM container providers can choose any kind of persistence mechanism to implement the SMP and CMP. Here we choose our StarPSS as the infrastructure to implement the component persistence of CCM.

3.3 Implementating the CMP

Since the StarPSS is used, the container manages all interactions with the persistence provider and the component developer need not use the persistence interfaces offered by the container. We provide an automatic code generation for the storage factories, finders, and some callback operations. These works are done by the *CIDL Compiler.* The Component Implementation Description Language (CIDL) is a superset of the PSDL. Component developers define the component's state information using CIDL, and the CIDL compiler parse this description file and generate all codes to automatically manipulate persistence of the component. As Figure 2 shows, home executor of component is bound to storage home of StarPSS, and correspondingly executor is stored as storage object which is managed by the former storage home.

Container-managed persistence is specified in CIDL and can be configured at deployment time to specify StarPSS-specific properties such as database server host, database name, authentication information and connection pool parameters, etc.

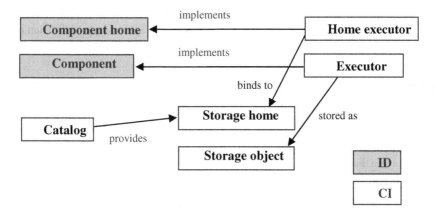

Fig. 2. Component Persistence

3.4 Guidelines for StarPSS-Based SMP

Like container-managed persistence, the component developer has two choices: to use the CORBA persistent state service or any other third-party or user-defined persistence mechanism. But since no declarations are available to support code generation, the component developer is responsible for implementing both the callback interfaces and the persistence classes.

If users choose StarPSS as the lower level persistence provider, the container supports access to a component persistence abstraction provided by the CORBA persistent state service, which hides many of the details of the underlying persistence mechanism from the component developer.

4 Conclusion and Future Work

In this paper, we introduced an implementation of CORBA persistent state service, which is named StarPSS. And based on this service, we analysis the CORBA component model's persistence framework and proposed a solution of CCM component persistence.The result of our work, which is named StarCCM, can be accessed at SourceForge site[12].All of the source files and docments can be found on the homepage as well as the binary packages.

The current version of StarPSS only supports one connector, which is for PostgreSQL databases. In the future, other prevailing database systems will be also supported.

Moreover, the performance is another important problem need to pay more attention to, and we plan to do some work on this aspect including evaluation and improvement.

Acknowledgement

This work was funded by National Natural Science Foundation of China Under grant No. 90104020 and China National Advanced Science & Technology (863) Plan under Contract No. 2001AA113020.

References

1. Atkinson, M. P., Bailey, P. J., Chisholm, K. J., Cockshott, W. P. , Morrison, R. An Approach to Persistent Programming, Computer Journal 26, 4 (1983) pp 360-365
2. Persistent Object Service Specification, OMG
3. Petr Tuma. Persistence in CORBA. PhD. Thesis. Charles University, 1997
4. Jan Kleindienst, Frantisek Plasil and Petr Tuma, Lessons Learned from Implementing the CORBA Persistent Object Service,OOPSLA'96
5. Persistent State Service Specification version 2.0,OMG,2001
6. CORBA 3.0 Specification , OMG,2001

7. Nanbor Wang, Douglas C Schmit, Carlos O'Ryan. Overview of CORBA Component Model. 2000
8. CORBA Component Model, OMG
9. Enterprise JavaBeans Specification 2.0 Final Release 2,Sun Microsystems
10. Keiron McCammon, Heiko Bobzin,Sameer Tyagi,Michael Vorburger, "Core JDO", Prentice Hall,2002
11. The Java Community Process, Java Data Objects Specification
12. StarCCM Project, http://starccm.sourceforge.net

Load Balancing Design Issues
on Prefetch-Based DSM Systems

Hsiao-Hsi Wang, Kuan-Ching Li, Kuo-Jen Wang, Ssu-Hsuan Lu,
and Chun-Chieh Yang

Parallel and Distributed Processing Center (PDPC),
Dept. of Computer Science and Information Management,
Providence University, Shalu, Taichung 43301, Taiwan
{hhwang, kuancli, g9134014, g9234024, g9234025}@pu.edu.tw

Abstract. In recent years, the cluster computing technology has become a cost-effective computing infrastructure, because it aggregates resources of computational power, communication and storage. It is also considered a very attractive platform for low-cost supercomputing. Software distributed shared memory (DSM) provides a convenient and effective solution for programming parallel applications. However, both page faults and communication are major sources of overheads in DSM systems. Prefetching strategy can overlap data transporting time with computation time, as also reducing page faults. Unfortunately, it conducts load imbalance during barrier synchronization. For solving such inconveniences, this research paper discusses the load balancing for barrier synchronization in DSM systems. We discuss that, leaving the loop when half of hosts have finished prefetching is the best method, and therefore, we modify the threshold of leaving loop. Experiments show that, by incorporating load balancing into DSM systems, the barrier synchronization has been improved.

Keywords: DSM system, prefetching strategy, home-based.

1 Introduction

In recent years, the cluster computing technology has become a cost-effective computing infrastructure because it aggregates resources of computational power, communication and storage. It is also considered a very attractive platform for low-cost supercomputing. Cluster of workstations are easy to build, cost effective and highly scalable. It consists of several workstations that are interconnected through a high-speed network (Gigabit Ethernet, SCI, Myrinet or Infiniband) for information exchange and coordination among them. With the advances in networking technology, connecting PCs and workstations is not a problem anymore. Despite of this fact, there is still much to do in the software domain.

People cannot solve many problems in a short time. People are finding a faster way to solve the problem when they are seeking the answer of the problem. The fastest way to solve the problem is through promoting the speed of computer hardware.

J. Cao, W. Nejdl, and M. Xu (Eds.): APPT 2005, LNCS 3756, pp. 234–243, 2005.
© Springer-Verlag Berlin Heidelberg 2005

However, computers progress so fast. Therefore, people start to find other methods to solve the problem. The most charming point of cluster computers is that can solve problems in a finite time as supercomputers by connecting many computers through network and using parallel technology. In addition, cluster systems also have advantages such as low price, high flexibility, and wide usage.

The way to make cluster computers to execute the same work is through message passing to transfer data between each other. Software distributed shared memory (DSM) provides a convenient and effective solution for programming parallel applications on cluster systems [2, 15, 16]. It does not need to change hardware architecture, and only needs to use software to achieve data consistency. DSM system provides the abstraction of shared address space among computers connected by a network. However, the performance of a DSM depends on the consistency scheme. Alleviating communication overhead that is induced by maintaining consistency is an important topic of investigation, because this overhead degrades program execution performance. Both page faults and communication increase overheads of DSM systems [11, 13]. Page faults increase communication overhead and affect other hosts when they execute on barrier synchronization. Prefetching strategy can overlap data transporting time with computation time, as also reducing page faults [8].

Liu and Hu [7] proposed two types of prefetch strategy. The former one is history-prefetching strategy. This strategy can be described as: when a host needs to get remote pages, remote hosts will record requester and requested page. The host will modify pages before barrier, because when all hosts update pages and send recent data to hosts that received invalid page at the barrier. The latter strategy is aggregate prefetching strategy. Essentially, the host takes remote pages from other hosts in occurrence of page faults. The aggregate prefetching strategy utilizes SIGSEGV to take pages for page faults and other related pages. Previous prefetching strategies have two drawbacks on latency issue. The first one is *accumulated waiting phenomenon*. When many hosts need to get remote pages of the same host n, host n needs to send several pages to requesters. Thus, host n spends much time to send data that increases prefetch time. The second drawback is *waiting synchronization phenomenon*.

The proposed Effective Prefetching Strategy solves shortcomings of previously proposed prefetching strategies, because previous developed strategies are affected by *accumulated waiting phenomenon* and *waiting synchronization phenomenon*, which cause system performance degradation or none-prefetching strategies system showing better performance than prefetching strategy. In previous experiments, the proposed Effective Prefetching Strategy shows the best performance when comparing with other existing prefetching strategies. Effective Prefetching Strategy adds three parts to software DSM. It uses Filter Unnecessary Prefetches to reduce misprefetch and Load Balance for Barrier Synchronization to improve *accumulated waiting phenomenon* and *waiting synchronization phenomenon*.

This research paper mainly discusses the implementation process of Load Balance for Barrier Synchronization. When executing a parallel application, the execution time of each node is different. In the process of prefetching, some nodes need to transfer many of prefetching data, but other nodes do not. Therefore, the time for processing prefetching of each node will be different. We use threshold to decrease differences between each other. We reduce idle time of nodes in the loop by threshold and try to find out the best threshold to promote performance of Effective Prefetching Strategy, as in [12]. We still discuss in this paper that, if leaving the loop when half of hosts have finished prefetching is the best method. Therefore, we modify the threshold of leaving loop. We compare it with other two types of threshold.

The remaining of this research paper is organized as follows. In section 2, we introduce Effective Prefetching Strategy. In section 3, we specify DSM system and system architecture. In section 4, the proposed strategy is evaluated with the execution of LU, IS and 3DFFT applications in a DSM system. Finally, a brief conclusion is presented in section 5.

2   Load Balance and Prefetching Strategy

We bring up the Effective Prefetching Strategy, by focusing on threshold in Load Balance with Barrier Synchronization of DSM systems. Effective Prefetching Strategy has three improving approaches that include Filter Unnecessary Prefetches, Distribute Prefetching Overhead and Load Balance with Barrier Synchronization. We discuss different values of threshold and the amount of Effective Prefetching Strategy performance they affect.

2.1   Load Balancing Interface in JIAJIA

Load balancing plays an important role in parallel and distributed systems in order to achieve good performance. In order to maximize performance based on available resources, the parallel system must not only optimally distribute the work according to the inherent computation and communication demands of the application, but also according to the available computation resources dynamically. In many scientific applications, loops are the richest source of parallelism, therefore, change the number of the loop iterations performed by each processor can balance the load [9, 10].

The basic idea of this scheme is keeping the processor affinity as close as possible. It provides auxiliary system calls named jia_lbarrier (&begin, &end) in JIAJIA system to support load balancing, where begin and end represent the upper and lower bound of the loop iterations of the calling processor. A pseudo code of load balancing interface is shown in Fig. 1.

After *STEP* iterations, we decide whether to redistribute the load according to the computing power of each participating processors. Here, *STEP* is an important parameter in our scheduling algorithm. Generally, the data locality will be changed after load redistribution. In comparison with other task queue based algorithms, this interface is simple and adds less programming burden to user since multiple threads must be used to represent the task in task queue based algorithm, which is difficult to use for application programmers [9, 10].

```
for (i=0; i<NUM; i++) {
  for (j=begin; j<end; j++) {
      execute iterations;
  }
  if (i%STEP==0) {
    jia_lbarrier (&begin,&end);
  } else {
    jia_mbarrier();
  }
}
```

Fig. 1. Pseudo-Code of Load Balancing Interface

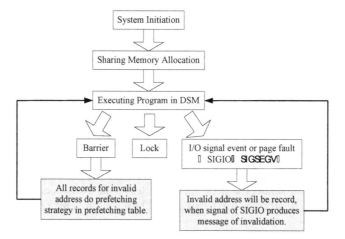

Fig. 2. Stages of History Prefetching Strategy in JIAJIA DSM System

2.2 History Prefetching Strategy

As in Fig. 2, we show procedures of history prefetching strategy in JIAJIA DSM system.

History prefetching strategy is built of two parts in a DSM system. The first part is to add part of it in I/O and page fault data structure. When receiving Invalid (INV) signals, nodes will record the invalidated addresses, if nodes have invalidation data. The second part to be added in is barriers. After all hosts at executing barrier step, they will send all memory addresses that had been recorded to page owners. Then home nodes will return data to nodes that have requested data.

2.3 Effective Prefetching Strategy

Effective Prefetching Strategy has three improving approaches that include Filter Unnecessary Prefetches, Distribute Prefetching Overhead and Load Balance with Barrier Synchronization. Effective Prefetching Strategy adds three parts into DSM software system.

The first part is Filter Unnecessary Prefetches. JIAJIA [3, 4, 5, 6, 14] manages cache pages by using the Read-Write (RW), Read-Only (RO), INV and UNMAP. We add new PREF status in cache pages. When each one of hosts receives prefetching pages from other hosts, we set the prefetching pages for PREF status and memory address status of none write and none read. When host accesses prefetching pages, it will induce local page faults. The prefetching page will change PREF status to RO status or RW status. We can know that prefetching page is reused page when PREF status changes to RO status or RW status. When hosts receive INV signal, all hosts will check status of invalid page. If page is RO status or RW status, host will record address of invalid page into the prefetching table. If the page is PREF status, host will ignore.

In Distribute Prefetching Overhead, we will develop Effective Prefetching Strategy based on history prefetching. History prefetching strategy has several overheads in executing prefetching, so we distribute some prefetching overheads to the requester. Each one of hosts will record memory address of invalid pages in barrier or lock. Each one of hosts collects requests of pages that be requested. Each one of hosts will send address of invalid pages to hosts of each home page before barrier or lock. Therefore, hosts of each home page just send prefetching data to remote request. The hosts of home pages are not managing GETP string and INV string, and other system overheads.

We will discuss Load Balance with Barrier Synchronization and ways to reduce accumulated waiting phenomenon and waiting synchronization phenomenon. When hosts were requested sending prefetching data, the receiving request hosts will execute conditional loop with threshold. When half of hosts finishing sending prefetching data, all senders will be forced to leave the conditional loop. When all hosts into the barrier, all hosts leave barrier synchronization in the same time.

2.4 The Selection of Threshold

In this research paper, we evaluate performance with three different thresholds. Values of threshold not only limit the leaving from loop, but also affecting nodes processing prefetching. Different thresholds will affect idle time of nodes finishing prefetching in the loop, as also affect the max time of node processing prefetching. We want to find out the best threshold for Effective Prefetching Strategy to balance between idle time and prefetching time.

3 Implementation

We use the JIAJIA DSM software to implement our DSM platform. The JIAJIA is a home-based DSM system. It records page status with invalid, read and write, and uses scope consistency [1, 5, 11] as memory consistency model. The JIAJIA provides instructions for barrier and lock to achieve data consistency.

3.1 Building DSM System

First, we execute JIAJIA as initial system. The initial system makes the environment for DSM, while the second step declares sharing memory in all of the hosts. The third step is to execute DSM application, and JIAJIA maintains data consistency in the I/O signal event or page fault. The barrier and lock are provided to achieve consistency in JIAJIA. The barrier stops all hosts on executing point and updates all data. The lock just stops the same locking number in all of hosts, and updates data of the same locking number.

3.2 Prefetching Strategy in JIAJIA System

Prefetching Strategy in JIAJIA is achieved by adding two parts into it. In the part of the I/O signal event or page fault, it is added first part to record all invalid addresses in SIGIO signal, and all recorded addresses will perform prefetching strategy. In barrier, it is added the second part for executing prefetching strategy. Executing prefetching strategy will send updating page to remote invalid page by each page owner. When all records for invalid page already send to remote requester, all hosts will continue executing program.

3.3 Effective Prefetching Strategy in JIAJIA DSM System

The Effective Prefetching Strategy has five additional models compared to JIAJIA DSM system, as shown in Fig. 3. In the part of I/O signal event or page fault it is added part of Filter Unnecessary Prefetches.

Filter Unnecessary Prefetches includes two conditions. When page status is PREF, it will not record addresses of page to prefetching table. Since that page is not used, we eliminate this to produce misprefetching. When page status is RO or RW, it will record address of page into prefetching table.

In the part of barrier, we add four parts of Load Balance with Barrier Synchronization. The first part, we add to count hosts that will send prefetching data to requester. Host 0 is the coordinator of all hosts. All hosts that sending prefetching page will inform host 0, and host 0 will count number of hosts for sending prefetching page. Host 0 sends message for number of hosts for sending prefetching page to other hosts. The second part, if host needs to send prefetching pages, it will be into the unlimited loop and sending prefetching pages. Because each one of hosts can send max prefetching pages for three pages of data, each one of hosts sends prefetching pages for the max number three pages in each loop. If host does not send prefetching pages, it will wait in barrier synchronization. The third part is when half of hosts that are in conditional loop finish sending prefetch pages, the hosts that are in loop will be forced to leave loop, which goal is to find out that if leaving loop when half of hosts finishing prefetching will has better performance. The fourth part is that the hosts that leave unlimited loop will clear the remained addresses of prefetching pages in prefetching table, and all hosts leave barrier synchronization synchronously. Then it will go back to execute application in DSM platform.

In next section, we will explore the performance of threshold in performing experiments with parallel applications LU, IS, and 3DFFT. The experimental results are compared when using one-eighth, one-fourth, and half of all computing hosts involved in the computation.

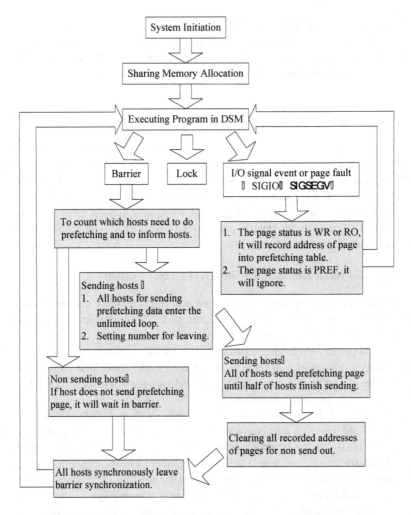

Fig. 3. Procedures of Effective Prefetching Strategy in JIAJIA System

4 Performance Analysis

The cluster-computing platform we used for our investigations is formed by 8 PCs, where each is AMD Athlon 2400+, 1GB DDR memory, interconnected via Gigabit Ethernet. We evaluated performance of Effective Prefetching Strategy with three thresholds when running three parallel applications, IS, LU, and 3DFFT. Table 1 is the data of execution time.

Fig. 4(a) shows the execution time of three different thresholds of IS parallel application. It is shown that, using half of prefetching nodes, as threshold of leaving loop is quicker than other two thresholds of about 23%. Fig. 4(b) is the execution time of LU parallel applications, while Fig. 4(c) is the execution time of 3DFFT on three thresholds. In these last two applications, using half of prefetching nodes as threshold of leaving loop is quicker than others are about 9%.

From above experiments, we can see that using half of prefetching nodes as threshold is more suitable, while other two thresholds are not suitable because they leave loop too fast that induces prefetching pages that has not been sent entirely. These two threshold values can let all nodes leave barrier quickly, but they also increase chance of occurring page faults. Since they add chance of occurring page faults, they also induce performance of Effective Prefetching Strategy can not be expressed.

Table 1. Execution Time of Each Application (sec.)

	1/2 (4 nodes)	1/4 (2 nodes)	1/8 (1 nodes)
IS	5.751	6.99	7.095
LU	25.75	27.71	28.01
3DFFT	30.12	32.04	32.77

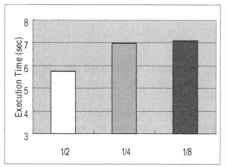

(a) IS application

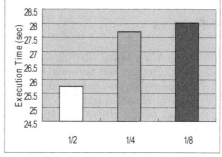

(b) LU application

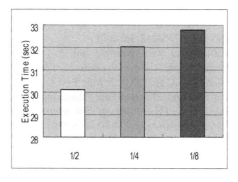

(c) 3DFFT

Fig. 4. Execution Time of Parallel Applications

5 Conclusion

Effective Prefetching Strategy uses Load Balance with Barrier Synchronization to balance prefetching system overload. Load Balance with Barrier Synchronization reduces waiting time among hosts. We use thresholds to achieve balance. From experimental results, we can see that we use half of prefetching nodes finishing prefetching as our threshold is ideal, while other two thresholds are slower than previous one, since it is too early to let prefetching nodes leave loop and that reduces effect of prefetching.

We find that we let all prefetching nodes leaving loop when half of prefetching nodes have finished prefetching indeed can reduce waiting time of prefetching nodes in loop and let most prefetching nodes finishing prefetching.

Acknowledgements

This research is partially supported by National Science Council, Taiwan, under grants no. NSC 93-2213-E-126-007 and NSC 93-2213-E-126-010.

References

[1] Benny Wang-Leung Cheung, Cho-Li Wang, and Francis Chi-Moon LAU, "Migrating-Home Protocol for Software Distributed Shared Memory," *Journal of Information Science and Engineering 18*, pp. 929-957, 2002.

[2] Jason A. Crawford and Clark M. Mobarry, "Hrunting: A Distributed Shared Memory System for the BEOWULF Parallel Workstation," in the *Proceedings of Aerospace Conference*, Vol. 4, 1998.

[3] M. Rasit Eskicioglu, T. Anthony Marsland, Weiwu Hu, and Weisong Shi, "Evaluation of the JIAJIA Software DSM System on High Performance Computer Architectures," in the *Proceedings of the Hawaii's International Conference On System Sciences*, January 5–8, 1999.

[4] Weiwu Hu, Weisong Shi, and Zhimin Tang, "Optimizing Home-Based Software DSM Protocols," *Journal of Networks, Software Tools and Applications, Baltzer Science Publishers*, Vol. 4, No 3, pp. 235-242, Jul 2001.

[5] Weiwu Hu, Fuxin Zhang, Li Ren, Weisong Shi, and Zhimin Tang, "Running Real Applications on Software DSMs," in the *Proceedings of High Performance Computing in the Asia-Pacific Region*, Vol. 1, pp. 148-153, May 14-17, 2000.

[6] Weiwu Hu, Weisong Shi, and Zhimin Tang, "Reducing System Overheads in Home-Based Software DSMs," in the *Proceedings of 13th International and 10th Symposium on Parallel and Distributed Processing*, pp. 167-173, April 12-16, 1999.

[7] Haiming Liu and Weiwu Hu, "A Comparison of Two Strategies Dynamic Data Prefetching in software DSM," in the *Proceedings of 15th International Parallel and Distributed Processing Symposium*, p. 62, April 23-27, 2001.

[8] Y. Roh, B. H. Seong, and D. Park, "Hiding Latency Through Bulk Transfer and Prefetching in Distributed Shared Memory Multiprocessors," in the *Proceedings of The Fourth International High Performance Computing in the Asia-Pacific Region*, Vol. 1, pp. 164-166, 2000.

[9]   Weisong Shi and Zhimin Tang, "Dynamic Computation Scheduling for Load Balancing in Home-based Software DSMs," in *Proceedings of the 1999 International Symposium on Parallel Architectures, Algorithms and Networks (I-SPAN 99)*, IEEE CS Press, Perth, Australia, pp.248-253, June, 1999.

[10] Weisong Shi and Zhimin Tang, "Load Balancing in Home-based Software DSMs," *Special Issue of International Journal of Foundations of Computer Scienc*e, World Scientic Publishing Co. Inc., Vol. 12, No. 3, pp. 307-324, USA, June, 2001.

[11] Andrew S. Tanenbaum, *Distributed Operating System*, PRENTICE HALL INTERNATIONAL EDITIONS, 1995.

[12] K. J. Wang, H. H. Wang, and K. C. Li, "On Design of a Prefetching Strategy for DSM System", in the *Proceedings of PDPTA'2004 International Conference on Parallel and Distributed Processing Techniques and Applications*, Las Vegas, USA, 2004.

[13] Barry Wilkinson, and Michael Allen, *Parallel Programming Techniques and Applications Using Networked Workstations and Parallel Computers*, PRENTICE HALL, Upper Saddle River, 1999.

[14] B. Yu, Z. Huang, S. Cranefield, and M. Purvis, "Homeless and home-based Lazy Release Consistency protocols on Distributed Shared Memory," in the *Proceedings of the 27th conference on Australasian computer science*, Vol. 26, pp. 117–123, January 2004.

[15] http://www.ict.ac.cn/chpc/dsm/index.html

[16] http://www.ics.uci.edu/~javid/dsm.html

Task Assignment
for Network Processor Pipelines Using GA

Shoumeng Yan, Xingshe Zhou, Lingmin Wang, Fan Zhang,
and Haipeng Wang

School of Computer Science, Northwestern Polytechnic University, Xi'an, China
yansm@mail.nwpu.edu.cn

Abstract. In several commercial network processors programming environments, programmer must manually assign many processing tasks to the processor pipelines which consist of many processing engines. Due to the large exploration space, this manual procedure is usually very tedious and inefficient and the optimal or even near-optimal assignment scheme may be difficult to obtain. This paper proposes an automated task-to-PE assignment algorithm based on genetic algorithm. Experimental results show that this method can quickly obtain near-optimal solutions from the large solution space and the algorithm execution time is decoupled with pipeline stages. These two features make this method very suitable to be used in a NP application development environment and provide a more efficient development experience for developers.

1 Introduction

Network Processor (NP), designed to take place of GPP and ASIC in networking applications, can combine the programming flexibility of GPP and high performance of ASIC through introduction of programmable processing engine (PE) and multiprocessor architecture. A popular NP-based application model is to group multiple PEs into a processing pipeline and assign the packet processing tasks to each stage of the pipeline. In this model, the system performance is determined by whether the assignment scheme is optimal. Due to the large exploration space, it will be a very tedious and inefficient procedure if the assignment is made manually; and the optimal or even near-optimal assignment scheme may be difficult to obtain. Thus, this paper proposes an automated task-to-stage assignment method for processing pipelines. Given partitioned task set and the number of processing stages, this method can rapidly get a near-optimal assignment solution.

There are some similar researches in traditional parallel computing domain. Compared with them, our work has a unique problem domain in that we consider multiple processing pipelines in NP. And instead of shortening program execution time, our goal is to maximize system throughput and improve developing efficiency through allocating the task set properly and automatically. Seema Datar etc. [2] have studied the same problem with ours and adopt a greedy algorithm called GreedyPipe, but they did not take into account the communications among tasks which make problem much more complicated. And their algorithm cannot be extended easily to

J. Cao, W. Nejdl, and M. Xu (Eds.): APPT 2005, LNCS 3756, pp. 244–252, 2005.
© Springer-Verlag Berlin Heidelberg 2005

add communication consideration because the algorithm requires the total execution time of a path is static. However, in fact the time varies with different assignment schemes when communication cost is considered. Even if without communication consideration, the time complexity of their method is greatly affected by the number of stages because they attack the problem in a stage-by-stage style. In ref. 7, our preliminary work, we have proposed a GA-based algorithm for this problem, but for simplicity we did not consider communication cost either. In this paper, we will investigate the problem further with communication cost consideration.

The remainder of this paper is organized as follows: Section 2 gives a formal description about the automated task assignment problem model. Section 3 proposes a solution to the problem with two steps, generic algorithm for a single path and greedy algorithm for the multiple paths. In this section, we illustrate how GA can be used in solving the problem, and provide details about encoding mechanism, generic operations, etc. Besides, it also presents the method to combine the assignment scheme for each path together. Section 4 discusses algorithm performance and section 5 summarizes the paper.

2 Problem Model

Packets from network can be classified into various flows based on their packet header. These flows may not necessarily processed by same sequential task set. For clearness, we define the packets that will be processed by the same ordered task set as a **path** which is a different concept from flow and is determined only by its task set. Packets from different flows may be processed by the same task set and thus we can say they belong to the same path. In a networking application system, there usually exist a number of paths and these paths may be multiplexed to the same processing pipeline. Evidently, the pipeline throughput is decided by its slowest stage. In this common scenario, we must map tasks for theses paths to each PE of the pipeline with a goal to maximize the system throughput.

Task partitioning problem is not considered here. We assume that networking application algorithms have been in advance converted to a pipelined implementation, i.e., a set of ordered tasks. ([5, 6] are efforts towards automatically pipelining the application).

We assume that every input path has a constant packet arrival rate and an unlimited temporal domain. For a processing pipeline, the data packets from network are classified into N different input paths, which can be expressed as a set:

$$PT = \{PT_1, PT_2, \cdots, PT_N\} \tag{1}$$

where PT_j means path j.

The corresponding processing task set T_j consists of M_j tasks:

$$T_j = \{T_{1j}, T_{2j}, \cdots, T_{M_j j}\}, (1 \le j \le N) \tag{2}$$

where T_{ij} means task i of path j.

There are L identical PEs in a processing pipeline P:

$$P = \{P_1, P_2, \cdots P_L\} \tag{3}$$

where $L < M_j$ is tenable in general situation.

Based upon the above symbols, one task assignment scheme can be defined by a three-dimensional binary assignment matrix:

$$A = \{a_{ijk}\}, a_{ijk} = \begin{cases} 0, T_{ij} \ not \ assigned \ to \ P_k \\ 1, T_{ij} \ assigned \ to \ P_k \end{cases} \tag{4}$$

Each task associated with a PE has an execution time:

$$t_j = \left(t_{1j}, t_{2j}, \cdots, t_{M_j j}\right), (1 \le j \le N) \tag{5}$$

Besides, task also has a communication time with its adjacent task and this time c_{ij} usually has different value (ca_{ij} and cb_{ij} in the following equations) determined by whether the two tasks are allocated to the same PE. It can be expressed as follows:

$$c_j = \begin{cases} \left(ca_{1j}, ca_{2j}, \cdots, ca_{M_j-1j}\right) if \ a_{ijk} + a_{(i+1)jk} = 2, (1 \le j \le N) \\ \left(cb_{1j}, cb_{2j}, \cdots, cb_{M_j-1j}\right) if \ a_{ijk} + a_{(i+1)jk} = 1, (1 \le j \le N) \end{cases} \tag{6}$$

Thus, the number of tasks assigned to processor k is denoted by:

$$N_k = \sum_{j=1}^{N} \sum_{i=1}^{M_j} a_{ijk} \tag{7}$$

And the total execution time for path j on processor k is given by:

$$e_{jk} = \sum_{i=1}^{M_j} a_{ijk}(t_{ij} + c_{ij}) \tag{8}$$

The processing latency for path j is defined as maximum execution time on all PEs and denoted by:

$$D_j = \max_{k=1}^{L}\{e_{jk}\} = \max_{k=1}^{L}\left\{\sum_{i=1}^{M_j} a_{ijk}(t_{ij} + c_{ij})\right\} \tag{9}$$

The processing latency for the pipeline is considered as the maximum delay time of all flows:

$$D = \max_{j=1}^{N} D_j = \max_{j=1}^{N}\left\{\max_{k=1}^{L}\left\{\sum_{i=1}^{M_j} a_{ijk}(t_{ij} + c_{ij})\right\}\right\} \tag{10}$$

Therefore, the overall throughput is $Th = 1/D$. Up to now, the automated task assignment problem becomes one problem about how to find a rational assignment matrix to maximizes the throughput Th. Such problem has been proven to be NP-Complete [3], which means it is difficult to find an optimal efficient solution by enumerating all possible assignments. And, heuristic methods like GA [1, 4] are very suitable to this kind of problem.

3   Design of GA-Based Assignment Method

To apply GA, solutions for task-to-stage assignment problem need to be encoded into a binary structure. Since it is difficult to directly encode the three-dimension assignment matrix, we solve the problem in two steps. In the first step, we aim at a single path, of which near-optimal assignments should be achieved after genetic operations on the initial population generated with Monte Carlo Method. Secondly, some of the near-optimal assignments of all the paths will be put together to constitute a candidate library. Then, greedy idea will be applied to the library through combination of library elements and the final assignment scheme will be obtained.

3.1   Encoding a Single Path Assignment Matrix

As far as a single path is considered, the assignment matrix is reduced to two dimensions. The formulation to decide this matrix is presented as the following.

$$A = \{a_{ij}\}, a_{ij} = \begin{cases} 0, & T_i \text{ not assigned to } P_j \\ 1, & T_i \text{ assigned to } P_j \end{cases} \tag{11}$$

It can be noticed that not all the binary planar matrixes are viable and the following constraints must apply:

Constraint 1: A task may only be assigned to a single PE, which indicates there is one "1" in every column of the matrix.

Constraint 2: The assignment must maintain the original order of tasks, which means "1" in the matrix is arrayed right-downwards.

Constraint 3: One of the reasonable assignments is to allocate the first task to the first PE.

The above three constraints are reflected in Fig. 1. In the figure, each candidate assignment matrix satisfying above constraints corresponds to a ladder in right-downwards and downwards direction.

If we directly encode the assignment matrix into binary string, we will get a very long chromosome. And we must check the validity of each chromosome after any genetic operation to ensure each chromosome represents a feasible assignment. Evidently, there exist lots of unreasonable assignments for this direct encoding mechanism, which means we may spend lots of time checking and repairing them.

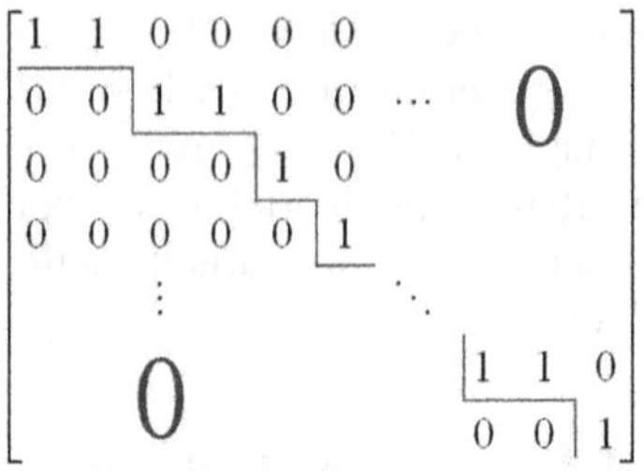

$$\begin{bmatrix} 1 & 1 & 0 & 0 & 0 & 0 & & & \\ 0 & 0 & 1 & 1 & 0 & 0 & \cdots & & 0 \\ 0 & 0 & 0 & 0 & 1 & 0 & & & \\ 0 & 0 & 0 & 0 & 0 & 1 & & & \\ & \vdots & & & & & \ddots & & \\ & & 0 & & & & & 1 & 1 & 0 \\ & & & & & & & 0 & 0 & 1 \end{bmatrix}$$

Fig. 1. A reasonable assignment matrix of a single path can be reduced to a ladder in rightwards and downwards direction. In this figure, row and column indicates PE and task individually.

The big length and this additional checking imposed on such long chromosomes will increase algorithm execution time greatly because GA involves large amount of genetic operations. Also, too many validity checking and repairing operations may make the solution space cannot be covered fully and thus good solutions may be not able to be obtained. Thus, we decide to pursue other ways to encode the matrix.

Considering the ladder distribution of "1" in Fig.1, it is easily to notice that a ladder corresponds to an assignment scheme meeting the three constraints, so one ladder can be considered as one possible solution to our problem. Since each ladder moves right-downwards and rightwards from the top left corner with $n-1$ steps if there are n tasks in the path, we can encode the ladder into a binary string following the rule: moving rightwards represents "1", moving right-downwards represents "0". Obviously, the length of the binary string is only $n-1$. However, the ladder in right-downwards and downwards direction still may not necessarily correspond a reasonable assignment scheme. In occasions where task number is greater than PE number, the times that a ladder move right-downwards should not be greater than PE number (we called this constraint 4.) because each move in right-downwards direction means assignment for a new PE stage. Thus, the encoding still needs some checking and repairing after genetic operations to ensure the number of "0" in the binary string should be no more than that of PE. Fortunately, due to the much smaller unfeasible solution space compared with direct matrix encoding, this encoding mechanism still can greatly shorten algorithm execution time and the repairing process is just a simple heuristics to convert excessive "0" to "1". The advantage of this encoding mechanism is twofold: firstly, the length of each chromosome is shortened greatly; secondly, it makes a much smaller probability to do the repairing process.

Here, we give an example to illustrate the encoding process. Suppose there is an assignment matrix for a pipeline with 5 stages and a path with 10 processing tasks. The matrix is:

$$\begin{bmatrix} 1 & 1 & 0 & 0 & 0 & 0 & 0 & 0 & 0 & 0 \\ 0 & 0 & 1 & 0 & 0 & 0 & 0 & 0 & 0 & 0 \\ 0 & 0 & 0 & 1 & 1 & 1 & 0 & 0 & 0 & 0 \\ 0 & 0 & 0 & 0 & 0 & 0 & 1 & 1 & 1 & 0 \\ 0 & 0 & 0 & 0 & 0 & 0 & 0 & 0 & 0 & 1 \end{bmatrix}.$$

According to the encoding mechanism, this assignment matrix can be encoded into a binary string as follows:

$$[1\ 0\ 0\ 1\ 1\ 0\ 1\ 1\ 0].$$

The length of this string is 9 (i.e., 10 - 1). As we can see, this encoding mechanism is very easy and convenient to apply.

We assume the number of processing pipeline stages is L, each stage is identical and the path contains M tasks. The initial generation can be randomly generated with number of "0" less than L. In our problem, the population size is determined by the number of processing tasks: the larger the task set, the larger the number of the individuals.

3.2 Fitness Calculation

When the target function D is applied to two-dimensional condition where a single path containing M tasks is considered, the maximum processing time becomes:

$$P = \max_{j=1}^{L}\{e_j\} = \max_{j=1}^{L}\left\{\sum_{i=1}^{M} a_{ij}(t_i + c_i)\right\} \tag{12}$$

Therefore, what we should do is to find a_{ij}, which minimize P. Thus, we define the fitness function as follows:

$$F = \frac{1}{P} = \frac{1}{\max_{j=1}^{L}\left\{\sum_{i=1}^{M} a_{ij}(t_i + c_i)\right\}} \tag{13}$$

3.3 Genetic Operations

The genetic algorithm creates a population of chromosome then applies crossover and mutation to the individuals in the population to generate new individuals. It uses various selection criteria so that it can pick the best individuals for mating.

The individuals that have chance to pass down their genes should be those of big fitness. We use roulette sampling to select individuals. The basic idea is in the whirling of the roulette, i.e., the selection probability of big-fitness individuals is larger than the others since they occupy more area in the roulette. In detail, we first sum up fitness of all the individuals in present generation, then generate a Pseudo Random Number r evenly distributed between zero and the sum value, and then individual k is selected meeting the following condition:

$$\sum_{i=0}^{k-1} f_i < r \le \sum_{i=0}^{k} f_i \tag{14}$$

where f_i is the fitness of individual i.

The primary purpose of the crossover operator is to get genetic material from the previous generation to the subsequent generation. In our implementation, a pseudo random number determines the crossover sites, which can be a single point or multiple points. In our problem, we choose multi-point crossover if the binary string is long, otherwise we choose single-point crossover. The mutation operator is used to introduce a certain amount of randomness to the search. It can help to find solutions that might not be encountered using crossover alone. To speed up the convergence of the algorithm and improve its performance, we introduce self-study feature into mutation. This feature helps to identify and keep good results of individual mutation operations by comparing the fitness of newly generated individuals with that before mutation. After crossover and mutation, a heuristics is applied to do some repairing as noted section 3.1.

3.4 Greedy Idea for Multiple Paths

The analysis above gives a good solution to single-path task assignment problem. It is difficult to extend this solution directly to multi-path problem in that the assignment matrix for multi-path has three dimensions. Due to the rapidly increasing complexity, it is hard to find a clear and viable encoding mechanism. Therefore, we need to look for other approaches to further study multi-path problem. Here, we adopt the greedy idea. Simply speaking, we first solve the task assignment problem for each single path using GA, then choose some good assignments per path, and finally combine them to get the best solution to the total system. Ref. 2 presents detailed description about greedy idea, which is utilized to process the task assignment problem stage by stage. Unlike its approach, we use greedy idea to process in a path-by-path manner here.

We take four good assignments per path. Then, we have 4^N different path combinations if there are N input path. Since in real applications N usually has a small value, the execution time in this step makes a relatively minor contribution to the total algorithm execution time. For each combination, we calculate the maximum execution time among all PE stages. Then, by comparing these values, we will find the combination minimizing maximum execution time among all PE stages (i.e., the combination maximizing the overall throughput). If there is more than one such combination, we choose the one that makes the standard deviation minimum. In summary, we use genetic algorithm to get near-optimal assignments for every path, and greedy idea to get the solution to the task assignment problem for multiple paths.

4 Performance Evaluation

We have implemented the algorithm and test it on a 2.6GHZ x86 PC. We have also implemented GreedyPipe in Ref.2 to make performance comparison and a time-consuming exhaustive search algorithm to get the optimal solutions. Since GreedyPipe can not be easily extended to include communication cost, the comparison with it is based on results of GA without communication consideration. Results show that it has a very small influence on our algorithm performance whether we consider communication cost or not. This implies that our GA-based assignment algorithm possess a good scalability in contrast to GreedyPipe.

The performance can be evaluated from two perspectives. The first aspect concerns how closely our results match the true optimal results, and the second is the algorithm execution time.

Extensive experiments have been done to evaluate how well this genetic algorithm approach is than other allocating techniques. And, we have got a lot of experiment data. Through the statistical analysis of the data, we have obtained the following overall results. When the number of input paths is no more than five, if the task number per path ranges between 10 and 15 and the number of PE ranges between 3 and 8, 90% of the time the result got by GA was within 20% of the optimal solution; if the number of task is more than 16 and PE number ranges between 3 and 15, 98% of the time the result was within 10% of the optimal. With the task number increase, our results approach the optimal solution closer and closer. We infer that the reason lies in the fact that longer chromosome strings (In our method, length of chromosome is determined by the number of tasks.) make a larger candidate space for GA operations.

In our experiments with systems containing no more than 5 input paths, 15 to 20 tasks per path, exhaustive searches took on the order of hours versus on the order of seconds with GA. This difference increases exponentially as the complexity of the system increases further. Experiments also show that the variance of pipeline stage number leads to little effect on our algorithm execution time (see Fig. 2), which is different from the method in ref. 2. We think the reason is that in our method, the length of chromosome has no relation with number of pipeline stage. On the other hand, GreedyPipe applies its algorithm in a stage by stage style, thus its execution time is linear with the number of pipeline stages.

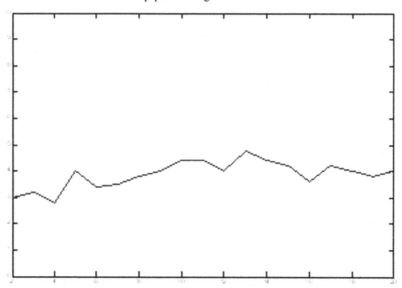

Fig. 2. In this figure, the execution time (Y axis) does not increase abruptly with the number of pipeline stages (X axis)

In summary, we find that we can obtain very near-optimal assignment schemes using GA. And unlike method in Ref.2, execution time of our method does not linearly increase with number of pipeline stages. These two features (near optimal assignment and short execution time) make this method very suitable to be used in NP application development environment and hence provide a better development experience for developers.

5 Conclusions

This paper proposes an automated task-to-PE assignment method to improve the development efficiency of NP based system. Given partitioned ordered task set and the number of available processing pipeline stages, this method can obtain a near-optimal assignment quickly. Our work features in combining the genetic algorithm and greedy idea and the interesting encoding mechanism applied in GA.

Currently, we are making efforts to further test this approach targeting the architecture of Intel's IXP2400, extend and improve it by considering program storage constraints of PEs, and other factors. Also, we want to integrate this algorithm into our system software platform for network processor [8]. Using the platform, programmer can develop applications and make efficient performance explorations with the underlying support from our automated assignment method.

Acknowledgements

This work is supported by the 863 project of China under contract no. 2003AA1Z2100. We gratefully acknowledge the financial and technical support from the committee of the 863 project. We also wish to thank the anonymous reviewers for their constructive comments.

References

1. Wang Xiaoping, and Cao liming, Genetic Algorithm: Theory, Application and Software Implementation, Xi'An, Press of Jiaotong University
2. Mark A. Franklin and Seema Datar, Pipeline Task Scheduling on Network Processors, in Workshop on Network Processors & Applications - NP3(Madrid, Spain), Feb. 2004
3. B. A. Malloy, E. L. Lloyd, and M. L. Souffa, Scheduling DAG's for asynchronous multiprocessor execution, IEEE Transactions on Parallel and Distributed Systems, 5(5):498–508, May 1994
4. David E. Goldberg, Genetic Algorithms in Search, Optimization, and Machine Learning (Hardcover), Addison-Wesley Professional, Jan. 1989
5. Ning Weng and Tilman Wolf, Pipelining vs. Multiprocessors – Choosing the Right Network Processor System Topology, In Proceedings of ANCHOR 2004, June 2004
6. Jianquan Dai, Bo Huang etc., Automatically Partitioning Packet Processing Applications for Pipelined Architectures, In Proceedings of ACM SIGPLAN PLDI 2005, June. 2005
7. Y. Shoumeng, and Z. Xingshe and W. Lingmin, GA-Based Automated Task Assignment on Network Processors, In Proceedings of ICPADS 2005, July 2005
8. Zhang Fan, and Zhou Xingshe and Yan Shoumeng, Design and Implementation of Network Processor Programming Model Based on Software Component, to appear in Computer Engineering and Applications (in Chinese)

Test-Suite Reduction Using Genetic Algorithm

Xue-ying Ma[1,2], Bin-kui Sheng[2], and Cheng-qing Ye[1]

[1] College of Computer Science and Technology, Zhejiang University,
Hangzhou 310027, P. R. China
hzmaxueying@hotmail.com, ycq@zjip.com
[2] Dept. of Information Management,
Zhejiang University of Finance & Economics,
Wenhua Road 269#, Hangzhou, 310012, P. R. China
sbkmxy@yahoo.com.cn

Abstract. As the software is modified and new test cases are added to the test-suite, the size of the test-suite grows and the cost of regression testing increases. In order to decrease the cost of regression testing, researchers have researched on the use of test-suite reduction techniques, which identify a subset of test cases that provides the same coverage of the software, according to some criterion, as the original test-suite. This paper investigates the use of an evolutionary approach, called genetic algorithms, for test-suite reduction. The algorithm builds the initial population based on test history, calculates the fitness value using coverage and cost information, and then selectively breeds the successive generations using genetic operations. This generational process is repeated until a minimized test-suite is founded. The results of studies show that, genetic algorithms can significantly reduce the size of the test-suite and the cost of regression testing, and achieves good cost-effectiveness.

Keywords: Test-suite reduction, Regression testing, Genetic algorithm, Gene modeling, Cost-effectiveness.

1 Introduction

Regression testing is frequently executed maintenance process used to revalidate modified software. As the software is modified and new test cases are added to the test-suite to test new or changed requirements or to maintain test-suite adequacy, the size of the test-suite grows and the cost regression testing increases. Improvements in the regression testing process would help reduce the cost of software.

Researchers have investigated two approaches for addressing the test-suite size problem: test-suite reduction and test selection. Test-suite reduction (also known as test set minimization) algorithms (e.g., [1][3][7][8][10][12]) identify minimized test-suite that provides the same coverage of the software as the original test-suite. Test selection algorithms (e.g., [2][9][11]) selects a subset of the test-suite that will execute code or entity changes; this test-subset, however, may not provide the same coverage as the original test-suite.

J. Cao, W. Nejdl, and M. Xu (Eds.): APPT 2005, LNCS 3756, pp. 253–262, 2005.
© Springer-Verlag Berlin Heidelberg 2005

Test-suite reduction problem can be stated as follows:

Given a test set TS $= \{t_1, t_2 \ldots t_n\}$ consisting of the test cases and a positive cost, c_j assigned to each test case measuring the amount of resources its execution needs, identify a minimal subset of test-suite that provides the same coverage of the software, according to some criterion, as the original test-suite.

This problem is NP-complete. The adaptive search technique has been used to find solutions to many NP-complete optimization problems and could often find a very good solution in a limited amount of time (Goldberg 1989[14]). As an adaptive search algorithm, genetic algorithm has been widely used in software testing, especially in test-case generation for software structural testing ([4][5][19][20][22][16]) or for functional testing ([13][21]). But there is few application of genetic algorithm to reduce test-suite. One of the important motivating factors for using genetic algorithm to reduce test-suite is to promote the effectiveness of test-suite reduction technique.

This paper investigates the use genetic algorithms for test-suite reduction. The main contributions of this paper are:

♦ Present a detailed description of the genetic algorithm used to reduce the test-suite for regression testing.
♦ Implement a prototype of this genetic algorithm using C++ language and perform a set of empirical studies to evaluate the performance and effectiveness of the reduction algorithm. The results of the studies show that our algorithm can find a minimal (or an approximate minimal) set of test cases, which maintains the same coverage of the software as the original test-suite, and achieves good cost-effectiveness.

The remainder of this paper is organized as follows: Section 2 describes some related work. Section 3 presents a genetic algorithm for test-suite reduction. In section 4, we present some studies' results that evaluate the reduction algorithm. Finally, Section 5 summarizes our results and discusses some future work.

2 Related Work

2.1 Test-Suite Reduction

In order to perform test-suite reduction, we should do something includes:

• Maintaining a testing pool where contains all the test cases used in previous test activities.
• Keeping the test coverage information which denotes how many and which parts of the program tested by each test cases during the previous tests.
• Recording the test-execution cost information that measures the amount of resources each test-case's execution needs.

A test-suite reduction technique should possess several qualifications listed in the following.

Adequacy, which means that the reduced test-suite must provide the same test coverage of the software, according to some criterion, as the original test-suite.

Precision, which means that the test-suite algorithm must be able to reduce the redundant test cases from the original test-suite, that is, be able to find a minimal or an approximate minimal subset of test cases.

Cost-effectiveness, which means that it is worth doing test-suite reduction only if the cost of the analysis necessary to do test-suite reduction is less than the savings realized by reducing the test-suite.

Generality, which means the technique must be general, that is, it must be applicable to a wide class of program and modification.

In order to analyze test-suite reduction techniques, we assume equivalent test coverage and test execution costs under regression test reduction and retest all.

2.2 Algorithms for Test-Suite Reduction

Test-suite reduction techniques have been extensively studied. Harrold, Gupta, and Soffa (1993, [7]) proposed a methodology for controlling the size of a test suite. Jones and Harrold (2003,[3]) presented an algorithm for test-suite reduction that can be tailored effectively for use with Modified Condition/Decision Coverage (MC/DC). TIBÖR CSONDES and BALÁZS KOTNYEK (2002, [6]) presented two greedy algorithms for the test-suite reduction problem in protocol conformance testing, namely ADD-DELTA and DROP-DELTA. The ADD-DELTA algorithm can be stated as follows:

*Step*1: Let T=φ;

*Step*2: For each $ti \in$ TS-T, calculate the increase in coverage and cost if it is added to T: $\Delta Cov(ti) = Cov(T \cup \{ ti \}) - Cov(T)$,

$\Delta Cost(ti) = Cost(T \cup \{ ti \}) - Cost(T)$

*Step*3: Find a test cast ti in TS-T for which $\Delta Cov(ti)/ \Delta Cost(ti)$ is minimal. If there are more, then choose the one with the lowest index. Let $T = T \cup \{ ti \}$;

*Step*4: If Cov(T)≥K, then STOP, otherwise go to Step 2.

All of these algorithms repeat the process that select test case which has most contributions to coverage from the remaining test-suite into the minimized test-suite until the minimized test-suite can provide the same coverage as the original test-set.

2.3 Test History

In earlier work [17], we developed computer-aided software testing tools for Visual Basic, Delphi, and C++ etc. Our software-testing tools presented a new block division mechanism and then extended some block-based test adequacy criteria [18]. According to the block division mechanism, there is only one kind of component: block in the program. Formally, a block is such a sequence of statements that if any one statement of the block is executed, all statements thereof are executed. we also extended some block-based test adequacy criteria: *SC0, SC1, SC1+, J-Coverage*[18].

Test history includes static analysis information and dynamic analysis information. The static analysis information includes all the structure information of the program to be tested. The dynamic analysis information includes the coverage information, the test execution costs that are associated with running a test case and so on. All the test histories are recorded in the test database of the testing tools. The logical view of the block coverage information can be derived, as shown in table 1:

Table 1. Block coverage information of a test

	Block$_1$	Block$_2$	...	Block$_j$	...	Block$_k$
Case$_1$	1	0	...	1	...	0
Case$_2$	0	0	...	1	...	1
...	...		...		...	
Case$_i$	0	1	...	0	...	1
....	...	...	...	...	...	...
Case$_n$	1	1	...	0	...	0

In the table, $block_1$, $block_2$, ..., $block_j$,..., and $block_k$ are the blocks sequence of the program, and $Case_1$, $Case_2$, ... , $Case_i$,..., and $Case_n$ are the test cases that have been executed. The information in the table is all the digits of '0' or '1' which denote the blocks-coverage information; here '1' in the row i and the column j means $case_i$ tested $block_j$, and '0' means $case_i$ did not test $block_j$. The number of '1' in row i means how many blocks tested by $case_i$; and the number of '1' in column j means how many times the $block_j$ executed in a test.

3   Genetic Algorithms for Test-Suite Reduction

3.1   Genetic Algorithms

Genetic algorithms (GAs) [14] have been first developed by John Holland [15], and are rooted in the mechanisms of evolution and natural genetics and manipulate a population of potential solutions to an optimization or search problem. A genetic algorithm computed following the process described Figure 1.

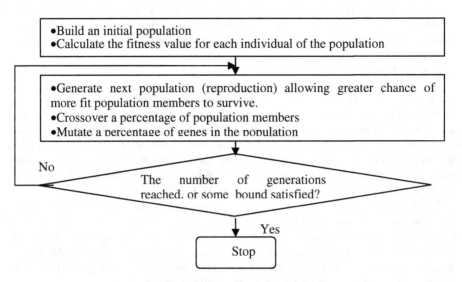

Fig. 1. The process of a genetic algorithm

3.2 Genetic Algorithm for Test-Suite Reduction

Gene modeling for test-suite reduction. For our problem of test-suite reduction, gene of an individual is modeled as a '0'-'1' string which is represented as a column in table 1. As we have mentioned above, each column in table1 denotes the information of which test cases tested the associated block of this column, and these test cases construct a subset of test-suite. Although in its primitive form each test-subset can't satisfy the test coverage bound, it is believed that after evolution of numbers of generations it can evolve to be a feasible solution to this problem. So, each test-subset is naturally to be considered as a potential solution or an ancestor of the solution to our reduction problem. The initial population consists of all the test-subsets denoted by the columns in table 1. That is to say if there are k test cases in the original test-suite and the program has n blocks, the size of population is k, and the length of gene code is n. So the gene code of an individual can be represented as follows:

$Gj=[g_{j1}, g_{j2}, ..., g_{jn}]$, $g_{ji} \in \{0,1\}$, $1 \leq j \leq k$ and $1 \leq i \leq n$, here if $g_{ji} = 1$, it means test case t_i has tested *block_j*, and if $g_{ji}=0$ means case t_i has not tested *block_j* according to the original test histories shown in Table 1. So, the gene of each individual denotes a subset of test cases that have tested *block_j* and this subset is named as *Tj*.

Fitness function. The fitness value for an individual is the combination value of its associated coverage and its cost. The fitness function for individual Gj can be computed as follows:

Fitness(Gj) = Cov(Gj) / Cost(Gj), $1 \leq i \leq n$, here Cov$(Gj)= \sum_{t_i \in Tj} c(t_i)$ is the coverage of the test case t_i according to some criteria, $C(t_i)$ is the test execution cost of t_i.

Let $X_i=[x_{i1}, x_{i2}, ..., x_{ik}]$ be a '0'-'1'string code that denotes the information of row i in Table 1. Consider the information in table, we conclude that if only there are one '1' in the column j, we could say the associated *block_j* is tested by test cases *Tj*. The function Cov() is shown in Figure 2 and the function Fitness() in Figure 3.

```
Double Population::Cov(int j)
{ int *covnum; double Cov;
Cov=0;
covnum=new int[sizePop];
for(int j=0;j<sizePop;j++) covnum[j]=0
for(int i=0;i<sizeCase;i++)
  { if G[j]. code[i]>0 /* G[j] is the individual Gj*/
    for (j=0;j<sizePop;j++)
    {covnum[j]+=X[i][j]; /*X[i] is the vector Xi*/}
  }
for (j=0;j<sizePop;j++)
{if (covnum[j] >0)
  Cov+=w[j]; /* w[j]is the weight of block_j*/ }
Cov=Cov*100/sizePop;
delete []covnum;
return Cov;
}//endCov
```

Fig. 2. The function Cov() for calculating coverage of test set *Tj*

```
void Population::Fitness()
{   double cost,cov;
    for(int i=0;i<sizePop;i++)
    { cost = P->Cost(kids[i].code);
      cov = P->Cov(kids[i].code);
      if(req<miniReq)
       kids[i].fitness= 0;
      else
       kids[i].fitness = cov*kids[i]/cost;}
    //endfor
}//endFitness
```

Fig. 3. The function Fitness() for calculating fitness value of our genetic algorithm

Consider the function Fitness(G_j), the bigger the function Fitness(G_j) value is, the more possible the individual can be selected, because the bigger value of F denotes that the subset of test cases T_j have tested more blocks with less cost. Now we should define the genetic operators for the particular problem of test case reduction.

Reproduction. This operator copies the individuals that are going to participate in crossover: they are chosen according to their fitness value. The choice can be seen as spinning a roulette wheel where each individual has a slot proportional to its fitness value. We spin the wheel as many times as the number of the individuals, and so we have a new population that is going to participate to crossover. This new population is made of individuals of the old one, and the number of each type of individual is proportional to its fitness.

Crossover. Let m be the size of individuals in a population, and let's select an integer i at random between 1 and m-1, then from two individuals ind_1 and ind_2, we can create two new individuals ind_3 and ind_4; one made of the i first genes of ind1 and the m-i last genes of ind2, and the other made of the i first genes of ind2 and m-i last genes of ind1, as shown in the following figure.

$$ind_1 = \{G_{11}, ..., G_{1i}, G_{1i+1}, ...G_{1m}\} \quad ind_2 = \{G_{21}, ..., G_{2i}, G_{2i+1}, ...G_{2m}\}$$
$$ind_3 = \{G_{11}, ..., G_{1i}, G_{2i+1}, ...G_{2m}\} \quad ind_4 = \{G_{21}, ..., G_{2i}, G_{1i+1}, ...G_{1m}\}$$

Mutation. Based on the gene model, the mutation operator consists in changing a '1' in to '0' and vice versa. The mutation operator chooses a gene at random in an individual as illustrated in the following figure: $G_j = [g1, ..., gi, ..., gk] \rightarrow G_{mut} = [g1, ..., \boxed{gi}, ..., gk]$, here, if gi=1 , $\boxed{gi}$=0; otherwise if gi=0, $\boxed{gi}$=1.

4   Empirical Studies

To investigate the quality of our algorithm presented in Section 3, we implemented prototype of the genetic algorithm for test-suite reduction, and the prototype is written in C++, named GeA. In our genetic algorithm, the ratio of crossover is 0.8, the ratio of mutation is 0.01, and the number of generations is 150. In order to evaluate the relative improvement of our reduction algorithm, we also implement the prototype of greedy algorithm ADD-DELTA [6].

We used two real VB subjects for our study: one is a 'CALCULATOR' program, and the other is a 'LITTLE-TV' program. CALCULATOR consists of 85 blocks and LITTLE-TV consists of 241 blocks. For CALCULATOR, we have a test pool of 174 test cases, and, for LITTLE-TV, we have a test pool of 238 test cases. From these test pools, 400 randomly sized, randomly generated test suites, for each subject program, were extracted. The test-suites for CALCULATOR ranged in size from 5 to 80 test cases, ranged in coverage from 65% to 95% (The average coverage is 90%). The test-suites for LITTLE-TV ranged in size from 5 to 85 test cases, ranged in coverage from 60% to 98% (The average coverage is 89%.).

We evaluated three characteristics of the reduction algorithm: size reduction of test-suites, cost reduction of test-suites, and time to perform the reduction.

To evaluate the reduction technique's ability to reduce test-suites, we plotted the size of the reduced test-suites as a function of the original test-suites size. Because there are several test-suites of each size, we calculated the average size of the reduced test-suites of each size. The study results are shown in Fig. 4 and Fig. 5. The horizontal axis represents the original test-suite size and the vertical axis represents the reduced test-suite size. The results show that the sizes of reduced test-suites achieved by GeA are slightly smaller than those reduced by ADD-DELTA. The average size of test suites reduced by GeA for CALCULATOR is 6.5, and the average size of test suites reduced by ADD-DELTA is 7.8. For LITTLE-TV, the average size of reduced test suites is 6.4 and 7.9, for GeA and ADD-DELTA, respectively.

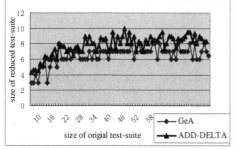 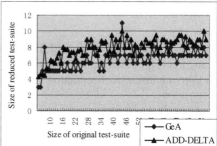

Fig. 4. Size reduction for CALCULATOR   **Fig. 5.** Size reduction for LITTLE-TV

To evaluate the technique's effectiveness to reduce test-suites, we plotted the percentage of cost-reduction as a function of the original test-suite size. Fig. 6~ Fig. 7 show the percentage reduction in cost of the reduced test-suite compared to the original test-suite for CALCULATOR and LITTLE-TV, respectively. The percentages of cost-reduction achieved by GeA are higher than those achieved by ADD-DELTA For CALCULATOR, the average cost-reduction is 81.19% and 71.00%, the maximum cost-reduction is 95.35% and 83.84%, and the minimum cost-reduction is 15.29% and

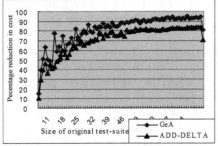

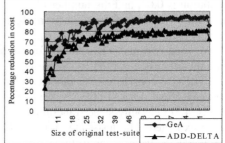

Fig. 6. Percentage reduction in cost for CALCULATOR

Fig. 7. Percentage reduction in cost for LITTLE-TV

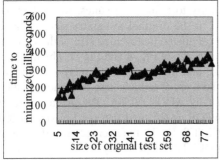

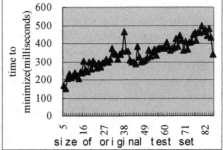

Fig. 8. Time to reduce test-suite for CALCULATOR

Fig. 9. Time to reduce test-suite for LITTLE-TV

10.50%, for GeA and ADD-DELTA, respectively. For LITTLE-TV, the average cost-reduction is 85.38% and 71.59%, the maximum cost-reduction is 94.85% and 80.91%, and the minimum cost-reduction is 29.52% and 22.92%, for GeA and ADD-DELTA, respectively.

To evaluate performance in reducing test-suites, we plotted the average time to reduce test-suites of each size as a function of the original test-suite size. Fig.8 and Fig.9 show the average time to reduce test-suite needed by GeA and ADD-DELTA algorithms, for CALCULATOR and LITTLE-TV, respectively. The vertical axis for both figures represents milliseconds. For CALCULATOR, the time to reduce test-suite needed is ranged form 150 to 401 milliseconds (the average time needed is 286.03 milliseconds), and for LITTLE-TV, the time to reduce test-suite needed is ranged from 140 to 521 milliseconds (the average time needed is 336.98 milliseconds). For CALCULATOR, the savings in test-execution cost achieved by GeA is ranged from 16 to 2262 seconds; for LITTLE-TV, the savings in test-execution cost is ranged from 16 to 1857 seconds. So, the costs in analysis for test reduction are significantly less than the cost-savings, and therefore our algorithm GeA is very cost effective. The time needed for reduction by GeA is much more than those needed by ADD-DELTA(339 milliseconds more in average), however the reduction in cost achieved by GeA is more than those achieved by ADD-DELTA (range 15 to 245 senconds more for CALCULATOR and 12 to 274 seconds more for LITTLE-TV). So, GeA achieves a higher cost-effectiveness than ADD-DELTA does.

5 Conclusion

This paper has presented a genetic algorithm for test reduction based on block-based test history. By modifying the function Cov(), which is used to calculate the coverage of test-suites, the presented reduction algorithms, can be conveniently modified to account for different coverage criteria. Furthermore, because our testing tool saves the test histories of all tested program of different languages in a uniform structure of the media database, our reduction algorithm can be applicable to a wide class of program and modification.

The paper presents the results of empirical studies that evaluate the reduction algorithm. We can conclude from the initial experimental results that our test-suite reduction technique has cost-effectiveness and generality. The results of studies also show that our genetic algorithm GeA is more effective than ADD-DELTA both in size and cost reduction.

Although our initial studies are encouraging, much more experimentation must be conducted to verify the effectiveness of our techniques in general and in practice. Another experiments we should do to further investigate the fault-detection capabilities of a block-based test adequate test-suite on software. These studies will also let us evaluate our algorithms and help us provide guidelines for test-suite reduction in practice.

Acknowledgments

This work was supported in part by project 'Study on Object-Oriented Software Testing Automation' (60073027) from National Natural Science Foundation of China, by project 'Study on Automated Structure Testing of Object-Oriented Software' (20030201) from Scientific Research Fund of Zhejiang Provincial Education Department.

References

1. Y. Chen and M. F. Lau, Dividing Strategies for the Optimization of a Test Suite. Information Processing Letters, Vol. 60, no. 3, pp.135-141, Mar, 1996.
2. Y. Chen, D. Rosenblum, and K. Vo, TestTube: A System for Selective Regression Testing. Proc. 16th Int'l Conf. Software Eng., pp. 211-222, May 1994.
3. James A. Jones and Mary Jean Harrold, Test-Suite Reduction and Prioritization for Modified Condition/Decesion Coverage. IEEE Trans. On Software Engineering, Vol. 29, no. 3, pp. 195-209, Mar. 2003.
4. Borgelt, K., Software Test Data Generation From a Genetic Algorithm, Industrial Applications of Genetic Algorithms, CRC Press 1998.
5. JC.Lin and PU. Yeh, Automatic Test Data. Generation for Path Testing using GAs, Information Sciences, 131: 47-64, 2001.

6. TIBÖR CSONDES,BALÁZS KOTNYEK, Greedy Algorithm For The Test Selection Problem In Protocol Conformance Testing, Journal of Circuit, System and Computers, Vol. 11, No. 3(2002),273-281, World Scientific Publishing Company.
7. M. J. Harrold, R. Gupta, andM.L. Soffa, A Methodlogy for Controlling the Size of a Test Suite. ACM Trans. Software Eng. And Methods, vol. 2, no. 3, pp. 270-285, July 1993.
8. J. Offutt, J. Pan, and J. M. Voas, Procedures for Reducing the size of Coverage-Based Test Sets. Proc. 12th int'l Conf. Testing Computer Software. PP. 111-123, June 1995.
9. G. Rothermel and M.J. Harrold, A Safe, Efficient Regression Test Selection Techinique. ACM Trans. Software Eng. And Methods, vol. 6, no. 2, pp. 173-210, Apr. 1997.
10. G. Rothermel, M.J. Harrold, J. Ostria, and C. Hong, An Empirical Study of the Effects of Minimization on the Fault Detection Capabilities of Test Suites. Proc. Int'l Conf. Software Maintenance, PP. 34-43, Nov. 1998.
11. Gregg Rothermel, Mary Jean Harrol, Selecting Tests and Identifying Test Coverage Requirement for Modified Software. Proceedings of the 1994 international symposium on Software testing and analysis, Seattle, Washington, United States, Pages: 169-184, 1994 .
12. W. E. Wong, J. R. Horgan, S. London, and A. P. Mathur, Effect of Test Set Minimization on Fault Detection Effectiveness. Software Practice and Experience, vol. 28, no. 4, pp. 347-369, Apr. 1998.
13. D. Berndt, J. Fisher, L. Johnson, J. Pinglikar, and A. Watkins, Breeding Software Test Cases with Genetic Algorithms, Proceedings of the 36th Hawaii International Conference on System Sciences (HICSS 36), Track 9, p. 338a, January 6-9, 2003.
14. D.E. Goldberg, Genetic Algorithms in Search, Optimization and Machine Learning, Addison-Wesley, 1989.
15. Holland, J. H. Adaptation in Natural and Artificial Systems. Ann Arbor, MI: University of Michigan Press, 1975
16. A. Watkins, The Automatic Generation of Software. Test Data using Genetic Algorithms, Proceedings of the Fourth Software Quality Conference, 2: 300-309, Dundee,. Scotland, July, 1995.
17. YANG Jian-Jun, CHEN Wei-Dong, YE Cheng-Qing, and PAN Yun-He, Design and Implementation of Testing Tools for Context-Free Languages, Journal of Computer Research & Development, vol. 37(11): p1375-1382, Nov. 2000.
18. Xueying MA, Weidong CHEN, Jianjun YANG, Chengqing YE, Block-based Test Data Adequacy Measurement Criteria and Test Complexity Metrics, Journal Of Computer Science, Vol. 29, No. 5(141-143), May 2002.
19. M. Roper, CAST with GAs (Genetic Algorithms) - Automatic Test Data Generation via. Evolutionary Computation, IEE Colloquium on Computer Aided Software Testing Tools, digest no. 96/096, April, 1996
20. Pargas, R., Harrold, M., Peck, R., Test data generation using genetic algorithms. Software Testing Verification & Reliability, vol. 9, no. 4, pp. 263-282,1999.
21. Michael, C., McGraw, G., Schatz, M., Generating Software Test Data by Evolution, IEEE Transactions On Software Engineering, 27(12),pp. 1085-1110, December 2001.
22. Wegener, J.; Baresel, A. and Sthamer, H., Evolutionary Test Environment for Automatic Structural Testing. Information and Software Technology, Special Issue devoted to the Application of Metaheuristic Algorithms to Problems in Software Engineering, vol. 43, pp. 841 - 854 (2001).

A Constellation Model for Grid Resource Management

Yinfeng Wang, Xiaoshe Dong, Xiuqiang He, Hua Guo, Fang Zheng,
and Zhongsheng Qin

School of Electronics and Information Engineering, Xi'an Jiaotong,
University, Xi'an, 710049,China
{wangyf, hexqiang, guohua}@mailst.xjtu.edu.cn
xsdong@mail.xjtu.edu.cn

Abstract. By analyzing the demand of the Grid resource management, this paper proposes a constellation model for dynamically organizing and managing Grid resources according to the integrated service capabilities of each node. The logical layer of the constellation model matches with the underlying physical organization. Some evaluation criterions for the integrated service capabilities are proposed in this paper, which can also be used as the criterion in selection of the standby fixed star node in resource management. By defining the minimum resource management unit, the model is more suitable to manage the resources dynamically and easier to realize uniform resource management. The services in the constellation model can be developed according to WSRF and the model conforms to OGSA.

1 Introduction

In essence, the Grid resources are heterogeneous, dynamic and distributed. So it is difficult for distributed and heterogeneous systems to organize and manage the grid resources dynamically. The complexity of managing distributed, heterogeneous large-scale systems increases the administration cost and the risk of human errors.

Therefore, this paper considers Grid resource management from the following two aspects: (1) To reduce the complexity of management and usage, the heterogeneous resources have to be encapsulated and standardized into the uniform logical and abstract service for management; (2) To improve the efficiency of management and usage, the distributed physical resources have to be extracted and standardized into uniform form to match the reality of underlying physical structure. In order to resolve the above problems, Grid resource management can focus on organizing and managing the resources dynamically. The dynamic coordinated sharing of resources is a fundamental capability of Grid. The basic requirement of dynamically organizing and managing resources includes scalability, security and QoS assurance.

As a widely accepted open standard, Open Grid Services Architecture (OGSA) [1] uses service to represent all the Grid resources, which wraps the complexity of management in the service interoperation.

The Web Service technology, uniform resource abstract and common configuration principles make the system homogeneous. As an influential tool for OGSA, GLOBUS makes the local management policy compatible with the Grid Resource Allocation &

J. Cao, W. Nejdl, and M. Xu (Eds.): APPT 2005, LNCS 3756, pp. 263–272, 2005.
© Springer-Verlag Berlin Heidelberg 2005

Mgmt (GRAM)[2]; warps the heterogeneity of local resources. However, providing GRAM implementation with each local resource management tool restricts the scalability of the resource model, which not only leads to some resources managed by the other resource tools failing to be used in Grid environment, but also weakens the general organization and management.

Condor [3] can be considered as a computational Grid that has multiple Condor pools with a flat organization. Condor uses an extensible schema with a hybrid namespace and has no QoS support. Resource discovery is centralized and queried with periodic push dissemination; the scheduler is centralized, too. Both of them tend to become the bottleneck and do harm to the scalability.

SpiderNet [4] can achieve load balance efficiently and select peers based on QoS requirement, but fails to offer an effective method to guarantee QoS once critical path or resource confliction occurs.

All the resource models mentioned above cannot resolve the mismatch problem existing between the distributed abstract layer and the underlying physical structure. Therefore, they can't satisfy the requirements for scalability, QoS assurance and so on simultaneously

We consider there are reliable nodes in Grid, which can provide long-term, reliable and stable services for users. These high available systems can be used as the building blocks to realize stable, highly reliable execution environments in OGSA.

With the publishing of the OGSA/OGSI, and later the WS-Resource Framework (WSRF), service becomes a more and more important concept in grid. The service-oriented solutions to the grid problems have been the common realization in the field of grid. In the constellation model, each resource is encapsulated as a standard WSRF service, with the assistance of the Globus Toolkit 4(GT4). The services can be accessed through uniform and standard interfaces, which make the heterogeneous resources easy to integrate. The grid nodes are classified into different categories, such as planet node, fixed star node, according to their ISC properties. The constellation model proposed in this paper can guarantee that the logical topology matches the underlying physical organizations.

2 Components of Constellation Model

Definition 1. *Planet* is the node that resides in a steady physical location, encapsulates self-resources and can provide service.

Definition 2. *Meteor* is the node that lies on the instable physical location, can provide service, and may be mobile computing objects.

Definition 3. *Integrated Service Capabilities* (ISC) *value* is the overall evaluation criterion of a node's service capabilities, including the node's computation resources, the network bandwidth and the availability (Mean Time To Failure, MTTF). The ISC value metric Φ can be defined as the weighed average for these values as in the equation 1.

$$\Phi = \sum_{i=0}^{m-1} w_i \times \frac{r_i}{rs_i} + w_m \times \frac{b}{bs} + w_{m+1} \frac{t_{MTTF}}{ts} \tag{1}$$

r_i: the available node's computing resources, such as a single CPU speed, number of CPUs, memory, disk capability and so on, $0 \leqslant i \leqslant$ m-1 ;

b: network bandwidth;

t_{MTTF}: node's Mean Time To Failure;

rs_i: benchmark for different computation resources, the units adopt Gflops, Gbit and so forth;

bs: benchmark network bandwidth, the unit adopts Mbps or Gbps;

ts: benchmark MTTF, unit is hour;

w_i: weight factor; $\sum_{i=0}^{m+1} w_i = 1$, administrator can define each object's weight factor

according to its relative significance.

Definition 4. *Fixed star* is the node that has the maximum ISC value in Solar system, or a fixed star can be designated by selecting one from the nodes whose ISC value exceed the criterion. Fixed star manages the planets, meteors around. The management includes registering, scheduling, monitoring, discovery, controlling, and accounting, etc.

Definition 5. *Solar system* owns one fixed star at most, the numbers of Planets and Meteor is x and y respectively, and then x, y $\geqslant$ 0. The system obeys the same sharing and management rules such as policy of security and etc. It can be represented as a fourtuple:

Solar system={Fixed star, Planet [x], Meteor [y], Rule}.

3   Construction of Solar System

3.1   Solar System Construction Strategy

Organization Strategy: Solar system can be organized according to geographical dispersal or service types. When it comes to the nodes that fail to find suitable specialty organizations, based on the geographic dispersal strategy, they are allowed to join the local solar system

Restriction 1. Each planet node belongs to one solar system at most.

As the management and sharing rules given as follows, if some planets were controlled by more than one fixed star nodes, once Rule of the solar systems changed, mutual conflict of management and sharing rules would occur, which would make the planet at loose ends. Hence, in order to decrease management complexity and uniform designing and management, it should be restricted that each planet belongs to only one solar system at most.

In figure 1, Rule 1 and 3 represent the rule sets of planet nodes, and Rule2 represents the rule sets of fixed-star node. The rule sets of solar system Rule = Rule1 ∩Rule2∩Rule3, which reflects the relaxation degree of the solar system management. Based on the management and sharing rules, Service Level Managers (SLMs)[1] perform the negotiation with Service level agreement (SLA) in solar system. If the intersection of Rule1 and Rule3 is relatively great, the fact accounts for that the two rule sets have much in common; the two planets are tightly coupled, and they can collaborate to perform some jobs.

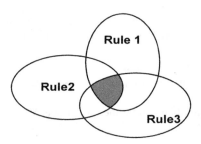

Fig. 1. Solar system Management and sharing rules

Community Authorization Service (CAS) [5] or PKI, Certification Authority (CA) can be used to in the security of solar system. Solar system owns Self-Management service, including self-configuring, self-healing and self-optimizing. Self-Management Services can reduce the cost and complexity of managing the system. Such as Simple Network Management Protocol and etc. enable the interoperability among the solar system node.

To avoid single points of failure and improve scalability, in the nodes, we should appoint a node whose ISC value exceeds the criterion as the standby of the fixed star, and the standby fixed star should keep updating to make sure of consistency.

3.2 Update Solar System Node

Employing WS-ResourceProperties and WS-ResourceLifetime of WSRF, we can regard nodes as a kind of service to manage. To improve stability, once the solar system comes into being, any fresh node can be taken as a standby fixed star at most.

The Solar-system Update Algorithm can be described as bellows:

Step1) Check the identity of incoming nodes; if ISC value $\geqslant$ the evaluation criterion of fixed star, then go to Step3); else go to Step4).

Step2) If quitting node's identity is fixed star, then go to Step5),else is standby fixed star go to Step6); Otherwise go to Step7).

Step3) In terms of the evaluation criterion, if the incoming node accords with the appointment requirement or its ISC value $\geqslant$ the standby fixed star, then the newcomer upgrades to the standby fixed star, backups and keeps updating managing data which is consistent with the fixed star. Meanwhile the former descends to planet. Else go to Step4).

Step4) If newcomer lies on the steady physical location, then node joins the solar system as a planet, else as meteor ; managed by the fixed star.

Step5) Fixed star quits, if there is standby fixed star, management Agent moves to evaluation criterion of fixed star; a planet is appointed as a backup fixed star; else, i.e. there is no standby fixed star, the solar system fails. Its meteors remain to join the neighbor solar system following the geographic dispersal strategy.

Step6) Standby fixed star quits, if there are planets satisfying the evaluation criterion of fixed star, upgrade one of them as the standby fixed star; else, the in-service fixed star updates the managing data.

Step7) A planet or meteor quits; the fixed star update managing data.

maximum response time is T_{max}, while the maximum delay is not allowed to exceed 50 percents, then the corresponding QoS range $<a_i^{min}, a_i^{max}>$ could be expressed as $<0, 1.5 \times (estimated\ response\ time)>$. Through monitoring, scheduling, and tuning the running services, The SLMs[1] deployed on the fixed star negotiate a SLA within the QoS range to realize the Service level attainment.

4.2 Inter-constellation Connections

In the constellation model, each constellation is self-managed and thus independent of the others. Routers, links and end systems may fail and their performance may fluctuate over time. The network transience occurred between solar systems are also independent. In dynamic grid environment, overlay network is required to be built to organize the redundant network links in the physical layers, thus the availability and scalability of the constellation will be improved, while the number of transience occurrence could be decreased.

In the overlay network of P2P systems, a node maintains an overlay routing table containing the IDs of a small set of other overlay nodes. Each such entry can be thought of as a virtual, direct link between the current node and the table entry. They are neighbors to each other. The P2P-overlay network is loosely coupled and dynamic, and easily to be constructed. But the overhead of the structured overlay network is relatively too high while organizing the dynamic resources.

In P2P, the inefficient overlay topology and the blind flooding is unnecessary, which makes the unstructured P2P systems far from scalable [8]. Topology mismatch between the P2P logical network and physical underlying network results in that the messages may be relayed multiple times between the same physical links incurring a large volume of unnecessary traffic.

The Location-aware Topology Matching (LTM) algorithm proposed in [9] can be used to disconnect the redundant slow connection between solar systems in constellation and construct the P2P overlay network, which matches the underlying physical network. The solar systems can be organized together through the overlay network reliably and efficiently. The resources in grid are dynamic and distributed, which requires the solar systems be able to collaborate. The performance can be improved and the provided peers are able to actively collaborate through the powerful overlay network.

4.3 Dynamic Resource Organization Based on Solar Systems

The VOs in the constellation model could be existed inner the solar systems or constellation, or inter the constellations. The minimum unit of the grid resources is a service, and the solar system is the minimum unit of grid resources to be managed. The reasons are the followings:

1. In Grid, all the resources are encapsulated as services. Users or VOs invoke the services through uniform interfaces. Only the services are visible to the outside. So the service is the minimum unit of grid resources.
2. If the dynamic resources are managed by direct managing the services, when the resources changes, the services must be modified and reconfigured to adapt to

the changes. This may makes the services unable to be shared if the modification information cannot be notified to all the users timely.

3. The dynamic resources are managed in the unit of solar system. The fixed stars are responsible for managing the resources securely and efficiently without changing the organization rules of the solar system.
4. The VOs in the solar system are only managed by the solar system.
5. According to Restriction 1, a planet is managed by only one solar system.

So the solar system is the minimum unit of grid resources managed in the constellation model.

The solar system, as the minimum management unit of grid resources, contributes to the formation of a uniform global name space. The WS-Addressing of a service can be extended to point out the fixed star node to which it belongs. It can be guaranteed that the name of a service is unique by referring the address of endpoint of the node where the service is deployed.

In VOs, sharing relationship among participants is peer-to-peer in nature [10]. In the forming of the VOs, peer members need to be found through the Metadata service while the VOs are cross-organization. Compared to the existing resource management models, the constellation model treats the solar system as the minimum unit of the resource management. The peer members could be found more easily, which could benefits the aggregation of the Metadata service, resolves some security problems existing in site autonomy, such as security Policy Exchange [1], and benefits the global uniform management and the formation of a single system image as well.

4.4 QoS Assurance in Dynamic Organization of Resources

In the dynamic organization of resources, the QoS assurance requirements include Service level agreement, Service level attainment, and Migration [1]. The QoS-Radius in the constellation model is negotiated by the SLMs in all the solar systems.

Take the QoS migration problem as an example, when a certain service in a planet or fixed star is being used by one user or VO, at the same time, another VO with a higher priority issues a request to occupy the service, then the fixed star will make the running tasks migrate to satisfy the higher-priority VOs. In the constellation model, the solar systems may share some common service rules and services. When the high-priority VO issues a request to the fixed star, the SLM of the fixed star will be responsible for the migration of the running tasks inner the solar systems or between different solar systems. If the migration cannot be performed, then the checkpoint of the running task will be retained.

Once the high-priority VO releases service, the SLM will decide whether the task should be retrieved back to the original fixed star if the task has not been finished, or continue the task from the checkpoint. Utilizing resource across organization introduces stricter accounting requirements. Mechanisms for collecting and exchanging necessary information for accounting are required [1]. In constellation model, the accounting is also managed by the SLM.

5 Relationships with OGSA

In constellation model, the basic web services (encapsulated resources) are deployed inside the solar systems. The fixed star, planet, and meteor nodes are the resident host of the basic services that support the base resources of OGSA.

The solar systems map to the infrastructure level of OGSA, shown in figure 3. The agent that is running in the fixed star monitors, accesses, and terminates the resources inside the solar systems, and performs the capabilities required by the base manageability model, such as discovering. The service instances can be created, monitored and destroyed through the generic manageability interface that the agents provide.

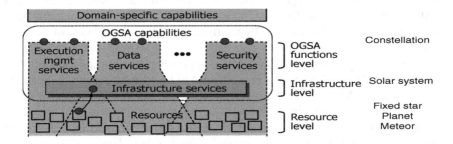

Fig. 3. Relationship between Constellation model and OGSA

The constellations correspond to the function level of OGSA. By composing the relatively simple self-management functions of lower-level solar systems, the complex self-management function of the constellation can be realized. The scalability of the constellation model can be improved by providing standard functional interfaces and manageability interface to the public services of OGSA. For example, the interface to create and destroy the tasks should enable to be invoked by the service that manages the operation process of services.

The solar systems in the constellation model can provide metadata services for the cross-organization behaviors, such as identity authentication/authorization, policy transaction or negotiation, which is necessary in OGSA. The self-management function of the constellation can provide standard mechanisms to collect or exchange the information between organizations.

6 Application Experiences

The National High Performance Computing Center (NHPCC) of Xi'an Jiaotong University (XJTU) is one of the eight nodes in the China National Grid. Based on the Xi'an node, a prototype system has been developed for dynamically sharing the Computational Fluid Dynamics (CFD) Fluent software licenses, and in this system the constellation resource management model is used. The purpose of the system is to integrate the Fluent software licenses [11] in Xi'an and Shanghai areas, and realize dynamic sharing of the licenses according to the demands of the users.

Each university in Xi'an area that has the Fluid Computing Environment (FCE) can join in as a solar system. All the FCE of universities in Xi'an area constitute the Xi'an CFD constellation, while the Shanghai constellation is composed of the FCEs in Shanghai area, such as Shanghai Jiaotong University (SJTU), Supercomputer Center (SSC) etc. The NHPCC can be designated as the fixed star, while other computing centers or clusters in XJTU can join in the solar system as planet or meteor nodes. Totally there are four licenses for Fluent software in XJTU solar system. Figure 4 shows the structure of the CFD system based on the Constellation model.

Once a certain task is submitted to system, the solar system will firstly choose a suitable resource for it according to the scale of task.

The solar system manages the software registry, scheduling of requests from users, the collection of resource information, the software resource search engine and the task execution. The fixed star monitors the loads of the planet nodes, and schedules the tasks according to the information. If the current Solar System is heavily loaded, new tasks will be migrated to another Solar System with light load in the same constellation.

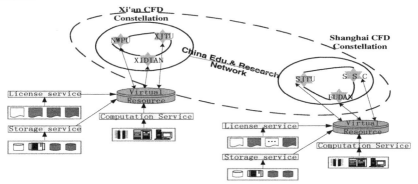

Fig. 4. The illustration of the Constellation model for CFD

When the Fluent licenses in Xi'an area are not enough in use, the Xi'an CFD Constellation will negotiate with Shanghai CFD Constellation to share the Fluent licenses resource, where more eight Fluent licenses can be shared. The licenses from Shanghai constellation will be transmitted through China Education and Research Network (CERNET) to active the Fluent software.

7   Summaries and Future Work

This paper proposes a resource model named Constellation for dynamically organizing the grid resources. By analyzing each component of the constellation model, the metrics for measuring the capabilities of the grid nodes are proposed. The demands of the overlay network used to connect the solar systems or constellations are also analyzed, and a P2P-like overlay network combined the LTM technology is proposed in this paper to make up the gaps between the logical network and the underlying physical network. The introduction of solar system, the minimum resource management unit, can contribute to better organizing the dynamic resources across multiple administrative domains and easily implement the global and uniform resource management.

There are several aspects in which constellation model could be advanced. Security functionality should be strengthened at all layers, the model should be more scalable and the QoS guarantees need to be improved. We will do some research in these issues in the future.

WSDM-WS has published the specification of MUWS (Management Using Web Service) and MOWS (Management Of Web Service) in March 2005. The related future works will accord with the two specifications.

Acknowledgments

This research is supported by China Education and Research Grid (Grant No.CG2003-CG008) and 863 projects of China (Grant No.2002AA104550)

References

1. I. Foster, Argonne & U.Chicago (Editor): Open Grid Services Architecture GWD-I (draft-ggf-ogsa-spec-019). http://forge.gridforum.org/projects/ogsa-wg.
2. http://www-unix.globus.org/toolkit/docs/development/4.0-drafts/GT4Facts/index.html.
3. http://www.cs.wisc.edu/condor/.
4. Xiaohui Gu, Klara Nahrstedt: QoS-Aware service composition for large-scale P2Psystem. Chapter 24.Grid Resource Management, Kluwer Publishing, Fall 2003.
5. L. Pearlman, et al.: A Community Authorization Service for Group Collaboration. IEEE 3rd International Workshop on Policies for Distributed Systems and Networks, 2002.
6. Charlie Catlett: The TeraGrid: Progress and Applications. February 2004.
7. http://lcg.web.cern.ch/LCG/.
8. J. Ritter: Why Gnutella Can't Scale. No, Really. http://www.tch.org/gnutella.html, 2001.
9. Yunhao Liu, Li Xiao, Xiaomei Liu, Lionel M. Ni, Xiaodong Zhang: Location Awareness in Unstructured Peer-to-Peer Systems. IEEE Transactions on Parallel and Distributed Systems, VOL. 16, NO. 2, February 2005, pp. 163-174.
10. J. Joseph, M. Ernest, C. Fellenstein: Evolution of grid computing architecture and grid adoption models, IBM SYSTEMS JOURNAL, VOL 43, NO. 4, 2004, pp.624-645.
11. Fluent RLicense Service Description. http://www.fluent.com/software/rlicense/descr.htm.

An Effective Information Service Architecture in Grid Environment

Huashan Yu, Yin Luo, Xingguo Zhu, and Xiaoming Li

School of Electronics Engineering and Computer Science, Peking University,
100871 Beijing, P.R. China
yuhs, lxm@pku.edu.cn, luoyin@ailab.pku.edu.cn
zxg@db.pku.edu.cn

Abstract. The information service is a vital part of any Grid software platform, providing the fundamental mechanism for discovering and monitoring services and resources in a Grid. This paper presents an effective information service architecture developed for ChinaGrid Support Platform (CGSP). The architecture is based on domains in CGSP, which are autonomous grid systems. The key issues are to collect information of diverse resources within a domain dynamically, and to share these collected data across domains effectively and securely by providing a unique interface based on a delegation method and an index aggregation mechanism.

1 Introduction

Grid technologies [3] have emerged as an attractive distributed computing paradigm over wide-area network, focusing on large-scale resource sharing and coordinated problem solving in a dynamic, efficient and secure way. The term "resource" refers to hardware, software and data, such as computers, networks, sensors, programs and databases. These resources usually are heterogeneous and contributed by different organizations. To realize the goal of Grid computing, mechanisms for describing, discovering and monitoring the diverse resources are required. Let's consider a BLAT problem for example. Among the numerous computers, only a few can provide the gene data required by the problem, and the executable version required by each candidate computer is platform-dependent. Some candidates are currently too busy to be appropriate for the problem. Moreover, these candidates are owned by different organizations, therefore it's important to provide a consistent way for problem submitters to monitoring job statuses on remote computers. To address the above challenges, we have designed and implemented an information-service architecture in ChinaGrid Support Platform (CGSP) [2], which is a service-oriented Grid middle-ware for the development of ChinaGrid [1] and its applications.

ChinaGrid aims to utilize resources distributed in Chinese education and research network with Grid technologies to provide high-quality services for scientific research. CGSP consists of five modules: Grid Portal, Grid Development Toolkits, Information Service, Grid Management and Grid Security. The Grid Management includes: Service Container, Data Manager, Job Manager, and Domain Manager [1]. CGSP has been

J. Cao, W. Nejdl, and M. Xu (Eds.): APPT 2005, LNCS 3756, pp. 273 – 281, 2005.
© Springer-Verlag Berlin Heidelberg 2005

successfully applied in the development of ChinaGrid's campus Grid in more than 20 universities. Every campus grid is a domain of ChinaGrid, and it has its own resource sharing policies and provides a mechanism to handle delegated request from other domains.

Information Service of CGSP (CGSP-IS) is a service-oriented architecture conforming to OGSA [4] and WSRF [17]. First, CGSP-IS takes the form of services, which means it is in fact a group of services providing several kinds of information about resources in the Grid; Secondly, CGSP-IS only takes care of services, which means we view all the resources in the grid as services and we describe, discover and monitor a resource via a service which represents it. To facilitate the development of CGSP-IS, we propose an architecture, named Global Information Service Architecture (GISA), which addresses the issues encountered in building information services for diverse, dynamic and distributed resources. The key issues are how to collect diverse resources within a domain dynamically and effectively, and then how to share these collected data across domains effectively and securely by providing a unique interface based on delegation method and index aggregation mechanism, and hence providing a global view of resources to users in any domain.

The rest of this paper is organized as follows: Section 2 presents GISA's architecture and related mechanisms. Section 3 gives a brief description of GISA's implementation in CGSP. Section 4 overviews some related works, and a conclusion about this paper is presented in section 5.

2 Information Service Architecture

In this section, we present the functional and architectural design of CGSP's information service in detail. All resources in CGSP are presented as Grid services conforming to WSRF. However, developing and running Grid applications directly on this abstract level is very complex, since the number of Grid services is very large and statuses of their hosting environments are dynamic. Information service in CGSP plays an intermediary among resource providers, Grid application developers and job submitters.

2.1 Overview

As the infrastructure of the information services in CGSP, GISA is built on the concept of domain. The term "domain" refers to an independent and autonomous grid environment, which provides services to the users and modules within the domain. People establish domain for many reasons. For example, a bio-informatics Grid is a domain established according to application domain; the campus grid of Peking University is also a domain. Although domain has autonomy in nature, this concept is introduced to integrate grid systems into a larger one, which provides larger scale of and multi-level policies for resource sharing. The integrating process occurs when a domain joins another as its child, and thus a recursive integration can lead to the integrated domains have tree-like structure. It is important to notice that for transparency and flexibility considerations, each domain knows nothing about the whole topology except its father and children.

Information Center (IC) is another important concept for GISA. It is an abstract information-serving module with the following three interfaces: 1) to collect information from all kinds of resources in a domain; 2) to provide information to the users and other modules in a domain; and 3) to provide information to ICs in other domain. We introduce this concept for two reasons: one is that although a grid system usually has multiple information services for diversity, scalability and fault-tolerance considerations, a logic single access point for these services will help to design a general framework; the other reason is that to address the challenge of integrating heterogeneous grid system, an abstraction of various information serving modules in different systems is necessary.

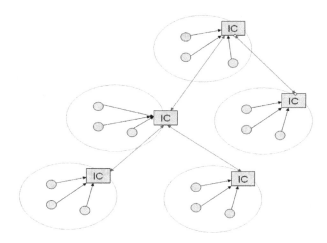

Fig. 1. GISA based on integrated domains

Based on these tow concepts, GISA has the same tree-like structure as the integrated domains, as illustrated in Fig. 1. Each domain has an Information Center to collect data, and the collected data can be shared with its parents and children. Correspondingly, GISA is separated into two parts: one is for monitoring and discovering resources within a domain, named Resource Discovery and Monitoring Architecture (RDMA); another is for sharing resource information by providing a global view of resources in multiple domains, named Global Resource Sharing Mechanism (GRSM).

2.2 Resource Discovery and Monitoring Architecture

Due to the diversity of Grid resources, it is difficult to develop a general resource monitoring and discovering system. The key issues are: 1) describing the state of every resource in a generic way; and 2) discovering appropriate resources among a large number of distributed resources.

Because of the considerable diversity and openness of resources in a grid environment, it is obvious that the language used to describe the state of a particular kind of resource should be customized. Therefore, mechanisms are required to keep the consistency of vocabulary, syntax and semantic between the resource provider and consumers. In fact, RDMA uses XML [21] to describe resource state, and utilizes **resource template** to assure the agreement between the providers and consumers. A resource template defines the content, format and other properties of the information used to describe a kind of resource. The definition should represent nature of the resource and satisfy the need of resource discovery. Generally, resource templates are defined by application experts. Every resource must be associated with a resource template for its discovery and monitoring. In RDMA, every resource template contains a XML schema to define the state of resource.

For the second issue, we prefer to aggregate the information of all the resources into one central place and make query upon the aggregated information. Thus, at conceptual level, RDMA has a classic tree-like structure, which comprises three levels of modules:

Information Provider (IP) is an agent who takes charge of feeding raw data to RDMA. The data's syntax and semantics are arbitrary, ranging from a single item about a workstation, such as the current load of CPU or the available storage capability, to a complex information model about a cluster with hundreds of nodes at which a rare operating system and other valuable software are installed. The way that an IP retrieves data is also arbitrary: it can query the machine's state by invoking some system APIs, or just copy data from a file output by other programs. An IP can be any program that is customized and provided by the contributor of the resource, using some predefined interface to communicate with the Resource Monitor with which it associated.

Resource Monitor (RM) is a service that represents a resource by monitoring and publishing its state. As mentioned above, each resource has a resource template to specify what pieces of information should be collected by RM. According to the associated template, RM starts monitoring via some appropriate IPs and publishing to IC and other users. The way a RM gets its associated resource template can be arbitrary.

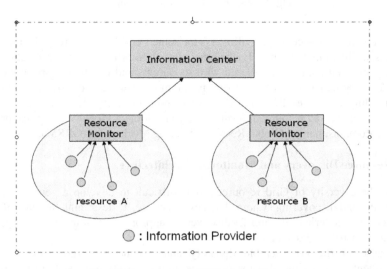

Fig. 2. The structure of RDMA in a domain

Information Center (IC) is a group of services that are responsible for aggregating information from RMs and providing user interfaces for discovering and monitoring resources. Each service is associated with one or more resource template, and every resource template represents a kind of resources that the service is responsible for. The contributor of a resource should register it to the corresponding service in IC, according to its resource template, with the address of its RM, and then the service communicates with RM to retrieve the state of resource. The way to retrieve data can be arbitrary.

As shown in Fig. 2, Resource Monitors, who has the resource template for the resource, know what data item should be collected, so it starts some providers according to a configure file, and starts reporting the state of the resource to IC.

2.3 Global Resource Sharing Mechanism

GRSM is based on the concept of domain. As mentioned in chapter 2.1, a domain refers to an independent and autonomous grid system. The grid system provides services to the users and modules within the domain, has its own necessary runtime modules such as portal, execution management, data Management and information service. Although domain has autonomy in nature, it is introduced to integrate grid systems into a larger one that provides larger scale of and multi-level policies for resource sharing. As far as information services is concerned, domain has the following features:

– **Autonomy:** The process of integration should not conflict with the autonomy of the involved domains. A user of a domain cannot log in other domain and utilize the services provided in the domain unless it has a legal mapping identity. To information service, this feature means that the information of a resource should not be exposed outside the domain except the resource sharing policies allow that.
– **Heterogeneity:** Since domains are autonomous grid systems, they could be heterogeneous. To build an information service based on heterogeneous domains, we must consider the ICs in different domains have different methods to express information and the query language is also different.
– **Uniformity:** The integrated grid system with multiple domains should provide a global view to the users of all the domains. Though, according to different local resource sharing policies in different domains, a user cannot get a complete view of resources in a remote domain, he/she should find all resources as possible as the resource sharing policies allow via a unique interface.
– **Flexibility:** Reconstructing the integrating relationship of domains is easy. Due to a recursive integrating process, the structure of all involved domains can be tree-like, and thus a domain can join the family at any level, except the root, and can quit at any time. To information service, this flexibility leads to a dynamic view of resources.

These features of domain make an information service, providing unique interface based on multiple autonomous and heterogeneous domains, hard to be a reality. However, we are willing to contribute our efforts on this problem.

Due to the feature of autonomy of a domain, a user cannot issue a query request to the IC in a remote domain, because the remote domain cannot accept the request according to its resource sharing policies that usually require a user identity within the system. To handle this problem, GRSM requires that each IC should be equipped with a mechanism to map a remote user identity to a local one and provide an interface to accept remote request from the ICs in other domain. The policy used in the identity mapping mechanism depends on each domain. Hence, when a user issues a query request to a specific domain, the local IC should delegate the query request to the remote IC and retrieves the result from it. For example, a user in domain A wants to issue a query to domain B. First, he/she issues a request to local IC, and the IC in domain A delegates the user's request to the IC in domain B with the user's identity in domain A; and then the IC in domain B maps the user's identity into a local one and makes a local query, and then returns the result of the query back to the IC in domain A which returns the result to the user.

However, because of absence of the whole topology, as shown in Fig. 3, a remote request can just delegate to the parent domain and child domains. A global query can be done by flood the query request to all integrated ICs by a recursive delegation process. In this case, the request will cost a lot.

To accelerate the query process based on a global view of resources in all domains, the only way is to expose the information in an IC to other IC directly. GRSM provides an interface to administrator to configure what part of information in an IC can be exposed directly and a mechanism to use these exposed information. In the mechanism, we utilize the following things to facilitate the global query process.

- **Index** is a mapping from the information that an IC has, which can be completely exposed to other ICs. In general case, an index is just a part of the whole piece of information and, in more sophisticated case, an index could be some data derived from the original information.
- **Query Operator** is an agent who can query upon the index. Due to the heterogeneity feature of domains, how to express information and how to do a query upon the information can be various. When the administrator of an IC exposes some index, he must provide its associated Query Operator. In practical sense, Query Operator can be Java class.

Once a domain is established, it is a trivial root node without any children. When this domain is integrating into another one for larger-scale resource sharing, the IC of this domain is also integrating into the counterpart of the parent domain. In GRSM, As shown in Fig. 3, once the parent-child relationship is set up, the index and its Query Operator in the child domain starts aggregating into the parent domain. In this way, an upper IC will have all the indices and Query Operators that its descendants have.

When a user issues a global query for a resource meets his needs, the IC in local domain first makes a local query on the information of its own and the indices aggregated from its descendants, and if no resource can meet the needs, it delegates the query to its parent and children who will do the same process.

In this pattern, the upper IC will have more load because of the aggregated indices, and the consistence between original data and its index can be a problem. However, we suppose that the index exposed should be relatively static and limited when the network does not have a good performance.

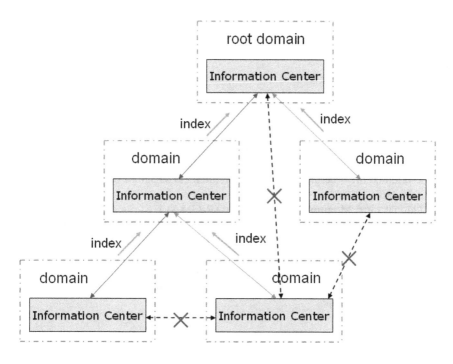

Fig. 3. Relationship between ICs in different domains

3 Implementation

GISA has been implemented in CGSP-IS (see Sect. 1). Based on RDMA and GRSM, CGSP-IS provides information services for three kinds of resources, namely, grid-service, hyper-service, and physical resource. A grid-service is a web-service or a WS-Resource. A hyper-service is an abstract grid-service defined by its port-type and can be dynamically bind to a physical grid-service, according to some QoS requirement. And a physical resource is a platform on which grid-services run. Based on these information services, CGSP supports execution management across domains, which makes the resource sharing and dynamic problem solving over multiple campus a reality.

In CGSP, RDMA is based on WSRF, where a RM is implemented as a WS-Resource [17]. For example, a RM for a computer can be WS-Resource providing resource properties such as disk capacity, CPU load and so on. The state of resources is aggregated into IC using the standard method GetResourceProperty as the pull approach and the mechanism provided by WS-Notification [17] as the push approach. And the definition of resource template is just consistent with the definition of ResourceProperties, but we do not remove resource template because we extend the template to specify the approach to update the state of resource for RMs.

For GRSM, the domains in CGSP 1.0 are actually all homogeneous, and the structure of integrated domains is a two-layer tree with a root domain and about 20 child domains. And each IC aggregates all information as index to the root IC due to the limited number of domains and the good stability and high band-width of the underlying network, CERNET.

4 Related Work

The Aggregator Framework of WS MDS [15] provides a general mechanism to collect data from distributed WS-Resource, which is similar to RDMA's implementation in CGSP 1.0, but it do not focus on the resource sharing across domains.

MDS-2 [6] is based on a configurable gateway Grid Resource Information Service (GRIS) and an aggregating component Grid Index Information Service (GIIS) in a service-oriented architecture [4]. It has a hierarchical structure based on VO, but it is not for domains either.

Relational Grid Monitoring Architecture (R-GMA) [18], as an implementation of GMA [14], is based on relational data model and provides a centralized architecture in which Producers feed data to a central registry and Consumers query the registry for locating a Producer.

Compared with these existing approaches, GISA focus on providing a unique interface for a global query of resources in domains that are autonomous and independent grid systems.

5 Conclusions and Future Work

In this paper, we propose an information service architecture GISA, serving for CGSP, which is based on the concept of domain and the way in which domains integrate, and provides a mechanism to provide resource monitoring and discovery services over domains. GISA consists of two parts: RDMA and GRSM, which are responsible for discovery and monitoring resources within a domain and resource information sharing across domains respectively.

RDMA utilizes resource template to define the format and other properties of state of resource. A resource template must be carefully designed to achieve the agreement between information provider and information consumer.

GRSM requires a user identity mapping and delegation mechanism and use index to accelerate a global query process.

In the future, we will focus on how to improve the performance of GISA. The key point is to remove the recursive delegation process for a global query, which requires the global view of domains. And we are going to provide a mechanism to generate resource template by inheriting from an existing one.

Acknowledgement

This paper is supported by ChinaGrid project from Ministry of Education of China (CG2003-CG/GP001), and National Science Foundation of China (NO. 60303001, 90412010). We would also like to give special thanks to LU Fakai, GAO Aiqiang at Peking University for their contributions to the initial design and implementation of our work.

References

1. H. Jin, "ChinaGrid: Making Grid Computing a Reality," Proc. Of The 7th International Conference of Asian Digital Libraries, Shanghai, P.R.China, Dec. 2004, pp.13-24.
2. ChinaGrid Support Platform, http://www.chinagrid.edu.cn/CGSP.
3. I. Foster, et al., "The Anatomy of the Grid: Enabling Scalable Virtual Organizations," International Journal of High Performance Computing Applications, vol. 15, 2001.
4. I. Foster, C. Kesselman, "The Physiology of the Grid: An Open Grid Services Architecture for Distributed Systems Integration", J. Nick, S. Tuecke, (2002).
5. I. Foster, C. Kesselman, "Globus: A Metacomputing Infrastructure Toolkit", International J. Supercomputer Application, (1997), 11(2), 115-128.
6. K. Czajkowski, S. Fitzgerald, I. Foster, and C. Kesselman, "Grid Information Services for Distributed Resource Sharing," in proceedings of the 10th IEEE International Symposium on High-Performance Distributed Computing (HPDC-10), 2001.
7. X. Zhang, J. L. Freshl, and J. M. Schopf, "A Performance Study of Monitoring and Information Services for Distributed Systems," in proceedings of the 12th IEEE International Symposium on High Performance Distributed Computing (HPDC-12), 2003.
8. I. Foster and C. Kesselman, The Grid 2: Blueprint for a New Computing Infrastructure: Morgan Kaufmann Publishers,2003.
9. J. Frey and T. Tannenbaum, "Condor-G: A computation Management Agent for multi-Institutional Grids," Journalof Cluster Computing, vol. 5, pp. 237, 2002.
10. N. Furmento, W. Lee, A. Mayer, S. Newhouse, and J. Darlington, "ICENI: An Open Grid Service Architecture Implemented with Jini," in Parallel Computing, vol. 28, 2002, pp. 1753-1772.
11. M. Litzkow, M. Livny, and M. Mutka, "Condor – A Hunter of Idle Workstations," in proceedings of the 8th International Conference of Distributed Computing Systems, California, 1988.
12. A. Medina, A. Lakhina, I. Matta, and J. Byers, "BRITE: An Approach to Universal Topology Generation," in proceedings of the International Workshop on Modeling, Analysis and Simulation of Computer and Telecommunications Systems (MASCOTS), 2001.
13. W. Smith, A. Waheed, D. Meyers, and J. Yan, "An Evaluation of Alternative Designs for a Grid Information Service," in proceedings of the 9th IEEE International Symposium on High Performance Distributed Computing (HPDC-9), 2000.
14. B. Tierney, R. Aydt, D. Gunter, W. Smith, V. Taylor, R. Wolski, and M. Swany, "A Grid Monitoring Architecture," The Global Grid Forum GWD-GP-16-2.
15. WS MDS, http://www.globus.org/toolkit/docs/4.0/info/.
16. Globous Toolkits, http://www.globus.org/.
17. The Web Services Resource Framework, http://www.globus.org/wsrf/.
18. "DataGrid Information and Monitoring Services Architecture: Design, Requirements and Evaluation Criteria, Technical Report.," DataGrid.
19. China Education and Research Network, http://www.edu.cn/
20. B. Tierney, B. Crowley, D. Gunter, M. Holding, J. Lee, and M. Thompson. A monitoring sensor management system for grid environments. In Proc. 9th IEEE Symp. on High Performance Distributed Computing, pages 97–104, 2000.
21. XML, http://www.w3c.org/xml/

An Efficient Data Management System with High Scalability for ChinaGrid Support Platform*

Hai Jin, Wenjun Gong, Song Wu, Muzhou Xiong, Li Qi,
and Chengwei Wang

Cluster and Grid Computing Lab.,
Huazhong University of Science and Technology,
Wuhan, 430074, China
hjin@hust.edu.cn

Abstract. There are a great number of data intensive applications in ChinaGrid. They require an efficient and high performance data management. The data management in *ChinaGrid Support Platform* (CGSP) supplies a data access mechanism with location transparency, name transparency, and protocol transparency as while as ensuring the transfer efficiency. The data management system consists of five parts: the storage data server based on Global Distributed Storage System to guarantee the reliability and performance of data transfer; the storage resource agent to discover, publish and catalog the storage resources; the data logical domain management to enable applications to select specific storage resources; the metadata management to publish, query and access metadata; and the uniform data access entry to organize grid users' data space. We present the design philosophy of the efficient data management system with high scalability for CGSP and also give preliminary performance results.

1 Introduction

Nowadays, a great number of data-intensive applications are emerging. These applications need many researchers to work together in one or more domains to analyze and process the shared data. We see collaborations of hundreds of scientists in areas such as gravitational-wave physics [1], high-energy physics [2], astronomy [3], and many others coming together and sharing a variety of resources with common goals. These applications require the efficient management and transfer of terabytes or petabytes of information in wide-area, distributed computing environments. The users should be able to move large datasets to local sites or other remote resources for processing. They may want to put their datasets only on several specified storage resources or share their data to others. The storage resources may be heterogeneous. They may just want to find a space to store their datasets. Grid technologies [4] enable efficient resource sharing in collaborative distributed environments.

ChinaGrid [5, 7] project integrates all kinds of resources in Chinese universities to make use of heterogeneous grid resources cooperatively. It provides transparent grid

---

* This paper is supported by ChinaGrid project of Ministry of Education of China, National Science Foundation of China under grant 60125208 and 90412010, Hubei Science Foundation under grant 2004ABA053.

J. Cao, W. Nejdl, and M. Xu (Eds.): APPT 2005, LNCS 3756, pp. 282–291, 2005.
© Springer-Verlag Berlin Heidelberg 2005

services with high performance, high reliability for all kinds of scientific computing and research. *ChinaGrid Support Platform* (CGSP) [8] is the core middleware of ChinaGrid, which also provides development environment for grid application. CGSP contains five building blocks [9]: Grid Portal, Grid Development Toolkits, Information Service, Grid Management, and Grid Security. Grid Management contains four parts: Service Container, Data Manager, Job Manager, and Domain Manager [5, 9].

Data management is the core service in CGSP, which manages heterogeneous storage resources and data in grid environment. It includes four key functionalities: 1) reliable and efficient transfer mechanism based on *Global Distributed Storage System* (GDSS) [6]; 2) Data Logical Domains based on physical storage resources which provide great flexibility for user to reorganize the resources; 3) transparent file accessing mechanism shielding the heterogeneous storage resources and transfer protocol; 4) and the storage resource management organizing the heterogeneous resources. Therefore, we design data management system five parts: 1) data storage server based on GDSS to guarantee the reliability and performance of data transfer; 2) storage resource agent to discover, publish and catalog the storage resources; 3) the Data Logical Domain management to enable applications to select specific storage resources to share their data; 4) the metadata [10] management to publish, query and access metadata; 5) and the Uniform Data Access Entry to organize grid users' data space and provide a series of data access API for users.

This paper is organized as follows. Section 2 presents the functionalities of data management system in details. Section 3 describes the design philosophy of data management in CGSP. Then we give two use cases studies in the context of data management in section 4. Section 5 evaluates the performance of the system. Section 6 gives some related works. And section 7 concludes this paper.

2 Functionalities of Data Management

Data management is one of the core services in grid system. Its main responsibilities are to manage the storage resources and user's data in the grid and to provide data service for users. Data management is divided to three levels: data service access, metadata management, and storage resources. Data management of CGSP can shield the heterogeneous storage resources and transfer protocols for users. It provides a uniform data assess entry. The data service provided by metadata management can shield the physical storage path of data by logical file path and organize the data space for every user to get a transparent data access. According to the requirements of the applications in ChinaGrid, data management provides four functions as follows:

2.1 Reliable and Efficient Data Transfer Mechanism

The data server based on GDSS processes the data transfer. It improves efficiency by parallel transmission using multiple file slices. For reliability it can restart the transfer tasks from the break point. The data server includes two kinds of resuming mechanism. One is that it continues the previous data transfer until the link of network recovered from failure. The other is that it automatic switches to another backup storage to resume data transfer when the original storage server fails to provide service.

2.2 Data Logical Domain Based on Physical Storage Resources

Physical storage resources consist of collections of resources in different geographical locations or owners. All the collections register to the data center. This hierarchy greatly enhances the scalability of storage resources in data management. Meanwhile, these storage collections should be able to be re-organized under different conditions such as the network latency, the storage capability. For example, three organizations want to share their data that only can be accessed by the users from these three organizations. They want the data to be stored in the specific storage resources owned only by them three. To achieve this, the concept of *Data Logical Domain* (DLD) is introduced. DLD is a logical storage resource sets created based on physical data collections for a specific application.

In the former example, the storage resource set shared by the three organizations is called a DLD. A DLD contains storage resources in multiple storage resources even multiple data management systems (Fig. 1). The data task executed in a DLD can only be run within the group of storage resources specified in the DLD. In this way, it guarantees that data will not be stored outside the DLD, and satisfies the requirements of data store security, efficiency, and so on.

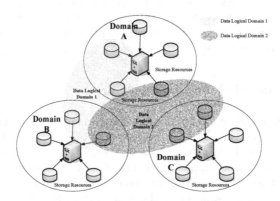

Fig. 1. Data Logical Domains

Users in ChinaGrid can join in one or more DLDs. There is a shared space in a DLD, which can be accessed by every user that participates in the DLD with definite privilege control. There is personal data space belonging to each individual user in the DLD controlled only by the relevant user.

2.3 Transparent File Access Mechanism

The users in ChinaGrid do not care which storage resource their data stored in. The transparent file access mechanism lets the users to access their data just by using the logical path in his user space. This mechanism requires the system be able to map and transform logical file name to physical file name. Every data file has a unique global identifier. If more than one user has the privilege to access this file, every user can name the file as a different logical name in his way. Meanwhile there may be several replicas for a file, each of which is stored in a different storage resource with a different physical path to access the file.

There are three file name spaces: logical file name, unique global identifier, and physical file name. The data management in CGSP provides a mechanism to map and transform these three names. A user gives the logical file name when he wants to access the file in his own space, which will then be transformed to the global identifier by the system. Then the system chooses one of the best replicas according to the global identifier, and returns the physical file path to user to access the data by the data transfer API provided by data management system.

2.4 Management of Storage Resources

The management of storage resources receives the registration of storage resources and monitors the running status of them. According to the status information, the system will choose a *nearest* resource to store the data. The management of storage resources provides a function of error resource detection. The storage resources report their running status to data center periodically. If the status recorded in the data center is not updated for a given time, the storage is then considered to be no longer available until it is recovered.

3 Design Philosophy

To achieve the functionalities above, the design of the data management system consists of five parts: Data Storage Server; *Storage Resource Agent* (SRA); DLD Management; Metadata Management; and Uniform Data Access Entry, shown in Fig. 2.

3.1 Data Storage Server

The storage resource here is not a single hard disk or a disk array. It is used as a file access server as well as a storage status collection sensor. The file access sever is a GDSS server, which supports parallel transfer, data channel reuse, partial file transfer, and failure task restarting. After a storage resource has registered to the SRA, the status collection sensor will report its status to the SRA periodically including available space, CPU load, available memory, status of network, and so on. The SRA allocates storage resources for data transfer according to the information collected from storage status collection sensor. The storage resources also have the responsibility to execute file deleting tasks.

3.2 Storage Resource Agent

All the storage resources wanting to join a CGSP domain must register to the SRA and report its status periodically. The SRA maintains the available storage list within its CGSP domain, records the size of available space and the performance of each storage resource. All the information is used to allocate a proper storage resource for a transfer task by the SRA. The SRA receives registration from the resources and collects their status information for allocating resources of a transfer task, which is implemented to be a *Web Services Resource Framework* (WSRF) [11] service.

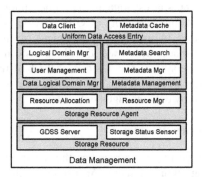

Fig. 2. Implementation of Data Management in CGSP

3.3 Data Logical Domain Management

Data Logical Domain (DLD) is a logical storage resources set created based on physical data collections for a specific application. The resource set in a DLD has a common characteristic such as network latency, storage capacity. DLD management obtains the physical storage resources information from SRA. The administrator creates, deletes and modifies the DLD through *Data Logical Domain Agent* (DLDA). Each user has a default DLD which can not be modified. The default DLD has no fixed storage resource but with limited capacity. When the user uploads data to his default DLD, the system will randomly choose him a proper resource for the task to satisfy the capacity demand.

Meanwhile, DLD management maintains user lists for each DLD. If a user joins in a DLD, he can store his data in the storage resources within DLD. He can also share his data with other users that join in the same DLD. One user can join one or more DLDs. The administrator can add or delete users for a DLD through DLDA. DLD management has also been implemented as a WSRF service for the administrator.

3.4 Metadata Management

The metadata refers to the data used to describe the physical data [12] including file length, file type, access privilege, logical file name, global identifier, and so on. The physical data is given as a URL which specifies the location of the file. While the attributes of the metadata, organized as a tree-structured directory, provide a uniform logical view of heterogeneous storage files. When a user uploads his file, he will publish his metadata after data transferring. He can list his data by directory name in his data space as well as move, copy, delete files and create directories. The user can search his data by the logical file path. The metadata is currently implemented as *Lightweighted Directory Access Protocol* (LDAP) directory.

3.5 Uniform Data Access Entry

To make the heterogeneous storage resources and transfer protocols transparent to users, a *Uniform Data Access Entry* (UDAE) is designed to organize the user's data space view in different DLDs, helping the user access the logical files in the DLDs. The user can upload, download, delete, move, and copy data by only giving the source logical file name and destination logical file name. UDAE also has a cache to keep the hot metadata recently be read or written. We choose the *Least Recently Used* (LRU) as the cache replace algorithm to greatly improve the efficiency of metadata access. We have implemented the UDAE as a WSRF service. A GUI data client has been implemented to help users communicate with the UDAE.

4 Use Cases Study

There are two types of users in data management, administrator and common user. The administrator manages the storage resources, DLDs and users. The common user accesses and modifies the data in his user space. The followings are two typical use cases in the data management of CGSP.

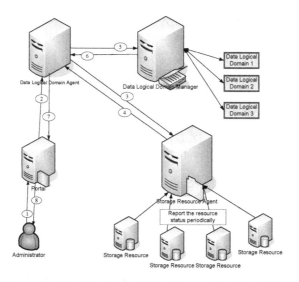

Fig. 3. The Working Flow of Administrator Creating a DLD

The working flow of an administrator creates a DLD is (see Fig.3):

1) An administrator requests for creating a DLD through portal;
2) The request is forwarded to DLDA;
3) DLDA queries usable storage resources from SRA;
4) SRA returns information of the storage resources list to DLDA;
5) DLDA registers the information of the new DLD to DLD manager;
6) DLD manager returns the result (true or false) of the operation to DLDA;

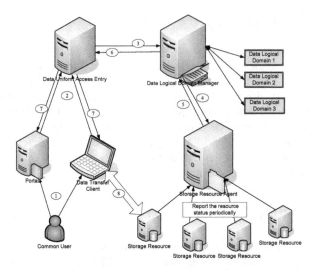

Fig. 4. The Working Flow of Common User Uploading Data

7) DLDA forwards the operation result to Portal;
8) The administrator gets the operation result from Portal.

The working flow of a common user downloading files is (see Fig.4):

1) The user commits the data request with data logical path through Portal or data client;
2) Portal or data client forwards the request to UDAE;
3) UDAE gets the metadata and gives it to DLD manager to find DLD;
4) DLD manager forwards DLD name to SRA;
5) SRA gets the physical location information of the requested data and returns it to DLD manager;
6) DLD manager return the result to UDAE;
7) UDAE forwards the result to Portal or data client;
8) Data client or Portal will connect to the storage resource returned by the result and download data.

Besides the administrator and common user, there is another special type of user: storage resource provider. This user provides storage resources for data manager to satisfy the storage requirement of common users. The working flow of this kind of users is very simple. They first deploy the data storage server on the storage device. Then they properly configure the storage resource and register the resource to SRA. After that the common users can use the storage resources.

5 Performance Evaluation

The testing experiments here tries to address two issues: 1) the performance of data transfer in DLD and default DLD; and 2) the response time of metadata writing with and without cache.

Fig.5 shows the file transfer performance in DLD and default DLD. We register 10 storage resources into the data management system distributed in 3 universities: 5 in Huazhong University of Science and Technology, 3 in Tsinghua University, and 2 in Peking University. The default DLD can select any of them to finish data transfer tasks. The DLD we select is consisted of 5 storage resources with similar network latency. We get data in DLD and default DLD with the size from 50 to 1000 MB, respectively. It is easy to draw the conclusion that the performance of data transfer speed in DLD is averagely 60% higher than that in default DLD. The data transfer speed in DLD is steadier because all the storage resources have similar network latency. For data in default DLD, data transfer speed may be high, but for the most circumstances, we just get poor performance. Because the system selects storage resources for data transfer tasks randomly in default DLD.

Fig. 6 shows the response time of uploading data with and without cache through UDAE. The cache is set to 4MB. UDAE replaces the metadata in cache by LRU algorithm. We have processed five groups of tests separately on UDAE with and without cache. Each group has 50 times data uploading with the same file length. We record the average respond time for each group. From the result, we find that setting a cache for UDAE can greatly reduce the response time of uploading data and improve the performance.

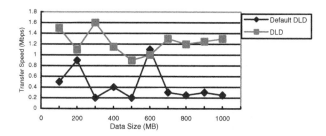

Fig. 5. The Performance of Data Transfer in DLD and Default DLD

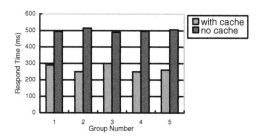

Fig. 6. The Response Time of Uploading Data with and without Cache through UDAE

6 Related Works

Storage Resource Broker (SRB) [13] is a middleware infrastructure to provide a uniform, UNIX-style file I/O interface for accessing heterogeneous storage resources distributed over the wide area network. Using its *Metadata Catalog* (MCAT) [14], SRB provides collection-based access to data based on high-level attributes rather than on physical filenames. SRB also supports automatic replication of files on storage systems. SRB uses an integrated architecture to access data via the SRB interface and MCAT and with SRB control over replication and replica selection. Data management in CGSP uses a layered architecture to supply different services for grid users. It can also be re-organized the storage resources for some special demand.

The Globus Toolkit [15] from Globus Alliance [16] provides a number of components for data management. GridFTP [17] and the Globus *Reliable File Transfer* (RFT) [18] service take care for data movement. The *Replica Location Service* (RLS) [19] is a tool to provide the ability keeping track of one or more copies, or replicas, of files in a grid environment. They do not provide metadata service in the Grid Toolkit and they have not integrated all the components to a single system to server as a data management system.

7 Conclusions and Future Work

In this paper we have presented the four key functionalities of data management for CGSP. The reliable and efficient transfer mechanism and the storage resource management guarantee the basic transfer demand for data management. The data logical domains based on physical storage resources give users great flexibility to reorganize the resources. The transparent file accessing mechanism shields the heterogeneous storage resources and transfer protocols for user. We also discuss our implementation and use cases. The performance of data transfer in DLD and with cache in UDAE is presented. In the future, we plan to implement a replica service to improve the efficiency and reliability of data transfer. We will implement security mechanism to guarantee the secure transfer.

References

1. B. C. Barish and R. Weiss, "LIGO and the Detection of Gravitational Waves", Physics Today, Vol.52, pp.44, 1999.
2. C.-E. Wulz, "CMS – Concept and Physics Potential", Proceedings II-SILAFAE, San Juan, Puerto Rico, 1998.
3. NVO, http://www.us-vo.org/.
4. Foster, C. Kesselman, and S. Tuecke, "The Anatomy of the Grid: Enabling Scalable Virtual Organizations", International Journal of High Performance Computing Applications, Vol.15, 2001.
5. H. Jin, "ChinaGrid: Making Grid Computing a Reality", Digital Libraries: International Collaboration and Cross-Fertilization - Lecture Notes in Computer Science, Vol.3334, Springer-Verlag, December 2004, pp.13-24.

6. H. Jin, L. Ran, Z. Wang, C. Huang, Y. Chen, R. Zhou, and Y. Jia, "Architecture Design of Global Distributed Storage System for Data Grid", High Technology Letters, Vol.9, No.4, December 2003, pp.1-4

7. ChinaGrid, http://www.chinagrid.edu.cn.

8. ChinaGrid Support Platform, http://www.chinagrid.edu.cn/CGSP.

9. CGSP Work Group, Design Specification of ChinaGrid Support Platform, Tsinghua University Press, Beijing, China, 2004

10. G. Singh, S. Bharathi, A. Chervenak, E. Deelman, C. Kesselman, M. Mahohar, S. Pail, and L. Pearlman, "A Metadata Catalog Service for Data Intensive Applications", Proceedings of Supercomputing (SC'03), November 2003.

11. The Web Services Resource Framework, http://www.globus.org/wsrf/.

12. E. Deelman, G. Singh, M. P. Atkinson, A. Chervenak, N. P. C. Hong, C. Kesselman, S. Patil, L. Pearlman, and M. Su, "Grid-Based Metadata Services", Proceedings of 16th International Conference on Scientific and Statistical Database Management (SSDBM'04), p.393, June 2004.

13. C. Baru, R. Moore, A. Rajasekar, and M. Wan, "The SDSC Storage Resource Broker", Proc. CASCON'98 Conference, 1998.

14. MCAT, MCAT – A Meta Information Catalog (Version 1.1), http://www.npaci.edu/DICE/SRB/mcat.html.

15. Globus Tookit, http://www.globus.org/toolkit/.

16. Globus Alliance, http://www.globus.org/alliance/.

17. B. Allcock, J. Bester, J. Bresnahan, A. Chervenak, I. Foster, C. Kesselman, S. Meder, V. Nefedova, D. Quesnel, and S. Tuecke, "Secure, Efficient Data Transport and Replica Management for High-Performance Data-Intensive Computing", Proceedings of IEEE Mass Storage Conference, 2001.

18. W. E. Allcock, I. Foster, and R. Madduri, "Reliable Data Transport: A Critical Service for the Grid", Building Service Based Grids Workshop, Global Grid Forum 11, June 2004.

19. M. Ripeanu and I. Foster, "A Decentralized, Adaptive, Replica Location Service", Proceedings of 11th IEEE International Symposium on High Performance Distributed Computing (HPDC-11), Edinburgh, Scotland, July 24-16, 2002.

CGSP: An Extensible and Reconfigurable Grid Framework[*]

Yongwei Wu[1], Song Wu[2], Huashan Yu[3], and Chunming Hu[4]

[1] Department of Computer Science and Technology,
Tsinghua University, Beijing, 100084, China
[2] Cluster and Grid Computing Lab, School of Computer,
Huazhong University of Science and Technology, Wuhan, 430074, China
[3] School of Electronics Engineering and Computer Science,
Peking University, Beijing, 100871, China
[4] School of Computer Science, Beihang University, Beijing, 100083, China

Abstract. ChinaGrid Support Platform (CGSP) is proposed to provide grid toolkit for ChinaGrid application developers and specific grid constructors, in order to reduce their development cost as greatly as possible. CGSP extensible and reconfigurable framework, which satisfies the expansion and autonomy requirement of ChinaGrid, is mainly discussed in the paper. In the framework, domain is presented to denote one unit which could provide grid service for end users by itself. Layered structure of domains and corresponding interactive relationship are paid much more attention. CGSP design motivation and simple execution management mechanism are also described in this paper.

1 Introduction

Grid computing [5,6,7] has emerged as an important new field by its focus on large-scale computing resource sharing and coordinated use of resources at multiple sites. It provides approaches to integrate widespread heterogeneous resources into one multi-institutional virtual organization and uniform application interface. It is important to recognize that the resources in this context include computational systems and data storage and specialized experimental facilities.

Based on existing network infrastructure, many grid computing projects have been launched, such as UK e-Science Program [12], Information Power Grid (IPG) [13], TeraGrid [14], China National Grid (CNGrid) [15]. As on national wide grid, China Education and Research Grid (ChinaGrid) [1,2] aims at constructing a public service system for Chinese education and research. Without exception, all of these grid projects paid much more attention on the middleware-the software that enables grid computing/services.

[*] This Work is supported by ChinaGrid project of Ministry of Education of China, Natural Science Foundation of China under Grant 60373004, 60373005, 90412006, 90412011, and National Key Basic Research Project of China under Grant 2004CB318000.

J. Cao, W. Nejdl, and M. Xu (Eds.): APPT 2005, LNCS 3756, pp. 292–300, 2005.
© Springer-Verlag Berlin Heidelberg 2005

Grid middleware is the kernel of constructing grid effectively. Great efforts have been made to develop scalable, secure, and highly available grid platforms, which transparently shield the heterogeneities and dynamic behaviors of participants, on top of local operating systems. As one grid middleware, ChinaGrid Support Platform (CGSP) [3,4]aims to provide grid toolkits for ChinaGrid application development and specific grid construction, in order to make them completed more easily and quickly.

Based on Open Grid Service Architecture (OGSA) [9], CGSP is developed according to the Web Service Resource Framework (WSRF) [10] specification for the construction of ChinaGrid from April, 2004. In Jan. 2005, CGSP version 1.0 (CGSP1.0) was released. CGSP1.0 is developed by 5 top universities including Tsinghua University, Huazhong University of Science and Technology, Peking University, Beihang University, and Shanghai Jiao Tong University. It aims to integrate all sorts of heterogeneous resources, especially education and research resources, distributed over China Education and Research Network (CERNET), to provide transparent, high performance, reliable, secure and convenient grid services for scientific researchers and college students.

In this paper, we begin with the design principles of CGSP. Then, CGSP function modules are described simply. In the fourth and fifth parts, CGSP extensible and reconfigurable framework, and execution management across domains are put forward respectively in detail. These provide a whole vision for CGSP framework. Future work and conclusion are discussed at last.

2 Design Principles

CGSP is a grid middleware developed for the construction and evolution of ChinaGrid. The design goal is to reduce the cost of grid construction and application development greatly. In addition to supplying the grid runtime environment of ChinaGrid, CGSP offers a whole set of tools for building portal, deploying grid services and developing various grid applications.

Based on CGSP, ChinaGrid can be constructed into one tree/layered structure. Each node of the tree is a domain. Each domain has the same logic structure and consists of same function modules. They interact through CGSP information center.

In moving forword, CGSP has been guided by a set of key design principles as follows.

- Support localized requirement of ChinaGrid. In fact, it still needs a long way to implement intensive message passing computation over Internet. It makes more proportional local users to use local resources in the grid in order to avoid the reduced performance caused by limitation of network bandwidth and latency over the Internet, and improves the service efficiency of the grid as well.
- Meet the autonomy requirement of ChinaGrid. Each grid application, such as bioinformatics grid [20], image processing grid [22], or computing fluent dynamics grid [21], has its own user and resource management mechanism.

CGSP makes it easier to construct various independent ChinaGrid applications with their own management protocols and mechanisms. Each application grid of ChinaGrid could be a solely domain which could provide specific service for its users by itself. At the same time, they could interact with each other through CGSP high level interactive regulation.

- Scalability of CGSP satisfies the demand of expansion of ChinaGrid. The tree or layered structure of CGSP ensures that ChinaGrid can link more and more universities. Actually our goal is to link up to 100 universities in China in the near future.
- Flexibility of CGSP makes it easier to rebuild the ChinaGrid. Layered structure of CGSP could be easily reconfigurable and rooted by any domain through converting its original parent domain to its child domain.
- Reconfiguration of the tree structure guarantees the integrity and uniformity of a grid system through constructing a global monitoring engine started from any node of the ChinaGrid tree.
- Different from normal grid middleware efforts (GT series [11,8], OMII [16], TinghuaGrid [17] et. al.), CGSP is a platform. Not only does it include the grid running components, such as portal, service container, service monitoring and discovery, file delivery and transformation, but it also provides the grid developing tools, such as programming API, portal constructing tool, service deploying and packaging tools and so on.

3   CGSP Function Modules

CGSP is a collection of cooperative software components. It contains several software modules which can run independently to support each step of development process, execution process, system installation process, and system management. In addition to supplying the components for building grid platform to reduce development cost of ChinaGrid applications, CGSP also offers a whole set of tools for developing and deploying various grid applications.

CGSP logically consists of 6 components showed in Fig. 1.

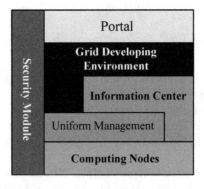

Fig. 1. Logic Diagram of CGSP

- **Portal**: Portal is a web based user interface for defining, submitting, monitoring jobs, and managing personal data, viewing resource information as well.
- **Grid Developing Environment/Toolkit (GDE)**: GDE provides a set of toolkits for grid application's construction and development. They are portal developing tool, service packaging tool, grid-enabled parallel programming tool [19], job definition tool etc.
- **Information Center (IC)**: Information Center, or Service Manager, manages the relatively static information of resources and services in the grid. It provides real-time grid information service for other CGSP function modules under a global uniform information framework. Dynamic information, such as status of grid job, is obtained from uniform management module.
- **Uniform Management(UM)**: UM aims to make heterogeneous resources, computing tools over grid, grid users, grid jobs and job operations managed in a uniform view. It consists of following four items.
  - **Service Container (SC)** is deployed to act as a runtime environment for the installing, deploying, running and monitoring of grid service in the specific node. It also provides support for real-time grid resource monitoring.
  - **Data Management (DM)** mainly provides data delivery and transformation for CGSP. It also implements one global file view and makes all sorts of grid data accessible transparently by grid end-users.
  - **Job Management (JM)** is responsible for the scheduling and monitoring of grid jobs. JM plays a key role in the execution management of grid job. It deploys and submits jobs to the service container of real computing node and starts up them through general running service of CGSP.
  - **Domain Management** ensures the autonomy of each domain of China-Grid with focus on user management, log and accounting, and interactive call with other domains for the user identity mapping.
  - **Grid Monitoring** It gathers status data from all the CGSP components and reports the results in structured and standardized documents through WSRF service to correlative modules or users. At the same time, it could notify the ChinaGrid system modules, administrators or users in time when the status is changed.
- **Grid Security (GS)**: GS is in charge of user identity authentication, identity mapping between different domains, service and resource authorization, and secure message passing.
- **Computing Nodes (CNs)**: CN provides real computing power for grid services. It could be a cluster, or PC server or workstation.

4   CGSP Framework Architecture

CGSP consists of a set of well-assorted software packages. Base on it, ChinaGrid can be constructed into a layered tree showed in Fig. 2 (left). Each node of the

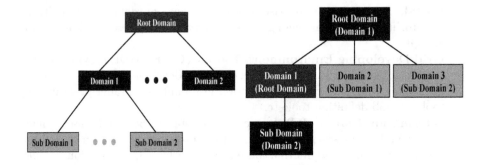

Fig. 2. Tree Structure of ChinaGrid Based on CGSP

tree is a domain. Each domain, maybe root domain, domain or sub-domain, has its own Portal, GDE, IC, UM, GS and CNs. It could provide independent grid service for end users by itself. Each domain can be a specific application grid (bioinformatics grid, image processing grid). It can also be a region grid (Shanghai Grid, Tsinghua Grid). All domains could share one certificate authorization (CA) center, and can also have their own CAs.

In Fig. 2, Root Domain, Domain and Sub-Domain are only used to note their parent-child relationship. The tree Fig. 2 (left) can be reconfigurable easily into another tree as Fig. 2 (right). Each domain could become the root domain of the CGSP if its original parent domain is converted to its child domain same as showed in Fig. 2. Root Domain in the left becomes the Domain 1 in the right. It is clear that the cost of such a reconfiguration is very low with a little effort.

Each domain has the same internal logic architecture and consists of 6 same function modules showed in Fig. 1. At the same time, the module interactive relationships in one domain can also be gotten from Fig. 1. Same as Portal, GDE must get the support from UM and IC, but no direct interaction with CNs. As the grid middleware kernel, UM and IC are called each other and interactive with the real CNs. GS always is closed to all components of CGSP.

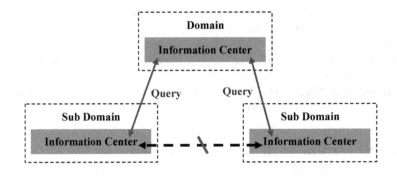

Fig. 3. Relationship between ICs in different domains

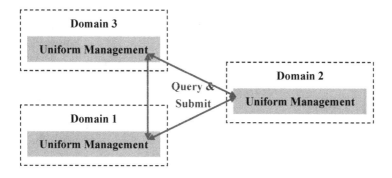

Fig. 4. Relationship between UMs in different domains

Different domains could access each other. But the interactive calls only happen between UMs and ICs of different domains. Fig. 3 shows the IC module relationship of CGSP. In order to see clearly and easily portrayed, Portal, GPE, Gs, and CNs are ignored in the figure. First, we can find that IC in domain has a bidirectional call relationship with the IC in sub domain. But, there is no direct relationship between ICs of different sub domains. That is to say, one IC only has a bidirectional relationship with its parent ICs or its child ICs and has no any relationship with its brother, grandson or grandfather ICs. From each IC, all resource and service information of the whole grid can be gotten through querying the whole tree. In fact, the tree structure of CGSP is held and determined totally by the IC module.

Different from the relationship between ICs, we can find that each UM could call UMs of any other domains directly from the Fig. 4. But which UM will be called by the UM? This is determined by the IC. That is to say, one UM wants to get the support from another UM in other domains, but it does not know which domain could supply such support. So, it must ask help from its own IC. Its IC will search one domain which could provide this requirement through querying its parent IC or child IC. We know that the ICs in a grid are constructed into a tree structure. So, if there is one domain that could provide such support, it can be found through querying the tree at least. But any UM could only call its own IC (in the same domain) and has no interactive relationship with other ICs (in other domains). Fig. 4 shows such relationship.

5   Execution Management Across Domains

Let's see the execution management of job which needs to be submitted from one domain to another one. That is to say that such job can not be completed by its own domain. It will be submitted to another domain to execute.

Fig. 5 shows such a job execution flow step by step. Before submitting job to the job manager, user must upload the input data required by the job to the personal data space in the data manager. Then, job manager will query the available computing nodes from the IC when it gets a computing job from

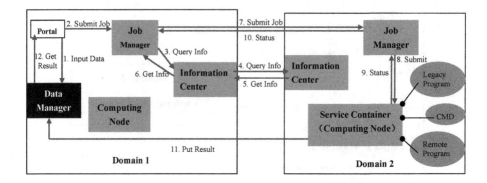

Fig. 5. Job Execution Flow across Domains

Portal. If its own domain can not complete the job, the IC will query other ICs of the grid. And then the job manager will submit the job to the job manager of another domain, which could complete the job. At last, the job will be sent to the service container of one real computing node and begins to execute it.

After the job is started up, the service container will report job status to the original job manager through local job manager in time. When the job is completed, the service container will put the computing results to the personal data area of the data manager of the original domain, which in turn sends the results to user through Portal at last.

There is a problem about user identity mapping during accessing and executing across domains in CGSP. Each domain has its own user management, so user identity should be converted when one job manager submits jobs or report job status to another job manager (such step 7 and 10 in the Fig. 5). The domain manager is in charge of the user identity mapping between different domains in CGSP. But for the interaction of ICs between different domains, it is another story. ICs complete the interaction through system user, who can only access its parent of child domain ICs

6   Conclusion and Future Work

Deferent from the other grid middleware efforts, CGSP supplies a grid platform to satisfy all kinds of requirements of grid constructors and grid application developers. The cost of grid application is reduced greatly based on the CGSP. Besides runtime environment, it also provides developing tools, such as portal constructing tool, programming environment, service packaging tools and so on.

Furthermore, CGSP extensible and reconfigurable framework makes nation-wide grid reality. Especially in China, there are many universities and research institutes. How to guarantee their respective autonomy is the key design principles. At the same time, the integrity of grid is also involved into the CGSP extensible framework.

At present, four ChinaGrid applications, Bioinformatics Grid, Image Grid, Computational Fluid Dynamics Grid and Mass Information Grid have completed the initial building over the CGSP and almost 30 most famous universities in China, as the members of ChinaGrid program, are building their campus grid on the basis of CGSP.

In the future, WS core 4.0 from Globus Toolkit will be replanted into CGSP. Job Submission Description language (JSDL) will be used to replace current Grid Job Description Language (GJDL). At the same time, open source ActiveBPEL [24] Engine will be used to implement the workflow component of job manager in CGSP. OGSA-DAI [23] based integration of heterogeneous database is one of the most important new functions which will be implemented in CGSP version 2. Media service, grid monitoring, accounting and management will be also strengthened in the near future.

Acknowledgement

We wish to express our sincere appreciation to the Prof. Hai Jin, Prof. Weimin Zheng, Prof. Xiaomin Li, Dr. Xiaowu Chen and other ChinaGrid experts for extending their generous support for the successful conduct of the CGSP. Special thanks to CGSP working group members for the CGSP success.

References

1. ChinaGrid, http://www.chinagrid.edu.cn.
2. H. Jin, *ChinaGrid: Making Grid Computing a Reality*, Proceedings of ICADL 2004, Lecture Notes of Computer Science, (2004), 3334, 13-24
3. ChinaGrid Support Platform, http://www.chinagrid.edu.cn/CGSP.
4. CGSP Work Group, *Design Specification of ChinaGrid Support Platform*, Tsinghua University Press, Beijing, China, 2004
5. I. Foster, C. Kesselman, S. Tuecke, *The Anatomy of the Grid: Enabling Scalable Virtual Organization*, International J. Supercomputer Applications, 15(3), (2001)
6. I. Foster, C. Kesselman, *The Physiology of the Grid: An Open Grid Services Architecture for Distributed Systems Integration*, J. Nick, S. Tuecke, (2002)
7. Baraglia, R., Laforenza, D., Lagana, A., *A Web-based Metacomputing Problem-Solving Environment for Complex Applications*, Proceedings of Grid Computing 2000, (2000), 111-122
8. I. Foster, C. Kesselman, *Globus: A Metacomputing Infrastructure Toolkit*, International J. Supercomputer Application, (1997), 11(2), 115-128
9. Open Grid Services Architecture (OGSA), https://forge.gridforum.org/projects/ogsa-wg, or http://www.globus.org/ogsa/
10. Web Service Resource Framework (WSRF), http://www.globus.org/wsrf/ and http://www.ggf.org/documents/GFD.30.pdf
11. Globous Toolkits, http://www.globus.org.
12. UK e-Science Program, http://www.rcuk.ac.uk/escience/;
13. NASA Information Power Grid, http://www.ipg.nasa.gov/;
14. TeraGrid, http://www.teragrid.org/;
15. China National Grid, http://www.cngrid.cn;

16. Open Middleware Infraxtructure Institute, http://www.omii.ac.uk/
17. Dazheng Huang, Fei Xie, Guangwen Yang, *T.G.: a Market-oriented Computing System with Fine-grained Parallelism*, 9th Workshop on Job Scheduling Strategies for Parallel Processing Seattle, Washington, (2002)
18. Tuecke, S., Czajkowski, K., Foster, I. , et.al.: *Open Grid Services Infrastructure (OGSI) Version 1.0*, Global Grid Forum Draft Recommendation. (2003).
19. Yongwei Wu, Guangwen Yang, Qing Wang, Weiming Zheng, *Coarse-grained Distributed Parallel Programming Interface for Grid Computing*, Lecture Notes in Computer Science, (2004), 3032, 255-258; Expanded Version is accepted by International Journal of Grid and Utility Computing;
20. ChinaGrid Bioinformatics Grid, http://166.111.68.168/bioinfo/tools/index.jsp
21. ChinaGrid Computational Fluid Dynamics (CFD) Grid, http://grid.sjtu.edu.cn:7080/grid/
22. ChinaGrid Image Processing Grid, http://grid.hust.edu.cn/ImageGrid/
23. http://www.ogsadai.org.uk/
24. The Open Source BPEL Enging, http://www.activebpel.org/

Early Experience of Remote and Hot Service Deployment with Trustworthiness in CROWN Grid*

Hailong Sun[1], Yanmin Zhu[2], Chunming Hu[1], Jinpeng Huai[1],
Yunhao Liu[2], and Jianxin Li[1]

[1] School of Computer Science, Beihang University, Beijing, China
{sunhl, hucm, huaijp, lijx}@act.buaa.edu.cn130
[2] Department of Computer Science,
Hong Kong University of Science & Technology, Hong Kong
{zhuym, liu}@cs.ust.hk

Abstract. CROWN Grid aims to empower in-depth integration of resources and cooperation of researchers nationwide and worldwide. In such a distributed environment, to facilitate adoption of services, remote and hot service deployment is highly desirable. Furthermore, when the deployer and the target container are from different domains, great security challenges arise when a service is deployed to the remote container. In this paper, we present ROST, an original scheme of Remote & hOt Service deployment with Trustworthiness. By dynamically updating runtime environment configurations, ROST avoids restarting the runtime system during deployment. Moreover, we adopt trust negotiation in ROST to assure the security of service deployment. We conduct experiments in a real grid environment, and evaluate ROST comprehensively.

Keywords: Service grid, CROWN, Remote and hot deployment, ROST, Trust Negotiation Agent (TNA).

1 Introduction

Grid computing promises to enable coordinated resource sharing and problem solving in dynamic, multi-institutional virtual organizations [1]. In recent years, service-oriented grid architecture is introduced, which is widely considered as the future of grid computing [2]. Built on web services, OGSA [3] is the de facto standard for building service grids, in which various resources are encapsulated as services with uniform user interfaces.

The main goal of our key project, CROWN (**C**hina **R**&D Environment **O**ver **W**ide-area **N**etwork) Grid, is to empower in-depth integration of resources and cooperation of researchers nationwide and worldwide. CROWN project was started in late 2003. A number of universities and institutes, such as Tsinghua University, Peking

---

* This work is partially supported by the National Natural Science Foundation of China under Grant 91412011, China Ministry of Education under grant CG2003-CG004 & GP004 & GA004 and Microsoft Research Aisa.

J. Cao, W. Nejdl, and M. Xu (Eds.): APPT 2005, LNCS 3756, pp. 301–312, 2005.
© Springer-Verlag Berlin Heidelberg 2005

University, Chinese Academy of Sciences, and Beihang University, have joined CROWN, with each contributing several computing nodes. More universities and institutes will be invited to join CROWN Grid by mid 2005.

In the past years, many key issues in grid computing have been extensively studied. However, remote and hot service deployment has not been fully addressed. Before a service is ready for invocation, it must be deployed in a service container which provides a runtime environment. A grid is a highly distributed environment, in which numerous domains could be involved. The domains are usually geographically dispersed. It is highly desirable for a user to deploy its services into remote service containers for multiple purposes. For example, in CROWN Grid for bioinformatics application, there are many computing intensive applications such as BLAST. A computing node could easily be over-loaded when multiple jobs arrive in a short period. The heavy load can be balanced if the node is able to deploy one or more BLAST service replica to remote nodes and then redirect some jobs. Similar requirements also exist in many other grid applications.

Traditionally, remote service deployment is supported in a cold fashion, which means, to deploy a new service, the runtime environment need to be restarted. This results in many disadvantages because previously running services must be stopped, and they may have to resume or even restart their jobs, causing significant overhead. Therefore, hot service deployment has become increasingly important, which does not need to restart the runtime environment while deploying services. With the availability of remote and hot service deployment, many applications will benefit, such as load balancing, job migration and so on.

Service deployment is actually not a new issue. Similar demands also exist in mobile agents [4] and active networks [5]. To the best of our knowledge, however, there is no successful solution to enabling remote and hot service deployment in grid systems. The most updated Globus Toolkit version 4 [6], the de facto standard for grid middleware, does not provide the function of remote and hot service deployment yet. This may be due to the great security challenges arising when a user deploys a service to a remote container. Here we call a node *deployer*, which intends to deploy a service, and the remote service runtime environment *target container*, which is responsible for running and managing services being deployed. Without proper security mechanisms, a service provided by a deployer may be malicious, and the target container may be rogue or fragile. Also, the security policies of the deployer and the container could be incompatible. In an open grid environment, we can not expect any deployer and the corresponding target container to set up required trust relationship in advance. Moreover, it is too costly to build the trust across domains based on the traditional PKI infrastructure every time during remote deployment.

In this paper, we present our original work, ROST (**R**emote and h**O**t **S**ervice deployment with **T**rustworthiness), which achieves its goal by dynamically updating the runtime environment configurations. ROST avoids restarting runtime systems during remote deployment. Moreover, we include trust negotiation in ROST scheme, which greatly increases flexibility and security of CROWN. Major contributions of this work are as follows:

- We identify the necessity of remote and hot service deployment in service grids, and their challenges.
- We propose an effective approach, ROST, to enable remote and hot service deployment. Also, we add trust negotiation into the scheme to meet general security requirements for grid environments.
- We implement ROST in CROWN Grid and evaluate the performance of ROST by comprehensive experiments.

The rest of this paper is organized as follows. We discuss related work in Section 2. In Section 3, we introduce the design and implementation experiences. We present experimental methodology and performance evaluation of ROST in section 4. And in section 5, we conclude this work.

2 Related Work

Globus Toolkit is the most famous grid middleware and it has begun to support service-oriented grid computing based on OGSA since version 3. But even in the updated release version 4, remote and hot service deployment is not supported. Grid service is actually built on Web service, and extended to include functions such as state and life cycle management. For Web services, several middleware, such as Apache Axis [7], JBOSS [8] and Microsoft .NET [9], have partly implemented dynamic service deployment, i.e., deploying a local service without restarting service containers. However, Web service is much simpler than grid service, e.g. web services are normally stateless, so web service middleware can not apply to grid environments. Also, most of them only consider local deployment.

Friese et al. [10] proposed a method for hot service deployment in an ad hoc grid environment based on OGSI which is now replaced by WSRF. To ensure security, they make use of sandbox which can restrict the service function. DistAnt [11] extends the Apache Ant build file environment to provide a flexible procedural deployment description, and provides a solution to remote and hot service deployment based on Globus Toolkit 3. It does not provide any security mechanism for remote deployment. Baude et al. [12] proposed a solution for deployment and monitoring of applications written using ProActive, which is a Java-based library for concurrent, distributed and mobile computing. It does not consider grid service deployment issues.

3 ROST Design and Implementation

CROWN consists of numerous organizations with each of them forming a domain, as illustrated in Figure 1. Domains are usually connected by the Internet. CROWN, as a service-oriented grid, encapsulates various resources as services. In CROWN, a computer must be installed a Node Server (NS), a CROWN middleware. An NS contains a service container which provides runtime environment for various services. Each NS usually belongs to a security domain. Every domain has at least one RLDS (Resource Locating and Description Service) to provide information service. RLDS maintains dynamic information of available services.

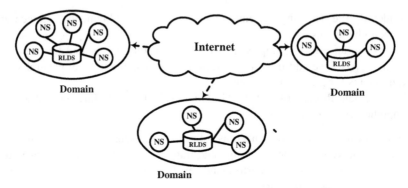

Fig. 1. Resource organization in CROWN

Remote service deployment is needed when a deployer needs to deploy a service on an NS in a different domain. In this paper, we refer to deploying a service to an NS and deploying a service to a container interchangeably, which means the same. A service is basically an entity that consists of an executable program, a description file, and several configuration files. Before a deployer's services can be ready for invocation in the remote NS, two key issues must be addressed. The first is security, namely, how to guarantee the service provided by the deployer is not malicious and the environment provided by the remote container is safe to the service. The second is how to enable the service to be available without restarting the remote container.

In CROWN, services follow the WSRF specifications [13]. A complete service consists of several files, as shown in the following.

- Executable programs. Such as Java classes, scripts, EJBs, etc.
- One or multiple WSDL files. Description of interfaces and access protocols of a service.
- A WSDD file. Web Service Description Descriptor, description of service configuration for the service container.
- BPEL files. Description of composed services which are described in BPEL4WS (Business Process Execution Language for Web Services).
- A JNDI configuration file. Description of WSRF resources of a service.
- A security configuration file. Description of authorization approach and other security related information.

To facilitate the transportation and protection of services, we compress a service into one single file. By far, we have adopted Globus Toolkit's GAR file format. In addition, we have extended GAR so that it is able to contain multiple types of executable programs and description files.

3.1 ROST Architecture

As shown in Figure 2, ROST is composed of several components while our discussions will focus on the two major ones, i.e., TNA and RHD.

TNA is responsible for trust establishment between a pair of deployer and container, and RHD is for remote and hot service deployment. The SCC (Service

Container Configuration) is the abstract of various configurations of service containers. Indeed, each deployment operation results in an update to SCC.

The procedure of service deployment can be divided into two phases: trust negotiation by TNA and deployment by RHD. To be more specific, the workflow of ROST is depicted as follows:

Step 1: the deployer sends a deployment request to a remote NS;

Step 2: the remote NS checks locally whether it can afford the new service; if yes, goes to Step 3;

Step 3: the remote NS checks whether the deployer has been trusted according to the local domain controller or the history information. If yes, sends a trusted notification; otherwise, initiates trust negotiation;

Sep 4: the deployer checks whether the remote NS has been trusted. If yes, sends a trusted notification, and goes to Step 5; otherwise, initiates trust negotiation;

Step 5: if the negotiation successfully sets up the desired trust, the deployer initiates service deployment by transferring the service to the remote NS;

Step 6: the remote NS performs hot deployment of the service.

Step 7: the remote NS acknowledge the success of the deployment.

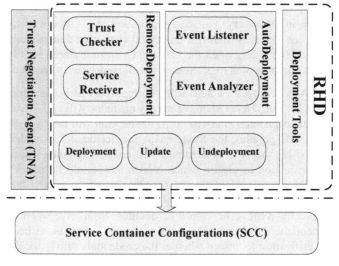

Fig. 2. ROST components

3.2 TNA: Trust Negotiation Agent

3.2.1 ATN Technology

Several security infrastructures have been proposed for grid computing. For instance, in Grid Security Infrastructure (GSI) [14], every user or computer is uniquely identified by a X.509 certificate, which is issued by a Certificate Authority (CA). This fashion provides very limited capability of security control and it is rarely possible to deploy such a global hierarchy of CAs in an open environment like CROWN.

ATN (Automated Trust Negotiation) [15-19] is a new approach to access control in an open environment, which, in particular, successfully protects sensitive information while negotiating a trust relationship. With ATN, any individual can be fully autonomous. Two individuals try to set up a trust relationship by exchanging credentials according to respective policies.

Based on the above observations, we solve the trustworthiness problem in ROST by adding a Trust Negotiation Agent (TNA) , which is generally based on ATN technologies.

3.2.2   Trust Negotiation in ROST

As illustrated by Figure 3, TNA has mainly four components as follows.

- **TrustTicket Manager**: The Access Mediator is responsible for issuing new Trust-Tickets for requesters and validating TrustTickets based on local Ticket Repository.
- **Strategy Engine**: The negotiation strategy [20] is used to determine when and how to disclose local credentials and policies. Also, it makes decisions to update the negotiation states, including success, failure or continuance.
- **Compliance Checker**: This component determines which local credentials satisfy the requester's policies or whether the requester's credentials satisfy local policies.
- **Credential Chain Discovery**: For trust negotiation in open networks, access control decision often involves finding a credential chain that delegate authority from the source to the requester, when the credentials are not stored locally. The main function of this component is to discover and collect necessary credentials.

In ROST, TNA is deployed on both sides of deployers and target containers. If a requestor has a valid *TrustTicket*, then the access mediator will call *TrustTicket* Manager to make access decision. Otherwise, trust negotiation will be triggered. When the requestor discloses its policies, the Strategy Engine decides whether the negotiation should continue. If so, the Access Mediator will call Compliance Checker to make corresponding verification to ensure which credentials should be provided, then responds with the necessary credentials and policies. In some cases, if the credentials are not available in local Credential/Policy Repository, Credential Chain Discovery is called to dynamically retrieve necessary credentials. Similarly, when the requester submits its credentials, the Access Mediator will call Compliance Checker to make corresponding verification to ensure whether the credentials satisfy local policies and make access decisions.

In TNA, we adopt refined RTML (Role-based Trust Management Language Markup Language) to represent both access control policy and attribute-based credentials. When credential storage is distributed, the goal-directed algorithm [16] ensures that all credentials available can be discovered and collected. In ROST design, the *TrustTicket* takes the form of *<subject, issuer, subject, valid date, expiration data, signature>*. It is an identity assertion represented with XML with short lifetime assigned by the issuer.

In addition, negotiation information exchange between participants must rely on a secure communication protocol such as SSL/TLS to prevent eavesdropping, man-in-the-middle attacks, replay attacks, etc. Our ROST implementation conforms to WS-Security and WS-Conversation specifications for SOAP message protection.

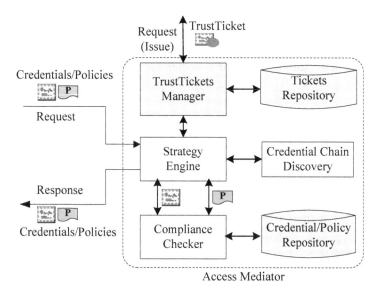

Fig. 3. TNA structure

3.3 RHD

After a negotiation successfully sets up desired trust, the container receives the service from the deployer and begins to deploy it.

RHD enables remote hot deployment as well as providing a convenient way for local hot deployment. RemoteDeployment and AutoDeployment, as shown in Figure 2, are respectively responsible for remote and local service deployment.

3.3.1 RHD APIs

We design APIs for both remote and local deployment, through which users are able to develop high level middleware and applications. There are basically three types of deployment operations: deploy, update, and undeploy. We define nine APIs to support these deployment operations as follows.

(1) *deploy* (String *garFilePath*)
(2) *deployByFTP* (URL *garFileURL*, String *user*,
    String *password*)
(3) *deployBySOAPAttach*(String *garFilePath*)
(4) *update* (String *garFilePath*)
(5) *updateByFTP* (URL *garFileURL*, String *user*,
    String *password*)
(6) *updateBySOAPAttach*(String *garFilePath*)
(7) *undeploy*(String *garFileName*)
(8) *undeploy*(String *serviceName*)
(9) *getAllDeployedServices*()

Note that (1)-(3) are three interfaces for deploying a service, while (1) is for deploying a service locally; and (2)(3) provide two different interfaces for remote deployment. The (4)-(6) defines three interfaces for updating deprecated services. The (7) and (8) defines two interfaces for removing services from service containers. We define (9) for querying all services deployed in a service container.

3.3.2 Remote Deployment

After mutual trust is successfully established, *ServiceReceiver* is called to receive the GAR file and uncompress it by *GARUnzipper*. Then the underlying deployment functions are called to perform corresponding operations.

A service container must include various configurations of the deployed services. Indeed, the key to hot deployment is to update the configuration of SCC dynamically. Relevant configurations include executable programs, WSDL description, WSDD, and JNDI configuration. For example, when a new service implemented with JAVA needs to be deployed, we have to let SCC load JAVA classes of the service.

For updating or un-deploying an existing service, it should be careful since other services or users might be using it. Simply updating or undeploying a service without adding special measures may lead to unexpected service interruption to users. To solve this problem, we add a reference counter for each deployed service. The initial value of a counter is zero, and the value increases/decreases by one each time when the service is invoked/completed. When an update or undeployment request comes, we first check the counter of the service. A service is ready to be updated or undeployed only if the reference counter is equal to zero.

3.3.3 Auto Deployment

Besides remote and hot deployment, RHD component also provide a convenient method to hot-deploy services to local containers.

A file folder is specified to receive GAR files and an *EventListener* keeps listening to the events associated to the folder. The *EventListener* is interested in two types of events: arrival of new files and deletion of existing files.

Suppose an event *e* caught by *EventListener* is passed to the *EventAnalyer* for analysis and further process. Based on contents of an event, the *EventAnalyzer* will call underlying different deployment functions. In the following, we provide the pseudo code of this process.

```
if (e is arrival of a new file){
  if (file type is GAR){
        if ( the file already exists){
             while(reference number > 0){
            sleep(2000 milliseconds);
      }
         update the corresponding service;
         }else{
         deploy the corresponding service;
         }
  }else{
        remove the file;
  }
  }else if ( e is file deletion){
```

```
    while( reference number > 0){
        sleep(2000 milliseconds);
    }
undeploy the corresponding service;
}
```

As a result, users may deploy/update/undeploy a local service by simply storing/replacing/removing its GAR file to/in/from a folder. They need not to care about underlying processes, and services are deployed/undeployed automatically and transparently.

4 Performance Evaluation

ROST is implemented as a core component of CROWN middleware. We evaluate the performance of ROST by comprehensive experiments in real grid environments.

4.1 Experimental Environment

The experiments are conducted across two domains connected by the Internet. The deployer resides in Tsinghua University, while the target NS's (i.e., target containers) are located in Beihang University. The deployer has a Pentium III 1.6Ghz CPU and 512M memory, with a 10M bps connection to the Internet. Remote NS's reside in a 32-node cluster with each has two Intel Xeon 2.8GHz CPUs and 2G memory. The cluster is connected to the Internet through a 100M bps connection. No other tasks are running on each node except the necessary CROWN middleware.

4.2 Performance Metrics

We use the following metrics to evaluate ROST.

- *Deployment response time.* It is important that a remote service deployment introduces shorter response time. When multiple concurrent deployment requests are sent to a single NS, the deployment response time increases.
- *Task execution time.* A task here means a collection of independent jobs, while a job means an invocation of a specific service. Given a task, we concern its total execution time.

4.3 Experimental Results and Analysis

We execute the experiment 100 times and report the average.

In the first experiment, we evaluate the performance of ROST in terms of deployment response time. The deployer in Tsinghua University issues concurrent deployment requests to a node server in Beihang University. We vary the ways of service GAR file transfer, FTP and SOAP attachment. Each GAR file has a size of about 6K bytes.

Figure 4 shows the average deployment response time as a function of the number of concurrent requests. When there is only one request each time, the response time of ROST is as short as seven seconds. In contrast, the cold deployment needs as long as 30 seconds to merely stop and restart the service container so as to load a new service. With increasing number of concurrent requests, the average response time increases roughly linearly. When the number of concurrent requests reaches 30, the average response time is about 52 seconds. We also observe that SOAP transfer has similar performance with FTP mechanism.

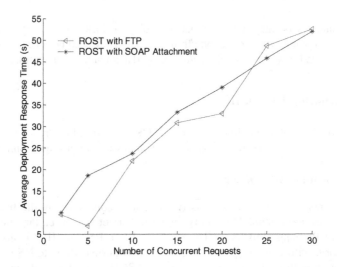

Fig. 4. Average deployment response time v.s. Number of concurrent requests

We then study how well ROST can help to achieve load balancing. In the second experiment, two schemes are compared, *with* and *without* ROST. There are 20 NS's available for processing jobs, while initially only a fraction of the nodes are deployed with the required service.

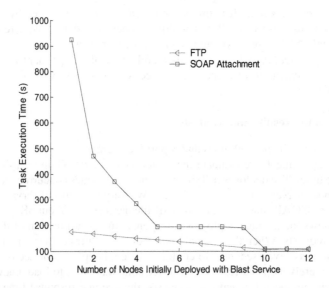

Fig. 5. Task execution time vs. number of nodes initially deployed with Blast service

Figure 5 plots the task execution time with different number of nodes initially deployed with the service. With ROST, the task execution time is significantly reduced, as a node may easily deploy its service to other relatively idle nodes. In some specific cases, the maximum improvement can be four times faster. When the fraction of nodes initially deployed with the service increases, the effect of time reduction becomes less.

5 Conclusions and Future Work

CROWN Grid aims to integrate nationwide and worldwide valuable Internet resources. In CROWN, remote and hot service deployment is highly demanded. In this paper, we present early design and implementation experience of remote & hot service deployment with trustworthiness (ROST). With ROST, services can be deployed to a remote container in a different security domain in a hot and secure fashion, which significantly improves service efficiency and quality. The experiments in real grid environment demonstrate the effectiveness of ROST.

In future work, we will perform more experiments, explore more relevant trust mechanisms, and further improve trust negotiation and deployment efficiency. Additionally, we will further integrate ROST with other CROWN middleware to handle real application problems such as load balancing and job migration.

References

1. I. Foster, C. Kesselman, and S. Tuecke, "The Anatomy of the Grid: Enabling Scalable Virtual Organization," The International Journal of High Performance Computing Applications, vol. 15, pp. 200-222, 2001.
2. I. Foster and C. Kesselman, The Grid 2: Blueprint for a New Computing Infrastructure. San Francisco: Morgan Kaufmann, 2003.
3. I. Foster, C. Kesselman, J. M. Nick, and S. Tuecke, "Grid Services for Distributed System Integration," IEEE Computer, vol. 35, pp. 37-46, 2002.
4. L. Bernardo and P. Pinto, "Scalable Service Deployment using Mobile Agents," presented at the Second International Workshop on Mobile Agents, 1998.
5. M. Bossardt, A. Muhlemann, R. Zurcher, and B. Plattner, "Pattern Based Service Deployment for Active Networks," presented at the Second International Workshop on Active Network Technologies and Applications, 2003.
6. "The Globus Toolkit: http://www.globus.org/toolkit/."
7. "Apache Axis: http://ws.apache.org/axis/."
8. M. Fleury and F. Reverbel, "The JBoss Extensible Server," presented at ACM/IFIP/USENIX International Middleware Conference, 2003.
9. "Microsoft.NET: http://www.microsoft.com/net/."
10. T. Friese, M. Smith, and B. Freisleben, "Hot Service Deployment in an Ad Hoc Grid Environment," presented at International Conference on Service Oriented Computing, 2004.
11. W. Goscinski and D. Abramson, "Distributed Ant: A System to Support Application Deployment in the Grid," presented at the Fifth IEEE/ACM International Workshop on Grid Computing, 2004.

12. F. Baude, D. Caromel, F. Huet, L. Mestre, and J. Vayssiere, "Interactive and Descriptor-based Deployment of Object-Oriented Grid Applications," presented at the 11th IEEE International Symposium on High Performance Distributed Computing, 2002.
13. WSRF Specifications, http://www.oasis-open.org/committees/tc_home.php.
14. I. Foster, C. Kesselman, G. Tsudik, and S. Tuecke, "A Security Architecture for Computational Grids," presented at the 5th ACM Conference on Computer and Communications Security, 1998.
15. W. H. Winsborough, K. E. Seamons, and V. E. Jones, "Automated Trust Negotiation," presented at DARPA Information Survivability Conference and Exposition, 2000.
16. N. Li, W. H. Winsborough, and J. C. Mitchell, "Distributed Credential Chain Discovery in Trust Management," presented at the 8th ACM Conference on Computer and Communications Security, 2001.
17. W. H. Winsborough and N. Li, "Towards Practical Automated Trust Negotiation," presented at the 3rd International Worshop on Policies for Distributed Systems and Networks(POLICY 2002), 2002.
18. W. H. Winsborough and N. Li, "Safety in Automated Trust Negotiation," presented at IEEE Symposium on Security and Privacy, 2004.
19. M. Winslett, T. Yu, K. E. Seamons, A. Hess, J. Jacobson, R. Jarvis, B. Smith, and L. Yu, "Negotiating Trust on the Web," IEEE Internet Computing, vol. 6, 2002.
20. T. Yu and M. Winslett, "A Unified Scheme for Resouce Protection in Automated Trust Negotiation," presented at IEEE Symposium on Security and Privacy, 2003.

Grid Developing Environment in CGSP System[1]

Weimin Zheng, Lisen Mu, Qing Wang, and Yongwei Wu

Department of Computer Science and Technology,
Tsinghua University, Beijing, 100084, China
zwm-dcs@tsinghua.edu.cn
{mulisen99, wangqing02}@mails.tsinghua.edu.cn

Abstract. Grid computing is becoming a mainstream technology for multi-institutional distributed resources sharing and system integration. Normally, the programmer's productivity in designing and implementing efficient parallel applications over grid remains a very time-consuming task, especially for the non-compute users. At the same time, the development of grid programming environments, which would enable programmers to efficiently exploit grid technologies, becomes an important and hot research issue too. In this paper, grid developing environment (GDE) based on ChinaGrid Support Platform (CGSP) is discussed. GDE supplies the portal building, job defining, programming interface, and administration tools for CGSP. The GDE motivations, architecture and corresponding implementation over CGSP are presented respectively.

1 Background

Grid computing is emerging as a main stream for the large-scale distributed resource sharing and system integration.

As one of the essential characteristics, a grid system aggregates all kinds of the low-level resources and shields heterogeneities and dynamic behaviors of those resources. Therefore, grid middlewares are developed to implement and generalize those characteristic functionalities to ease the construction of the grid systems. Furthermore, agreements and standards across the grid systems and the grid middlewares are established to improve the interoperability between the grid systems. Corresponding achievement includes OGSA [4], WSRF [5] and Globus Toolkits (GT) [3]. But there is still a little achievement at the grid developing tools for the end users, especially for of non-computer developers.

In 2002, China Ministry of Education (MoE) launched the largest grid computing project in China, called ChinaGrid project [1] , aiming to provide the nationwide grid computing platform and services for research and education purpose among 100 key universities in China.

The underlying common grid computing platform for ChinaGrid project is called ChinaGrid Supporting Platform (CGSP).[1] CGSP integrates all kinds of resources in education and research environments, makes the heterogeneous and dynamic nature

---

[1] This Work is supported by ChinaGrid project of Ministry of Education of China, Natural Science Foundation of China under Grant 60373004, 60373005, 90412006, 90412011, and National Key Basic Research Project of China under Grant 2004CB318000.

J. Cao, W. Nejdl, and M. Xu (Eds.): APPT 2005, LNCS 3756, pp. 313 – 322, 2005.
© Springer-Verlag Berlin Heidelberg 2005

of resource transparent to the users, and provides users various ways to access and monitor resources within the grid system constructed with CGSP. GDE in CGSP focuses on providing sufficient programming interfaces and rich developing and monitoring tools to enable all levels of users, from expert grid developers to grid administrators without programming background.

2 Design Considerations

The CGSP middleware system is designed to reduce the cost of constructing grid systems for grid system developers. Application development and resource accessing and monitoring are also important concerns for grid system deployers, application developers and administrator users. Thus, the design goal of the grid developing environment is to provide sufficient user interfaces and instruments to cover all functionalities supported by CGSP and to satisfy different requirements from different roles.

Therefore, the key design considerations of Grid Developing Environment can be summarized as follow:

1. GDE interfaces must provide sufficient functionalities. They must enable users to access all necessary functionalities of the inner modules of CGSP.
2. GDE interfaces must meet the requirements of users of all levels. Grid system developers may need low level API to write their own meta services. Application developers may need both low level and high level API to develop applications. Grid system deployers may need various tools to register resources and deploy services. Administrator users may also need various client tools.
3. GDE must provide functionalities to simplify or auto-complete some complex developing and deploying procedures during the whole lifetime of the grid development and management. According to the design of CGSP, such procedures include software packaging, application deploying, workflow defining, etc.

3 Architecture

According those design considerations, the architecture of GDE can be described as follow.

As shown in Fig.1, from the perspective of implementation, the GDE has a layered structure. In this structure, upper layer is implemented based on the lower layer. The lower layer has more precise operations in the API, while the upper layer provides friendlier interface and more integrated functionalities. The main modules in the GDE architecture are introduced as below:

3.1 Meta Service Access Layer

This layer provides a basic API which contains primitive operations provided by the CGSP functional modules. These functional modules are Container, Information Center, Job Manager, Storage Manager and Domain Manager. This interface has the most complete functionality, and could be used by any kind of developers who are interested. MSA is the base layer of all other GDE modules.

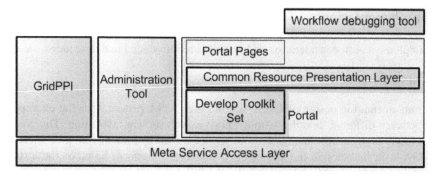

Fig. 1. Architecture of Grid Developing Environment in CGSP

3.2 Grid Parallel Programming Interface

From the user's aspect, it is a MPI-like interface granting users the ability to develop parallel applications. GridPPI [2] encapsulates the logics of operations in the MSA layer, provides users interface to customize and execute grid jobs, and provides a synchronization control mechanism on the job level. Other necessary functions like resource query and data transfer are also supported by this interface. Application developers can develop parallel programs via GridPPI interface.

3.3 Development Toolkit Set

It provides a set of tools which encapsulate some complex and frequently used procedures during the development and deployment of the grid service. Currently such procedures include software packing and workflow defining, thus the software packing tool and workflow defining tool are implemented in this toolkit set.

3.4 Common Resource Presentation Layer

The concept of this interface mainly comes from the demand of the Portal and a variety of monitoring client applications. In the most case, such applications aggregate and integrate information from the grid system, provide a certain form of view of the information to the user, and keep the view up to date by frequently retrieving information from the grid system. To ease the development of applications and to lower the performance cost of the query operations, Common Resource Presentation layer is presented on the purpose of providing a relatively uniformed query interface.

3.5 Administration Tool Set

Administration functionalities, like resource registering, are provided by administration tool set.

4   Implementation

The implementation consideration and features are introduced in this section.

4.1 MSA Layer

The implementation design of CGSP follows the OGSA pattern, thus the interoperation between different modules within CGSP is web service invocation. The goal of MSA Layer is to encapsulate all the web service SOAP operations into java interface, so MSA layer includes all the service client stubs and interfaces based on these stubs. The main functions of MSA layer fall into the following category:

1. Container Functions:
   Hot deploy, Remote deploy: deploy a grid service archive (.gar file) into a given container.

2. Information Center Functions:
   Resource query and management: Query computing resources registered in the InfoCenter; register a new computing resource into the InfoCenter; remove a computing resource record from the InfoCenter; update information of a computing resource in the InfoCenter.
   Service query and management: Similar to resource query and management. Provide query/add/remove/update operations on the grid service registrations.
   Deploy configuration query and management: Query and update domain configuration settings in the InfoCenter.

3. Data Storage Management Functions:
   File management: provide copy/move/delete operations on file or directory in the corresponding user's file space.
   File import and export: import a file from the client; export a file to the client.

4. Domain Manager Functions:
   User management: provide add/remove/update operations on user and user group information in the domain manager;
   Authorization and authentication: send authorization and authentication request to the domain manager and retrieve the result.

5. Job Manager Functions:
   Job submition: submit grid jobs to the Job manager.
   Job monitoring: query job states; send commands to a specified job.

4.2 GridPPI

The main design goal of GridPPI is to provide a parallel programming model to the users. The interface of GridPPI should be easy to use and highly abstract from MSA elayer. So the GridPPI would be relatively independent from the underlying grid

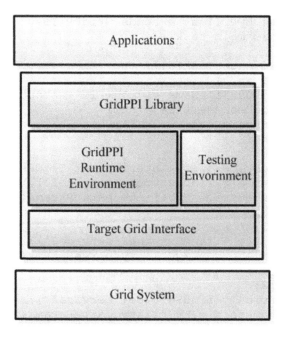

Fig. 2. Architecture of GridPPI

system. GridPPI can be transplanted onto different grid middleware other than CGSP without much work.

The inner structure of GridPPI is shown and described as follow:

1. GridPPI Runtime Environment

It is the kernel functional module of GridPPI. Runtime environment supports the parallelism provided by GridPPI. It schedules the different paralleled processes, handles synchronization between processes, invokes the underlying grid services accordingly, and monitors the status of the services. It is implemented as a lightweighted client-side runtime.

2. Target Grid Interface

It is the adaptor to the underlying grid system. Different OGSA-based gird systems may have different architecture and invocation paradigm, yet some basic meta services or functional modules can be abstracted in common, like data management, resource registration management and job management etc. Target System Interface makes the difference between different grid systems transparent and provides a standard interface to the GridPPI Runtime Environment.

3. GridPPI Library

It is the programming library providing API directly invoked by the developer's application code. There are 4 main aspects of interfaces included:

```
public class MyPPITask extends GridPPITask {
  public void main_task() {
    try {
      int id = getTaskId();
      if (id >= 3)   return;
      String outFileName = "r" + id + ".out";
      ServiceDesc service = null;
      Vector parameters = null;
      switch (id) {
      case 0:
        //service query interface of GridPPI
        service = findService("service1");
        parameters = MyUtility.parseParams(file1);
        break;
      case 1:
        service = findService("service2");
        parameters = MyUtility.parseParams(file2);
        break;
      case 2:
        service = findService("service3");
        parameters = MyUtility.parseParams(file3);
        break;
      }
      //Job Submition interface of GridPPI
      GridJob gridJob =
        executeJob (service, parameters);
      //Job Synchronization interface of GridPPI
      waitJob(gridJob);
      String result = MyUtil-
ity.parseResultMsg(gridJob);
      if(id != 0) {
        sendSignal(0, "sig", result);
      } else { //id == 0
        String r1 = result;
        String r2 = (String) waitSignal(1, "sig");
        String r3 = (String) waitSignal(2, "sig");
        String sum = MyUtility.postProcess(r1, r2, r3);
        reportStatus("result=" + sum);
      }
    }catch(Exception e) {
      reportError(e.getMessage);
    }
  }
}
```

Fig. 3. Sample Code of GridPPI

Job Synchronization: GridPPI supports task-level synchronization between processes. The invocation pattern of synchronization API is similar to the MPI interface, as shown in the {sample code}.

Job Submition: Submits a job to the underlying grid system.

Data Transfer: Specifies data transfer operations. In this interface, the source and destination of the data transfer is represented in the form of URL, thus both transfer between client and server and third-party transfer are supported by this interface.

Service Query: Searches through the target grid system registration and retrieves appropriate services according to the query condition. Different grid systems may have different forms of service registration. However, as stated above, heterogeneities of the service registration mechanisms of the grid middleware is dealt and encapsulated within the Target Grid Interface module.

To illustrate the basic coding paradigm, a sample code of GridPPI task is shown below in Fig. 3.

From Fig. 3, it can be seen clearly that the invoking paradigm of the GraidPPI interface, especially of the process synchronization functions, is quite similar to that of MPI libraries which parallel application developers are quite familiar with.

4. Testing Environment

Flawed program code may cause unpredictable performance expenditure or damage to the grid system. To avoid, or at least to minimize, the fault or bugs in the user program code, a Testing Environment is designed to simulate the real grid environment for testing of user's program code. In consideration of unpredictablility and heterogeneities existing in various real grid environments, it is impossible to perfectly simulate a real grid system. However, some simpler testing like deadlock detection or approximate performance prediction can be done in the Testing Environment.

4.3 Administration Tool Set

All the Administration tools are developed based on MSA layer. The main goal of these tools is to provide a graphic user interface to system administrators or application deployers to register resource, import files, deploy services etc.

4.3.1 Data Storage Client Tool

Data storage client tool provides data management user interface. In the design of GDE, data management UI is also provided in the portal page. However, large file transfer through portal has mainly 2 disadvantages:

The file data must be transferred to the portal server before the file is imported into the data storage server, thus 50% waste of bandwidth is inevitable;

Long period data transfer from client to portal server over http protocol will cause unbearable idle of the web browser.

Therefore, it is recommended to import large data files into the data storage manager through the client tool.

4.3.2 Domain Manager Client Tool

In the design of CGSP architecture, the composition unit of the entire grid is called domain. Each domain represents a virtual organization. A whole set of CGSP components are deployed within each domain and make the domain fully operational independently from the other domains. Each domain has its local security mechanism. Authentication across domains is also supported to support interoperation between domains.

Both the internal and external user information is administrated by the Domain Manager. Therefore, the Domain Manager Client Tool is a GUI for the domain administrators to maintain the account information of the domain.

4.4 Development Toolkit Set

The development toolkits ease the development of the portal and various client tools by encapsulating frequently used procedures. As stated above, currently 2 of such procedures are encapsulated in development toolkit:

Software Packing. In the architecture of CGSP, software applications can be packed into a general running service with the arguments of the application being mapped to the input message of the service. Current implementation of this design is to generate a special wsdl file with extensional elements describing mapping information from the input message to the application arguments and to zip the application executable file into the .gar file which will be interpreted and deployed by CGSP GRS(general running service). This implementation procedure is denoted as 'software packing', and is encapsulated in the software packing tool. Currently, it is used in the development of CGSP Portal to implement software packing user interface.

Job Defining. The terminology 'job' in CGSP denotes a single grid service invocation or an invocation of a well defined workflow, which is described in BPEL[7] and is also deployed in CGSP container as a service. The job defining tool allows the developer or administrator user to construct such workflow jobs by specifying inputs and outputs, defining control flow structure of the job and exporting BPEL scripts and other related meta information descriptions. In the current release of the CGSP, the job defining tool is also used in the development of Portal.

4.5 Common Resource Presentation Layer

In the second release of CGSP, the CRP layer is deployed in the Portal server. Portal server receives and responses http requests. The request URL of a specific pattern defined in CRP is forwarded to the CRP handler which performs the actual query operation via MSA interface and returns the query result in the form of xml document. The advantage of CRP layer in performance consideration is that the cache mechanism based on the URL indexing of the Portal engine can be easily utilized to minimize the actual query operation submitted to the underlying grid system.

4.6 Portal Pages

Utilizing functionalities of MSA layer, Development toolkits and CRP layer, Portal pages provide user interfaces to all administrative and common user operations of the CGSP system.

4.7 Workflow Debug Tool

In the design of CGSP authentication policy, after a workflow is defined and deployed by the application developer (or administrator), it can be invoked by common users who have sufficient access right. Thus the application developer may wish to debug the workflow being constructed to verify its correctness. In the second release of CGSP, breakpoint and debug mechanism is to be supported by the CGSP Job Manager and Workflow debug tool.

The workflow debug tool is to be implemented as a standalone client application. Its main function is to load a workflow definition, specify breakpoints in the control flow, start workflow execution in the Job Manager, and to monitor and control the runtime status of the workflow. It uses CRP layer to update the status of the current workflow, and submits controlling commands to the Job Manager via MSA library.

5 Conclusions

The current architecture design of the Grid Developing Environment is based on the CGSP middleware system. GDE provides various programming interfaces and administrative utilities to allow users to develop grid services, develop applications, deploy software, register resources, construct workflow, submit jobs and monitor many kinds of information within the grid. Compared to the other client-side developing environments or client tools, CGSP GPE offers the following advantages:

1. It provides API in different abstraction levels to satisfy demands of different users;
2. It provides both API and various client tools to enrich the measure to develop and deploy services and applications;
3. Modules and applications in the higher layer of GDE architecture is developed based on the lower layer, thus higher-layer modules themselves provide sufficient examples to the developers on how to utilize the lower-layer modules.

Reference

1. Jin, H., ChinaGrid: Making grid computing a reality. Digital Libraries: International Collaboration and Cross-Fertilization, Proceedings 2004
2. Yongwei Wu, Guangwen Yang, Qing Wang, Weiming Zheng, Coarse-grained Distributed Parallel Programming Interface for Grid Computing. Lecture Notes in Computer Science, (2004), 3032, 255-258
3. I. Foster and C. Kesselman: Globus: A Metacomputing Infrastructure Toolkit. International Journal of Supercomputer Applications, Vol.11, No.2, pp.115-128, 1997
4. Open Grid Services Architecture,
   http://www.ggf.org/Public_Comment_Docs/Documents/draft-ggf-ogsa-specv1.pdf

5. Web Service Resource Framework (WSRF),
   http://www.globus.org/wsrf/ and http://www.ggf.org/documents/GFD.30.pdf
6. ChinaGrid, http://www.chinagrid.edu.cn
7. Business Process Execution Language for Web Services,
   http://www-128.ibm.com/developerworks/library/specification/ws-bpel/
8. G. Andronico, R. Barbera, A. Falzone: Grid portal-based data management for lattice QCD
   data. Nuclear Instruments and Methods in Physics Research A 534 (2004) 76–79
9. A. Andronico, R. Barbera, A. Falzone, P. Kunszt, G. Lo Re, A. Pulvirenti, A. Rodolico:
   GENIUS: a simple and easy way to access computational and data grids. Future Generation
   Computer Systems 19 (2003) 805–813

Grid Job Support System in CGSP[*]

Jinpeng Huai[1], Yu Wan[1], Yong Wang[1], and Haifeng Ou[2]

[1] Advance Computer Technology Institution, Computer Science and Technology,
Beihang University 7−28#, 100083 Beijing, China
`huaijp@buaa.edu.cn`, {`wanyu, wangy`}`@act.buaa.edu.cn`
[2] The Key Laboratory of Virtual Reality Technology, Ministry of Education, China,
Beihang University 7−28#, 100083 Beijing, China
`ouhf@vrlab.buaa.edu.cn`

Abstract. As Grid computing becomes more and more practical, User needs a steady computing environment and an effective way to assemble simple services into complicated service to meet their requirements. Based on BPEL4WS, this paper introduces a Grid Job Description Language (GJDL) to combine correlation services into composited service. With GJDL, we design and implement a job support system which has been already applied successfully in China Grid Support Platform (CGSP).

1 Introduction

Grid computing turns out to be an effective solution for large scale computing resources sharing and coordinated use of resources at multiple sites. In grid architecture OGSA service is a key conception, every thing in grid is a Grid Service. WSRF which heavily depends on web service specifics defines and refers a set of specifications which are used to implement resource lifetime management, discovery, notification and so on to form a Grid Service. It makes Grid Service compatible with current web service technology.

As grid jobs became more and more complex, the categories of jobs evolute from simple job and batch job to complicated jobs described by workflow languages that are called grid workflow. In most common grid systems, such as Globus Toolkit[10], do not support this kind of job. In service oriented grid architecture, a simple job is the request to a service; a batch job is the request to a set of services that are simply called one by one. Complicated job, in its nature, is the request to a set of correlation services. It includes the control of the execution sequence, which makes constructing a more complex service by simple services a reality. To define a complicated job, user needs to define the execution sequence of the element services, correlation data handling, exception handling mechanism, identification of job instance, roles of element services and so on.

In our grid middleware, China Grid Support Platform (CGSP) [1, 2, 3], in order to support both simple jobs and complicated jobs, we defined a Gird Job Description

---

[*] This Work is supported by ChinaGrid project of Ministry of Education of China, Natural Science Foundation of China under Grant 60373004, 60373005,90412006, 90412011.

J. Cao, W. Nejdl, and M. Xu (Eds.): APPT 2005, LNCS 3756, pp. 323 – 331, 2005.
© Springer-Verlag Berlin Heidelberg 2005

Language (GJDL), which is simplified from BPEL4WS [12] and added some feature for grid characteristic. We designed and implemented a job support system that supports the jobs described by this language.

In this paper, we begin with goals and functions of job support system. Then, we introduce our job description language GJDL. In fourth part, we describe our job support system architecture and its implements in detail. Conclusion and future work are discussed at last.

2 Goals and Functions

CGSP is a grid middleware developed for the construction and evolution of ChinaGrid. Different from normal grid middleware efforts (GT series [10], OMII [17]), CGSP is a platform. Not only does it include the grid running components, such as portal, service container, service monitoring and discovery, file delivery and transformation, but it also provides the grid developing tools, such as programming API, portal constructing tool, service deploying and packaging tools and so on.

2.1 Goals

Job support system, as a key part of this platform, must meet the following goals.

- Easy to use. Most users of ChinaGrid are not computer scientists, nor are they computer engineers. They are most likely the experts of different fields who need to take advantage of the grid computing. Keeping our job support a simple model will helps them to easily define and submit jobs for use.
- Robust. Since many grid jobs are time-consuming mission, if we cannot keep our job support system running in a steady status, we cannot really provide services for grid jobs.
- Effective. Grid jobs are often time critical. For an instance, weather forecast. To perform user's jobs as soon as possible, we must schedule jobs in an effective way.
- Throughput. Grid platform does always support many jobs concurrent execution. As the nature of grid, it tries to make full use of the large scale computing resources.

2.2 Functions

In the first stage of CGSP job support system, the main functions provided as follow.

- Defining a job. User chooses suitable Grid Services (or let job support system to do this for him) to make up his job. This procedure makes a job definition that is used for job execution.
- Submitting a job. After user defines a job, he should submit it to job support system. Job support system will process the job later.
- Scheduling a job. During job support system running procedure, it interacts with information center(IC) to get the suitable Grid Service to perform user's job.

– Workflow management. In CGSP, jobs are often workflow. Job support system provides flow management of this kind of jobs. It includes the management of execution process, data and lifetime.
– Job status management. Job support system provides job execution information to grid administrators and the job owner. And it allows administrators and the job owner to control the job's execution.

3   Grid Job Describing Language (GJDL)

With the progress of the grid computing, grid jobs become more and more complex. Simple job, which is a simple request to a Grid Service or resource, cannot meet user's requirements. In service oriented grid system, we need to give job support system the ability to compose correlated services into a complicated job. Traditional job description languages (ClassAD [14], RSL [18]) do not have this capacity.

BPEL4WS, which is based on both XLANG [16] and WSFL [15], is designed for web service oriented workflow. BPEL4WS, for its nature, is an effective web service workflow description language. It is the standard recommended by OASIS.

As WSRF refers, Grid Service now relies heavily on web service specifications and is an extension from web service. In order to take advantage of web service environment and workflow technology, we defined our grid job description language based on BPEL4WS. So our GJDL has the ability to describe complicated jobs, which means to describe complicated work flow of grid applications. Our jobs described by GJDL, is whole service oriented, according to the fact that grid technology and web service technology integrate daily. Job in our system is composed by correlated Grid Services, and job itself is a Grid Service.

Although BPEL4WS is a good language for service composition, its nature is for business process described by web service. It doesn't suit for describing Grid Service composition from the beginning. It does not support WSRF, which means it is impossible to use a WS-Resource [13]. And the BPEL4WS language is too complex and too large for Grid Service composition. For grid computing, job support system needs the ability to make full use of the large scale computing resources, which means job support system should balance the load between different computing nodes to make a well throughput. That is done by dynamic binding element services in runtime according to information center. This late binding technology also helps job support system has the capacity of fault tolerance. We simplified the useless features and extended some elements in BPEL4WS to form our GJDL. Now we only keep four core atom activities (invoke, receive, assign and reply) and four structure activities (sequence, flow, while, switch) in our GJDL. We extends invoke activity for supporting WS-Resource, dynamic element service selection and binding, fault tolerance.

Here are brief differences between BPEL4WS and GJDL.

Features	BPEL4WS	GJDL
Support WSRF specifics	No	Yes
Simple structure	Yes	No
Dynamic element service binding	No	Yes
Fault tolerance	Must be defined by user	Job support system automatism

4 Job Support System Architecture

4.1 Whole Architecture

As the goals and functions described in second part of this paper, we designed our job support system architecture as figure 1 shows.

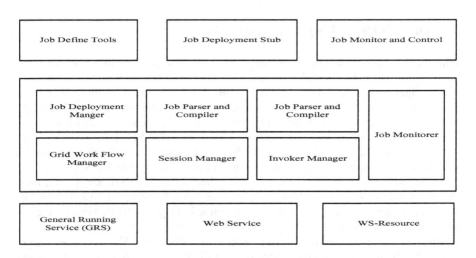

Fig. 1. Job support system whole architecture

There are three layers in job support system. The top layer is for client to define a job, submit a job to job manager, and control and monitor stubs for user to take full control to his a job. We will discuss the job definition in other papers. The middle layer is the core of our job support system, which we will describe in detail. It is named job manager, which takes charge of deploying users' jobs, grid workflow instances management, monitoring jobs' running information and control jobs execution under users' instructions. There is only a job manager in each domain of CGSP. Domain administrator deploys it, and Information Center [4] remembers its location, which will be used by CGSP Portal [4] or CGSP Grid PPI [4]. The bottom layer is the foundation of job support system, which includes support for web service, WS-Resource and General Running Service (GRS). GRS, in brief, adapts user's local binary runnables to normal WS-Resource. Its work mechanism likes GRAM [10].

4.2 Job Manager Architecture

Main functions of job manager is handling user's request based on job definition. Its core functions are described as follow.

– Job deployment. User defined jobs in user space, which can be a simple service (job) or a complicated service (job). Job manager should have the ability to dynamic deploy jobs for future execution.

- Session management. During execution, a job may interact with other job instances. To delivery the requests between job instances correctly, job manager should have a session manager in job instances level.
- Scheduling. For better throughput in domain, job manager need to schedule both job instances and element services.
- Lifetime management. Job manager maintains jobs' status and full control of their lifetime. This part also includes job status monitoring and control.
- Execution. Job manager translates job definition into job instance, controls element services execution sequences, copes with related data, and handles exceptions in execution procedure.
- Element service binding. For load balance in domain, job manager should inquire information center for a set of services that meets user's expectation deployed at different computing nodes in domain and select the computing node with a low load to perform the request to the service.

The architecture is shown in figure 2.

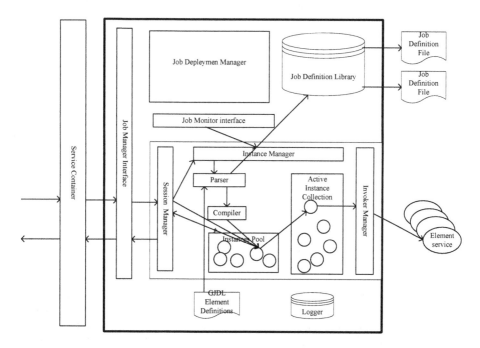

Fig. 2. Job manager architecture

Modules and the relations of modules are described as follow.

- Job manager interface. Service container (in this version, it's Globus container [10]) deliveries user's request of deploying a job or executing a job to job manager with this interface. All the interactions between user and job manager are through this interface. It receives the user's request, deliveries it to related module, blocked the request procedure until the related module answer back. Then it transforms the answer message with the right format, and answers service container with it. Service container then does the dirty job to interact with user.

- Deployment manager. User defined complicated job should be known by job manager. Deployment manager is responsible for receiving user deploying file which is called gar file, putting the job definition file into job definition library and adding description information and other useful stuff into service container to help future user to use.

- Session manager. Its main function is picking up correct job instance for the incoming request message from job instances pool and identifying the session information to the in-coming and out-going messages so that related job instance knows which job instance it deals with.

- Instance manager. It takes charge of job instances management. For effective reason, it caches job definitions and their execution related structure and data for future requests for same jobs. It also monitors job instances status and manages instances' lifetime by user's order. It can start, pause, resume and terminate a job instance. Following figure (figure 3) shows the exactly translation between job instance states.

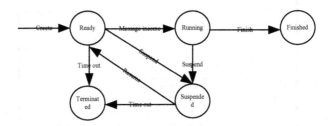

Fig. 3. Job Instance status

- Invoker manager. Invoker manager handles the request to element service. As we discussed formerly, considering load, the invoker manager inquiries information center for a set of services that meets user's expectation deployed at different computing nodes in domain. Based on specific load algorithms, it invokes the best fit onet and handles the exceptions during the service invoking procedure.

- GJDL parser and compiler. When user request for complicated job incomes, if there is no suitable job instance for that request, the instance manager will pick up the right job definition file from job definition library and call GJDL parser to parse the job definition file, it translates the GJDL describing job into a definition entity. Then instance manager uses compiler to make the entity to a runtime entity, which is called a job instance. Instance manager will then put this new job instance into instance pool and take care of its lifetime.

Other parts of job manager are introduced briefly here. Job definition library is the place where job manager puts the job definition files. Job monitor interface replies user's request for job instance status information, which is collected by instance manager during the job instance execution.

4.3 Execution Diagram

To make the execution procedure straight, we will show a typical job request execution.

1. User sends a request to service container via SOAP message.
2. Service container unpacks the message. For example, logging. Then service container deliveries it to job manager interface.
3. Job manager interface finds out it is a request for a job. It deliveries the request to session manager.

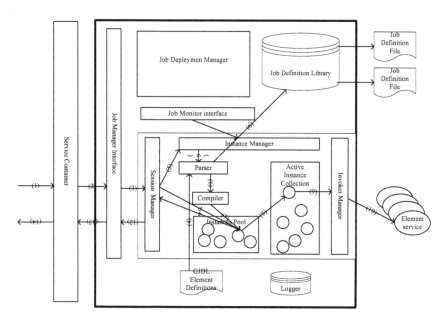

Fig. 4. Job execution diagram

4. Session manager depends on the identification in the request to pick up the right instance to delivery the request to. If success, execution procedure goes to step 8.
5. Instance manager receives that request and picks up the right job definition file from job definition library.
6. Instance manager uses parser and complier to make out a new job instance for that request.
7. Instance manager puts the newly created job instance into instance pool.
8. Instance manager begins or resumes the execution of the job instance. If the job instance is not an active one, that is to say, it has been paused by user or job manager for scheduling. Instance manager actives the job instance to execute.
9. During the job instance execution procedure, instance manager records the changing status of the job instance.
10. When the job instance needs to call a service, the invoker manager inquiries information center for a set of services that meets user's expectation deployed at different computing nodes in domain and selects the right computing node with low load to perform the request to the service.

11. During the job instance execution, if some exception occurs, instance manager handles the exception and make the job instance finished with useful error information.
12. When job instance finishes, instance manager returns the result to job manager interface, takes out the job instance from instance pool and destroys the job instance.
13. Job manager interface then returns the result to service container. Then do some extra work such as recording the process period and so on.
14. Service container ships the result to user. So a job instance execution finishes.

5 Conclusion and Future Work

Grid jobs became more and more complex. Simple job or service could not satisfy user any more. User needs an effective way to assemble simple services into complicated service to meet their requirements. In OGSA, a service-oriented architecture, composing correlation services into a complicated job is an effective and smart way to resolve this issue. In many important grid systems, such as Globus Toolkit, VEGA[5], there is no such part to match this up. We use complicated job, which describes the relationship between related services, to stand for user's requirement for complicated service.

Job support system is a core part of a grid platform. It takes charge of defining a job, executing a job and scheduling jobs in domain to get the largest throughput and best use of large-scale computing resources shared in domain.

Supporting complicated job is an important part of the job support system in CGSP. Based on BPEL4WS, which suits describing service composition about web service, we simplified BPEL4WS for easy usage and extended some activities for grid feature such as WSRF and throughput and load balance for domain computing resources effective use. This is our most valuable contribution to grid computing field. Our job manager, which is the core of our job support system, has following strongpoint. First of all, it supports the execution procedure management of complicated job. Secondly, it handles the job request message in layers, which helps us to focus on the main issue we try to resolve. Also that makes it easy to add some new features to existing process procedure for user's interests. The third is that enabling our platform with the capacity of load balance and throughput. Finally, our job support system is service container independent, so other gird platform can easily replant our job support system to their environments.

In the future, we will support Job Submit Description Language (JSDL) to support job submitting between inhomogeneous grid systems and CGSP. We will enrich our GJDL that has been proven to be an effective language to describe complicated job to meet up other key feature of grid computing fields such as large data management between element services and so on.

References

1. ChinaGrid, http://www.chinagrid.edu.cn.
2. H. Jin, "ChinaGrid: Making Grid Computing a Reality", Proceedings of ICADL 2004, Lecture Notes of Computer Science, (2004), 3334, 13-24
3. ChinaGrid Support Platform, http://www.chinagrid.edu.cn/CGSP.

4. CGSP Work Group, Design Specification of ChinaGrid Support Platform, Tsinghua University Press, Beijing, China, 2004
5. Z. Xu, W. Li, "Vega Grid: A Computer Systems Approach to Grid Research", Keynote speech paper at the Second International Workshop on Grid and Cooperative Computing (GCC 2003), Shanghai, China, December 2003.
6. Hu CM, Huai JP, et al. "WebSASE4G: A Web Services-based Grid Architecture and its Supporting Environment". Journal of Software, 2004 (in Chinese)
7. Foster I, Kesselman C, Nick J, Tuecke S. "The Physiology of the Grid: An Open Grid Services Architecture for Distributed Systems Integration". http://www.globus.org/ research/papers/ogsa.pdf, January 2002.
8. Foster I, Kesselman C, Tuecke S. "The Anatomy of the Grid: Enabling Scalable Virtual Organizations". International J. Supercomputer Applications, 2001, 15(3). 1~10.
9. Open Grid Services Architecture Roadmap. 7th Global Grid Forum Draft, 2003. http://www.gridforum.org/ogsa-wg/papers/ ogsa_roadmap.0.4.pdf.
10. 10.Globus Toolkit, http://www.globus.org
11. Curbera, F., Goland, Y., Klein, J., Leymann, F., Roller, D., Thatte, S., and Weerawarana, S. (2002) 'Business Process Execution Language for Web Services', http://msdn.microsoft.com/webservices/default.asp?pull=/library/en-us/dnbiz2k2/html/bpel1-0.asp, accessed 22 September 2002.
12. BPWS4J, http://www.alphaworks.ibm.com/, 2004-3-15
13. Karl Czajkowski. "The WS-Resource Framework", http://www.globus.org/wsrf/specs/ws-wsrf.pdf, 2004-9-2
14. ClassAD, http://www.cs.wisc.edu/condor/classad/refman/, 2004-11-7
15. 15. Leymann, F.(2001) "Web service flow language (WSFL) 1.0", http://www-4.ibm.com/software/solutions/webservices/pdf/WSFL.pdf, accessed 22 September 2002.
16. Thatte, S. (2001), "XLANG: Web Services for Business Process Design", http://www.gotdotnetcom/team/xml_wsspecs/xlang-c/default.htm, accessed 22 September 2002.
17. Open Middleware Infraxtructure Institute, http://www.omii.ac.uk/
18. RSL, http://www.globus.org/gram/rsl_spec1.html , 2004-11-7

JFreeSim: A Grid Simulation Tool Based on MTMSMR Model[*]

Hai Jin, Jin Huang, Xia Xie, and Qin Zhang

Cluster and Grid Computing Center, School of Computer Science and Technology,
Huazhong University of Science and Technology, Wuhan, 430074, China
hjin@hust.edu.cn

Abstract. Due to the non-repeatability of the grid environment, limitation comes out when conducting grid performance analysis in real environment. Therefore, grid simulation tool is used extensively as an important research tool. This paper proposes JFreeSim, a new grid simulation tool based on *multiple tasks, multiple schedulers and multiple resources* (MTMSMR) model. As a modular and extensible simulation tool, JFreeSim realizes many kinds of entity modeling and communication mechanism between all entities, and makes system simulation accord with the characteristics of the grid environment. Experiments indicate that JFreeSim can provide users with flexibility in configuring and meet requirements of different system architectures and applications, and the simulation results are as expected.

1 Introduction

For using resources that belong to different organizations and are geographically distributed, the grid system is the realization infrastructure and techniques. Grid system has great potential to help people solve various problems. In the grid environment, the execution of the task needs to be adjusted dynamically according to the change of the run-time environment. Though the performance of applications can be evaluated by executing the applications on actual resources, this approach usually has very large limitation. This is because not only the resources needed may not be able to satisfy, but the experiments are difficult to repeat in grid environments. In addition, the scale of current experimental grid is relatively small, and cannot totally reflect the characteristics of the actual grid systems. Therefore, grid simulation tool is used extensively as an important research tool to evaluate the performance of various fields of grid systems [1, 2].

This paper proposes JFreeSim, a new grid simulation tool base on *multiple tasks, multiple schedulers and multiple resources* (MTMSMR) model. The following characteristics make JFreeSim feasible to simulate various grid environments:

1. JFreeSim models the main parts of the grid system, and provides flexible simulation according to the needs of the user. MTMSMR model that multiple

---

[*] This paper is supported by National Science Foundation under grant 60273076 and 90412010, ChinaGrid project from Ministry of Education of China.

J. Cao, W. Nejdl, and M. Xu (Eds.): APPT 2005, LNCS 3756, pp. 332–341, 2005.
© Springer-Verlag Berlin Heidelberg 2005

tasks are scheduled by different schedulers and executed on multiple resources is supported. This model makes JFreeSim convenient to simulate various complex execution situations in grid environments.

2. The dynamic characteristic of grid is fully considered. In the course of simulation, the status of resources is described accurately, which makes the execution status of application correspond to the actual situations. In addition, the simulation of layered information directory also reflects the dynamic characteristic of grid environments.

3. Various overheads during running are estimated because they have a significant impact to the performance of an application. JFreeSim estimates the control and communication overhead incurred in the execution, thereby providing an effective approach to describe internal feature of real system accurately.

4. The design of JFreeSim follows the principles of making it modular and extensible [3]. Each part is independently modeled, and interacts by a certain message mechanism. The extensible design allows the user to construct different simulation script for various applications and execution processes.

The rest of the paper is organized as follows. Related works are discussed in section 2. Section 3 describes the entity models used in JFreeSim. Section 4 discusses the architecture and implementation of JFreeSim. Section 5 illustrates a use example of JFreeSim. Section 6 concludes the paper.

2 Related Works

For the grid system, one approach used to study the system characteristics is to emulate a grid system on a real computing system. This approach is used by MicroGrid [4] which emulates multiple computing resources on a real resource to increase grid size using limited resources. MicroGrid is suitable for testing real application on real, controllable environment. However, emulating real application may take a substantially long time. Hence, in order to minimize the turnaround time for the extensive study of grid environment, a simulator is needed.

At present, the main related simulation tools include Bricks [5], SimGrid [6], GridSim [7], ChicSim [8], HyperSim [9]. Bricks is a Java-based discrete event simulator. It is designed to maximize modularity of reconstructing system model based on client-server architecture. One may run Bricks to evaluate scheduling heuristics or to evaluate data movement algorithms on grid. Status of each component is the estimation of real world system trace.

SimGrid is a C language based toolkit for the simulation of task scheduling in a distributed environment. In SimGrid, resources are assumed to be time-shared with other applications. The performance of a resource can be a constant number or specified through a trace file containing real run time load variations. Tasks are described using a *Directed Acyclic Graph* (DAG). The task scheduling algorithm is defined by the SimGrid user, and SimGrid provides highly accurate network model for TCP and non-TCP transport. Thus, SimGrid offers the flexibility to simulate various applications and scheduling algorithms.

GridSim is a Java-based discrete event grid simulation toolkit. It provides high extensibility and portability through Java and thread technologies. GridSim supports modeling and simulation of heterogeneous grid resources, gird users and applications. Designed for a market-like grid computing environment, GridSim can simulate multiple computing grid users, applications, and schedulers, each with their own objectives and policies.

ChicSim is a Parsec-based simulator for concurrent job and data scheduling. System model is fixed. The user just needs to specify resources, networks, and workload characteristics to the simulator by a list of files.

HyperSim is a general-purpose discrete event simulation library developed on C++. It provides comprehensive classes for constructing a simulator. The experimental evaluation shows that HyperSim can be used to simulate the same grid environment with a much faster simulation speed [10].

3 Entity Models in JFreeSim

In this section, we discuss various entity models in JFreeSim. These entities include user entity, application and task entity, resource entity, scheduler entity, information directory entity, network entity, statistics and analysis entity. These entity models are used to describe the process of application executed and resource changed in grid environments [11, 12].

3.1 User Model

User entity is used to simulate the behavior of the user in JFreeSim. The user's behaviors involve selecting scheduler, dispatching the requests of the user, and receiving the processing results of the tasks. In the user model, the user request to implement one or several tasks. In the case of those involvements with more than one task, their submission and execution are based on the order of the tasks.

3.2 Task Model

In JFreeSim we assume that the simulations are performed at the task level. The static view of an application consists of two components: the problem and the scheduling algorithm used to solve this problem. A problem is defined as a set of related tasks, and the algorithm of the scheduler describes the process of resource selection and task scheduling.

Usually, task modeling defines various basic attributes needed for its simulation, including execution code quantity, data quantity before and after executing, etc. In addition, the conditions under which tasks are executed on resources are also defined, including resource type, computing capability, storage capacity, etc. In JFreeSim, DAGs are used as internal representations of the problem. These DAGs are generated by the user's input configuration files and the script tool.

3.3 Resource Model

A typical grid system consists of various resource nodes connected by a network. In JFreeSim, the resources are characterized by the computing resources. Each computing resource can own one or more computing nodes, and each computing node can own a lot of CPU units. We assume that the performance of a computing node is denoted by its computing power and storage capacity.

JFreeSim describes the performance characteristics of computing node through constants, random variables with predefined distributions, trace files, or system level abstraction based on random variables. In JFreeSim the storage capacity of a computing node is described through available storage space. When the execution of the tasks has constraints for the storage capacity of the computing resource, these constraints will be considered by the scheduler while selecting resources.

In grid environments, resources may fail and new resources may become available during computation. This scenario can be usually used to simulate hardware related events such as power shutdown. In JFreeSim, user can directly configure the available time range of resources when defining these entities, which can offer support to describe the dynamic characteristics of resources.

Each computing node has its own local scheduling policy. In JFreeSim, possible choices provided include *Round Robin*, *First In First Out*, and *Shortest Job First*. In addition, the users can extend new scheduling policies by themselves. When multiple tasks are ready to run on a single computing node, the designated local scheduling algorithm is used to determine the order of their execution.

3.4 Information Directory Model

The information about computation process and dynamic performance of resources is maintained by the information directory entity. In order to reflect accurately the collecting and maintaining process of the information directory in the real system, we analyze the existing typical grid information service systems, abstract and partition their function modules and construct the four independent modules.

The bottom layer is the *information collector* (IC) used to collect the status of the object monitored. In resource layer, the *information server* (IS) classifies and combines the information from the lower layer, and forms the usage information of the local resources. A layer above IS is the *aggregate information server* (AIS) which is responsible for aggregating variety of information from different IS layers and organizing the information directory. AIS offers resource information in a large scope. The top layer is the user-oriented *directory server* (DS). DS provides the function of querying, finding and allocating the resources for the user, and offers other functions according to the needs of the system. In addition, DS also sends queries to IS directly to obtain the resource information.

Due to the different architectures of information directory and deployment of resources and applications, each application has different resource view from others.

3.5 Scheduler Model

Scheduler entity is used to select resources and assign tasks to those resources after the tasks are submitted. The scheduler obtains dynamically useful information in the system according to the scheduling strategy, and finds out suitable resources and assigns tasks to them. The scheduling strategy during simulating can be specified by the user. So the simulation process can perform various scheduling strategies containing plenty of constraints to tasks, users and resources. The scheduler entity responds to the performance variations through querying information directory. In addition, the scheduler also monitors execution status of the tasks that have been assigned. Thus, it can control the process of simulation execution.

3.6 Network Model

Another important aspect of the grid simulation is the simulation of the network. Assuming a fully connected network would minimize the difficulty of implementing the simulator. However, the major drawback of a fully connected model is that it may not accurately describe the situation in actual grid system. In an actual grid system, a network link may be shared by several computing nodes.

Network model consists of some network entities, including network link, router and network traffic generator. Network link joins two entities that need to communicate with each other, and a user can configure the bandwidth and the latency of the link. Router is responsible for joining links that have different communication performance and determining the accurate transfer path of data. The data transferred in the network is split into transport units with specified size. Transport units are transmitted according to the routing policy of the router entity. In an actual environment, a network link may be shared. So a network traffic generator is used to simulate the impact of other network traffic for the data transfer in the system. User can predefine the network traffic generator with specified distribution, which benefits us in evaluating the transfer performance affected by traffic congestion.

3.7 Statistics and Analysis Model

During the whole simulation, obtaining, coordinating and analyzing the simulation data produced must be supported as an important function. It makes sense only when the simulator is designed to reflect the simulation data produced during the simulation. In JFreeSim, processing simulation data is classified into three stages: acquisition, statistics and analysis process.

The process of obtaining the simulation data is the base of further study in this part, and it goes through the whole simulation. The simulator needs to obtain various data, such as the status information and the running information, at any time. After obtaining the simulation data, the simulator will classify them. The statistical measurements include system's build-in measurements such as average response time of the task, success rate of the request, utilization rate of the resource, and user-defined measurements as well. The various statistical results are displayed through a comprehensive GUI tool. Finally, JFreeSim analyzes the whole course of simulation according to the relevant statistical measurements and evaluation algorithms. The process of analysis may be directed at the performance of a single function as well as many interactive functions and even the whole system. As the analysis results, the simulator may provide a series of evaluation results and possible references for further improvement.

4 Architecture and Implementation of JFreeSim

4.1 JFreeSim Architecture

The modular architecture of JFreeSim is shown in Fig. 1. The simulation is performed on various real platforms. The discrete event simulation infrastructure is the base of various entity models. The GUI of the system includes the definition of the users, tasks and resources and the configuration of the information directory, scheduler, network, statistics and analysis entities. JFreeSim provides a variety of GUIs of the output results [13], including many graphic displays and simple animation of whole simulation process. Thus, users can immediately understand the process and results of the simulation.

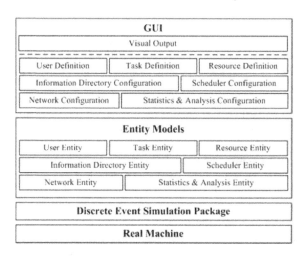

Fig. 1. Modular architecture of JFreeSim

In JFreeSim, the simulation of an application is performed as follows. First, system receives global configuration parameters sent by user entity, initializes other entities and starts the process of simulation. Meanwhile, information directory receives the register information from various resources available. User entity passes the tasks to scheduler entity according to DAG configuration. Scheduler queries the information directory and assigns the tasks to suitable resources according to scheduling strategy. Each resource maintains a local task queue. When multiple tasks are ready to run, the execution of those tasks is simulated according to the local scheduling policy. During the simulation, new resources may become available, some resources may fail, and the performance of the resources is variable. These performance variations are updated in the information directory. By querying the information directory, the scheduler determines the scheduling situation. The above process is repeated until all the tasks complete execution.

4.2 Implementation of JFreeSim

To implement JFreeSim, SimJava [14] is used as the discrete event simulation infrastructure. SimJava is a process-oriented discrete event simulation package, and has a series of API based on Java language to define and execute simulation process.

The behavior of each entity in the system is implemented by an independent Java thread. Each thread is looping continuously to process the messages from other entities. The way of thread implementation is able to well support the MTMSMR model and makes JFreeSim convenient to complete various complex simulations.

The instance of the user entity is started first and it is main procedure in the course of simulation. User instance is responsible for creating other entities according to the user's configuration. In addition, user instance creates task instances as well, which are represented by certain class structures. User instance submits the tasks to corresponding schedulers after simulation starts.

After information directory is created and initialized, it receives the register information from the resources and organizes global status information. Information directory is responsible for maintaining the performance variations of resources, and responds to the queries from the scheduler. The above process is repeated until finish signal is received.

Each computing node receives tasks from the scheduler, executes the tasks according to its local scheduling policy. This loop is terminated by the simulator once it receives finish signal from main procedure. The execution time of a task on a computing node is determined by the number of floating point operations in the task, the local scheduling policy, the computing power, and the I/O capacity. The computing node may also receive messages from the scheduler to remove or cancel a task.

The kernel code of the scheduler is the scheduling algorithm. The scheduling algorithm queries the information directory for the current status of the resources in the system. JFreeSim provides the flexible scheduler configuration tool for the user, which makes it easy to perform different scheduling algorithms.

Network entities including network link, router and network are all created and initialized by the main procedure. Network links are assumed to be time-shared if multiple data transfers are performed concurrently. The data transfer time over a network link is determined by the size of the data, the bandwidth and the latency of the network link at the time of the transfer.

During the implementation of JFreeSim, control and communication overheads are considered. Control overhead refers to the execution time used to discover and select the resource, and communication overhead refers to the execution time spent to transfer control message among the entities. Control overhead is estimated by considering the complexity of the scheduling algorithm in terms of the number of operations. As JFreeSim models various entities independently, it is able to trace each individual message transfer among these entities and to estimate the related communication overheads. The above overheads may also be directly specified by the user.

JFreeSim displays various build-in and user-defined measurements with the comprehensive graphical interface through the special statistics entity. Furthermore, JFreeSim is also able to analyze the activities of entities for a simulation and to determine possible performance bottlenecks.

5   Use Example of JFreeSim

In this section, we take hierarchical resource selection and task scheduling model as an example to illustrate the use of JFreeSim.

To solve the problem of resource management and control of the application with a large number of tasks in grid, a hierarchical resource selection and task scheduling model is proposed and applied to control the execution of a large number of tasks. This model solves the participant application problems, increases parallelism of task scheduling, accelerates the speed of resource allocation, shortens the average response time of the tasks, and therefore, offers an effective means to improve the system performance.

In the course of simulation, the scenario that the tasks only request computing resources is considered. Table 1 lists the main configuration used during the simulation.

Table 1. Configuration used during the simulation

Parameter	Configuration
Length of instructions	5000*80% ~ 5000*120% (million instructions)
Size of data file before execution	12.5*80% ~ 12.5*120% (kilobytes)
Size of data file after execution	30*80% ~ 30*120% (kilobytes)
Computing power of resources	Each resource contains three computers: computer 1 contains four CPU units, computer 2 contains four CPU units, computer 3 contains two CPU units; computing power is 377MIPS for each CPU unit
Number of resources	6 resources
Sharing mode of CPU	Time-shared mode
Algorithm of assignment of center scheduler	Round Robin algorithm for dispatching the tasks to lower schedulers
Algorithm of resource selection for lower scheduler	Random algorithm for selecting resources
Information directory configuration	Default mode: simple layered mode
Network configuration	Simple network mode, bandwidth: 10Mb/s
Statistics and analysis configuration	Default configuration

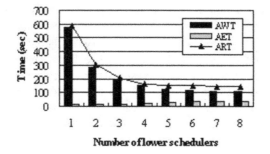

Fig. 2. Situation of simulating 1000 tasks

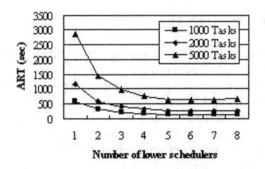

Fig. 3. Impact of the number of tasks to the execution time

The situation of simulating 1000 tasks execution is shown in Fig. 2. When the number of lower scheduler increases from 1 to 8, *average waiting time* (AWT) of the tasks is decreasing, but *average execution time* (AET) is increasing. Generally speaking, *average response time* (ART) of the tasks is decreasing gradually.

In order to illustrate the impact of the number of tasks to the scheduling performance, simulation is executed under the circumstances that the number of tasks is 1000, 2000, and 5000, respectively. The simulation result is shown in Fig. 3.

From the above results, JFreeSim has flexible configuration strategies, and is able to perform various simulations in the condition of different architectures and applications. The results of experiments are as we expected.

6 Conclusions

In this paper, we have proposed a new simulation tool, JFreeSim, based on MTMSMR model. As a modular and extensible simulation tool, JFreeSim realizes many kinds of entity modeling and communication mechanism between all entities, and makes system simulation accord with the characteristics of the grid environment. Examples indicate that JFreeSim can provide users with flexibility in configuration and meet requirements of various system architectures and applications, and the simulation results are as expected.

Although JFreeSim is a useful tool for simulating the execution of various entities in grid system, several limitations need to be addressed. Currently, the performance of a computing node refers only to its computing capability. We are going to extend the model with more information, such as the memory availability of each node. The model of the network entity needs to be designed exactly the same as that in the actual environments. More overheads should be estimated during data transfer. In addition, JFreeSim suffers from the high overhead of Java thread management. The number of entity thread is limited because of the capability of operating system. We are investigating these issues and expanding the related entity models to overcome these limitations.

References

1. A. Sulistio, C. Shinyeo, and R. Buyya, "A Taxonomy of Computer-based Simulations and its Mapping to Parallel and Distributed Systems Simulation Tools", *International Journal of Software: Practice and Experience*, Vol.34, No.7, pp.653-673, June 2004.
2. R. M. Fujimoto, "Parallel and Distributed Simulation Systems", *Proceedings of the Simulation Conference*, Vol.1, pp.147-157, Dec. 2001.
3. B. Hong and V. K. Prasanna, "A Modular and Extensible Simulator for Performance Evaluation of Adaptive Applications in Heterogeneous Computing Environments", *Proceedings of the Algorithms and Architectures for Parallel Processing*, pp.453-461, Oct. 2002.
4. MicroGrid: Online Simulation Tools for Grids, Distributed Systems and the Internet, http://www-csag.ucsd.edu/projects/grid/microgrid.html.
5. Bricks Project. http://www.is.ocha.ac.jp/~takefusa/bricks/.
6. SimGrid: A Toolkit for the Simulation of Application Scheduling, http://gcl.ucsd.edu/simgrid/.
7. GridSim: A Grid Simulation Toolkit for Resource Modelling and Application Scheduling for Parallel and Distributed Computing, http://www.gridbus.org/gridsim/.
8. ChicSim Project, http://people.cs.uchicago.edu/~krangana/ChicSim.html.
9. HyperSim Project, http://hpcnc.cpe.ku.ac.th/moin/HyperSim/.
10. S. Phatanapherom, P. Uthayopas, and V. Kachitvichyanukul, "Fast Simulation Model for Grid Scheduling Using HyperSim", *Proceedings of the Simulation Conference*, Vol.2, pp.1494-1500, Dec. 2003.
11. A. Sulistio and R. Buyya, "A Grid Simulation Infrastructure Supporting Advance Reservation", *Proceedings of the 16th International Conference on Parallel and Distributed Computing and Systems*, pp.1-7, Nov. 2004.
12. R. Buyya and M. Murshed, "GridSim: A Toolkit for the Modeling and Simulation of Distributed Resource Management and Scheduling for Grid Computing", *The Journal of Concurrency and Computation: Practice and Experience*, Vol.14, No.13-15, Wiley Press, Nov.-Dec. 2002.
13. A. Sulistio, C. Shinyeo, and R. Buyya, "Visual Modeler for Grid Modeling and Simulation (GridSim) Toolkit", *Proceedings of the 3rd International Conference on Computational Science*, June 2003.
14. SimJava: a discrete event simulation package for Java with applications in computer systems modeling, http://www.dcs.ed.ac.uk/home/simjava/.

OOML-Based Ontologies and Its Services
for Information Retrieval in UDMGrid[*]

Xixi Luo and Xiaowu Chen

The Key Laboratory of Virtual Reality Technology, Ministry of Education,
School of Computer Science and Engineering, Beihang University,
Beijing 100083, P.R. China
{luoxixi, chen}@vrlab.buaa.edu.cn

Abstract. In order to effectively integrate and share the enormous dispersed resources of various digital museums, University Digital Museum Grid (UDMGrid) has been developed to provide one-stop information services about kinds of digital specimens in the form of grid services. To eliminate the heterogeneity between the information resources, shared concepts for these digital museums are indispensable. This paper studies OOML-based ontologies and its services for information retrieval in UDMGrid, including the object oriented ontology construction and ontology mapping, in which a novel inheritance mechanism is proposed to eliminate logic confusion. On the basis of OOML-based ontologies, ontology services are developed to assist the information retrieval by transforming global concepts to local concepts.

1 Introduction

Eighteen featured university museums have been digitized mainly relating to Geology & Geography, Archaeology, Humanities & Civilization, and Aeronautics & Astronautics [1]. These digital museums play an important role in the fields of education, scientific research, as well as specimen collection, preservation, exhibition, and intercommunication. However, these digital museums dispersed on different nodes in CERNET (China Education and Research Network) [2] confront a problem that the multi-discipline resources at these digital museums are isolated and dispersed without sufficient interconnection. Hence it is necessary to propose a digital museum resources integration solution, through which these eighteen digital museums would be incorporated as a more comprehensive virtual one.

University Digital Museum Grid (UDMGrid) [3] [4] is proposed to using the grid technology to integrate and share the distributed digital museum information resources. From the user's perspective, UDMGrid should perform as a virtual digital museum, in which users can browser the digital specimen information in multiple manners on only one UDMGrid portal instead of eighteen separate homepages, with-

---

[*] This paper is supported by China Education and Research Grid (ChinaGrid)(CG2003-GA004 & CG004), National 863 Program (2004AA104280), Beijing Science & Technology Program (200411A), National Research in Advance Fund (51404040305HK01015).

J. Cao, W. Nejdl, and M. Xu (Eds.): APPT 2005, LNCS 3756, pp. 342–352, 2005.
© Springer-Verlag Berlin Heidelberg 2005

out rushing about among these digital museums. However, the information resources of digital museums are constructed by different domain experts, who use special metadata to describe information, thus the heterogeneity among these metadata brings difficulties for users to locate, organize and integrate the information resources. Therefore, shared concepts for these digital museums are indispensable.

Ontology defines a set of representational terms that we call concepts, among which the interrelationship describes a target world [5]. In grid environment, Ontology becomes increasingly crucial to operations about the analysis and integration of information resources [6].Ontology is widely applied to information system. Knowledge Sifter is a scaleable agent-based system that supports access to heterogeneous information sources such as the Web, open-source repositories, XML-databases and the emerging Semantic Web, in which the concept of ontology is central to this approach. By the Ontology Agent, user can pose queries to those data sources without needing to know the location of the supporting data, nor how the ontological concepts are materialized through the integration and ranking process [7]. Infosleuth is a retrieval agent system that provides access to information form multiple domains, regardless of its heterogeneity or distribution, in which domain ontologies and an evolutionary model of the use's interests are some of the basic concepts used by the system to help users identify and retrieve relevant domain information [8]. Meanwhile, the application of ontology in grid is emerging. Earth System Grid, a project of the U.S. Department of Energy Scientific Discovery through Advanced Computing (SciDAC) program, used and modified ontological concepts for its domain area to provide a basis for classifying and retrieving data files, collections and information about the files and collections based on content for use in a grid context [6]. Furthermore, some ontology development methodology comes out. ROD, a rapid ontology development methodology that can be used to build ontology for underdeveloped domains, make the development process more efficient so that domain concepts and relations can be automatically discovered from large-scale semi-structured and/or unstructured textual resources[4]. Additionally the ontology modeling is up-and-coming. OIL [9], DAML [10], SHOE [11], RDF(S) [12] and OOML (improved Object Oriented Markup Language) [13] are introduced as ontology modeling. Compare to former four kinds of makeup language, OOML are suitable for the UDMGrid ontology modeling, since the better support of generalization and specialization goes well with the ontology internal structure of UDMGrid.

This paper studies the OOML-Based Ontologies and its services for information retrieval in UDMGrid, through which a unified view of available resources is provided to users. In this paper, OOML is adopted for object oriented ontology modeling, in which a novel inheritance mechanism is proposed to eliminate logic confusion. Based on the mechanisms of ontology mapping in UDMGrid, ontology services are developed to assist the information retrieval.

The remainder of this paper is organized as follows. Section 2 presents the modules concerning ontology in UDMGrid. Section 3 elaborates the OOML-based ontologies in UDMGrid, including the hierarchy architecture and the mapping strategy. On the basis of OOML-based ontologies, the ontology services are introduced in Section 4, and Section 5 briefly issues the application workflow. Section 6 ends this paper with conclusions and future work.

2   Ontology-Modules in UDMGrid

The framework and function modules of UDMGrid are shown as Fig.1, in which the modules concerning ontology are in gray, including User & Applications, Ontology and Data management, whose collaboration is depicted in Fig.2. The other modules are indispensable for constructing a grid environment, such as job scheduler, and security etc.

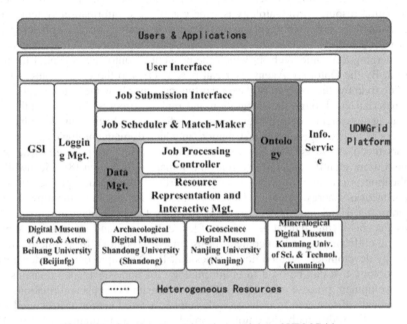

Fig. 1. The framework and function modules of UDMGrid

UDMGrid supplies an "on-stop" information service to users and applications. A global ontology, which is tree structured, is presented to users & Application, so that query statements can be constructed using the terms that appear in it. After the construction, the query statements are sent to query agent which is responsible for submitting the queries to the ontology services.

Ontology is used to integrate information from heterogeneous domains to provide a unified view of resources to users & application layer. In UDMGrid, it serves as a bridge for application to utilize digital museum resources. Ontology Repository contains a collection of concepts and their interrelationships, which provides an abstract view of an application domain [5]. Ontology services make the mapping between different ontology be available.

Data Management offers uniform interface to access heterogeneous resources, including the database resources and web resources, and Grid Services Architecture Data Access and Integration (Grid-DAI) [14] is a middleware to provide a component library for accessing and manipulating data in Grid environment especially for UDMGrid.

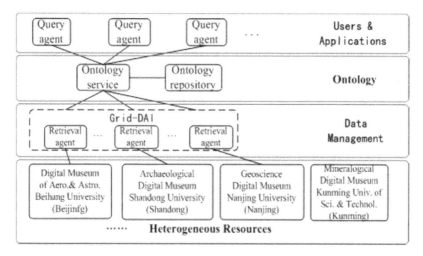

Fig. 2. The collaboration of ontology-modules in UDMGrid

3 OOML-Based Ontologies of UDMGrid

3.1 OOML-Based Ontologies

OOML is an improved object oriented markup language which is outstanding in reusability and expansibility suitable for the distributed environment. By means of object oriented paradigm, such as abstract class [15], OOML can simplify the concept of domain of interest. Moreover, it is also capable of solving frequently occurring conflicts in the distributed environment, thus enhance reusability and expandability of the ontology [13]. In the field of object-oriented modeling, Unified Modeling Language (UML) has become a standard modeling language and is widely supported by robust commercial tools [16]. Therefore, we utilize the UML to implement the object oriented ontology model.

There are two kinds of ontology to describe information resources of UDMGrid. One is global ontology, including generic ontology and domain-dependent ontology, both of which are stored in global ontology repository. The other one is local ontology which are also called source-specific ontology, and this ontology is stored in the local resources repository and might be keyword in web, attribute in database, or some other terms used to index the local resources.

3.2 Global Ontology

Global ontology, the key to integrate resources, consists of a set of concepts used to describe entities and relationships in digital museum. In principle, the global ontology can be tailored to suit the needs of each user or each group of users that share a common vocabulary [17].

As mentioned above, there are two kinds of global ontology which are generic ontology and domain ontology. Generic ontology is characterized by information-rich and flexibility, and with coarse granularity, thus suitable for kinds of domains, for example, "name", "person" can be used to describe a entity of aviation, archeology and so on. Domain ontology is constructed by domain experts, who hold professional and specific information of a domain. Compare to generic ontology domain ontology has finer granularity and more specific, for example, "airplane name", "aviator" of aviation domain.

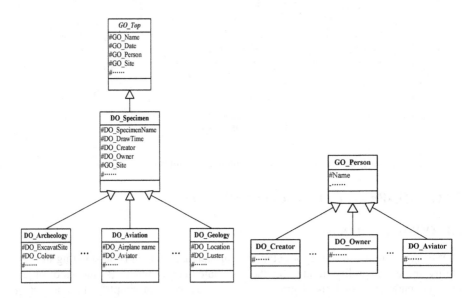

Fig. 3. UML for relationship of global ontology in UDMGrid

Fig. 4. UML for relationship of attribute classes

OOML-based ontology in UDMGrid express the conception of global ontology by "class" which can be divided to node class and attribute class, just as show in Fig.3, the node classes construct the basic hierarchical structure, such as GO_TOP, DO_Specimen, DO_Archeology, DO_Aviation, DO_Geology ("Go" means generic ontology, and "DO" means domain ontology). Each node class has a list of attribute classes. For example, GO_Name, GO_Date, GO_Person .etc. are attribute classes of node class GO_Top. The classes maintain the hierarchically structure by a inheritance, which not only for the node classes, but also for the attribute classes, just as shown in Fig.4, GO_Person, an attribute class of GO_Top, is the sup class of DO_Creator, DO_Owner, and DO_Aviator which belong to different node classes.

Traditional inheritance mechanism means that the sub class should have all the attributes of the sup class, which means that DO_Specimen should inherit all the attribute class of GO_TOP, then the GO_Person and DO_Creator, DO_Owner are the attributes of DO_Specimen, however, among which imply a inheritance relationship (DO_Creator, DO_Owner inherit GO_Person), the concurrence of these three attribute class in one node class will bring logic confusion. Thus we proposed a special inheritance mechanism, during which, the sub node class will automatically omit the attribute class, which is the supper class of that node class's attribute classes. Following this special inheritance mechanism, the GO_Person would not appear in the attribute list of DO_Specimen, because the DO_Creator, DO_Owner has substitute the GO_Person by finer granularity.

3.3 Local Ontology

Local ontology is source-specific, might be keyword in web, attribute in database, or some other terms can used to index the local resources. As resource provider, each digital museum should publish their digital specimens accompanying by local ontology in the form of metadata, which is utilized to support search, access and explanation. In addition, the mapping relationship between global ontology and local ontology should also be provided. In UDMGrid, there are three kinds of local ontology: access metadata, database resource metadata and web resource metadata.

3.3.1 Access Metadata
Access metadata contains information on the location, structure, access rights and ownership of the distributed data sources. For example, the database name, user name and password, which are dispensable to access database are metadata of this kind.

3.3.2 Database Resource Metadata
Database is the main carrier of digital museum information resources. Database resource metadata describes the table information and attribute information of digital museum databases, which are described by two designed object classes: Table Object, Attribute Object. The detail design of these two object classes is illustrated in Table 1.

Table 1. Database resource metadata object classes

Object class	Attributes describing the object class
Table Object	Name, Database, Description, etc.
Attribute Objec	Name, Type, Database, Table, Description, etc.

Table 2. Web resource metadata related object classes

Object class	Attributes describing the object class
Keyword Object	Name, URL, Description, etc.

3.3.3 Web Resource Metadata
Besides database, there are a large mount of web pages of digital museums which are mainly non-structured or semi-structured. In order to utilize keywords to annotate the contents of web pages, one object class is designed to record the information about the keywords: Keyword Object. Table 2 shows some attributes belonging to this object classes.

3.4 Ontology Mapping

Ontology mapping is to establish the mapping relationship between global concepts of global ontology and local concepts of local ontology. There are two mapping strategy for two kinds of information resource. For database resource, every attribute object will be mapped to an attribute class of global ontology with the database and table information, for web resource, the keyword object will be mapped to an attribute class of global ontology with the web page's URL.

The ontology mappings are done by resource publishers, by which each local ontology is mapped to one or more than one global ontology. For example, an attribute object "MingCheng" (Chinese for name) which is the local ontology of an aviation database would be mapped to DO_Airplane_Name, a keyword "别名" of web page http://digitalmuseum.buaa.edu.cn/store/aircraft.jsp?aircraftid=12020201 （ as Fig.5 shows） would also be mapped to DO_Airplane_Name. The mapping information will be stored in the ontology repository in the form of Table 3 or Table 4.

Originally, the information resources in UDMGrid is expressed by local conceptions, after mapping, all the information resources can be described by the global conceptions, on the basis of which the heterogeneity between the information resources can be eliminated.

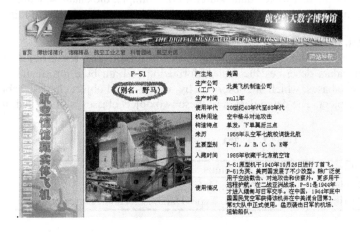

Fig. 5. Web page with keyword "别名"

Table 3. An example of the mapping information from database attribute "MingCheng" to DO_Airplane_Name

Local ontology	Global ontology	Type	Database	Table
MingCheng	DO_Airplane_Name	String	dm	feiji

Table 4. An example of the mapping information from keyword " 别 名 " to DO_Airplane_Name

Local ontology	Global ontology	URL
别名	DO_Airplane_Name	http://digitalmuseum.buaa.edu.cn/store/air craft.jsp?aircraftid=12020201

4   Ontology Services

Ontology services assist the information retrieval by transforming global concepts to underlying local concepts. As shown in Fig. 2, users select interested global concepts of global ontology from portal and offer the corresponding value, and then query agents construct an abstract query with these parameters, which is comprehensible but can not be executed on local data source. Therefore, it is required for ontology services to transform the abstract query to executable queries on relevant data source which is comprised of local ontology metadata. Then the retrieval agent can utilize the executable query to collect desirable information resources.

Two steps form the transformation. Firstly, the global concept should be extended according to the inheritance relationship between attribute classes, after that the original global concept is extended to more than one global concept which are the sub classes of the original global concept. Secondly, these global concepts will be transformed to the local concepts according to the mapping information.

We will take an example to detail the procedure. Suppose users select the global concept and the corresponding value in the form of name = "horse" (name is a global concept, and "horse" is the expected value of the name). Firstly, the original global concept should be extended, and the original global concept GO_name will be extended to DO_Airplane_Name, DO_Ore_Name, and so on. Secondly, these global concepts will be transformed to the local concepts. From Table 3 and Table 4, the DO_Airplane_Name would be transformed to "MingCheng" , "别名" and so on.

Every local concept is from a specific data source, which may contain the corresponding expected value, for example, the web page: http:// digitalmuseum.buaa.edu.cn/store/aircraft.jsp?aircraftid=12020201 with the local concept "别名" may contain the expected value "horse". If that is true, then the URL will be returned to the user as the grid service result.

5   Application

The prototype has been developed to integrate the information resources of several university digital museums, and provide integrative and intelligent information retrieval service through grid portal. The university digital museums involved include the Digital Museum of Aeronautics and Astronautics (BUAA) [18], the Archaeological Digital Museum (SDU) [19], the Geoscience Digital Museum (NJU) [20], and the Mineralogical Digital Museum (KUST) [21], etc.

Kinds of grid application services have been designed and developed, such as AbstractSQLService, OntologyService and DoSearchService (as shown in Fig.6). AbstractSQLService is to form an abstract query based on the users' request parameters which are shown on the portal, and it is important to note that the definition of users' request parameters is according to the definition of global ontologies. OntologyService is responsible to transform the abstract query to a few executable ones, and during the process of transformation, OntologyService completes the transformation between global ontologies and local ontologies using the mapping information in ontology repository. The grid service DoSearchService has two functions. One is to

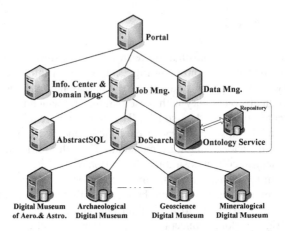

Fig. 6. UDMGrid deployment

execute these executable queries with the assist of Grid-DAI to get information resources stored in database, and the other function is to get information resources from webs, and both of the two functions need to transfer the retrieved result to the portal.

Here is an example to demonstrate the workflow. Suppose that a global ontology and local ontology have been constructed, and the mapping relationship is also established.

1. A user access to the UDMGrid Portal, and submit his/her request to UDMGrid through the Portal
2. The request then will be sent by Portal to Job manager, and the Job manger invokes specific services to response the request
3. The first service is AbstractSQL, which is to form an abstract SQL query statement that is composed of terms in global ontology. The formed query statement is ASQL(SELECT all FROM specimen WHERE name='horse'), and then this query statement is sent back to Job manager
4. The second service is OntologyService that transforms abstract SQL query statement (ASQL) to executable ones (RSQL). Two steps form the transformation. In the first step the OntologyService extends ASQL to ASQLs using global ontology, while in the second is to convert ASQLs to RSQL. In this example, the query statements after first step are ASQLs1 (SELECT all FROM Aviation WHERE DO_Airplane_Name ='horse'), ASQLs2 (SELECT all FROM Archaeology WHERE DO_Ore_Name ='horse') and so on, the terms like 'Aviation' and 'Archaeology' are sub classes of 'specimen', the terms like 'aero name' and 'cultural relic' are sub classes of 'name'. The final executable query statements after second step are RSQL1 (SELECT all FROM aircraft WHERE MingCheng = 'horse'), RSQL2 (SELECT all FROM wenwu WHERE 别名='horse') and so on, the terms like "MingCheng" and "别名" are database metadata or web metadata, and the mapping information between global ontology and local ontology is used in second step
5. These RSQLs in addition with the access information (based on access metadata) will be sent to the last service DoSearchService that is response for the execution

of the query statements generated by OntologyService. During this process, Grid-DAI will be invoked to retrieve data from distributed data source, and then the result will be sent back to Portal

6. Finally, as Fig.7 shows, this job totally retrieved 1304 specimens from three digital museums except for the Mineralogical Digital Museum of Kunming University of Science & Technology, and 65 of which are from databases, 1249 from webs. In Fig.7, the left part is a picture of an aircraft named "wild horse" in the Aeronautics and Astronautics Digital Museum, and the right picture is about some 2000 years ago carriage equipments in the Archaeological Digital Museum of Shandong University. Moreover users can download the results from portal.

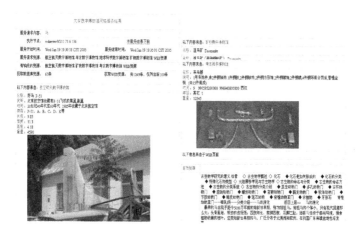

Fig. 7. Parts of grid service result in UDMGrid portal

6   Conclusion and Future work

This paper presents an ontology-based digital museum resources integration solution in Grid, its purpose is to solve the information island problem of established university digital museums, provide easy resource sharing, and offer intelligent information retrieval service. Much research and development has been done on ontology. In this paper, not only the OOML-based ontology hierarchical structure is explicated, but also the object oriented ontology mapping. Furthermore, this paper proposed a novel inheritance mechanism, which can eliminate the logic confusion in the ontology. In addition, Ontology services, which can assist the information retrieval by transforming global concepts to underlying local concepts, are also discussed.

Recently, our work is regular keyword based information retrieval, in order to make retrieved results more reasonable, and the potential relationship between specimens can be found, the research on the data mining technology in UDMGrid is one of the key topics of study work in the future, Moreover, distributed ontology need to be developed to suit the grid environment.

Reference

[1]   University Digital Museums, http://www.edu.cn/20020118/3018035.shtml.

[2]   China Education and Research Network, http://www.edu.cn/.

[3]   Xiaowu Chen, Xixi Luo, Zhangsheng Pan, Qinping Zhao. A CGSP-based Grid Application for University Digital Museums. Third International Symposium on Parallel and Distributed Processing and Applications (ISPA'2005), Nanjing, China, 2005.

[4]   Xiaowu Chen, Zhi Xu, Zhangsheng Pan, Xixi Luo. UDMGrid: A Grid Application for University Digital Museums. Grid and Cooperative Computing (GCC 2004), pp. 720~728, Wuhang, China, 2004.

[5]   Latifur Khan, Feng Luo Ontology construction for information selection, ICTAI'02.

[6]   Line Pouchard, Luca Cinquini, Bob Drach. An ontology for scientific information in a Grid environment-the earth system Grid, CCGRID.03.

[7]   L. Kerschberg, M.Chowdhury, A. Damiano. Knowledge Sifter: Ontology-Driven Search over Heterogeneous Database, SSDBM'04.

[8]   Tomasz Ksiezyk, Gale Martin, Qing Jia Infosleuth: Agent-Based System for Data Integration and Analysis.

[9]   D. Fensel, I. Horrocks, F. Van Harmelen, S. Decker OIL in a nutshell. Proceedings of the 12th European Workshop on Knowledge Acquisition, Modeling, and Management.

[10] Dan Connolly, Frank van Harmelen, Ian Horrocks, Annotated DAML+OIL Ontology Markup, http://www.w3.org/TR/2001/NOTE-daml+oil-walkthru-20011218/.

[11] The SHOE Specification http://www.cs.umd.edu/projects/plus/SHOE/spec.html.

[12] Decker. S., Melnik. S., van Harmelen The Semantic Web: the roles of XML and RDF, IEEE Internet Computing.

[13] Kangchan Lee, Myonghwan Yoo, Injeong Chung Improved markup language for semantic Web using object oriented technology.

[14] Zhangsheng Pan, Xiaowu Chen, Xiangyu Ji. Research on Database Access and Integration in UDMGrid. Third International Symposium on Parallel and Distributed Processing and Applications (ISPA'2005), Nanjing, China, 2005.

[15] Fowler Martin, Scott Kendall, UML Distilled second edition, Addidon-Wesley.

[16] Wei Liu, Zong-Tian Liu, Kun Shao UML-based domain ontology modeling for multi-agent system.

[17] Jaime A Reinoso Castillo, Adrian Silvescu, Information Extraction and Integration from Heterogeneous, Distributed, Autonomous Information Sources - A Federated Ontology-Driven Query-Centric Approach.

[18] The Digital Museum of Aeronautics and Astronautics (Beihang University, BUAA) http://digitalmuseum.buaa.edu.cn/.

[19] The Archaeological Digital Museum (Shandong University) http://museum.sdu.edu.cn/index/index.asp.

[20] The Geoscience Digital Museum (Nanjing University) http://202.119.49.29/museum/default.htm.

[21] The Mineralogical Digital Museum (Kunming Univ. of Sci. & Technol.) http://www.kmust.edu.cn/dm/ index.htm.

A Hybrid Integrated QoS Multicast Routing Algorithm in IP/DWDM Optical Internet[*]

Xingwei Wang, Jia Li, and Min Huang

College of Information Science and Engineering,
Northeastern University, Shenyang, 110004, P.R.China
wangxw@mail.neu.edu.cn

Abstract. An integrated QoS multicast routing algorithm in IP/DWDM optical Internet is proposed in this paper. Considering load balancing, given a multicast request and flexible QoS requirement, to find a QoS multicast routing tree is NP-hard. Thus, a hybrid algorithm based on simulated annealing and tabu search is introduced to construct the cost suboptimal QoS multicast routing tree, embedding the wavelength assignment procedure based on segment and wavelength graph ideas. Hence, the multicast routing and wavelength assignment is solved integratedly. Simulation results have shown that the proposed algorithm is both feasible and effective.

1 Introduction

Dense Wavelength Division Multiplexing (DWDM) is a key technology to exploit enormous bandwidth of optical fibers to meet the explosive growth of bandwidth demand in the Internet. However, DWDM has been traditionally used just to increase the transport capacity, leading to lack of scalability, high cost and low efficiency. Thus, running IP directly over DWDM to eliminate one or more of these intermediate layers (e.g., SONET/SDH, ATM) is considered to be the right solution [1].

From the architecture viewpoint, there are three models for IP over optical networks: peer, overlay and augmented [2], and we call that with peer model as IP/DWDM optical Internet. Three routing approaches exist, namely integrated routing, overlay routing and domain-specific routing [2], and the integrated routing approach supporting the peer model is adopted in this paper.

There is no doubt that multimedia will become the dominate traffic in IP/DWDM optical Internet. Many multimedia applications, such as video conferencing, distance education, telemedicine, and etc. often have strict quality of service (QoS) requirements (e.g., bandwidth and delay) and also multicast demands [3]. In order to support such kinds of multimedia group applications, QoS based multicast routing tree should be established, spanning the source and all destinations of the group. It has been proved

---

[*] This work is supported by the National Natural Science Foundation of China under Grant No. 60473089 and No. 70101006; the Natural Science Foundation of Liaoning Province in China under Grant No. 20032018 and No. 20032019; the Modern Distance Education Engineering Project of China MoE.

J. Cao, W. Nejdl, and M. Xu (Eds.): APPT 2005, LNCS 3756, pp. 353 – 362, 2005.
© Springer-Verlag Berlin Heidelberg 2005

that finding such a tree is NP-hard [4]. Thus, a lot of heuristic and intelligent solutions are proposed in the literature [5-9]. Leung [5] proposed an algorithm that develops a new GA to solve the multiple destination routing problem without constraints. Jia [6] proposed a QoS multicast routing algorithm to minimize the number of used wavelengths under rigid delay constraint. Ran [7] considered multicast routing and wavelength assignment in multihop optical networks under a model in which multicast communication requests are made and released dynamically over time. Cui [8] discussed the problem of multi-constrained routing based on simulated annealing. Li [9] proposed a QoS-guaranteed multicast routing protocol operating on top of the unicast routing protocol, to find a minimum cost solution under multiple QoS constraints.

In general, most of the existing algorithms aim simply at minimizing the cost of the tree and often only cope with the rigid QoS constraints, i.e. the QoS requirements have strict upper or lower bounds. However, due to the difficulty in accurately describing the user QoS requirements and the inaccuracy and dynamics of the network status information, we believe flexible QoS should be supported. This motivates our work.

End-to-end delay has been widely considered as an important measure of QoS. To support flexible QoS, delay interval [10] instead of bounded delay is introduced to describe the requirement. By defining the user QoS satisfaction degree, the proposed algorithm could achieve a better tradeoff between QoS and network cost.

Considering load balancing, given a multicast request and flexible QoS requirement, the proposed algorithm could find a QoS multicast routing tree. Due to its NP-hard nature, a hybrid algorithm [11] based on simulated annealing and tabu search is introduced to construct the cost suboptimal multicast routing tree, embedding the wavelength assignment procedure based on segment [12] and wavelength graph [13] ideas. Hence, the multicast routing and wavelength assignment is solved integratedly, which could optimize both the network cost and QoS.

The rest of this paper is organized as follows. In Sec. 2, the network model and the mathematical model of the problem are described. Sec. 3 is devoted to the design of the hybrid integrated QoS multicast routing algorithm. In Sec. 4, the proposed algorithm is applied to several networks by simulation. The paper concludes with a summary of the results in Sec. 5.

2   Model Description

IP/DWDM optical Internet can be considered to be composed of optical nodes (such as wavelength routers or OXCs) interconnected by optical fibers. Assume each optical node exhibits multicast capability, equipped with optical splitter at which an optical signal can be split into an arbitrary number of optical signals. In consideration of the still high cost of wavelength converter, assume only some optical nodes are equipped with full-range wavelength converters. Assume the conversion between any two different wavelengths has the same delay. The number of wavelengths that a fiber can support is finite, and it may be different from that of others.

Given a graph $G(V, E)$, where V is the set of nodes representing optical nodes and E is the set of edges representing optical fibers. If wavelength conversion happens at node $v_i \in V$, the wavelength conversion delay at v_i is $t(v_i) = t$, otherwise, $t(v_i) = 0$. The set of available wavelengths, delay and cost of edge $e_{ij} = (v_i, v_j) \in E$ are denoted by $w(e_{ij}) \subseteq w_{ij} = \{\lambda_1, \lambda_2, \cdots, \lambda_{n_{ij}}\}$, $\delta(e_{ij})$ and $c(e_{ij})$ respectively, where w_{ij} is the set of supported wavelengths by e_{ij} and $n_{ij} = |w_{ij}|$.

A multicast request is represented as $R(s, D, \Delta)$, where $s \in V$ is the source node, $D = \{d_1, d_2, \cdots, d_m\} \subseteq \{V - \{s\}\}$ is the set of destination nodes, and Δ is the required end-to-end delay interval of users. Suppose the set $U = \{s\} \cup D$. The proposed algorithm is to construct a multicast routing tree from the source to all the destinations, i.e. $T(X, F)$, $X \subseteq V$, $F \subseteq E$.

The total cost of T is defined as follows:

$$Cost(T) = \sum_{e_{ij} \in F} c(e_{ij}) . \tag{1}$$

To balance the network load and thus to reduce the call blocking probability, those edges with more available wavelengths should be considered with priority. Thus, the edge cost function is defined as follows:

$$c(e_{ij}) = n - |w(e_{ij})| . \tag{2}$$

$$n = \max_{e_{ij} \in E} \{n_{ij}\} . \tag{3}$$

Therefore, for the edge on which available wavelengths are more, the cost takes smaller value, otherwise, takes larger value.

Let $P(s, d_i)$ denote the path from s to d_i in T. The delay between s and d_i along T, denoted by PD_{sd_i}, can be represented as follows:

$$PD_{sd_i} = \sum_{v_i \in P(s, d_i)} t(v_i) + \sum_{e_{ij} \in P(s, d_i)} \delta(e_{ij}) . \tag{4}$$

The delay of T is defined as follows:

$$Delay(T) = \max\{PD_{sd_i} | \forall d_i \in D\} . \tag{5}$$

Let $\Delta = [\Delta_{low}, \Delta_{high}]$, and the user QoS satisfaction degree is defined as follows:

$$Degree(QoS) = \begin{cases} 100\% & Delay(T) \leq \Delta_{low} \\ \dfrac{\Delta_{high} - Delay(T)}{\Delta_{high} - \Delta_{low}} & \Delta_{low} < Delay(T) < \Delta_{high} \\ 0\% & Delay(T) \geq \Delta_{high} \end{cases} . \tag{6}$$

3 Algorithm Design

Simulated annealing (SA) [11] is well suited for solving combinatorial optimization problems because it can avoid local optima effectively, but it requires excessive computation time. Tabu search [11] can improve convergence efficiency by using flexible memory structures and reasonable tabu criteria. Integrating simulated annealing with tabu search could not only alleviate local optima but also improve runtime performance.

3.1 Solution Expression

A solution is denoted by binary coding. Each bit of the binary cluster corresponds to one node in $G(V,E)$. The graph corresponding to the solution S is $G'(V',E')$. Let the function $bit(S,i)$ denote the ith bit of S, $bit(S,k)=1$ iff $v_k \in V'$. The length of the binary cluster is $|V|$. Construct the minimum cost spanning tree $T_i'(X_i',F_i')$ of G'. T_i' spans the given nodes in U. However, G' may be unconnected, thus S corresponds to a minimum cost spanning forest, also denoted by $T_i'(X_i',F_i')$. It's necessary to prune the leaf nodes not in U and their related edges in T_i', the result is denoted by $T_i(X_i,F_i)$, and assign wavelengths to T_i.

3.2 Neighborhood Structure

Choose one node not in U randomly, and take the reverse value for the corresponding bit in its binary coding solution.

3.3 Wavelength Assignment Algorithm

The objective is to minimize the delay of the multicast routing tree by minimizing the number of wavelength conversions, making $Degree(QoS)$ high. If T_i is a tree, assign wavelengths; otherwise, the solution is unfeasible.

3.3.1 Constructing Auxiliary Graph AG.
The method is described as follows:
    (1) Dividing T_i into segments.
    Locate the intermediate nodes with converters in T_i, and divide T_i into segments according to them, i.e., the edges having wavelength continuity constraint should be merged into one segment. Number each segment. The source node is considered to be equipped with wavelength converter.
    (2) Creating nodes in AG.
    In AG, add node a_0 as the source node, and create node a_j according to segment j, where $j=1,2,\cdots,m$, m is the number of segments. A mapping table is created to record the relationship between a_j and segment j. Each node in AG can be considered to be equipped with wavelength converter.
    (3) Creating edges in AG.
    Assume a_k $(1 \le k \le m)$ corresponds to the segment that the source node in T_i belongs to. Add a directed edge (a_0,a_k) between a_0 and a_k, making the intersection of the available wavelength set on each edge in segment k as its available wavelength set. For each node pair a_{j_1} and a_{j_2} $(1 \le j_1, j_2 \le m, j_1 \ne j_2)$, if segments j_1 and j_2 are connected in T_i, add a directed edge (a_{j_1},a_{j_2}) between a_{j_1} and a_{j_2}, making the

intersection of the available wavelength set on each edge in segment j_2 as its available wavelength set.

By now, AG is constructed. In the extreme, if all the intermediate nodes of T_i have been equipped with wavelength converters, AG is T_i.

3.3.2 Constructing Wavelength Graph WG.

Transform AG to WG. In WG, create $N * w$ nodes, named b_{ij}, for $i = 0,1,\cdots,w-1$ and $j = 0,1,\cdots,N-1$, where N is the number of nodes in AG, and w is the number of wavelengths available at least on one edge in AG. All the nodes are arranged into a matrix with w rows and N columns. Row i represents a corresponding wavelength λ_i' and column j represents a node a_j' in AG. A mapping table is created to record the relationship between i and λ_i', and another one is created to record the relationship between j and a_j'. The two tables will help map the paths in WG back to the paths and wavelengths in AG. Create edges in WG, where a vertical edge represents a wavelength conversion at a node, assigning 1 as its weight, and a horizontal edge represents an actual edge in AG, assigning 0 as its weight. The WG construction method is shown in [15].

3.3.3 Wavelength Assignment.

Treat WG as an ordinary network topology graph. Find the shortest paths from the source node column to each leaf node column in WG using Dijkstra algorithm [15], and construct the multicast routing tree T_{WG}. Map T_{WG} back to AG, and denote the resulting subgraph in AG by T_{AG}. Since in WG all the nodes in one column correspond to the same node in AG, those shortest paths that are disjoint in T_{WG} may intersect in T_{AG}. Thus, pruning some edges is needed. If these edges are pruned in T_{AG} directly, it dose not consider wavelength conversion cost, thus doing it in T_{WG} is better. For example, for column h in T_{WG}, there are two nodes b_{ij} and b_{kj} $(i \neq k, j \neq h)$ with the edges (b_{ij}, b_{ih}) and (b_{kj}, b_{kh}), indicating that two wavelengths λ_i' and λ_k' are selected on edge (a_j', a_h') in T_{AG}, thus b_{ij} or b_{kj} must be pruned. Here, only the node with the shortest distance from the source node is reserved.

Map the paths in WG back to the paths and wavelengths in AG, and then map them back to the paths and wavelengths in T_i, thus wavelength assignment is completed.

3.4 Generating Initial Solution

A destination-node-initiated joining algorithm [14] is adopted to find an initial feasible solution, leading the algorithm to be more robust with fewer overheads.

3.5 Fitness Function

Fitness function $f(S)$ is determined by $Cost(T_i)$ and $Degree(QoS)$ together:

$$f(S) = \frac{Cost(T_i) + [count(T_i) - 1] * \rho}{Degree(QoS)}. \tag{7}$$

$count(T_i)$ is the number of trees in T_i, ρ is a positive value. If T_i has more than one tree, add a penalty value to the cost of T_i and take a smaller value for $Degree(QoS)$.

3.6 Cooling Schedule

Set the initial annealing temperature $t_0 = K\eta$, K is a sufficiently large number, and $\eta = \max\{f(j) \mid j \in Sp\} - \min\{f(j) \mid j \in Sp\}$, Sp denotes the solution space, η can be estimated simply by $\eta = C_g - C_u$, C_g is the total cost of the current graph, and C_u is the cost of the subgraph composed of all the nodes in U. Based on the idea of "controlling iteration times by the ratio of acceptance to rejection", the allowed upper bound of iteration times and that of continuous non-improved current optimum solution is increased gradually with annealing temperature descending. Both "stopping after definite iteration times" and "controlling the change of the object value" are adopted as termination rules.

3.7 Tabu List

Constructing tabu list involves designing tabu object, tabu length, aspiration criterion and so on. According to the neighborhood structure, tabu object is 0-1 exchange of the components of solution vectors. Tabu length is the constant t. Aspiration criterion is as follows: if a tabued solution in the candidate set could give a better solution than the current optimum one, meaning that a new region has been reached, make it free. For convenience, make the size of the candidate set is larger than t and smaller than num, avoiding the phenomenon that all the elements in the candidate set are tabued. If num is too small to satisfy the above prescription, it is unsuitable to the proposed algorithm, however, it is easy to be solved because of its tiny searching space.

3.8 Algorithm Description

In the hybrid algorithm, tabu search is integrated into simulated annealing to help improve runtime performance. The proposed algorithm is described as follows:

Step 1.   Run the destination-node-initiated joining algorithm to get an initial feasible solution x, evaluate its fitness value $f(x)$ according to Sec. 3.5. Let the current optimum solution x_{best} be x. Clear the tabu list. Set initial annealing temperature to be t_0, and temperature-descended times k to be 0.

Step 2.   If the termination rule is satisfied, x_{best} is the final solution, the algorithm ends; otherwise, perform the following operations:

   Step 2.1.   Clear the candidate set and the exchanged bit set which records the exchanged bit of each element in the candidate set according to the current solution.

   Step 2.2.   If the candidate set is full or iteration times at the current temperature have exceeded the allowed upper bound, go to Step 2.4; otherwise, according to x, among the neighbors whose exchanged bits are not in the exchanged bit set, select a solution x' randomly, and add x' and its exchanged bit to the candidate set and the exchanged bit set respectively.

Step 2.3.   If x' is not in the tabu list or has been freed by the aspiration criterion, evaluate the fitness value of x' according to Sec. 3.5; otherwise, reject x', go to Step 2.2. If $f(x') < f(x_{best})$, $x_{best} \leftarrow x'$, accept x', i.e. $x \leftarrow x'$, go to Step 2.2; otherwise, let Δf equal $f(x') - f(x)$, if $\Delta f \leq 0$, accept x', otherwise, accept x' at the probability $e^{-\Delta f / t_k}$. Go to Step 2.2.

Step 2.4.   If the candidate set is full, among the solutions not in the tabu list in the candidate set, add the exchanged bit of the best one to the tabu list. Free the tabu object whose tabu term has become 0.

Step 2.5.   If iteration times at the current temperature have exceeded the allowed upper bound, go to Step 3, otherwise; go to Step 2.1.

Step 3. $k \leftarrow k+1$. Modify the annealing temperature: $t_k \leftarrow \mu t_{k-1}$ $(0 < \mu < 1)$, go to Step 2.

4   Simulation Research

Simulation research has been done over some actual network topologies and mesh topologies with 20 nodes to 100 nodes. Several example topologies are shown in Fig.1.

4.1   Auxiliary Graph Effect Evaluation

Compared with the wavelength graph based algorithm (WG-based) [15], the proposed wavelength assignment algorithm based on segment and wavelength graph (S&WG-based) needs one more time of graph transformation. To evaluate the auxiliary graph effect, compare the runtime of S&WG-based with that of WG-based. The results are shown in Fig. 2. The runtime of S&WG-based is less than that of WG-based in most cases. When all the intermediate nodes have wavelength conversion capabilities, the runtime of S&WG-based is just a little more than that of WG-based. Hence, the introduction of auxiliary graph could reduce the problem complexity effectively.

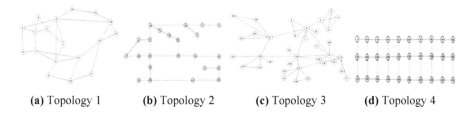

(a) Topology 1   (b) Topology 2   (c) Topology 3   (d) Topology 4

Fig. 1. Example topologies

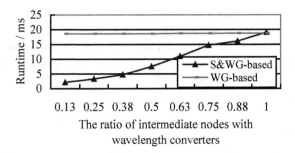

Fig. 2. Evaluation on auxiliary graph effect

4.2 Multicast Routing Tree Cost Evaluation

Comparing solutions obtained by the proposed algorithm with the optimal ones obtained by exhaustive search, the quantitative analysis is made. The results are shown in Table 1. ≤ 1% means that the ratio of the difference between the cost of the obtained solution and the optimal cost (i.e. the deviation of the obtained tree cost from the optimal one) is not more than 1%. ≤ 5% means that the deviation ratio is more than 1% and not more than 5%, and others have the similar meanings. The value under each ratio interval refers to the lower bound of the percentage of the solutions of which deviation ratios are within this interval. Apparently, the solutions obtained by the proposed algorithm are rather satisfied, sometimes are even optimal.

4.3 QoS Evaluation

The delay of the tree obtained by the proposed algorithm and its counterpart obtained without considering *Degree(QoS)* are compared. Take Topology 1 of Fig. 1 (a) as example, simulation results are shown in Fig. 3. The QoS of the multicast routing tree is improved effectively and efficiently.

4.4 Runtime Evaluation

The runtime of the proposed algorithm is compared with that of the simulated annealing based algorithm. Simulation results are shown in Table 2, indicating that the proposed algorithm could improve the searching efficiency effectively.

Table 1. Evaluation on multicast routing tree cost

Topology	Obtained tree cost vs. optimal tree cost			
	≤ 1%	≤ 5%	≤ 10%	>10%
Topology 1	0.84	0.01	0.1	0.05
Topology 2	0.7			0.3
Topology 3	1			
Topology 4	0.85	0.15		

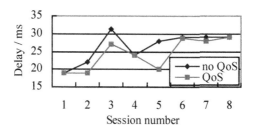

Fig. 3. Evaluation on QoS

Table 2. Evaluation on runtime

Topology	Runtime ratio (hybrid / SA)
Topology 1	0.70 / 1
Topology 2	0.98 / 1
Topology 3	0.83 / 1
Topology 4	0.52 / 1

5 Conclusions

An integrated algorithm for flexible QoS multicast routing and wavelength assignment in IP/DWDM optical Internet is proposed. Given a multicast request and flexible QoS requirement, a hybrid algorithm based on simulated annealing and tabu search is introduced to construct QoS multicast routing tree, embedding the wavelength assignment procedure based on segment and wavelength graph ideas. Hence, the multicast routing and wavelength assignment is solved integratedly. Simulation results have shown that the proposed algorithm is feasible and effective.

The multimedia group communications we considered here are static, that is, the group membership does not vary over time. Further study is needed on the dynamic multimedia group communications. Besides, additional QoS parameters (such as bandwidth, delay jitter, etc.) than only the delay should be considered, thus making the model more general. In this regard, we propose to utilize the fuzzy information processing methods.

References

1. Ghani N., Dixit S., Wang T. S.: On IP-over-WDM Integration. IEEE Communications Magazine. Vol. 38. No. 3 (2000) 72-84
2. Rajagopalan B.: IP over Optical Networks: a Framework. IETF-RFC-3717 (2004)
3. Carlos A. S. O., Panos M. P.: A Survey of Combinatorial Optimization Problems in Multicast Routing. Computers & Operations Research, Vol. 32. No. 8 (2005) 1953-1981
4. Ramaswami R., Sivarajan K. N.: Routing and Wavelength Assignment in All-Optical Networks. IEEE/ACM Transactions on Networking. Vol. 3. No. 5 (1995) 489-500

5. Leung Y., Li G., Xu Z. B.: A Genetic Algorithm for the Multiple Destination Routing Problems. IEEE Transactions on Evolutionary Computation. Vol. 2. No. 4 (1998) 150-161
6. Jia X. H., Du D. Z., Hu X. D., *et al*: Optimization of Wavelength Assignment for QoS Multicast in WDM Networks. IEEE Transactions on Communications. Vol. 49. No. 2 (2001) 341-350
7. Ran L. H., Rami M.: Multicast Routing and Wavelength Assignment in Multihop Optical Networks. IEEE/ACM Transactions on Networking. Vol. 10. No. 5 (2002) 621-629
8. Cui Y., Xu K., Wu J. P., *et al*: Multi-Constrained Routing Based on Simulated Annealing. Proc. IEEE ICC (2003) 1718-1722
9. Li L., Li C.: A QoS-Guaranteed Multicast Routing Protocol. Computer Communications. Vol. 27. No. 1 (2004) 59-69
10. Dean H. L., Ariel O.: QoS Routing in Networks with Uncertain Parameters. IEEE/ACM Transactions on Networking. Vol. 6. No. 6 (1998) 768-778
11. Jeon Y. J., Kim J. C.: Application of Simulated Annealing and Tabu Search for Loss Minimization in Distribution Systems. Electrical Power and Energy Systems. Vol. 26. No. 1 (2004) 9-18.
12. Aijun D., Gee S. P.: A Survey of Optical Multicast over WDM Networks. Computer Communications. Vol. 26. No. 2 (2003) 193-200
13. Chlamtac I., Farago A., Zhang T.: Lightpath (Wavelength) Routing in Large WDM Networks. IEEE Journal on Selected Areas in Communications. Vol. 14. No. 5 (1996) 909-913
14. Wang X. W., Cheng H., Huang M.: A Tabu-Search-Based QoS Routing Algorithm. Journal of China Institute of Communications. Vol. 23. No. 12A (2002) 57-62
15. Wang X. W., Cheng H., Li J., *et al*: A Multicast Routing Algorithm in IP/DWDM Optical Internet. Journal of Northeastern University (Natural Science). Vol. 24. No. 12 (2003) 1165-1168

An Efficient Distributed Broadcasting Algorithm for Ad Hoc Networks[*]

Qiang Sun and Layuan Li

School of Computer Science, Wuhan University of Technology,
Wuhan, Hubei 430063, China
chyang_sun@163.com, jwtu@public.wh.hb.cn

Abstract. In this paper, an efficient distributed heuristic-based algorithm is presented, which is based on joint distance-counter threshold scheme. It features a distributed manner by each node in the network needing no global information. Each node in an ad hoc network receives the message from its neighbors and decides whether to operate retransmitting or not according to the signal strength and times of the receiving messages. The algorithm has superiority such as reliability, rebroadcast saving, less communication overhead for broadcasting task, localized and parameter-less behaviors, so it is easy to operate and possesses a good performance in mobile ad hoc communication environments. A comparison with several other existing algorithms is conducted. It shows by simulation results that the new algorithm is more efficient than others.

1 Introduction

A mobile ad hoc network is a self-organizing network without any existing fixed communication infrastructure support. Because of its independence of a fixed infrastructure, instant deployment, and easy reconfiguration capabilities, the ad hoc wireless networking technology shows great potential and importance in many situations, such as in military and disaster-relief applications.

In mobile ad hoc networks, the research of routing is still under way and considerable routing protocols have been put forward [1, 2, 3, 4, 5, 6, 7]. But nearly all these protocols depend on a broadcasting mechanism [8, 9, 10]. An efficient distributed broadcast algorithm based on joint distance-counter threshold is proposed in this paper. It drastically reduces the effect of the mobility and no exchanged messages or control messages are needed. Joint distance-counter threshold is to provide both a satisfied coverage, less broadcast and average latency, unlike joint distance threshold just guaranteeing a high coverage, and counter threshold just guaranteeing a high saved broadcast and less latency.

The rest of this paper is organized as follows. Section 2 provides an overview of key efficient broadcast protocols proposed for ad hoc networks. Major related definitions and assumptions are presented in Section 3. Joint distance-counter threshold broadcast

---

[*] This work is proudly supported in part by the Grand Research Problem of the National Science Foundation of P.R.C. under Grant No. 90304018.

J. Cao, W. Nejdl, and M. Xu (Eds.): APPT 2005, LNCS 3756, pp. 363 – 372, 2005.
© Springer-Verlag Berlin Heidelberg 2005

algorithm is introduced in Section 4. Simulation results are in Section 5, thus conclusions are drawn in Section 6.

2 Related Work

A straightforward broadcast by flooding is usually very costly and will result the broadcast storm problem [8]. Various efficient flooding schemes have recently been proposed in the mobile ad hoc networks.

[12] proposed a broadcast protocol for ad hoc networks based on a distributed hierarchical framework i.e. cluster structure. With a cluster platform, the broadcast protocol in [12] can choose a dominating sets, which is only composed of clusterhead nodes and gateway nodes to rebroadcast packets. In [13], using a passive clustering, clustering algorithm is suspended until the data traffic commences. Thus it reduces setup latency and control overhead caused by active clustering. The main drawback of using cluster structure is its significant communication overhead for maintaining the structure in a moving environment [14]. Ivan Stojmenovic et al. [15] proposed a simple and efficient distributed algorithm for calculating connected dominating sets enhanced by neighbor elimination scheme and highest degree key. Nevertheless location information of each node should be available to implement such algorithm.

The Multipoint Relay (MPR) method [16, 18] is presented for efficient flooding in mobile wireless networks. In [16], a node periodically exchanges the list of adjacent nodes with its neighbors, each node collects *2-hop* neighbor information, and selects the minimal subset of forwarding neighbors that covers all neighbors within *2-hop* away. Only nodes chosen as forwarding neighbors rebroadcast the flooding packet. In [17], two practical heuristics for selecting the minimum number of forwarding neighbors are proposed. In order to choose forwarding neighbors, a node needs to know its *1-hop* and *2-hop* neighbors. Any changes from any neighbors can cause a reselection of forwarding neighbors. Thus, the exchanged messages are main overhead to the algorithm.

Note that the global information is difficult and sometimes is impossible to get in ad hoc networks, broadcast algorithms based on global network information cannot provide scalability and is unsuitable for the qualifications of mobile ad hoc networks. In [19], a generic distributed broadcast scheme is proposed, which depends on correct *k-hop* neighborhood information to reduce the broadcast storm problem and guarantee full coverage. If the neighborhood information cannot be acquired correctly, the algorithms in [12, 13, 14, 15, 16, 17, 18, 19, 20, 21] cannot run properly. In order to acquire correctly neighborhood information, more exchanged messages are needed, which will deteriorate the performance of networks and consume many of limited resources in ad hoc networks.

The distance-based algorithm and the counter-based algorithm in [8] give a simple scheme, in which each node, when receiving packets from its neighbors, decides whether to forward the broadcast packet or not according to distance or counter threshold. But all these methods are in dilemma about how to provide both high reachability, less rebroadcast and average latency.

3 Notations and Assumptions

An ad hoc network can be mapped to a unit disk graph $G(t) = (V, E)$, where V is the set of nodes and E is the set of logical edges at which two nodes are connected if their geographical distance is within a given transmission range r. Considering the effect of the mobile nodes, $G(t)$ is a time-relevant function.

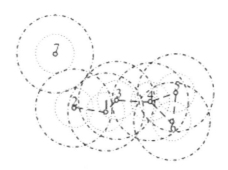

Fig. 1. An ad hoc network

The symbols and definitions used in this paper are defined as follows:

Definition 1: $d(x, y)$ is defined the distance between node x and y within their transmission range r, $d(x, y) \le r$.

Definition 2: $c(x)$ denotes the number of received messages in node x during broadcasting.

Definition 3: D_{Th} denotes a distance threshold, where $0 \le D_{Th} < r$.

Definition 4: C_{Th} denotes a counter threshold, where $C_{Th} \ge 0$.

Definition 5: $N(x)$ is a set of neighbors of node x, $y \in N(x), d(x, y) \le r$. See Fig.1, the neighbors of node 1 consist of node 2, 3 represented as $N(1) = \{2, 3\}$.

Definition 6: $I(x)$ is a subset of $N(x)$, $y \in I(x), d(x, y) \le D_{Th}$. In Fig.1, $I(1) = \{3\}$.

Definition 7: $E(x)$ is a subset of $N(x)$, $y \in E(x), D_{Th} < d(x, y) \le r$. See Fig.1, $E(1) = \{2\}$.

Definition 8: $Rt(x)$ is a set of nodes that retransmit the message from source node, $Rt(x) = \{x \mid x \in \{\{E(x_1) \cap E(x_2) \cap \cdots E(x_k)\} \cup \{x \notin \{E(x_1) \cap E(x_2) \cap \cdots E(x_k)\}, c(x) < C_{TH}\}\}, x_i \in V\}$.

Definition 9: R is defined as the coverage ratio of a broadcast algorithm, the number of mobile nodes receiving the broadcast message divided by the total number of mobile nodes.

The following assumptions are made in our system model.

1) Mobile nodes in an ad hoc network share a single common channel.
2) The maximum transmitting radius of each node in the network is the same.
3) There are no unidirectional links.
4) The explicit acknowledgement to confirm the reception is not needed.

4 Joint Distance Counter Threshold Broadcasting Algorithm

In this section, an efficient distributed heuristic-based algorithm is presented. The proposed algorithm aimed at solving the broadcast storm problem without consuming additional network resources, such as bandwidth and energy.

4.1 Details of the Algorithm

In order to alleviate the broadcast storm problem, $Rt(x)$ has to be found by an efficient distributed heuristic-based algorithm. The relationship between the redundancy of a broadcast and the additional coverage is revealed in [8] that the more additional coverage a node gets, the less broadcast redundancy it has. Moreover, it also shows the relationship among the additional coverage, $d(x, y)$ and $c(x)$ i.e. the farther $d(x, y)$ is, the larger additional coverage of a node can be acquired, and the larger $c(x)$ is, the smaller additional coverage of a node can be acquired.

Based on these relationships, joint distance-counter threshold broadcasting algorithm is proposed. It works as follows. When a node x sends a broadcast message M, its neighbors will receive M and compute $d(x, y)$ according to the signal strength [22]. If $y \in \{E(x_1) \cap E(x_2) \cap \cdots E(x)\}$ or $y \notin \{E(x_1) \cap E(x_2) \cap \cdots E(x)\}, c(x) < C_{Th}$, $\{x_1, x_2, \cdots x\} \subseteq N(y)$, it easy to get that $y \in Rt(x)$ and node y waits for a short time. The delay helps to avoid nodes transmitting M all at once. If node y didn't receive any messages during this short delay, it will transmit M when time expires. Otherwise, it will compute the distance from the sending node and wait again. If $c(y) \geq C_{Th}$, node y stops transmitting. The algorithm can be described as follows:

```
PROCEDURE  Broadcast (M )
while timer is expired
   do
      if M has not been transmitted then send (M)
             end if
   delete(M)
end while
PROCEDURE  HandleMess age(M )
while v receives M, v∈V
   do
             if  M has been retransmitted then delete(M)
             else
               cancelTimer ()
                c←c+1
               push d(u,v) to DisList
             end if
             if  d(u,v)>D_Th and no values in DisList less than D_Th
        then  t← setDelay(d(u,v),1)
             else
                if (c <C_Th) then  t← setDelay(d(u,v),c)
                 else
                   delete (m)
```

```
        end if
      setTimer(t)
      end if
end while
```

The *DisList* is a list to save distance value in each node. As nodes are mobile in ad hoc networks, in order to get a less latency, counter threshold will be introduced and distance threshold will be forbidden when there exists $d(x,y) \le D_{Th}, y_i \in N(x)$. In order to reduce collision, a distance and counter relevant function, *setDelay(d, c)*, is used to set the waiting time for each node.

An analysis and comparison with the distance-based algorithm, the counter-based algorithm, and the JDCT Algorithm will be given through an example in Fig. 2. The result shows that through using joint distance-counter threshold scheme, JDCT algorithm gets a better performance than the other two algorithms. First, some assumptions are given as follows,

(1) The sequence of retransmission for the neighbors of a source node is decided by its distance with the source node. The farther the distance is, the earlier a node may retransmit M. In Fig. 2, as $d(S,1) > d(S,7) > d(S,3) > d(S,5) > d(S,4) > d(S,6) > d(S,2)$, the sequence of retransmission is 1-7-3-5-4-6-2.

(2) There are no collision and contention during broadcasting, and $C_{Th} = 3$.

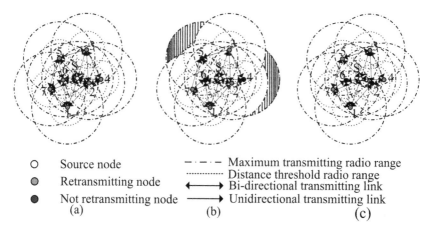

○ Source node	$- \cdot - \cdot -$ Maximum transmitting radio range	
◉ Retransmitting node	$\cdots\cdots\cdots$ Distance threshold radio range	
● Not retransmitting node	$\longleftrightarrow$ Bi-directional transmitting link	
(a)	$\longrightarrow$ Unidirectional transmitting link	
	(b)	(c)

Fig. 2. Comparison for redundant to broadcast among (a) distance-based algorithm, (b) counter-based algorithm, and (c) JDCT algorithm

In Fig. 2, node S is a source node which broadcasts a message M, node 1-7 are neighbors of S which will receive M all at once. Form above definitions and according to [8], the retransmitting node set of distance-based algorithm (Fig. 2 a) and counter-based algorithm (Fig. 2 b), $Rt_{DB}(x)$ and $Rt_{CB}(x)$, are,

$$Rt_{DB}(x) = \{1,2,3,4,5,6,7\}.$$

$$Rt_{CB}(x) = \{1,2,3,7\}.$$

The sum of transmitted M in distance-based algorithm and counter-based algorithm is 34 times and 25 times. In Fig. 2 b, the shaded region is the uncovered area caused by using counter-based algorithm. The results are in accordance with the analysis in [8], that is, the distance-based algorithm provides a better reachable but worse saved broadcast and more latency time than the counter-based algorithm.

In Fig. 2 c, each node uses JDCT algorithm to decide whether to retransmit or not. The node decide as follows,

S1: node 1 is the farthest node among the neighbor nodes of the source S, so node 1 retransmits M first.

S2: As $N(1) = E(1) = \{2,6,7,S\}, I(1) = \Phi, c(6) = c(2) = 2, d(S,3) > d(1,7)$, node 3 retransmits M.

S3: As $N(3) = E(3) = \{2,4,5,S\}, I(3) = \Phi, c(2) = 3, d(S,7) > d(3,4), d(3,5)$, node 7 retransmits M, node 2 quits retransmitting M.

S4: As $N(7) = \{1,5,6,S\}, E(7) = \{1,5,S\}, I(7) = \{6\}, c(6) = 3$, from definition 8, the retransmitting node set is $Rt_{JDCT}(x) = \{1,3,4,5,7\}$. The sum of transmitted M is 28 times.

Form above analysis, in Fig. 2, the JDCT algorithm can get the same reachable as distance-based algorithm with less cost, only 1 additional retransmitting node and 3 additional transmitted M, than counter-based algorithm.

5 Simulation Results

Simulations are performed to evaluate the new broadcasting algorithm and compare with other existing algorithms. A Mobility Framework for OMNeT++ [23] is used as a tool. The network possesses 100 nodes in a 1000×1000 meter square. Nodes in the simulation move according to "random waypoint" model [24]. The transmitting radius of each node is 230 meters and channel capacity is 10Kbits/sec. The mobility speed of a node is set from 0 m/s to 30 m/s. The CSMA/CA is used as the MAC layer in our experiments. Four distributed broadcasting algorithms are compared, as following.

- SB: straightforward broadcasting algorithm
- DB: distance-based broadcasting algorithm
- CB: counter-based broadcasting algorithm
- JDCT: JDCT broadcasting algorithm

The performance measures of interest are:

- Average latency: defined as the interval between its arrival and the moment when either all nodes have received it or no node can rebroadcast it further.
- Ratio of Saved Rebroadcast (RSR): the total number of saved rebroadcast nodes is divided by the total number of nodes receiving the broadcast message.
- Ratio of Collision (RC): the total number of collisions is divided by the total number of packets supposed to be delivered during broadcast.
- Total Number of Contention (TNC): the total number of contention during broadcast.
- Total Number of Received Messages (TNRM): the sum of the number of messages heard by each node during broadcast.

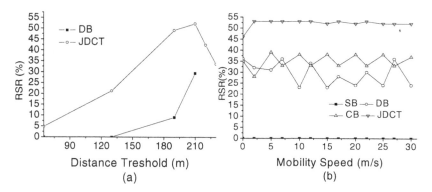

Fig. 3. (a) Distance threshold D_{Th} vs. RSR (R=1) (b) RSR vs. Mobility speed

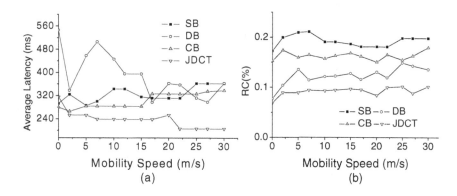

Fig. 4. (a) Average Latency vs. Mobility speed (b) RC vs. Mobility speed

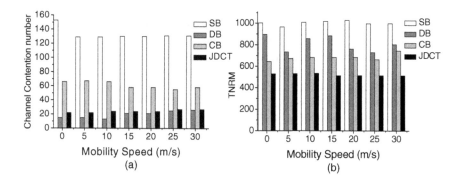

Fig. 5. (a) TNC vs. Mobility speed (b) TNRM vs. Mobility speed

The first set of experimental results (Fig. 3 a) demonstrates the relationship between average RSR and distance with R=1. The result shows that JDCT algorithm obtains higher SRS than the distance-based algorithm. When $D_{Th} \approx 0.9r = 210m$, the RSR of both distance-based algorithm and JDCT algorithm get their maximum value of SRS, about 52% in JDCT, while 28% in DB. When $D_{Th} > 0.9r$, the distance-based algorithm can't cover the whole network, and the RSR of JDCT decreases. This is understandable, because with the threshold value increasing, the number of retransmitting nodes decreases. When $0.9r \leq D_{Th} < r$, by using joint distance-counter threshold scheme, the JDCT can always find enough nodes to retransmit the packet. Thus a high RSR value can be acquired.

In Fig. 3b-Fig. 5, the results are obtained with the parameters of $R = 1$, $D_{Th} = 0.9r$, $C_{Th} = 3$. Fig.3b shows the ratio of saved rebroadcast using different algorithms with varying node speeds (from 0 to 30m/sec). The average latency using different broadcast algorithms with varying node speeds is reported in Fig. 4a. Fig. 4b gives the ratio of collision using different algorithms with varying node speeds. Fig.5 a presents the total number of contention in different node speeds using different algorithms. Fig. 5b shows the total number of received messages in different algorithms with varying node speeds. In ad hoc networks, the key to lessen the broadcast storm problem is to reduce redundant rebroadcasts, collision, and contention. Thus, a high and stable ratio of saved rebroadcast is desirable to reduce the broadcast storm problem. As shown in Fig. 4b-Fig. 5, the broadcast storm problem becomes serious with mobility increment. In Fig.3b, the RSR value of DB and CB fluctuates seriously when mobility increases, whereas the JDCT algorithm keeps a high stable ratio of saved rebroadcast (more than 50%). The results show the robustness and resilience of the JDCT. This makes JDCT a good choice for the mobile environment.

As the topology is dynamic in ad hoc networks, a low average latency is desirable for scalability of the protocol. In addition to the algorithm process time, the average latency is influenced by many factors such as collision, channel contention and the number of nodes not delivered broadcast packets. The other important factor that affects the average latency is related to the schedule algorithm that decides which node to rebroadcast the message. For example, in the distance-based algorithm, a node can rebroadcast the message only if none of transmission distances are below a given distance threshold, thus, the node may consume more time to hear the messages from its neighbors and the average latency may be higher than other algorithms. In Fig. 4a, the average latency of the JDCT algorithm is the lowest among the four broadcast algorithms because the JDCT algorithm gets the lowest collisions, lower contention, and the least received messages with varying node speeds by using joint distance-counter threshold scheme (Fig. 4b-Fig. 5).

6   Conclusion

It is challenging to build efficient broadcast algorithm for ad hoc networks, due to the dynamic nature of the nodes. In ad hoc networks, centralized algorithms are not suitable for the global information is extraordinarily difficult to get. While many existing broadcast algorithms depended on local information such as k-hop neighbor information can not get a desirable performance when local information cannot be acquired correctly because of existing hidden/exposed terminals. As we know, ad hoc networks are resource-limited networks, algorithms based on exchanging control messages such as hello message as well as that based on GPRS device are both unsuitable.

In this paper, an efficient distributed heuristic-based algorithm named JDCT algorithm is presented. The algorithm is based on joint distance and counter threshold scheme. It runs in a distributed manner by each node without needing any global information. The simulation experiments have demonstrated the efficiency of proposed broadcast algorithm meanwhile the broadcast storm problem is significant alleviated. It's a good choice for mobile ad hoc networks considering the algorithm's efficiency and robustness. Our future work includes a performance evaluation of the JDCT broadcast algorithm in realistic simulation environments with packet collision and node mobility. In addition, our JDCT algorithm will be embed in some routing protocols such as AODV protocol to investigate its efficiency in ad hoc networks.

References

1. Li Layuan, Li Chunlin: A QoS multicast routing protocol for dynamic group topology. Information Sciences, Vol. 169(1-2). Elsevier, UK (2005) 113-130
2. S. Ramanathan, M. Streenstrup: A Survey of Routing Techniques for Mobile Communication Networks. Mobile Networks and Applications, pp. 89-104, 1996
3. C. E. Perkins, E. M. Royer: Ad hoc on-demand distance vector (AODV) routing. In Proceedings of the 2nd IEEE Workshop on Mobile Computing Systems and Applications, Feb. 1999
4. Li Layuan, Li Chunlin: A routing protocol for dynamic and large computer networks with clustering topology. Computer Communications, Vol. 23(2), Elsevier, UK (2000) 171-176
5. David B. Johnson, David A. Maltz, Yih-Chun Hu, and Jorjeta G. Jetcheva: The Dynamic Source Routing Protocol for Mobile Ad Hoc Networks (DSR). http://www.ietf.org/internet-drafts/draft-ietf-manet-dsr-09.txt, April 2003
6. Li Layuan, Li Chunlin: Acta Informatica. A distributed QoS-aware multicast routing protocol. Computer Science, Vol. 40 (3). Springer-Verlag GmbH (2003) 221-233
7. J. Cartigny, D. Simplot: Border Node Retransmission Based Probabilistic Broadcast Protocols in Ad-Hoc Networks. In Proc. 36th International Hawaii International Conference on System Sciences (HICSS'03), Hawaii, USA. 2003
8. Y.-C. Tseng, S.-Y. Ni, Y.-S. Chen and J.-P. Sheu: The Broadcast Storm Problem in a Mobile Ad Hoc Network. MOBICOM'99, ACM Press New York (1999) 151-152
9. E. Royer and C-K. Toh: A Review of Current Routing Protocols for Ad-Hoc Mobile Wireless Networks. IEEE Personal Communications, Vol. 6(4). IEEE Communications Society (1999) 46-55
10. J. Broch et al: A performance comparison of multi-hop wireless ad hoc network routing protocols. Proc. ACM MOBICOM, ACM Press New York (1998) 85-97
11. Mike Burmester, Tri van Le and Alec Yasinsac: Weathering the storm: managing redundancy and security in ad hoc networks. Proceedings of the 3rd International Conference on AD-HOC Networks & Wireless. LNCS 3158, Springer(2004) 96-107
12. E. Pagnani, G. P, Rossi: Providing reliable and fault tolerant broadcast delivery in mobile ad-hoc networks. Mobile Networks and Applications, Vol. 5(4). DBLP (1999) 175-192
13. M. Gerla, TJ Kwon, G. Pei: On demand routing in large ad hoc wireless networks with passive clustering. Proceedings of the IEEE WCNC 2000, Chicago, IL, September 2000

14. Wu and H. Li: A dominating-set-based routing scheme in ad hoc wireless networks. Telecommunication Systems, Vol. 18(1-3). DBLP (2001) 13--36
15. Ivan Stojmenovic, Mahtab Seddigh, Jovisa Zunic: Dominating Sets and Neighbor Elimination-Based Broadcast Algorithms in Wireless Networks. IEEE Transactions on Parallel and Distributed Systems, Vol. 1(13). IEEE Computer Society (2002) 14-25
16. A. Qayyum, L. Viennot, A.Laouiti: Multipoint relaying for flooding broadcast messages in mobile wireless networks. In Proceedings of the 35th Annual Hawaii International Conference on System Sciences (HICSS'02), Hawaii, 2002
17. G Calinescu, Ion I. Mandoiu, P J. Wan, and A. Z. Zelikovsky: Selecting forwarding neighbors in wireless ad hoc networks. Mobile Networks & Applications, Vol. 9(2). ACM Press (2004) 101-111
18. ITSI STC-RES10 Committee. Radio Equipment and Systems: High Performance Radio Local Area Network Type 1, Functional Specifications. 1999, 7
19. Jie Wu, Fei Dai: A Generic Distributed Broadcast Scheme in Ad Hoc Wireless Networks. IEEE TRANSACTIONS ON COMPUTERS, Vol. 10 (53). IEEE Computer Society (2004) 1343-1354
20. B. Chen, K. Jamieson, H. Balakrishnan, and R. Morris: Span: An Energy-Efficient Coordination Algorithm for Topology Maintenance in Ad Hoc Wireless Networks. ACM Wireless Networks, Vol. 5 (8). ACM Press (2002) 481-494
21. H. Lim, C. Kim: Multicast Tree Construction and Flooding in Wireless Ad Hoc Networks. In Proc. of the ACM Int'l Workshop on Modeling, Analysis and Simulation of Wireless and Mobile Systems (MSWIM) (2000) 61-68
22. R. Dube, C. D. Rais, K.-Y. Wang and S. K. Tripathi: Signal stability based adaptive routing (SSA) for ad hoc mobile networks. Technical Report CS-TR-3646. University of Maryland, College Park (1996)
23. A. Vargas: OMNET++Discrete Event Simulation System. version 3.0 edition, 2005
24. Christian Bettstetter, Christian Wagner: The Spatial Node Distribution of the Random Waypoint Mobility Model. Mobile Ad-Hoc Netzwerke, 1. deutscher Workshop über Mobile Ad-Hoc Netzwerke WMAN 2002, March 25-26, 2002 41-58

Chaos-Based Dynamic QoS Scheme
and Simulating Analysis

Qigang Zhao and Qunzhan Li

School of Electrical Engineering, Southwest Jiaotong University,
Chengdu, Sichuan, 610031, China
qgzhao@vip.sina.com

Abstract. This paper takes use of chaos related theories to analyze the real network traffic about its chaotic properties and prediction attributes. Owning to the good performance of chaos-based prediction in short term forecasting, the prediction-based DiffServ framework and the Dynamic QoS scheme are firstly given in the paper. The OPNET-based simulating result shows that the QoS performaces and the network's throughputs in heavy-load environment are all improved remarkably, comparing with the traditional static QoS configuring and measure-based dynamic QoS setting methods.

1 Introduction

To guarantee the QoS (Quality of Service) in multi-service applied environment of Internet, IETF put forward both RSVP (Resource Reservation setup Protocol) based on InerServ (Integrated Services) model and coarse granularly control based on DiffServ (Differentiated Services) model. InerServ requires that all routers in the network have to support RSVP and keep the transmitting status for every service stream while service transports. Owning to the present networks needing huge upgrade at vast cost and the additional status data requiring too much network resources, InterServ is generally only applied in small-scale network or access network and difficulty to be used in large-scale networks. In DiffServ, Most of working loads are put to edge routers and the core routers only need classifying the different services according to predefined PHB (Per Hop Behavior). Since its simplicity and modifying the present network little, DiffServ is attached more importance [1-3] than InterServ.

The main idea of DiffServ model is that firstly to classify different services and define as different behavior aggregates (labeled by service code: codepoint) in edge route, then routing equipments to transmit the service streams according to its codepoint, wherein resources such as the cache and the bandwidth are allocated based on the service priority. The present resource allocated schemes of router are mainly static resource allocating method (SRAM), in which the services' resource allotting regulations and priorities are fixedly pre-set in both edge and core router, and the regulations and priorities shall not be changed in service transmitting. SRAM generally takes use traditional QoS parameters estimating method, such as to estimate the requirements of peak and average bandwidth of different services based on ON/OFF model, Markov modulating passion process or Markov modulating hydro-process. The limitations of SRAM are as following: (1) the statistic characteristics of services

J. Cao, W. Nejdl, and M. Xu (Eds.): APPT 2005, LNCS 3756, pp. 373 – 381, 2005.
© Springer-Verlag Berlin Heidelberg 2005

source generally are difficulty to be described by one service model; (2) the services' statistic properties will change owning to the queues; (3) different service usually has different stream characteristics. For the reasons, SRAM is difficulty to guarantee QoS of the precedence service while the network is busy and hard to improve the network throughput when the network is idle.

In view of the limitations of SRAM, the article [4] has given a measure-based parameters estimating method (MPEM). The method periodically collects the network traffic value in edge router, estimates the QoS and dynamically modify the resource allocating parameters. In paper [5], the method, that to allocate resources for different service is implemented by modify the WFQ (Weight Fair Queue) value of service based on the traffic measurement, is put forward. However, the references [6-7] have proved that Internet traffic has characteristics of long-term correlation and short-term multi-fractal, which just matches the chaotic attractor's macroscopical and microscopical characteristics. Thanks to the properties of bursting and complex dynamics of network traffic, the value of MPEM usually loses its availability. The present traffic property and distributing characteristic generally do not represent several hours, or several minutes later traffic properties, so MPEM also has its limitations.

Due to the limitations of SRAM and MPED, as well as chaos and fractal properties of network traffic, this paper puts forward a framework of prediction-based DiffServ model and gives the chaos-based dynamic QoS scheme. The OPNET-based simulating shows that the method given in this paper has the best performances among the three methods and the network throughput is also improved remarkably by using the scheme.

2 Network Traffic Prediction

Based on Lyapunov Exponent Methodology, the references [6-7] have given the analysis results of that both LAN and WAN's traffic have the characteristics of chaos and fractal. The Primary Component Analysis (PCA) method is used in this paper to analyze the real traffic data and the primary components of the data are shown as figure 1. According to PCA method, the curve of noise or period signal's primary components is a line paralleled with X-axis, but that of chaos signal's is a crossing fixed point, negative slope bias. In figure 1, the PCA slopes of both video and data traffic are all negatives, so the traffic of real network has distinct characteristic of chaos.

The traditional methodologies of network traffic forecasting are mainly based on dynamics or statistics, wherein the subjective model of data sequences is firstly created and then the prediction results are calculated by basis of the model. However, if the chaos theory is applied, the results of prediction may be directly calculated through the data sequences themselves and the subjective model may be not necessary to build. In this way, the man-made subjectivity is avoided and the accurateness and reliability are all improved. Because of "butterfly effect" property of chaos system, the chaotic data sequences can not be forecasted in long term. In short term, however, the motion path of the chaotic system changes little, so the traffic can be short-term predicted by basis of collected data sequences.

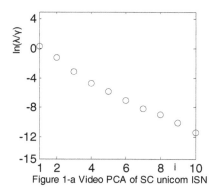

Figure 1-a Video PCA of SC unicom ISN

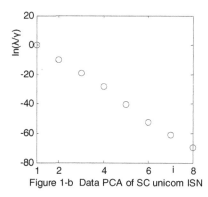

Figure 1-b Data PCA of SC unicom ISN

Fig. 1. The PCM of IP services traffic

As a measure value to estimate degree of chaotic motion divergence, Lyapunov exponent is an excellent traffic forecasting parameter. The reference [9] has employed it to predict short-term electric power load in power system and gotten good performance. We firstly take it as prediction parameter to forecast IP traffic in this paper.

Let Y_M as prediction central point, the nearest neighbor point of Y_M in phase space is Y_k, the distance between them is $d_M(0)$, and the largest Lyapunov exponent is λ_1. Here,

$$d_M(0) = \min_j \left\| Y_M - Y_J \right\| = \left\| Y_M - Y_K \right\| \tag{1}$$

$$\left\| Y_M - Y_{M+1} \right\| = \left\| Y_K - Y_{k+1} \right\| e^{\lambda_1} \tag{2}$$

In equations, for vector Y_{M+1} only the last component $x(t_{n+1})$ is unknown in reconstruction phase space, so Y_{M+1} can be computed based on equation (2). The algorithm of how to calculate the largest Lyapunov exponent λ_1 can refer to the paper [10].

Based on the historical data sequences of 2003 January to March (collected from telecom operator Sichuan Unicom , the sample rate is 1/5Min), we reconstruct the phase space by delaying the coordinate system, and build the short-term traffic prediction model. To guarantee the forecasting accuracy, only one step prediction method is used, that is, to search reference vectors over again after every prediction step. The prediction result and the original data of February 6 are compared and shown in figure 2. The statistics of prediction are shown in table 1.

As shown in table 1, the errors of chaos prediction are mainly smaller than 3%, and the amounts of those smaller than 2% exceeds 75% of total. In view of that the packet loss rate of the present operator's network has reached or exceeded 10% when the network is busy, and taking into account the reality that the packet loss rate of voice service smaller than 3% can be acceptable, the performance of chaos prediction can excellently satisfy the requests of IP dynamic QoS control scheme.

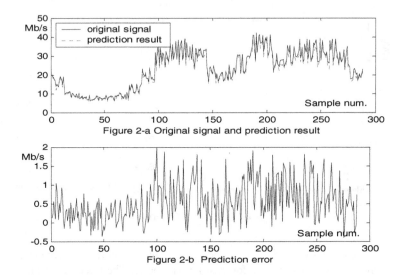

Figure 2-a Original signal and prediction result

Figure 2-b Prediction error

Fig. 2. Original data and prediction result comparison

Table 1. the Statistics of chaos prediction result

Month	Average Error	Max Error	<2%
1月	1.62%	8.34%	75%
2月	1.57%	7.23%	83%
3月	1.60%	7.91%	79%

3   Prediction-Based DiffServ Model

3.1   The Framework of Prediction-Based DiffServ

The QoS performance of IP services can be described as two layers parameters: network and service layer QoS. The QoS of two layers influence each other, but not only one layer affects another. The QoS parameters of network layer, such as delay, delay quiver, packet loss rate and bandwidth, will affect the service layer performance parameters for instance voice clearness and voice quiver etc. And, the admission control of service layer will influence the total traffic of network, so the QoS performance of network will be affected. Consequently, the correct IP QoS control scheme should be following as:

- The admission control of service layer must be based on the current states of the network's resources;
- Every service's traffic control of network layer must be based on the transmitting requirement of the service;
- The QoS of two layers should be controlled separately through the information feedback from another layer, so the communication scheme between service layer and network layer must be created.

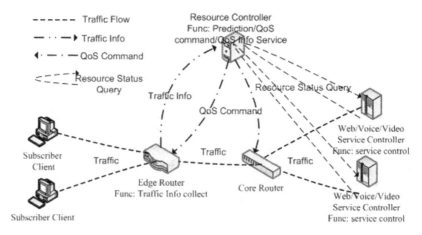

Fig. 3. Prediction-based DiffServ Model

By basis of before-mentioned network traffic prediction model and methodology, the service admission controller can get the guideline information about how to control the network traffic of the service, so the exchange scheme of QoS control information between two layers can be efficiently created. Accordingly, the prediction-based DiffServ model is built and shown as Fig.3.

As shown in Fig.3, being different with traditional model, one component—Resource Controlling Server (RCS) is added to the DiffServ framework. In the model, the edge router is responsible to collect the traffic data of every service, to upload to RCS, and to accept the QoS commands about resource assignment from RCS; RCS is responsible to predict next period traffic of every service, to send out the QoS commands to edge and core router, and to answer the information query about resource statues for service controller such as Softswitch, Web server and video server etc.

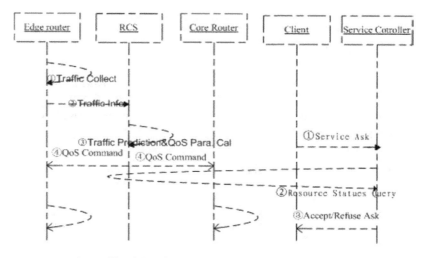

Fig. 4. Prediction-based QoS control process

3.2 Prediction-Based QoS Control Process

Based on the Prediction-based DiffServ model, the QoS control process is followed as Fig.4.

Therein, the steps to control the network traffic are as following:

1. The edge router collects the current traffic data of every service periodically (the interval time might be two or five minutes), and send to RCS by using UDP message;
2. The RCS predicts the next period traffic of every service by basis of the current traffic information from edge routers and the historical data in database;

Based on the next phasic traffic prediction and the values of the current network bandwidth, service priority etc, the resource allotting parameters are calculated;
The following are the formulas to compute the bandwidth allotting parameters:

$$\text{If, } S_{pt}(1) + S_{pt}(2) + \cdots + S_{pt}(n) \leq B_T$$

$$\text{then } B_A(i) = B_T \times \frac{S_{pt}(i)}{\sum_j S_{pt}(j)} ; \tag{3}$$

$$\text{else, } B_A(i) = B_T \times \frac{W(i)}{\sum_j W(j)} . \tag{4}$$

In equations, $S_{pt}(i)$, $B_A(i)$, W_i are separately the prediction traffic, allotted bandwidth and priority right of service i, and B_T is the total bandwidth of the router.

1. To send the QoS command to edge and core router, in which the resource allotting parameters are included;
2. The router modifies its PHB value and allot transmitting resource such as bandwidth, queues etc to every service according to the newly PHB, and do the transmitting and routing based on the service's priority and the modified PHB.

The steps of service admission control are as following:

1. The client sends the service request to the service controller;
2. The service controller queries the RCS about the traffic status of the next period after obtaining the service request;
3. The service controller accepts or refuses the service request based on the traffic changing tendency.

4 Simulating and Analysis

To evaluate the performance of prediction-based dynamic QoS control scheme, we have simulated the model (labeled as scheme 3) with OPNET modeler produced by OPNET Corp, and compared with SRAM (labeled as scheme 1) and MPEM (labeled as scheme 2). For convenience, the router queues algorithms of the three methods have all taken WFQ. The simulation topology of the model is as figure 5.

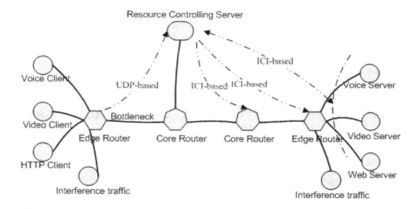

Fig. 5. Network simulation topology

Table 2. Service type and its priority

Service Type	Priority	Remark
_NETWORK_CONTROL	7	Network control signal
_GUARANTEED	5	Guaranteed service(Voice Client)
_CONTROLLEDLOAD	3	Load controlled (Video Client)
_QUANLITATIVE	2	
_BESTEFFORT	1	Best effort (HTTP Client)
NOR_CONFORMING	0	Interference Client

In simulating model, there have including three types traffic sources (Http, Video conference and voice) and a traffic interference client, which separately are counterparts with three types servers and another interference client. The traffic model used in simulating is Psedo-self-similar model. The service type and priority of four types traffic can referred to table 2. One VC++ designed module used to collect traffic data is added to the edge router node. The additionally designed node RCS (not necessary in SRAM and MPED simulating model) is communicating with edge and core router based on ICI. The link between edge router and core router is the bottleneck of the simulating network, whose bandwidth is 2Mb/s. The simulating is continued 20 hours and under heavy-load.

The corresponding Packed Delay curves of three methods are shown in figure 6-a, Packet Delay Jitter in figure 6-b, Packet loss in figure 6-c,traffic throughput in figure 6-d, and performance statistics in table 3.

As shown in figure 6 and table 3, when there exists bottleneck in network and the communications are under heavy load, the values of Packet Delay, Packet Delay Jitter and Packet Loss of SRAM is the biggest in three simulating, those of MPED smaller, those of our method smallest and the throughput of our method is biggest, thanks to applying the prediction-based service admission control and resource allocation in our model.

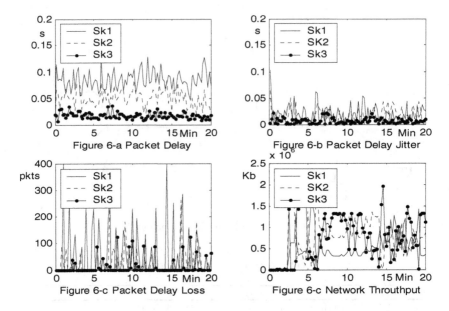

Fig. 6. Network simulating comparison

Table 3. Performance comparison of three schemes

Sch	Type	Packed Delay		APL	ATT (MB/S)
		APD(s)	MPD(s)		
Sk1	Voice	0.0839	0.1286	55.6917	0.51280
	Video	0.1025	0.1856	213.3478	0.45121
	HTTP	0.1531	0.2134	823.3532	0.39082
Sk2	Voice	0.0497	0.0844	24.1333	0.72692
	Video	0.0712	0.1567	163.3478	0.47133
	HTTP	0.1601	0.2029	845.3532	0.40185
Sk3	Voice	0.0170	0.0354	0.1222	0.76259
	Video	0.0623	0.1437	12.1213	0.51264
	HTTP	0.1453	0.2109	113.1568	0.48127

Notice: APD, Average Packet Delay; MPD, Max Packet Delay; Average Packet Loss(packets);APL, Average Traffic Throughput; ATT, Average Traffic Throughput; Sk1—Static Setting; Sk2—Measure Based; Sk3—Prediction Based

5 Conclusions

The bandwidth allotting and QoS control of IP network must be a complex and dynamic process due to the bursting, chaos and multi-fractal properties of IP multi-service. SRAM-based DiffServ can not guarantee the whole process QoS of IP integrated services, and the MPED-based improves some but not remarkable owning to not taking chaos characteristics of traffic into account. Based on the chaotic nature of network traffic, the chaos-based traffic prediction method is evaluated and the prediction-based DiffServ model is given in the paper. The simulating result shows that the QoS of services and throughput of network all have been improved remarkably.

Acknowledgments

We wish to thank Operation Managing Department of Sichuan Unicom for its providing the IP traffic data of the NGN for the experiment.

References

1. IETF working Group:Resource Reservation Protocol(RSVP)– version 1Functional Specification.RFC 2205.
2. IC.J. Cheng, M.Anthony, C.Armando et al.: QoS Architecture Based on Differentiated Services for Next Generation. IETF Internet Draft, July 2000.
3. K.Isoyama, M.Yoshida et al.: Policy Framework QoS Information Model for MPLS. IETF Internet Draft, 12 July 2000.
4. Bin Zhao, Zengji Liu: Estimation of QoS Parameters Based on Measurement and Its Apllication, Journal of Software, Vol.13, No.7 2002.
5. Xiaohui Jin, Jiandong Li: Using Measurement-Based WFQ to Implement PDS and Its Performance Analysin, ACTA Electronica Sinica, Vol.30 No.3 2002.
6. Leland W. E., TAQQU M. S., WILLINGER W.: On the self-similar nature of Ethernet traffic. IEEE/ACM Transactions on Networking,1995,2:1-15.
7. FELDMANN W, Gilbert A.C, Willinger W.: Data networks as cascades: investigating the multigractal nature of Internet WAN traffic. ACM SIGCOMM 98 Conference. Vancouver, BC, Canada,1998.
8. G. Sugihara and R. M. May: Nonlinear forecasting as a way of distinguishing chaos from measurement error in time series, Nature, 1993,344:734-741.
9. Zhishuan liang, Liming Wang: Short-term load prediction for power system based on Lyapunov exponent, Chinese Electrical Engineering Transaction, 2002, 18(5): 368-371.
10. D. Kugiurmtzis: State space reconstr uction parameters in the analysis of chaotic time series-the role of the time window length, Physica D, 1998,95:13-28.

Dynamic Delaunay Triangulation
for Wireless Ad Hoc Network

Ming Li, XiCheng Lu, and Wei Peng

School of Computer Science, National University of Defense Technology,
Changsha 410073, Hunan,China
liming_cs@hotmail.com

Abstract. Geometric routing protocols benefit from localized Delaunay triangulation, which can guarantee the delivery of packet and bound the length of route. In this paper we propose a localized algorithm to build Delaunay triangulation in wireless ad hoc network. The algorithm considers not only stationary situation but also dynamic situation in which nodes can dynamically join and leave the network. The communication cost of the algorithm is O($nlogn$). Therefore, our algorithm is applicable in wireless sensor network, in which nodes dynamically join and leave network. We also prove the correctness of the algorithm.

Keywords: Topology control, ad hoc network, Delaunay Triangulation, geometric routing protocols.

1 Introduction

Geometric routing protocol[1] is an attractive problem in wireless ad hoc network, which exploits the underlying geometry of ad hoc network locations to keep routing overhead small. Geometric routing does not require the nodes to maintain routing tables, a distinct advantage given the scarce storage resources and the relatively low computational power available to the wireless nodes. More importantly, given the numerous changes in topology expected in ad hoc networks, no re-computation of the routing tables is needed and therefore we expect a significant reduction in the overhead. Thus, geometric routing is scalable. Geometric routing is also uniform, in the sense that all nodes execute the same protocol when deciding to which other node to forward a packet.

But the design of geometric routing protocol is a challenging problem. Guaranteeing the successful arrival at the destination of the packet requires the underlying topology to be planar. Guaranteeing the route traveled by the packet is at most t times longer than the shortest path requires the underlying topology to be t-spanner of UDG. Delaunay triangulation[2] satisfies these two requirements. However, build Delaunay triangulation is not a viable solution, because edge may be longer than the communication range and it cannot be built locally.

In this paper, we present an algorithm to build Delaunay triangulation in wireless ad hoc network. The algorithm can build and maintain Delauany triangulation in dynamic

J. Cao, W. Nejdl, and M. Xu (Eds.): APPT 2005, LNCS 3756, pp. 382–389, 2005.
© Springer-Verlag Berlin Heidelberg 2005

environment efficiently. The communication cost of the algorithm is O(*nlogn*), where *n* is the number nodes in the network. Our algorithm is the first algorithm can ensure such low communication cost in dynamic environment.

The rest of the paper is organized as follows. We provide some definitions and notions in section 2. In section 3 we review some related works. In section 4 we describe our algorithm and prove its correctness. Finally, we draw conclusion in section 5. The proofs of theorems are in appendix.

2 Preliminaries

2.1 Definitions and Notions

Given a set of nodes *V* positioned in a two-dimensional space, a wireless ad hoc network is modeled as a Unit Disk Graph (UDG), which is composed of all vertices *V* and all possible edges connecting pairs of nodes of *V* whose distance is not longer than the maximum transmission range *R*. Small letters are used to represent nodes. The 1-hop neighbors of node u and itself is represented as $N(u)$. Edges are represented by the two nodes define them, for example, node *u* and *v* define edge *uv*; a triangle defined by node *u*, *v* and *w* is represented as $\triangle uvw$. The disk or circle that has edge *uv* as its diameter or has $\triangle uvw$ as its inscribed triangle is represented as $D(uv)$ or $D(uvw)$ respectively. An edge *uv* is called GG edge (Gabriel Graph) [3]if $D(uv)$ does not contain any other node of *V*. A triangulation of a node set *V* is called a Delaunay triangulation and represented as Del(*V*) if the circumstance of each of its triangles does not contain any node of *V*. Unit Delaunay Graph has all the edges in Del(*V*) except those longer than *R*, represented as UDel(*V*).

2.2 Spanner

A graph *H* is a *t*-spanner of a graph *G* if $V(H)=V(G)$ and, for any two nodes *u* and *v* of $V(H)$, $\| \Pi_H(u,v) \| \le t \| \Pi_G(u,v) \|$, where $\Pi_G(u,v)$ is the shortest path connecting node *u* and *v* in graph *G*, $\| \Pi_G(u,v) \|$ is the length of the shortest path. Constant *t* is called the length stretch factor of the spanner *H*. There are several geometrical structures that are proved to be t-spanners for the Euclidean complete graph For example, Yao graph [6], θ-graph [7] and Delaunay triangulation [8][9] have been shown to be *t*-spanners. However, the first two geometrical structures are not guarantee to be planar. Though Delaunay triangulation is planar *t*-spanner, it has edges that are longer than *R*. As UDel(*V*) is a *t*-spanner of UDG [4][5], it is much more suitable for topology of wireless ad hoc networks.

3 Related Work

Geometric routing protocols have received much attention for its excellent performance. It does not need to maintain routing tables and impose little routing computation overhead. Unfortunately these algorithms are not guarantee to converge, such as compass routing[10]. When fails they need to use methods, such as right-hand rule, to guarantee to converge, which require the underlying topology is planar. Under the assumption of the planar topology, many geometric routing protocols guaranteed to converge have been proposed, such as GPRS[11](Greedy Perimeter Stateless Routing), FACE routing [12].

As density is important to routing performance, many efforts have made to create good planner spanners of UDG. Delaunay triangulation is a planar spanner of UDG. Gao[4] use Delaunay triangulation to build a planar graph called RDG (Restricted Delaunay Graph), which is a spanner of UDG. The communication of their algorithm is $O(n^2)$.Li et el[5] use Delaunay triangulation to build another planar graph called PLDel(Planarized Delaunay triangualtion) with communication cost $O(nlogn)$.

Our algorithm improves Li's algorithm. Although the asymptotic communication cost is the same, $O(nlogn)$. The communication cost of our algorithm is about 50% less. And our algorithm can accommodate to node entrance and departure.

4   Dynamic Delaunay Triangulation Algorithm

4.1   Overview

In this section, we propose an algorithm, called DynDel, to dynamically build planar spanner of UDG. DynDel algorithm considers not only stationary situation but also dynamic situation.

DynDel algorithm is a localized algorithm, in which nodes only need to exchange information with 1-hop neighbors. And the topology generated by DynDel algorithm is planar spanner of UDG. DynDel algorithm comprises of three steps.

Neighbor Discovery. In the neighbor discovery step, nodes exchange HELLO message periodically. HELLO message contains the ID and position of nodes. As soon as the neighbor discovery step is complete, node can compute all GG edges incident to it. As GG edges belong to Delaunay triangulation, node puts these GG edges into its incident edges set.

GG Edge Broadcasting. After the computation of GG edges, nodes broadcast incident GG edges. All nodes collect the information of incident GG edges of its 1-hop neighbors.

Delaunay Triangulation. Assume node gathered all the information of its 1-hop neighbors, node computes the Delaunay triangulation of its 1-hop neighbors. If $uv, uw, vw \in Del(N(u))$ and $\angle vuw \geq \pi/2$ and edge vw does not intersect with any GG edge in node u's GG edge list, then node u sends message proposal(uvw). When node u receives message proposal(vuw), if $uw \in Del(N(u))$, then node u sends message accept(vuw), else node u sends message reject(vuw). When node u receives message accept(vuw), if $uw \in Del(N(u))$, then node u puts edge uw into incident edge set.

Condition $\angle vuw \geq \pi / 2$ is used to limit the number of proposal message. Without this condition, as shown in fig 1, node u will broadcast arbitrary number of wrong proposal messages. If edge uv belongs to $\mathrm{Del}(N(u))$, and exists node w so that $\angle vuw \geq \dfrac{\pi}{2}$, then node u waits for the proposal message from node w to put edge uv into its incident edge set.

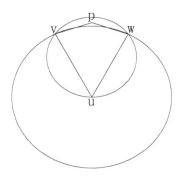

Fig. 1. Arbitrary number of wrong proposal messages

Our algorithm can support node join and leave network at any time at anywhere. So the neighbor discovery step is a continuous process. When detected some node joins and leaves network, the algorithm is triggered.

4.2 Dynamic Aspect of the Algorithm

Node departure or fail is detected by lost HELLO message. When node u detects some node v failed, which is 1-hop neighbor of u, node u computes $\mathrm{Del}(N(u))$ and broadcasts the disappearance and appearance of GG edges and proposal messages.

When node u detects some node v active, node u computes $\mathrm{Del}(N(u))$ and broadcasts all incident GG edges and proposal messages.

4.3 Final Algorithm

In this section we summarize the description of our algorithm. The algorithm is divided into two separated components. The first component reacts to asynchronous events, such as detection of new nodes or detection of failed nodes reception of messages. The action of this component is depicted in fig 2. The second component uses information collected in the first one to keep the Delaunay triangulation update. The second component is a synchronous component. The action of this component is depicted in fig 3.

Let $i+1$ be the current iteration of the second component of the algorithm. The algorithm keeps the following data structures:

---

case detects node x join the network

$N^{i+1}(u) = N^{i+1}(u) \cup \{x\}$

NewNode^{i+1} =TRUE

case detects node x leave the network

$N^{i+1}(u) = N^{i+1}(u) - \{x\}$

FailedNode^{i+1} = TRUE

case receives message proposal(wuv)

if $uv \in TRI^i(u)$ **then** sends message accept(wuv)

   else sends message reject(wuv)

case receives message proposalAdd(wuv)

   if $\triangle wuv \in TRI2^i(u)$ **then** sends message accept(wuv)

   else send message reject(wuv)

case receives message proposalDel(wuv)

   deletes edge uv from node u's incident edge set

case receives message GG(xy)

   $GG2(u) = GG2(u) \cup \{xy\}$

case receives message GGAdd(xy)

   $GG2(u) = GG2(u) \cup \{xy\}$

case receives message GGDel(xy)

   $GG2(u) = GG2(u) - \{xy\}$

case receives message accept(wuv)

   if $\triangle wuv \in TRI2^i(u)$ **then** puts edge uv into node u's incident edge set and marks $\triangle wuv$ as accepted

   case receives message reject(wuv)

   if $\triangle wuv \in TRI2^i(u)$ **then** deletes edge uv from node u's incident edge set and marks $\triangle wuv$ as rejected

---

Fig. 2. Asynchronous component

$N^i(u)$—1-hop neighbors of node u,

NewNode^{i+1}—Boolean variable, TRUE if detected new nodes,

FailedNode^{i+1} – Boolean variable, TRUE if detected failed nodes,

$GG^i(u)$ – GG edges incident to node u,

$GG2^i(u)$ – GG edges broadcasted by 1-hop neighbors of node u,

TRI$^i(u)$ – Triangles proposed by node u in ith iteration,

TRI2$^i(u)$ – Triangles not proposed by node u in ith iteration but in Del($N^i(u)$).

4.4 Correctness

Theorem 1 proves that DynDel builds a planar graph. Theorem 2 proves that UDel(V) is a sub-graph of the graph built by DynDel. Therefore, DynDel builds a planar t-spanner of UDG(V). See the proof in appendix.

Function Iteration

if $N^{i+1}(u) == N^i(u)$ and $GG2^{i+1}(u) == GG2^i(u)$ **then** return

Calculate Delaunay($N^{i+1}(u)$)

Compute $GG^{i+1}(u)$

Assume $uv_0, uv_1, uv_2, ..., uv_k(v^k == v_0)$ is the list of edges that belong to Del $(N^{i+1}(u))$
in clockwise order.

for each v_i **do**

if edge uv_i is a GG edge **then** puts edge uv_i into $GG^{i+1}(u)$

else if $\angle uv_{i+1}v_i \geq \pi / 2$ **then** put $\triangle v_{i+1}uv_i$ into $TRI^{i+1}(u)$

else puts $\triangle v_{i-1}uv_i$ into $TRI^{i+1}(u)$

if $\angle v_{i+1}uv_i \geq \pi / 2$ **then** put $\triangle v_{i+1}uv_i$ into $TRI2^{i+1}(u)$

endfor

if NewNode^{i+1} == TRUE

then broadcasts all the edges in $GG^{i+1}(u)$

if NewNode^{i+1} == FALSE and FailNode^{i+1} == TRUE

then broadcasts GGAdd message of all the edges in $GG^{i+1}(u) - GG^i(u)$

broadcasts GGDel message of all the edges in $GG^i(u) - GG^{i+1}(u)$

collects all GG edges broadcast by 1-hop neighbors into $GG2^{i+1}(u)$

for each $\triangle v_{i+1}uv_i$ in $TRI2^{i+1}(u)$ **do**

if edge $v_{i+1}v_i$ intersected with some edge in $GG2^{i+1}(u)$

then deletes $\triangle v_{i+1}uv_i$ from $TRI2^{i+1}(u)$

endfor

for each $\triangle v_{i+1}uv_i$ in $TRI2^{i+1}(u) - TRI2^i(u)$ **do**

broadcasts proposalAdd($uv_{i+1}v_i$)

endfor

for each $\triangle v_{i+1}uv_i$ in $TRI^i(u) - TRI^{i+1}(u)$ **do**

broadcasts proposalDel($uv_{i+1}v_i$)

endfor

$i = i+1$

NewNode^{i+1} = FALSE

FailNode^{i+1} = FALSE

$GG^{i+1}(u) = \Phi$

$TRI^{i+1}(u) = \Phi$

Fig. 3. Synchronous component

Theorem 1: Graph built by algorithm DynDel is planar.

Theorem 2: Graph built by algorithm DynDel $\supseteq$ UDel(V).

4.5 Communication Cost of Algorithm

Under stationary situation, where no nodes join and leave the network, the communication cost of the algorithm DynDel is O($nlogn$), where n is the number of nodes in the network. In fact, the cost of HELLO message is O($nlogn$). For the condition $\angle vuw \geq \pi / 2$, each node broadcast at most 4 proposal messages. And two nodes reply each proposal message. So the cost of proposals and their replies is O($nlogn$). For there are only O(n) GG edges, the cost of broadcasting GG edges is O($nlogn$). Therefore the total communication cost of algorithm is O($nlogn$), where n is the number of nodes in the network.

Under dynamic situation, communication cost of the algorithm is also O($nlogn$). For the communication cost is not more than the communication under stationary situation. Under favorable condition, the communication cost is O($logn$).

5   Conclusions

Geometric routing protocols take much advantage of localized Delaunay triangulation. Therefore, in this paper we propose an algorithm to build localized Delaunay triangulation, which is a planar spanner of UDG. Unlike previous algorithms, the algorithm considers not only stationary situation but also dynamic situation, in which nodes can join and leave network at any time and any moment. The communication cost of node's entrance and departure is O($nlogn$), where n is the number of nodes in the network. Like previous algorithm, the communication cost of setup the network is O($nlogn$), where n is the number of nodes in the network. In future work, we want to improve the algorithm to accommodate to network with mobile nodes.

References

[1] J. Urrutia. Routing with Guaranteed Delivery in Geometric and Wireless Networks. In I. Stojmenovic, editor, Handbook of Wireless Networks and Mobile Computing, chapter 18, pages 393-406. John Wiley & Sons, 2002.
[2] P. Bose and P. Morin. Online Routing in Triangulations. In Proc. 10th Int. Symposium on Algorithms and Computation (ISAAC), volume 1741 of Springer LNCS, pages 113-122, 1999.
[3] P. Bose, L. Devroye, W. Evans, and D. Kirkpatrick, "On the spanning ratio of gabriel graphs and beta-skeletons", in Proceedings of the Latin American Theoretical Infocomatics (LATIN), 2002.
[4] J. Gao, L. J. Guibas, J. Hershberger, L. Zhang, and A. Zhu. "Geometric spanners for routing in mobile networks. MobiHoc 2001.
[5] X.-Y. Li, G. Calinescu, and P.-J. Wan. "Distributed construction of a planar spanner and routing for ad hoc wireless networks," IEEE Infocom 2002,New York, June 2002
[6] A. C.-C. Yao, "On constructing minimum spanning trees in k-dimensional spaces and related problems," *SIAM J. Computing*, vol. 11, pp. 721–736,1982.
[7] J. M. Keil and C. A. Gutwin, "Classes of graphs which approximate the complete euclidean graph," *Discrete Computational Geometry*, vol. 7, 1992.
[8] J.M. Keil and C.A. Gutwin, "The Delaunay triangulation closely approximates the complete euclidean graph," in *Proc. 1st Workshop Algorithms Data Structure (LNCS 382)*, 1989.

[9] D.P. Dobkin, S.J. Friedman, and K.J. Supowit, "Delaunay graphs are almost as good as complete graphs," *Discrete Computational Geometry*,1990.

[10] E. Kranakis, H. Singh, and J. Urrutia. Compass Routing on Geometric Networks. In Proc. 11th Canadian Conference on Computational Geometry, pages 51-54, 1999.

[11] Brad Karp and H.T. Kung. GPSR: Greedy perimeter stateless routing for wireless networks. In Proceedings of the 6th Annual International Conference on Mobile Computing and Networking (MOBICOM-00). page 243-254. N.Y. August 6-11 2000. ACM Press

[12] P. Bose, P. Morin, I. Stojmenovic, and J. Urrutia. Routing with guaranteed delivery in ad hoc wireless networks. ACM/Kluwer Wireless Networks, 7(6):609–616, 2001. 3rd int. Workshop on Discrete Algorithms and methods for mobile computing and communications, 1999, 48-55.

Appendix

Lemma 1. If edge uv belongs to Del($N(u)$) and Del($N(v)$), edge uv intersects edge st, $\| uv \| \le R$, $\| st \| \le R$, $\| ut \| \le R$, $\| vt \| \le R$, $\| su \| > R$ and $\| sv \| > R$, then there is a GG edge intersects edge uv.

Proof is omitted for space limited.

Theorem 1: Graph built by algorithm DynDel is planar

Proof: Suppose there exist two edges in DynDel intersected each other.

For edges in DynDel do not intersect GG edges, neither of two edges are GG edges.

Assume these two edges are edge uv and edge st. As quadrangle is convex, there is at least one obtuse angle. Assume it is $\angle utv$. $\| uv \| \le R$ implies $\| ut \| < R$ and $\| vt \| < R$.

In addition, $\| us \| > R$ and $\| vs \| > R$. Otherwise, assume $\| us \| \le R$, so u,v,s,t are all in $N(u)$ and $N(s)$. So there exist at most one edge of uv and st in Del($N(u)$) $\cup$ Del($N(s)$). Therefore $\| us \| > R$ and $\| vs \| > R$. So lemma 1 implies there exists GG edge intersects uv, which contradicts the property of DynDel. Therefore, DynDel is planar.   □

Theorem 2: Graph built by algorithm DynDel, DynDel(V) $\supseteq$ UDel(V)

Proof: Graph built by algorithm DynDel satisfies the following property

$uv \in DynDel(V)$ iff $uv \in Del(N(u))$, $uv \in Del(N(v))$, uv does not intersect with any GG edges.

For each edge $uv \in UDel(V)$,it is obvious that $uv \in Del(N(u))$, $uv \in Del(N(v))$. For $GG(V)$ is sub-graph of $UDel(V)$ and $UDel(V)$ is planar, edge uv does not intersect with any GG edges.

Therefore, $DynDel(V) \supseteq UDel(V)$ □

Energy Efficient Multipath Routing in Large Scale Sensor Networks with Multiple Sink Nodes

Yuequan Chen, Edward Chan, and Song Han

Department of Computer Science, City University of Hong Kong
csedchan@cityu.edu.hk

Abstract. Due to the battery resource constraint, it is a critical issue to save energy in wireless sensor networks, particularly in large sensor networks. One possible solution is to deploy multiple sink nodes simultaneously. In this paper, we propose a protocol called MRMS (Multipath Routing in large scale sensor networks with Multiple Sink nodes) which incorporates multiple sink nodes, a new path cost metric for improving path selection, dynamic cluster maintenance and path switching to improve energy efficiency. MRMS is shown to increase the lifetime of sensor nodes substantially compared to other algorithms based on a series of simulation experiments.

1 Introduction

Recent advance in micro-electromechanical system technology has made it possible to develop low-power and low-cost sensors with at a much reduced cost, so that large wireless sensor networks with thousands of tiny sensors are well within the realm of reality. These large sensor networks are able to support many new applications, including habitat monitoring and agricultural monitoring . In such wireless sensor networks (WSN), sensors send data packets to sink nodes through multi-hop wireless communication. As the size of the network increases, the sensors near the sink nodes will dissipate energy faster than other sensors as they need to forward a larger number of messages, and prolonging the lifetime of whole network becomes a critical problem. One promising approach is to deploy multiple sink nodes in WSN, since it can decrease the energy consumption of sensors and improve the scalability of the networks.

In this paper, we propose a protocol called MRMS, which stands for "Multipath Routing in wireless sensor networks with Multiple Sink nodes". MRMS includes three parts: topology discovery, cluster maintenance and path switching. The topology is constructed based on the TopDisc algorithm [1] using our own path cost metric (which is described in a later section). Next we rotate the cluster head within a cluster and change delivery node between clusters to balance energy consumption in the cluster maintenance process. Finally, when some of the sensors in the original primary path have dissipated too much energy, we re-select the primary path to connect to an alternate sink node. Simulation shows that MRMS can improve energy efficiency significantly.

J. Cao, W. Nejdl, and M. Xu (Eds.): APPT 2005, LNCS 3756, pp. 390–399, 2005.
© Springer-Verlag Berlin Heidelberg 2005

The main contributions in our paper are as follows: First, we introduce a new path cost metric which is based on the distance between two neighbor nodes, hop count to sink node and the residual energy of sensor node. This metric is very useful in path selection and improve energy efficiency. Secondly our scheme uses stateless clusters in which all the ordinary sensors in the cluster maintain only the previous hop and corresponding sink. This means the cluster head does not need to maintain information on its children in its cluster, which simplifies cluster maintenance considerably. Finally, we introduce mechanisms for path switching when the energy of the sensors in original primary path has dropped below a certain level. This allows us to distribute energy consumption more evenly among the sensor nodes in the network. By combining these techniques, we are able to construct a strategy that outperforms existing algorithms, as shown in the extensive simulation experiments that we have carried out.

The rest of the paper is organized as follows. A summary of related work is presented in section 2. Section 3 describes the design of the MRMS protocol in detail. The performance of MRMS is examined in Section 4 and compared with other protocols using simulation. The paper concludes with Section 5 where some possible improvements to MRMS are pointed out.

2   Related Work

WSN is an area of much research recently. Since routing is a major issue, a large number of routing protocols such as Direct Diffusion [2] and LEACH [3] have been proposed by researchers [4]. In some of these protocols, cluster-based routing is used to decrease energy consumption, such as in TopDisc [1] and LEACH. However, only a few of these protocols deal explicitly with the multiple sink nodes problem, which is the key focus of our paper. A number of recent papers dealt with the optimal placement of sink nodes in multiple sink sensor networks [5] but do not deal directly with routing issues.

One of the earliest cluster-based routing algorithms is the TopDisc algorithm [1], which is based on the three-color or four color algorithm. TopDisc finds a set of distinguished nodes, using neighborhood information to construct approximate topology of the network. These distinguished nodes logically organize the network in the form of clusters comprised of nodes in their neighborhood. However, TopDisc neither considers the residual energy of sensor networks nor the possibility of path switching.

Dubois-Ferries and Estrin proposed an algorithm based on Voronoi clusters to handle multiple sink nodes [6]. This Voronoi algorithm designates a sink for each cluster to perform data acquisition from sensors in cluster. Each node keeps a record of its closet sink and of the network distance to that sink. A node also re-forwards the message if the distance traversed is equal to closest distance and the message came from the closet sink. A drawback of this algorithm is that it does not consider residual energy of sensor node.

A. Das [7] provides an analytical model of multiple sink nodes. However, it also does not also consider path switching or how to handle excessive energy dissipation among the sensors on the original paths. The Two-Tier Data Dissemination (TTDD) scheme [8] provides data delivery to multiple mobile base stations. However, this scheme requires precise position of the sensor nodes, which may be difficult to attain in many cases. It is also designed primarily for mobile sinks, and is not as efficient when the sink nodes are stationary.

3   The MRMS Algorithm

3.1   System Model

In this section we will present the system model used in our work. First we state our major assumptions. We assume there are multiple sink nodes in the wireless sensor networks, each of which has an infinite amount of energy. Every sensor, whose location is randomly distributed, has the same initial energy and radio range. Both the sensor nodes and the sink nodes are stationary. A perfect MAC layer and error-free communication links are assumed because MRMS focuses on routing algorithm, but no communication is possible once the energy of a sensor node has been depleted. A transceiver exhibits first order radio model characteristics in free space i.e. energy spent in transmitting a bit over a distance d is proportional to d^2.

A wireless sensor networks (WSN) is modeled as a graph G(V, E). where V is the set of all sensor nodes and all sink nodes, E is the set of all links.

$$V = V_{sink} \bigcup V_{sensor}, E = \{(i,j)|i,j \in V\} \tag{1}$$

Every sensor's initial energy is E_{init} and its residual energy is E_{RE}. The path is defined as $\{V_1, V_2, \cdots V_i, V_j, \cdots, S\}$, $V_i, V_j \in V_{sensor}$, $S \in V_{sink}$; the cost is defined as the cost of one link $\langle V_i, V_j \rangle$.

$$Cost_{ij} = \alpha \times d^2 + \beta \tag{2}$$

Now we define the path cost as follows:

$$path\_cost = \sum cost_{ij} \times E_{RE}^{\gamma} \tag{3}$$

where α is the energy/bit consumed by the transmitter electronics, β is energy/bit consumed by the transmitting and receiving signal operation overhead for amplifying, and d is the distance between two sensor nodes,which is based on modern RF technology. γ is the coefficient of residual energy and it is a none-zero negative value. From formula (1) and (2), it is clear that the longer the transmitting distance or the larger the overhead, the higher the cost. So the increase in the hop count between the sensor nodes and sink node will increase path cost. In addition, if the residual energy for each sensor decreases, the path cost will also increase. Hence, after a path has been used excessively, the residual energy of the sensors in the path will decrease, driving up the path cost and triggering the path-switching process. The role of the path cost metric in energy-efficient routing will be shown in greater detail in a later section.

3.2 Details of the MRMS Algorithm

There are three phases in MRMS: topology discovery, cluster maintenance and path switching.

MRMS Topology Discovery Algorithm. MRMS topology discovery is partially based on the three color algorithm used in TopDisc [1], which is derived from the simple greedy log (n)-approximation algorithm for finding the set cover. At the end of the TopDisc topology discovery process, the sensor network is divided into n clusters and each cluster is represented by one node, which is called the cluster head. The cluster head is able to reach all the sensor nodes in the cluster directly because they are all within its communication range. Each cluster head knows its sink, but they can not communication with each other directly. Instead, a delivery node (the grey node) acts as an intermediary which delivers messages between each pair of head node.

In the MRMS topology discovery mechanism, unlike TopDisc, the cluster is stateless because the cluster head will not maintain any children. Instead every sensor will note its previous hop and corresponding sink in its routing table at the initial topology discovery's broadcast phase, and topology discovery only occurs at the initial phase; this approach reduces the complexity of cluster reconstruction described in the next section. Thus each cluster can be considered a virtual node as far as the topology is concerned. A sensor node may keep information for more than one cluster heads and sinks in the routing table, as it can keep track of different paths from different sink nodes. However only one of these paths can be designated the primary path in the table, and this is the path with the minimum path cost, hence ensuring the topology will be an energy efficient one.

MRMS Cluster Maintenance. As explained in a previous section, MRMS cluster maintenance includes two parts: energy monitoring and cluster reconstruction. Energy monitoring in MRMS is relatively straight-forward. A cluster header will check its energy periodically. If the sensor's residual energy is below some threshold, it will invoke the cluster reconstruction process. A special case is that of a delivery node, in which case it needs to inform its child cluster head to change its delivery node as well.

In cluster reconstruction, when the residual energy of the cluster head (CH) is below some threshold, it will broadcast the SELECT_NEW_CH message to its neighbors,when CH can't communicate with other children directly, most of sensors' energy in this cluster has be below some low threshold based on our method, as shown in Figure 1(a). Any sensor that receives this message will checks its routing table and reports its residual energy to the CH if the previous hop in its primary path is the current CH. After combining all incoming information, the current CH will select the child with the maximum residual energy as the new CH and pass control to the new CH, if several nodes have the same maximum residual energy, old CH will select the new CH randomly,as shown in Figure 1(b). The new CH will probe new delivery node based on the path cost, and broadcast the NEW_CH to all its children in its cluster, as shown in

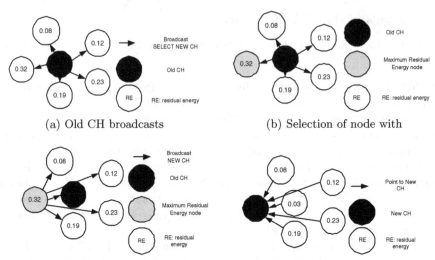

(a) Old CH broadcasts

(b) Selection of node with

(c) Node with maximum residual energy broadcasts NEW_CH Message to
all nodes in the cluster

(d) New Cluster after intra-cluster
re-construction

Fig. 1. Intra-Cluster Reconstruction

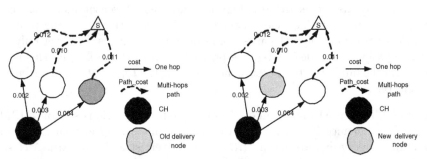

(a) Old delivery node before inter-
cluster reconstruction

(b) New delivery node after inter-
cluster reconstruction

Fig. 2. Changing Delivery Node in Cluster Reconstruction

Figure 1(c), and the result is shown in Figure 1(d). If a sensor node is the delivery node of some cluster head and its residual energy is below some threshold, it will notify these cluster heads to change their delivery nodes, as shown in Figure 2.

MRMS Path Switching. As discussed in a previous section, when the path to the original sink has been used for an extended period of time, there is a need to switch to another sink. There is a problem in determining the suitable sink, however, is that the value of the path cost is not always current. Although it is possible to have use a periodical update approach to refresh the path cost, this technique is expensive and quite unnecessary since switching to a different sink does not occur frequently for stationary sensors. In MRMS, we use an

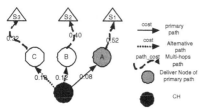
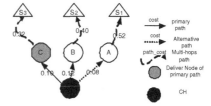

(a) Primary path before path switching

(b) Primary path after path switching

Fig. 3. Path switching in different sink node

event-based approach where path switching is triggered when during the cluster reconstruction process it is discovered that the current path is no longer the best path.

We will now describe the details of path switching in MRMS, using Figure 3 to illustrate the process. In the original path, because the sensor nodes which are near the sink node consume energy more rapidly than the sensor nodes which are far away from the sink node, these nodes close to the sink will invoke cluster-reconstruction. For example, in Figure 3, sensor A will probably see many of its upstream nodes invoke cluster reconstruction first. Since the path cost in the primary path will be updated whenever cluster reconstruction occurs, there is no need for sensor A to do anything explicitly to refresh the path cost in the primary path ($\{CH \to A \to S1\}$ in Figure 3(a)).

However, if in due time the cluster containing A undergoes cluster reconstruction, then there is a possibility that a new primary path will be chosen. The first task is to determine whether there is a need to refresh the path cost to the alternate path, which has not been updated for quite a while. The approach we have adopted is for the source CH to send a probe message to confirm another sink only if the path cost in the original primary path exceeds the path cost in the alternate path by a certain threshold η. The value of this threshold is dependent on the hop count of the CH to the sink, since the larger the hop count, the further away it is from the sink and the larger the interval between cluster reconstruction – and hence the more outdated the path cost of the alternate path is likely to be Figure 3(b) shows the CH sending a probing message to sink S_3 after the above condition has been met. After S3 has received the probing message, it will compare current path cost $(path\_cost_{CH \to S_3})$ to the original one $(path\_cost_{CH \to S_1})$. If the current path cost is larger than some threshold, the sink node will send fresh message to all sensor nodes in this path, and if the new calculated path cost in the new sink is less than the path cost of the original primary path, which is the case in Figure 3(a), then source CH will switch to the new sink node, otherwise S_3 will simply return its later path cost to the CH. Either way, the path cost will be broadcast by the CH to all its children in the cluster, and each child sensor node will update its routing table entries accordingly. In case the CH does not receive a reply from S_3, then topology discovery will be invoked again.

4   Performance Evaluation and Simulation Result

To evaluate the performance of the MRMS Algorithm, we have implemented it
in GloMoSim [9] which is based on Parsec [10]. In our simulation, we assume
the energy model is based on first order radio model in free space, that is, the
energy dissipation is:

$$e^T(d) = (\alpha \times d^2 + \beta) \times r$$

where α, β are real numbers. α is the energy consumed at the output transmitter
antenna for transmitting one meter, β is the overhead energy representing the
sum of the receiver, sensing and computation energy which is independent of the
distance d; r is the number of bits transmitted.

In our simulation, the sensor network consists of 250 nodes which are distributed randomly over a planar square region of 100m by100m. There are up
to 3 sink nodes, with positions (33.33, 33.33), (66.67, 33.33), (50.00, 66.67). The
initial battery capacity of the sensors is set as 0.5J, α is set to 0.1 $nJ/bit/m^2$ and
β set to 50 nJ/bit [3]. There are 10 stimulus which generate data flow randomly
in simulation, and the position of stimulus is random. And the sensor whose
distance to stimulus in 10m can receive the data flow. The data flow is based
on Poisson distribution model with an arrival rate of 0.5 packet per second. The
data packet size is 2000 bits and all control packets size is assumed to be 100 bits.
We set up the simulation time to range from 10 to 150 minutes for evaluating
the performance of the various protocols.

4.1   Performance Criteria

The main objective of our simulation is to evaluate the energy efficiency and
the lifetime of sensor networks. However, researchers have proposed different
definitions of lifetime. In our experiments, we use the following metrics.

- Time to first node failures: This metric indicate the duration for which the
  sensor network is fully functional i.e. no sensor failure due to battery outage.
- Number of dead nodes: We measure the number of dead nodes as time goes
  on; this metric provides an indication of the expected lifetime and the reliability of sensor networks.
- Mean Energy Consumption of one packet: the metric indicates the energy
  consumption of transmitting a packet to sink successfully.
- Average hop count to sink: This metric is useful since the larger the number
  of hops a packet has to traverse before it reaches the sink, the higher the
  aggregate energy consumption.
- Packet delivery ratio: this metric is defined as a ratio of the number of
  received packets at the sink to the number of packets transmitted by the
  source sensors. The higher the delivery ratio, the higher the reliability of the
  network. Uneven energy consumption in the network will lead to premature
  failure of sensors and reduced reliability, hence the packet delivery ratio is
  also a good indirect measure of the lifetime of the network.

4.2   Experiments and Result Analysis

In this section we discuss the performance of MRMS with the Voronoi Algorithm [6], TopDisc Algorithm and Direct Flooding algorithm. The Direct Flooding algorithm is a simplistic algorithm used as a base case; in its topology discovery phase the sink node simply floods its information to its neighbor sensor nodes without any optimization. After receiving the topology discovery request, the sensor nodes broadcast it again directly without any attempt to optimize the process. Similarly there is no optimization for sending packets to sink nodes. A single sink is used for Direct Flooding.

From Figure 4, we see that MRMS outperforms other protocols significantly, with MRMS close to doubling or tripling the time to first sensor node failure in some cases. In Direct Flooding, the first node dies quicker than the other protocols, because all packets are sent to only one sink and there is no cluster reconstruction and path switching. The TopDisc Algorithm uses clustering to decrease energy consumption which can improve the lifetime of sensor nodes and the Voronoi Algorithm uses the multiple sink nodes which improve the load-balance of data which is sent to sink nodes. However, MRMS by combining multiple sink nodes, cluster reconstruction and path switching, can best balance sensor energy consumption and prolong the duration for sensor network which is fully functional.

From Figure 5, it can be seen once again that MRMS decreases the number of dead nodes significantly, indicating that MRMS is indeed more energy-efficient than the other algorithms. The same conclusion can be reached by looking at Figure 7, which displays the average hop count to sink node for the various algorithms. The effect of using multiple sink nodes is seen clearly in this experiment, as both MRMS and Voronoi Algorithm decrease the hop counts by 1.5-2 hops compared to the Direct Flooding and the TopDisc algorithm. This result is quite obvious since multiple sink nodes will decrease the average distance from the sensor nodes to sink node and hence the hop count will drop accordingly.

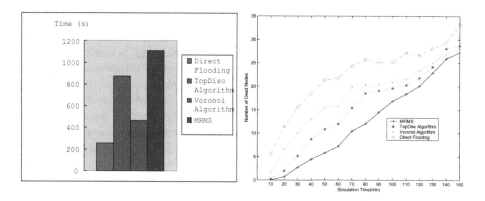

Fig. 4. Time to First Node Failure   **Fig. 5.** Number of Dead Nodes

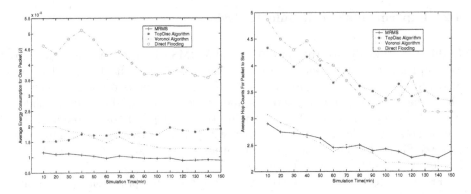

Fig. 6. Average Energy Consumption for packet

Fig. 7. Average Hop Count vs Time

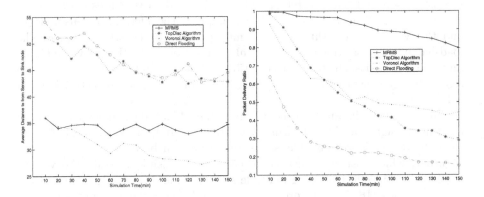

Fig. 8. Average Distance to Sink vs Time

Fig. 9. Packet Delivery Ratio

From Figure 6, it can be seen that MRMS decreases the energy consumption considerably compared with the Voronoi algorithm, TopDisc algorithm and Direct Flooding. As simulation time increases, the average energy consumption for one packet in MRMS and the other algorithms remain relatively stable. There are actually several factors at work. With path switching and cluster reconstruction, the average hop count decreases (as seen in Figure 7). However, the actual distance from (which greatly affects the energy consumption of the packet, as seen in Figure 8) may stay relatively the same because with some of the original best paths no longer available, more and more of the outlying sensors becomes unreachable meaning that the remaining sensors tend to be closer to the sink nodes. Figure 9, shows that MRMS outperforms significantly the other three algorithms significantly based on the packet delivery ratio, indicating the MRMS is indeed more energy efficient and reliable, since most of the packets are actually able to reach the final destination, unlike the other algorithms.

5   Conclusion and Future Work

In this chapter, we proposed the MRMS algorithm which includes topology discovery, cluster maintenance and path switching. Since MRMS uses multiple sink nodes, cluster maintenance and path switching which can distribute the energy consumption in sensor networks more evenly, it enjoys significant improvement in key metrics compared to other approaches. We plan on exploring the effect of a lossy MAC layer on the MRMS, as well as how to construct node-disjoint multipaths for multiple sink nodes.

References

1. B.Deb, S.Bhatangar, and B.Nath. A topology discovery algorithm for sensor networks with applications to network management. Proc. IEEE CAS Workshop on wireless communication and networking, Passadena, USA, 2002.
2. C. Intanagonwiwat, R. Govindan, and D. Estrin. Directed Diffusion: a Scalable and Robust Communication Paradigm for Sensor Networks. Proc. ACM Mobicom, boston, MA, 2000.
3. W. Heizelman, A. Chandrakasan and H. Balakrishnan. Energy-efficient communication Protocol for wireless microsensor networks. Proc. 33rd International Conference on System Sciences, 2000.
4. J.N. Al-Karaki and A.E. Kamal. Routing Techniques in wireless sensor networks: A survey. IEEE Wireless Communication. **11**(2004)
5. E.I. Oyman and C. Ersoy. Multiple Sink Network Design Problem in Large Scale Wireless Sensor Networks. Proc. International Conference on Communication, 2004
6. H. Duboris-Ferries and D. Estrin. Efficient and Practical Query Scoping in Sensor networks. Tech Rep.2004-39, CENS/UCLA Tech Report, 2004.
7. A. Das and D. Dutta. Data Acquisition in Multiple Sink Sensor Networks. Proc. 2nd International Conference on Embedded Sensor Systems, 2004
8. F. Ye, H. Luo, J. Cheng, S. Lu and L. Zhang. A two-tier Data Dissemination Model for large Scale Wireless Sensor networks. Proc. ACM Mobicom, 2002
9. GloMoSim: A Scalable Simulation Environment for Wireless and Wired Network Systems. UCLA Parallel Computing Laboratory and Wireless Adaptive Mobility Laboratory.
10. R. Bagrodia, R. Meyer, M. Takai, Y. Chen, X. Zeng, J. Martin and H.Y. Song. PARSEC: A Parallel Simulation Environment for Complex Systems. IEEE Computer. **10**(1998)

FLC: A Novel Dynamic Buffer Tuner for Shortening Service Roundtrip Time over the Internet by Eliminating User-Level Buffer Overflow on the Fly

Wilfred W.K. Lin[1], Allan K.Y. Wong[1], and Tharam S. Dillon[2]

[1] Department of Computing, The Hong Kong Polytechnic University,
Hung Hom, Kowloon, Hong Kong S.A.R. P.R.C.
{cswklin, csalwong}@comp.polyu.edu.hk
[2] Faculty of Information Technology, University,
University of Technology Sydney, Broadway, Sydney, Australia
tharam@it.uts.edu.au

Abstract. The proposed Fuzzy Logic Controller (FLC) is for dynamic buffer tuning at the user/server level. It eliminates buffer overflow on-line by ensuring that the buffer length always cover the queue length adaptively. The FLC contributes to prevent the AQM (active queue management) resources dished out at the system level from being wasted and to shorten the service roundtrip time (RTT) by reducing retransmission caused by user-level buffer overflow at the reception side. Combining fuzzy logic and the conventional PIDC model creates the FLC that operates with the $\{0,\Delta\}^2$ objective function. The fuzzy logic maintains the given safety margin about the reference point, which is symbolically represented by "0" in $\{0,\Delta\}^2$. The short execution time of the FLC makes it suitable for supporting time-critical applications over the Internet.

1 Introduction

Applications running on the Internet are basically object-based, and the constituent objects collaborate in an end-to-end client/server relationship. This relationship is an asymmetric rendezvous because the server can serve many clients at the same time. Figure 1 shows the two levels of the end-to-end path (i.e. EE-path) in a client/server interaction. The system/router level involves all the activities inside the TCP (Transmission Control Protocol) channel, and the user-level includes only the client and the server that make use of the channel for communication. In general, it is difficult to harness the roundtrip time (RTT) of an EE-path in a time-critical application because the sheer size and heterogeneity of the Internet. Practically it is impossible to monitor every one of the overwhelming number of network parameters in order to control the RTT. If the EE-path error probability for retransmissions is ρ, the average number of trials (ANT) for a successful transmission is $\sum_{j=1}^{\infty} j[\rho^{j-1}(1-\rho)] \approx \lim_{j\to\infty} \frac{1}{(1-\rho)}$. The ρ value encapsulates all the possible faults and errors along the EE-path, and one of them is the chance of user-level buffer overflow at the receiver/server end.

J. Cao, W. Nejdl, and M. Xu (Eds.): APPT 2005, LNCS 3756, pp. 400–408, 2005.
© Springer-Verlag Berlin Heidelberg 2005

Those methods devised to prevent network congestion at the system level are known as active queue management (AQM) [1] approaches. The only known method from the literature that can eliminate user-level buffer overflow is dynamic buffer size tuning. One of the user-level dynamic buffer size tuners is the GAC (Genetic Algorithm Controller [2]). This controller is basically the "genetic algorithm (GA) + PIDC + $\{0,\Delta\}^2$ objective function" combination. The GA moderates the PIDC control process for more dynamic buffer size tuning precision. The GAC approach, however, produces occasional buffer overflow because the GA does not guarantee the global-optimal solution of the solution hyperplane [3]. Nevertheless it has demonstrated that soft computing is potentially a powerful technique for achieving dynamic buffer size tuning. The GAC has eliminated all the PIDC shortcomings but also produces undesirable occasional buffer overflow. The desire to keep the GAC merits and eliminate buffer overflow at the same time motivates the FLC proposal.

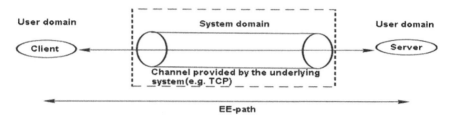

Fig. 1. The EE path

2 Related Work

In the practical sense, it is illogical for a user-level receiver to discard new requests in order to prevent local buffer overflow. Such an act not only increases retransmission and network congestion but also wastes the AQM effort already dished out by the system. It is sensible therefore to install a user-level dynamic buffer tuner/controller that auto-tunes the size of the server's queue buffer so that it always covers the queue length [4]. User-level buffer tuning and system-level AQM together provide a unified solution for reducing the chance of buffer overflow along the whole EE-path.

The "P+D" (Proportional + Derivative) is one of the earliest models for user-level dynamic buffer tuning. The P control element is the current, statistically computed "queue length over buffer length (QOB)" ratio, which is also known as the   . The D control element is the current rate of changes in the queue length (Q)). The "P+D" model worked well in simulations but failed frequently in real-life applications. The cause of failure is the unrealistic expectation of using a static set of control parameters to cover the whole spectrum of channel and buffer dynamics. The quest for a better user-level overflow controller led to the proposal of the conventional/algorithmic PID controller (PIDC), which incorporates integral (I) control to enhance the anticipative power of the "P+D". The PIDC effectively eliminates user-level overflow [5], even though it has two distinctive shortcomings: a) it locks unused buffer memory and this could lead to poor system performance, b) it does not have a safety margin to prevent possible overflow under serious system perturbations. The desire to eliminate these shortcomings and preserve the PIDC merits at the same time motivated the GAC research. Although the GAC has eliminated all the PIDC shortcomings, it produces occasional buffer overflow. The desire to keep the control merits of the GAC without buffer overflow has motivated the FLC proposal.

If {(dQ/dt > prescribed_positive_threshold) OR [(dQ/dt is_ positive) AND (QOB_i > prescribed_positive_threshold)]} then Lnow = Lnow +ICM; Lnow ≥ Lminimum

Else If {(dQ/dt < prescribed_negative_threshold) OR [(dQ/dt is_ negative) AND (QOB_i < prescribed_negative_threshold)]} then Lnow = Lnow-- ICM; Lnow ≥ Lminimum

Fig. 2. The basic PID controller (PIDC) algorithm

3 The Fuzzy Logic Controller (FLC) Framework

The FLC framework is basically the combination: "PIDC + fuzzy logic + $\{0,\Delta\}^2$ objective function". The fuzzy logic augments the algorithmic PIDC by keeping the latter's operation within the given safety margin   about the chosen reference symbolically represented by "0" in $\{0,\Delta\}^2$. The PIDC, which is abstracted by the pseudocode in Figure 2, basically works with the proportional (P), integral (I) and derivative (D) control elements and two range thresholds, namely, Th1 (i.e. the Δ value of $\{0,\Delta\}^2$ for the P or QOB_R control) and Th2 (i.e. the Δ value for the D or dQ/dt control). The PIDC working alone therefore has only four control regions, defined by different $\pm Th1$ and $\pm Th2$ combinations. The fuzzy logic in the FLC model divides these thresholds into finer membership functions, with range-thresholds among them for more refined control actions. For example, in Figure 3, which is the FLC[6x6] design, Th1 covers the range-thresholds for finer QOB control regions: ML, SL, L, G, SG and MG regions, and similarly Th2 for that of the NL, NM, NS, PS, PM and PL regions. When the FLC tuner is operating in a specific fuzzy control region there will be an intrinsic time delay before a corrective action. For example, if the increased value is less than the range-threshold of the current "don't care" region, no immediate action is taken. By the time the action is triggered there could be significant overshoot or undershoot already. The overshoot/undershoot accumulation contributes to the oscillations in the FLC convergence process in the steady state. In the experimental FLC[6x6] design matrix shown in Figure 3, the "dot" defines the chosen   reference and X the "don't care" state. In this case, the   reference ratio is equal to 0.8 for the objective function. The linguistic variables for the FLC design are as follows:

a) For QOB: ML for Much Less than QOBR, SL for Slightly Less than QOBR, L for Less than QOBR, G for Greater than QOBR, SG for Slightly Greater than QOBR, and MG for Much Greater than QOBR. The QOB membership function is shown in Figure 4.

b) For the current : NL for Negative and Larger than the threshold, NM for Negative and Medium than the threshold, NS for Negative and Smaller than the threshold, PS for Positive and Smaller than the threshold, PM for Positive and Medium than the threshold and PL for Positive and Larger than the threshold. The QOB membership function is shown in Figure 5.

The control decisions, which depend on the current QOB and dQ/dt values, include: Addition (buffer elongation) or "+", Subtraction (buffer shrinkage) or "- ", and don't care. For example, the FLC prototype based on Figure 4 and 5 has the following fuzzy rules to adjust the I control or ICM (integral control mechanism) proactively (Lnew and Lold denote the adjusted buffer length and the old buffer length respectively):

Rule 1: If (QOB is MG) AND (dQ/dt is PL) Then Action is "+"(Addition) AND Lnew = Lold + ICM
Rule 2: If (QOB is ML) AND (dQ/dt is NL) Then Action is "-"(Subtraction) AND Lnew = Lold - ICM
Rule 3: If (QOB is G) AND (dQ/dt is PS) Then Action is "X"(Don't care) AND Lnew = Lold

		dQ/dt					
		NL	NM	NS	PS	PM	PL
	ML	-	-	-	-	-	-
	SL	-	-	-	-	-	-
QOB	L	-	-	X	X	+	+
	G	-	-	X	X	+	+
	SG	+	+	+	+	+	+
	MG	+	+	+	+	+	+

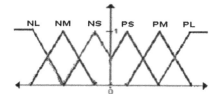

Fig. 3. A FLC example

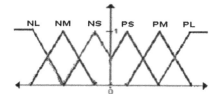
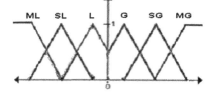

Fig. 4. Membership function of dQ/dt   **Fig. 5.** Membership function of QOB

4 Experimental Results

The FLC prototypes of different design matrices were verified by simulations on the Aglets mobile agent platform [6], which is chosen because: a) it is stable, b) it is rich in user experience, and c) it makes the experimental results scalable for the open Internet. The set up for the experiments is shown in Figure 6, in which the driver and the server are aglets (*agile applets*) that collaborate in as client/server relationship within a single computer. The driver picks a known waveform (e.g. Poisson) or a trace that embeds an unknown waveform from the table. It uses the pick to generate the inter-arrival times (IAT) for the simulated merged traffic into the server buffer. Self-similar traffic waveforms [7] in the table are generated by using the tool proposed by Glen Kramer [8]. A distribution F is heavy-tailed if and only if (1-F) is varying regularly with index α, for $\lim_{t\to\infty} \dfrac{(1-F(tx))}{(1-F(t))} = x^{-\alpha}, x > 0$. [9]. The

Selfis tool [10] is used in the simulations to identify the specific traffic pattern so that it can be matched with the corresponding FLC performance for detailed analysis. Figure 7 shows how the Selfis tool uses the Hurst exponent/effect to differentiate SRD (short-range dependence) and LRD (long-range dependence) traffic [11]. The $0 < H < 0.5$ and $0.5 < H < 1$ ranges indicate SRD and LRD respectively.

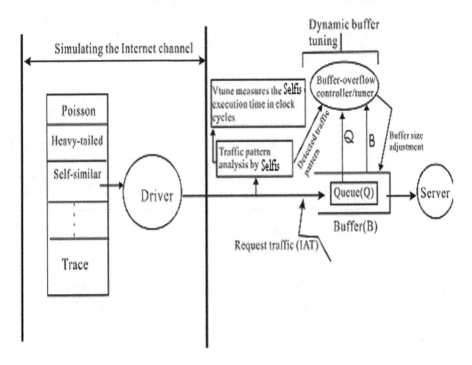

Fig. 6. Setup for the experiment

Figure 8 compares the FLC[6x6] performance with that of the PIDC working alone with the same trace. The buffer length controlled by the FLC follows the profile of the changing queue length more closely than the PIDC. The FLC trend line in Figure 9 settles quickly to the given QOB_R reference of 0.8. The trend-line for the PIDC working alone, however, lags behind and does not settle to QOB_R reference.

It was observed in the different experiments that traffic patterns did produce negative impacts on the FLC control accuracy. These negative impacts can be expressed in terms of the mean deviation (MD) values from given QOB_R reference. Figure 10 shows the calibrations of negative impacts by different traffic patterns (e.g. Poisson) on different FLC designs/configurations. The self-similar traffic consistently produces the large MD values for all the designs and Poisson the least. Figure 10 is by no means exhaustive, and the calibration table will expand if there is a necessity to include the negative impacts by new distinctive traffic patterns yet to be identified.

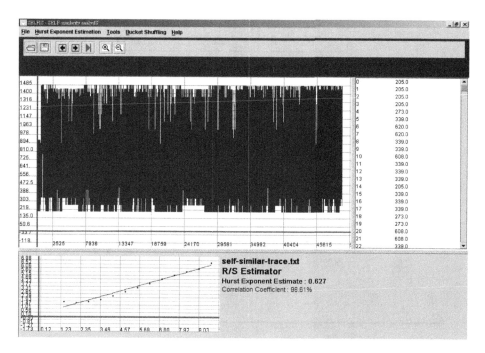

Fig. 7. Trace analysis/identification with Selfis [10]

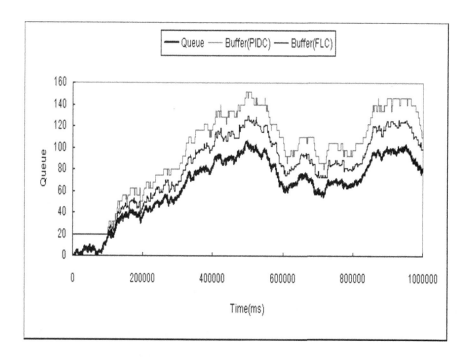

Fig. 8. The performance of the FLC and the PIDC

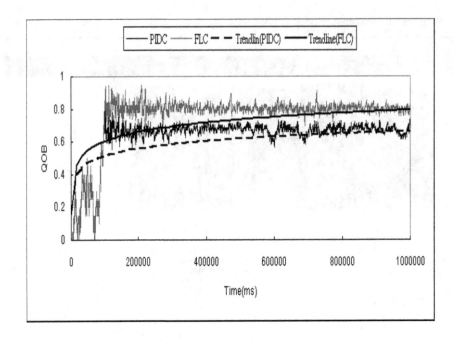

Fig. 9. Faster convergence of the FLC than the PIDC

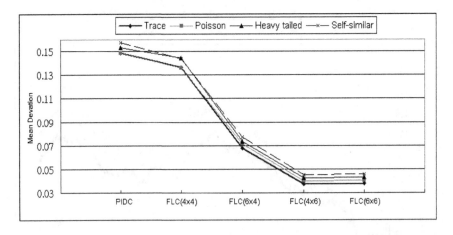

Fig. 10. Mean deviation errors of different basic FLC designs versus traffic patterns

The intrinsic execution time of any FLC design can be measured by using the *Intel's VTune Performance Analyzer* [12] in neutral clock cycles. These clock cycles can be converted for the platform of interest by the formula $PT = {CC}/{PS}$, where PT is the physical time in seconds, CC the measured number of clock cycles by

VTune, and PS is the platform speed in mega hertz (MHz). For example, the intrinsic execution time of the FLC[6x6] is around 275 clock cycles (Figure 11). It is intrinsic because when VTune measured its average execution (or control cycle) time the FLC prototype worked with traces from which IAT values were immediately usable (no actual delays). In real-life operations, however, FLC has to sample IAT values/delays one by one. Therefore the actual (versus intrinsic) FLC execution time includes the IAT delays and is therefore much longer than the average 275 clock cycles. For a platform that operates at the speed of 100 MHz, the intrinsic PT for FLC[6x6] is

$$PT = \frac{275}{100 * 10^6} = 2.75 \text{ micro seconds.}$$

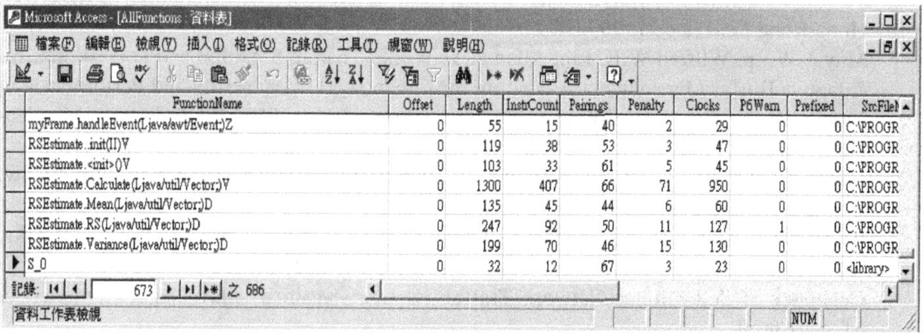

FunctionName	Offset	Length	InstrCount	Pairings	Penalty	Clocks	P6Warn	Prefixed	SrcFileN
myFrame.handleEvent(Ljava/awt/Event;)Z	0	55	15	40	2	29	0	0	C:\PROGR
RSEstimate..init(II)V	0	119	38	53	3	47	0	0	C:\PROGR
RSEstimate.<init>()V	0	103	33	61	5	45	0	0	C:\PROGR
RSEstimate.Calculate(Ljava/util/Vector;)V	0	1300	407	66	71	950	0	0	C:\PROGR
RSEstimate.Mean(Ljava/util/Vector;)D	0	135	45	44	6	60	0	0	C:\PROGR
RSEstimate.RS(Ljava/util/Vector;)D	0	247	92	50	11	127	1	0	C:\PROGR
RSEstimate.Variance(Ljava/util/Vector;)D	0	199	70	46	15	130	0	0	C:\PROGR
S_0	0	32	12	67	3	23	0	0	<library>

Fig. 11. R/S execution time by Intel's VTune Performance Analyzer

5 Conclusion

The preliminary experimental results confirm that the Fuzzy Logic Controller indeed eliminates user-level buffer overflow efficaciously. The fuzzy rules in the FLC tune the integral control (i.e. ICM) adaptively. As a result the dynamic buffer tuning process always maintains the Δ safety margin of the $\{0, \Delta\}^2$ objective function successfully. In this way the FLC effectively preserves the GAC merits minus its overflow shortcoming. The FLC performance is, however, affected negatively by different traffic patterns. This was observed in different experiments. As a result the negative impacts, in terms of mean deviations from QOB_R, by different traffic patterns were calibrated. Therefore, the next step in the research is to investigate how the FLC framework can make use of the calibrated negative impacts (in MD values as shown in Figure 10) to self-tune and mitigate/nullify these impacts.

Acknowledgement

The authors thank the Hong Kong Polytechnic University for the research grants: A-PG51, A-PF75 and HZJ91.

References

1. B. Braden et al., Recommendation on Queue Management and Congestion Avoidance in the Internet, RFC2309, April 1998
2. Allan K.Y. Wong, Wilfred W.K. Lin, May T.W. Ip and Tharam S. Dillon, Genetic Algorithm and PID Control Together for Dynamic Anticipative Marginal Buffer Management: An Effective Approach to Enhance Dependability and Performance for Distributed Mobile Object-Based Real-time Computing over the Internet, Journal of Parallel and Distributed Computing (JPDC), vol.62, Sept. 2002, 1433-1453
3. E. Mitchell, An Introduction to Genetic Algorithms, MIT Press, 1999
4. Allan K.Y. Wong and Tharam S. Dillon, A Fault-Tolerant Data Communication Setup to Improve Reliability and Performance for Internet-Based Distributed Applications, Proc. of the 1999 Pacific Rim International Symposium on Dependable Computing (PRDC'99), Hong Kong SAR, Dec.1999, 268-275
5. May T.W. Ip, Wilfred W.K. Lin, Allan K.Y. Wong, Tharam S. Dillon and D.H. Wong, An Adaptive Buffer Management Algorithm for Enhancing Dependability and Performance in Mobile-Object-Based Real-time Computing, Proc. of the IEEE ISORC'2001, Magdenburg, Germany, May 2001, 138-144
6. Mitsuru, O., Guenter, K., and Kouichi, O.(1998), IBM Aglets Specification, http://www.trl.ibm.com/aglets/spec11.htm
7. B. Tsybakov and N.D. Georganas, Self-Similar Processes in Communications Networks, IEEE Transactions on Information Theory, 44(5), September 1998, 1713-1725
8. Generator of Self-Similar Network Traffic, http://wwwcsif.cs.ucdavis.edu/~kramer/code/trf_gen1.html
9. Pareto Distribution,
10. http://www.cs.northwestern.edu/~agupta/ppts/dynamics/selfsimilarity/
11. T. Karagiannis, M. Faloutsos, M. Molle, A User-friendly Self-similarity Analysis Tool, ACM SIGCOMM Computer Communication Review, 33(3), July 2003, 81-93 (http://www.cs.ucr.edu/~tkarag/Selfis/Selfis.html)
12. S. Molnar, T.D. Dang and A. Vidacs, Heavy-Tailedness, Long-Range Dependence and Self-Similarity in Data Traffic, Proc. of the 7th Int'l Conference on Telecommunication Systems, Modelling and Analysis, Nashville, USA,18-21, 1999
13. Intel's VTune Performance Analyzer, http://ww.intel.com/support/performancetools/vtune/v5

Intelligent Congestion Avoidance
in Differentiated Service Networks

Farzad Habibipour, Ahmad Faraahi, and Mehdi Glily

Iran Telecom Research Center and Payame Noor University, Tehran, Iran
afaraahi@pnu.ac.ir

Abstract. Active Queue management (AQM) takes a trade-off between link utilization and delay experienced by data packets. From the viewpoint of control theory, it is rational to regard AQM as a typical regulation system. Although PI controller for AQM outperforms RED algorithm, the mismatches in simplified TCP flow model inevitably degrades the performance of controller designed with classic control theory. The Differentiated Service (Diff-Serv) architectures are proposed to deliver Quality of Service (QoS) in TCP/IP networks. The aim of this paper is to design an active queue management system to secure high utilization, bounded delay and loss, while the network complies with the demands each traffic class sets. To this end, predictive control strategy is used to design the congestion controller. This control strategy is suitable for plants with time delay, so the effects of round trip time delay can be reduced suing predictive control algorithm in comparison with the other exciting control algorithms. Simulation results of the proposed control action for the system with and without round trip time delay, demonstrate the effectiveness of the controller in providing queue management system.

1 Introduction

The rapid growth of the Internet and increased demand to use the Internet for voice and video applications necessitate the design and utilization of new Internet architectures with effective congestion control algorithms. As a result, the Differentiated Service (Diff-Serv) architectures were proposed to deliver Quality of Service (QoS) in TCP/IP networks. Diff-Serv architecture tries to provide QoS by using differentiated services aware congestion control algorithms. Recently several attempts have been made to develop congestion controllers [1,2], mostly using linear control theory. In this paper, the traffic of the network is divided into three types: Premium, Ordinary and Best Effort Traffic Services [3]. For very important people, there are VIPs passes. VIP passes get preferential treatment. This category is likened to our premium traffic Service. For ordinary people, there are common passes. To purchase these tickets, people may have to queue to get the best possible seats, and there is no preferential treatment, unless different prices are introduced for better seats. This category may be likened to our Ordinary Traffic Service. For reasons of economy, another pass may be offered, at a discount price for the opportunists at the door (Best Effort Traffic Service).

J. Cao, W. Nejdl, and M. Xu (Eds.): APPT 2005, LNCS 3756, pp. 409–416, 2005.
© Springer-Verlag Berlin Heidelberg 2005

In this paper, we will make use of predictive control strategy [4-7] to congestion control in differentiated services networks. Using the proposed control action, congestion control in Premium and Ordinary classes is performed. Best effort class is no-controlled. Some computer simulations are provided to illustrate the effectiveness of the proposed sliding mode controller.

2 Dynamic Network Model

In this section, a state space equation for M/M/1 queue is presented. The model has been extended to consider traffic delays and includes modeling uncertainties then three classes of traffic services are introduced in a Diff-Serv network.

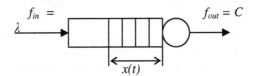

Fig. 1. Diagram of sample queue

2.1 Fluid Flow Model

A diagram of a sample queue is depicted in Fig.1. Let $x(t)$ be a state variable denoting the ensemble average number in the system in an arbitrary queuing model at time t. Furthermore, let $f_{in}(t)$ and $f_{out}(t)$ be ensemble averages of the flow entering and exiting the system, respectively. $\dot{x}(t) = dx(t)/dt$ can be written as

$$\dot{x}(t) = f_{in}(t) - f_{out}(t) \tag{1}$$

Equation of this kind of model has been used in the literature, and is commonly referred to as fluid flow equation [8,9]. To use this equation in a queuing system, C and λ have been defined as the queue server capacity and average arrival rate respectively. Assuming that the queue capacity is unlimited, $f_{in}(t)$ is just the arrival rate λ. The flow out of the system, $f_{out}(t)$, can be related to the ensemble average utilization of the queue, $\rho(t)$, by $f_{out}(t)=\rho(t)C$. It is assumed that the utilization of the link, ρ, can be approximated by the function $G(x(t))$, which represents the ensemble average utilization of the link at time t as a function of the state variable. Hence, queue model can be represented by the following nonlinear differential equation [3,8]

$$\dot{x}(t) = -CG(x(t)) + \lambda \tag{2}$$

In this model input and service rates both have Poisson distribution function. For M/M/1 the state space equation would be [9]

$$\dot{x}(t) = -C\frac{x(t)}{1+x(t)} + \lambda \tag{3}$$

The validity of this model has been verified by a number of researchers [3,8,10].

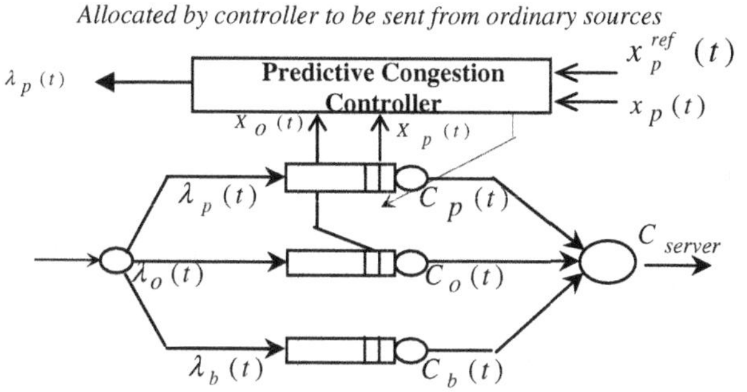

Allocated by controller to be sent from ordinary sources

Fig. 2. Control strategy at each switch output port

2.2 System Structure and Controller Mechanism

Consider a router of K input and L output ports handling three differentiated traffic classes mentioned above (Fig. 2). The incoming traffic to the input node includes different classes of traffic. The input node then separates each class according to their class identifier tags and forwards the packets to the proper queue. The output port could transmit packets at maximum rate of C_{server} to destination where

$$C_{server} < C_p + C_r + C_b \qquad (4)$$

2.3 Premium Control Strategy

Premium traffic flows needs strict guarantees of delivery. Delay, jitter and packet drops should be kept as small as possible. The queue dynamic model can be as

$$\dot{x}_p(t) = -C_p(t)\frac{x_p(t)}{1 + x_p(t)} + \lambda_p(t) \qquad (5)$$

The control goal here is to determine $C_p(t)$ at any time and for any arrival rate $\lambda_p(t)$ in which the queue length, $x_p(t)$ is kept close to a reference value, $x_p^{ref}(t)$, specified by the operator or designer. So in (5), $x_p(t)$ would be the state to be tracked, $C_p(t)$ is the control signal determined by the congestion controller and $\lambda_p(t)$ is the disturbance. Note that we are confined to control signals as

$$0 < C_p(t) < C_{server} \qquad (6)$$

2.4 Ordinary Control Strategy

In the case of ordinary traffic flows, there is no limitation on delay and we assume that the sources sending ordinary packets over the network are capable to adjust their rates to the value specified by the bottleneck controller. The queue dynamic model is as follows

$$\dot{x}_O(t) = -\frac{x_O(t)}{1 + x_O(t)} C_O(t) + \lambda_O(t - \tau) \qquad (7)$$

where, τ denotes the round-trip delay from bottleneck router to ordinary sources and back to the router. The control goal here is to determine $\lambda_o(t)$ at any time and for any allocated capacity $C_o(t)$ so that $x_o(t)$ be close to a reference value $x_o^{ref}(t)$ given by the operator or designer. There are two important points that must be considered, first, $C_o(t)$ is the remaining capacity, $C_o(t)=C_{server}-C_p(t)$ and would be considered as disturbance which could be measured from the premium queue. In our controller scheme we would try to decouple the affect of $C_o(t)$ on the state variable $x_o(t)$, and the another point is that λ_o is limited to a maximum value λ_{max} and no negative λ_o is allowed i.e.

$$0 \le \lambda_o(t) \le \lambda_{max} \le C_{max}$$

2.5 Best-Effort Traffic

As mentioned in the previous section, best effort traffic has the lowest priority and therefore could only use the left capacity not used by Premium and Ordinary traffic flows. So, this class of service is no-controlled.

3 Intelligent Predictive Congestion Controller Design

A predictive control anticipates the plant response for a sequence of control actions in future time interval, which is known as prediction horizon [4]. The control action in this prediction horizon should be determined by an optimization method to minimize the difference between set point and predicted response. Predictive control belongs to the class of model based controller design concepts. That is, a model of the process is explicitly used to design the controller, as is illustrated in Fig. 2. Usually, predictive controllers are used in discrete time. Supposed the current time is denoted by sample k, $u(k)$, $y(k)$ and $w(k)$ denote the controller output, the process output and the desired process output at sample k, respectively. More details about this strategy can be found in [4-7].

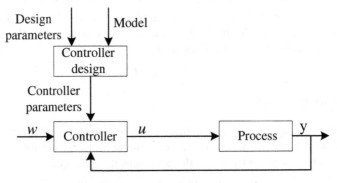

Fig. 3. Scheme of model based control

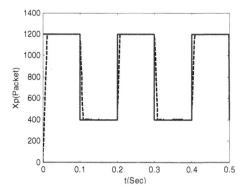

Fig. 4. $x_p^{ref}(t)$ and $x_p(t)$

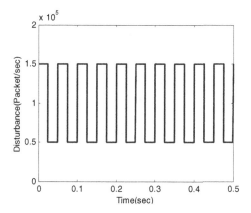

Fig. 5. Input rate of Premium's buffer

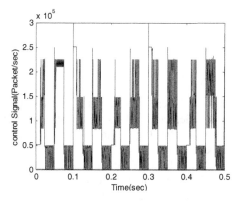

Fig. 6. Output rate of Premium's buffer

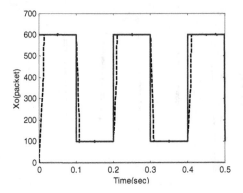

Fig. 7. $x_o^{ref}(t)$ and $x_o(t)$

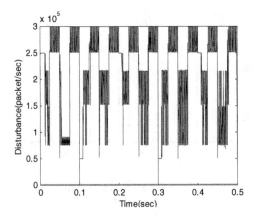

Fig. 8. Input rate of Ordinary's buffer

To design the controller, we have made the following assumptions for controller design throughout this paper

C_{max}=300000 Packets Per Second

$\lambda_{max} = 150000$ *Packets Per Second*

In addition at first is assumed there is not any delay in system (τ =0).

The simulation results are depicted in Figs. 4, 5 and 6 for premium traffic, and in Figs. 7, 8 and 9 for ordinary traffic. Figs. 4 and 7 show $x(t)$ with $x_{ref}(t)$ for Premium and Ordinary classes, respectively where good behavior for rising and settling of $x(t)$ is clear. The input and output rates of Premium buffer are shown in Figs. 5 and 6, respectively. Figs. 8 and 9 shows the input and output rates for the Ordinary buffers as well. To investigate the robustness of proposed controller, the round trip time delay and uncertainty is applied to the system as follows:

$$G(x(t)) = (1 + \frac{10}{100}) \frac{x(t)}{1 + x(t)}, \tau = 3 \ m\sec \qquad (8)$$

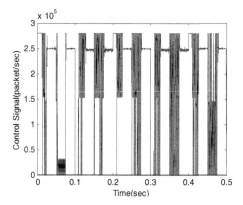

Fig. 9. Output rate of Premium's buffer

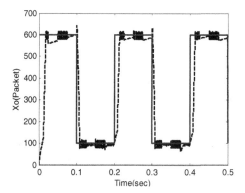

Fig. 10. $x_p^{ref}(t)$ and $x_p(t)$ with delay

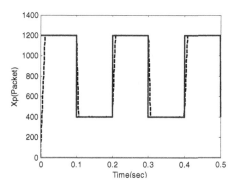

Fig. 11. $x_p^{ref}(t)$ and $x_p(t)$ with delay

Figs. 10 and 11 shows the set point tracking behavior of $x_o(t)$ and $x_p(t)$, respectively with above conditions. It is evident that the performance of $x_p(t)$ with the proposed control action does not vary much; so the above uncertainty does not effect on the closed-loop system very much. The performance of $x_o(t)$ is a little worst than the case of without delay. It means that our proposed robust controller still needs to be improved to compensate the effect of round trip time delay.

4 Conclusions

In this paper, predictive Controller was applied to congestion control in Differentiated-Services networks. A differentiated-services network framework was assumed and the control strategy was formulated for three types of services: Premium Service, Ordinary Service, and Best Effort Service. The proposed control action demonstrated robust performance against round trip time delay. Some computer simulations showed good and satisfactory performance for the proposed controller.

References

1. Kolarov and Ramamurthy G., A control theoretic approach to the design of an explicit rate controller for ABR service, *IEEE/ACM Transactions on Networking*, October 1999.
2. Pitsillides and Lambert J., Adaptive congestion control in ATM based networks: quality of service with high utilization, *Journal of Computer Comm.*, 20, 1997, pp. 1239-1258.
3. Pitsillides A.and Ioannou P., Non-linear Controllers for Congestion Control in Differentiated Services Networks, *TR-99-1, Dept. CS, University of Cyprus*, 2001.
4. Camacho, E.F. *Model predictive control*, Springer Verlag, 1998.
5. Garcia, C.E., Prett, D.M., and Morari, M. Model predictive control: theory and practice- a survey, *Automatica*, 25(3), pp.335-348, 1989.
6. Parker, R.S., Gatzke E.P., Mahadevan, R., Meadows, E.S., and Doyle, F.J. Nonlinear model predictive control: issues and applications, *In Nonlinear predictive control theory and practice, Kouvaritakis, B, Cannon, M (Eds.)*, IEE Control Series, pp.34-57, 2001.
7. Jalili-Kharaajoo, M. and Araabi, B.N. Neural network control of a heat exchanger pilot plant, *to appear in IU Journal of Electrical and Electronics Engineering*, 2004.
8. Sharma, S., D. Tipper, Approximate models for the study of nonstationary queues and their applications to communication networks, IEEE ICC 93, May 1993.
9. Tipper D., Sandareshan M. K., Numerical Methods for modeling Computer Networks Under Non-stationary Conditions, *IEEE Journal SAC*, Dec. 1990.
10. Rossides L., Pitsillides A. and Ioannou P., Non-linear Congestion control: Comparison of a fluid flow based model with OPNET simulated ATM switch model, TR-99-1, Dept. Computer Science, University of Cyprus, 1999.

Rule-Based Anomaly Detection of Inter-domain Routing System

Peidong Zhu, Xin Liu, Mingjun Yang, and Ming Xu

School of Computer, National University of Defense Technology,
Changsha 410073, China
National Laboratory for Modern Communications,
Chengdu 610041, China
pdzhu@nudt.edu.cn,
meteor5460@21cn.com,
ymjnudt@163.com,
xuming64@public.cs.hn.cn

Abstract. Inter-domain routing (IDR) system is a critical part of the Internet infrastructure. However, anomalies exist in BGP routing behaviors because of BGP misconfigurations, router malfunctions or deliberate attacking. To help secure the IDR system, this paper presents a rule-based framework and a rich set of detection rules to identify the abnormal routing behaviors. The detection rules are categorized into General Anomaly-detection Rules (GADRs) and Special Anomaly-detection Rules (SADRs), and they work together with the Basic Models and the Generated Models of the Internet respectively. Under the proposed framework, a prototype system, ISP-Health, is implemented, which can find out various abnormal routes and complex hidden routing attacks.

1 Introduction

The Internet is composed of thousands of independent networks, which are glued together using BGP (Border Gateway Protocol). BGP routing system is a critical part of communications infrastructure, and the correctness and stability of its operation are vital for the valid functioning of the Internet [1]. However, BGP is vulnerable to a variety of attacks [2, 3, 4], and a single misconfiguration or a malicious BGP speaker could result in large-scale service disruption [5, 6].

It is very difficult to find out the security problems of BGP. To diagnose BGP problems, a network operator is confronted with many challenges. Firstly, he needs high experience from many years' network operation; secondly, he cannot resolve problems that originate beyond the network's administrative boundary, and the situation gets even worse if a problem originates further beyond the neighbor networks; thirdly, current practices are mainly based on SNMP-based monitoring of routers, or writing scripts to process the output of "show ip bgp" command. In a word, it is an ineffective and laborious work without the aid of proper tools.

Because of BGP misconfigurations, router failures or deliberate attacking, many anomalies exist in BGP routing behaviors. To identify the routing anomalies and possible attacks, we present a rule-based framework and define a rich set of detection rules. The detecting engine uses these rules together with the Basic Models of the

J. Cao, W. Nejdl, and M. Xu (Eds.): APPT 2005, LNCS 3756, pp. 417–426, 2005.
© Springer-Verlag Berlin Heidelberg 2005

Internet and the Generated Models to find out numerous routing problems and security threats.

The remainder of the paper is organized as follows. Section 2 reviews the related work. Section 3 describes our rule-based framework. Section 4 describes GADRs (General Anomaly-detection Rules). Section 5 discusses SADRs (Special Anomaly-detection Rules). Section 6 depicts ISP-Health, a prototype system of the proposed framework. And in section 7 we conclude the paper.

2 Related Work

The security of inter-domain routing system is a hot topic in ISP (Internet Service Provider) meetings and network conferences.

From the point of view of protocol design and function enhancement, some solutions, e.g. S-BGP, soBGP, pSBGP and MOAS-list [7] have been proposed. Among these proposals, S-BGP is a relatively comprehensive solution, but it uses strict hierarchical public key infrastructure (PKI) for both AS (Autonomous System) number authentication and IP prefix ownership verification, and it is far from practical deployment due to high overhead and deployment difficulty [2]. Some other efforts have been made to help understand the state of BGP routing table in backbones, among which is the famous website by G. Huston [8]. [6] launches the first systematic study of BGP misconfigurations, and [9] focuses on route-missing issues.

Despite extensive research work in the literatures, network operators in practice still lack effective tools to diagnose BGP security problems. In this paper, we design a powerful rule-based framework and define a rich set of detection rules to detect various route anomalies and suspect routing attacks.

3 A Rule-Based Security-Monitoring Framework for IDR

The rule-based framework for monitoring IDR system is shown in Fig. 1. It consists of five components: BGP Data Collectors, Rule Repository, Internet Model Pool, BGP Security Report Generator, and Detecting Engine. Obviously, the *Detecting Engine* is the core part of the whole framework.

BGP routing table and BGP update messages are gathered from the monitored networks. The *Detecting Engine* checks if there are anomalies in these BGP data using the rules in the *Rule Repository* together with the network models in *Internet Model Pool* .The detection rules are categorized into two classes: *the General Anomaly-detection Rules (GADRs)* and *the Special Anomaly-detection Rules (SADRs)*. The *Internet Model Pool* contains the Basic Models of the Internet and the Generated Models. A model is made up of normal BGP behavior schemes or global routing restrictions.

Fig. 2 shows the relations between the detection rules and the network models. Rules work together with the corresponding models to check the routing behaviors.

The *Basic Models* of the Internet are constructed from large amount of basic information about the Internet, such as allocated AS numbers, allocated IP prefixes, and

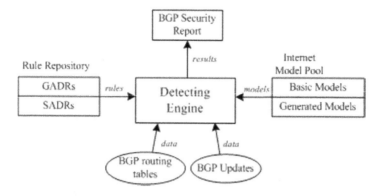

Fig. 1. A rule-based security-monitoring framework for IDR

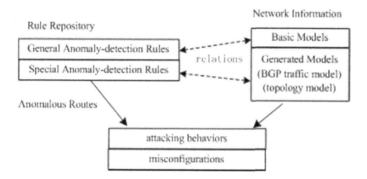

Fig. 2. Relations between the rules and the network models

the mapping from AS number to IP prefix. Such information can be acquired by inquiring IRR (Internet Route Registry) and RIR (Regional Internet Registry), or extracted from RFCs and formal bulletins. The *Generated Models* are not built so directly, and need be generated using specially devised algorithms. Such algorithms analyze volumes of BGP routing tables from numerous typical BGP speakers in backbone networks and try to discover more normal routing schemes that should be obeyed by ISPs. The *Generated Models* include ISP Commercial Relationship Model, Internet Hierarchy Model, ISP Geographical Relationship Model, Small-world Characteristic Model, and Power-law Characteristic Model, etc. Comparing the BGP data from monitored networks against the *Generated Models*, the Detecting Engine can find out more hidden routing misbehaviors.

4   General Anomaly-Detection Rules (GADRs)

In general, a route which violates *GADRs* is not allowed to spread in the Internet. Compared with *SADRs*, *GADRs* are relatively simple, and work with the Basic Models to detect bogus routes. For example, if a BGP route has a private prefix, it is assumed to be anomalous (according to *Rule a1*). Using *GADRs* and the Basic Models, the Detecting Engine can easily judge whether one route is problematic [10]. Here are ten *GADRs* that are extracted from ISP operation experience, BCP (Best Current Practices) guidelines and theoretical analysis of routing threats.

Rule a1: *If a route from an exterior network contains an IP prefix within any private address blocks, then it is anomalous. This rule is derived directly from RFC-1918.*

Rule a2: *If a route from an exterior network has unallocated addresses, then it is anomalous. Such a route contains an IP prefix that doesn't belong to any organization.*

Rule a3: *If a route contains an unauthorized IP prefix, then it may be anomalous. Usually, this kind of routes implies a network attack using forged or hijacking routes.*

Rule b1: *If a route comes from an exterior network and its AS-PATH contains private AS number, then it is anomalous. The private AS numbers are listed in RFC-1930.*

Rule b2: *If a route is from an exterior network and its AS-PATH has unallocated AS number, then it is anomalous. Such a route contains an AS number that doesn't belong to any organization.*

Rule b3: *If a route's AS-PATH has discontinuous repeated AS numbers, then it is AS-Loop anomalous.*

Rule c1: *If an IP prefix has more than one origin AS, then the relevant routes imply a conflict of MOAS (Multi-Origin Autonomous System).*

Rule c2: *If some routes satisfy Rule-c1 and the conflicting ASes haven't any affiliation relationship or the IP prefixes are not unauthorized by the legal AS, then they are judged as hijacked routes.*

Rule c3: *For two MOAS-conflicting ASes , e.g., A and B, if A indicates that B is the origin of an IP prefix in A's BGP routing table, while B's routing table shows that B isn't the origin, then it can be concluded that the routes are forged.*

Rule d: *If a route gathered from monitored network A indicates that its origin or the former hop is AS B, but the route table of B shows that B isn't the origin or has not propagated the route , then the route is forged.*

The above ten rules should be used together with the Basic Models. For example, to decide whether one route conforms to *Rule b2*, the Detecting Engine should inquire the Basic Models pool to know which AS numbers are unallocated.

Rules c2, c3 and *d* use routing data from different monitored networks, and compare them comprehensively to further verify the anomalies or find more hidden routing attacks. Such a multiple-view detecting ability is a special feature and an advantage of our system.

5   Special Anomaly-Detection Rules (SADRs)

Using *GADRs* and the Basic Models, the *Detecting Engine* can only find some fairly straightforward routing anomalies. To disclose more complex and hidden abnormal behaviors, it is necessary to make in-depth data-mining into the Internet data and find more orders in routing behaviors. We have tried to discover more normal behavior schemes of the network and use them to check the routing data from monitored networks. The normal behaviors schemes are dug out using specially devised algorithms, so it's reasonable to call them *Generated Models*. In this section, we first depict two Generated Models, i.e., *Internet Hierarchy Model* and *ISP Commercial Relationship Model*, and then define corresponding anomaly-detecting rules.

5.1 Internet Hierarchy Model and Corresponding Anomaly-Detection Rules

5.1.1 Building Hierarchy Model of the Internet

The Internet can be described in a hierarchy model made up of mult-levels of *ASes*. In this paper, we use a *core-transit-stub* three-level hierarchy to model the Internet. Generally speaking, it is not known which level an AS belongs to. We want to know the level information for all the active ASes. Therefore, we devise algorithms to achieve such a goal by analyzing the routing table available from Routeviews, RIPE-NCC and some other sources. The following *Method-A,B* and C are used to obtain the AS-sets that constitute core-level, transit-level and stub-level of the Internet hierarchy respectively. Some ideas of the methods are borrowed from the literature [12].

Method-A: The backbone networks of the top-level ISPs are the core of the Internet. To achieve the global connectivity, the Tier-1 ISPs build peer-peer commercial relationship between each other, and they are formed into a fully-connected mesh. Therefore, the problem of inferring the core-level of the Internet can be defined as: to acquire the node set of maximum fully-connected sub graph in the Internet graph at AS level. Obviously, it's an NP-hard problem. A heuristic algorithm is given in this paper. The algorithm uses the whole AS-PATH as input, which is obtained by combining all the AS-PATH attribute values in all the BGP tables used for the generation of the Internet Hierarchy Model.

Input: the whole AS-PATH set
Output: the core AS set (Tier1_AS_SET)
Step 1. Tier1_AS_SET = Φ;
Step 2. Compute the degree for every node in graph G, and store it into a table T;
Step 3. Get the node set with the max degree in G:
     max_degree_nodes(G) = $\{v| d(v) = \max(d(v_1),d(v_2),...), v_1,v_2,... \in V\}$
Step 4. If |max_degree_nodes (G)| = 1, suppose z is the only one item in
    max_degree_nodes(G);
Step 5. If |max_degree_nodes(G)| $\neq$1, look up the table T and choose one item z,
    such as z$\in$max_degree_nodes(G) and its historical degree in the table T is not
    smaller than the others;
Step 6. Tier1_AS_SET = Tier1_AS_SET $\cup$ {z};
Step 7. Neighbor_set $\leftarrow$ the neighbor node set of z in G;
Step 8. Get the induced sub graph G' from G, such that the node set of G' is
    Neighbor_set;
Step 9. G = G';
Step 10. Quit, if G satisfies the condition

$$|E(G)| >= \alpha \frac{(|V(G)|-1)*|V(G)|}{2}$$;

otherwise, go to the step 2. Here, $|E(G)|$ is the edge number of G, $|V(G)|$ is the node number of G, and α is a coefficient used to control the connection density of Tier1 set, e.g., G is a fully- meshed graph if $\alpha = 1$.

Method-B: An *AS* is a stub one if it does not transit traffic between ISPs. Such an *AS* lies in the bottom of the Internet hierarchy and only appears at the tail of an AS-PATH. Using the following algorithm, we can identify stub *ASes* by scanning the whole AS-PATH set.

Input: the whole AS-PATH set
Output: the stub AS set (STUB_AS_SET)
Step 1. STUB_AS_SET = Φ;
Step 2. Get the AS list , AS_LIST, which contains all the ASes in the whole AS-PATH set;
Step 3. Repeat Step 4, 5, 6, for every node v in AS_LIST;
Step 4. Flag = 0;
Step 5. Examine every AS-PATH, if node v isn't its last AS, set Flag = 1;
Step 6. If Flag = 0, add node v into STUB_AS_SET.

Method-C: After identifying the top level by *Method-A* and the bottom level by *Method-B*, we can construct the transit-level using all the left *ASes*. The method is described formally as follows.

Input: the whole AS-PATH set
Output: transit AS set (TRANSIT_AS_SET)
Step 1. Get the core AS set (Tier1_AS_SET) using Method-A;
Step 2. Get the stub AS set (STUB_AS_SET) using Method-B;
Step 3. Get the whole active AS set of the Internet, AS_SET;
Step 4. TRANSIT_AS_SET ← AS_SET - Tier1_AS_SET - STUB_AS_SET.

5.2 ISP Commercial Relationship Model and Relevant Rules

5.2.1 ISP Commercial Relationship Model
In general, there are three kinds of commercial relationships between two connected ASes, i.e. "provider-customer", "customer-provider" and "peer-peer" [11]. If we could get the business contracts between ISPs or learn commercial policies of all ISPs, the commercial relationship model between ISPs will be built easily. However, all these information are top secrets and inaccessible. Fortunately, there is an algorithm to approximately infer the ISP relationships. And it is described as Method-D in the paper, which borrows some idea from the literature [11]:

Method-D:

Input: The whole AS-PATH set, every AS-PATH is made up of an AS sequence, it can be denoted as $p = \alpha_1\alpha_2...\alpha_i...\alpha_n$, $1 \le i \le n$
Output: the commercial relationship set (Relastion_SET) of AS pairs <α, β>, where α and β are any AS that occurs in the whole AS-PATH set

Step 1. Use Method-A to get the ASes in Tier1_AS_SET; Label the peer-peer relationship to the AS pair $<\alpha, \beta>$, if $\alpha, \beta \in$ Tier1_AS_SET;

Step 2. Extract Core_AS- PATH that contains AS in Tier1_AS_SET from the whole AS-PATH set;

Step 3. If $p \in$ Core_AS- PATH, suppose $\alpha_i \in$ Tier1_AS_SET

3-1. Label customer-provider relationship to the AS pairs $<\alpha_{j-1}, \alpha_j>$ ($j \leq i$), which are on the left side of α_i in p;

3-2. Label provider-customer relationship to the AS pairs $<\alpha_j, \alpha_{j+1}>$ ($j \geq i$), which are on the right side of α_i in p;

Step 4. If $p \in$ AS-PATH - Core_AS- PATH

4-1. If the AS pairs in p, $<\alpha_{i-1}, \alpha_i>$ and $<\alpha_j, \alpha_{j+1}>$ ($i<j$), are customer-provider relationships, label customer-provider relationship to the pairs between them, $<\alpha_r, \alpha_{r+1}>$ ($i \leq r < j$);

4-2. If the AS pairs in p, $<\alpha_{i-1}, \alpha_i>$ and $<\alpha_j, \alpha_{j+1}>$ ($i<j$), are provider-customer relationships, label provider-customer relationship to the pairs between them, $<\alpha_r, \alpha_{r+1}>$ ($i \leq r < j$);

4-3. Repeat step 4-1 and 4-2, until no new customer-provider or provider-customer relationship need be labeled;

Step 5. If $p \in$ AS-PATH - Core_AS- PATH

5-1. If the AS pair $<\alpha_{i-1}, \alpha_i>$ in p has customer-provider relationship, and $<\alpha_j, \alpha_{j+1}>$ has provider-customer relationship, ($i<j$), label peer-peer relationship to the pairs between them, $<\alpha_r, \alpha_{r+1}>$ ($i \leq r < j$);

5-2. If the AS pairs $<\alpha_j, \alpha_{j+1}>$ ($j \geq i$) in p, which are on the right of α_i, are provider-customer relationship, and the AS pairs $<\alpha_{k-1}, \alpha_k>$ ($k \leq i$), which are on the left side of α_i, haven't been labeled, label peer-peer relationships to them;

5-3. If the AS pairs $<\alpha_{j-1}, \alpha_j>$ ($j \leq i$) on the left of α_i in p are customer-provider relationship, and the AS pairs $<\alpha_k, \alpha_{k+1}>$ ($k \geq i$), which are on the right side of α_i, haven't been labeled, label peer-peer relationships to them.

5.2.2 Anomaly-Detection Employing the ISP's Commercial Relationship Model

Routing activities are restricted by the defined commercial relationship models, and one route should conform to the normal schemes. For example, ISP A and ISP B are connected to their customer, ISP C, respectively. In this case, ISP A can not reach ISP B via ISP C, for C as a customer does not provide transit services between A and B. If a route indicates that A can reach B via C, or vice versa, it can be concluded that the route violates the commercial relationships between A, B and C [10] . Therefore, if a route is inconsistent with the commercial relationship model of related ISPs, it is suspected to be fraud, forged or abnormal. According to the above model of ISP Commercial Relationships, four SADRs are defined as follows:

Rule s3: *If a route passes from a provider-customer edge to a peer-peer edge, then it is anomalous.*

Rule s4: *If a route passes from a provider-customer edge to a customer-provider edge, then it is anomalous.*

Rule s5: *If a route passes from a provider-customer edge to a customer-provider edge, then it is anomalous.*

Rule s6: *If a route passes by a peer-peer edge and another peer-peer edge again, then it's anomalous.*

6 ISP-Health - A Monitoring System Based on Rule-Based Detection Framework

ISP-Health is an implementation of our proposed rule-based framework. It detects the abnormal behaviors of the routing system and finds out susceptible routing attacks using the detection rules together with the network models. The architecture of ISP-Health is shown in Fig. 3.

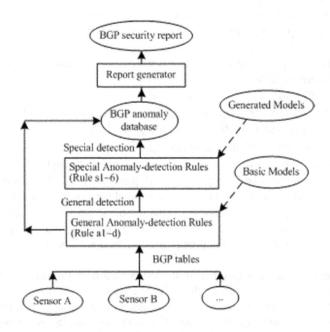

Fig. 3. The architecture of ISP-Health

The detecting procedure is mainly divided into two steps. At the first step, the General Anomaly-detection Rules are applied to the BGP data from the monitored networks; then the Special Anomaly-detection Rules are employed to detect more complex routing anomalies.

Compared with other detection tools, ISP-health can not only detect the average bogus routes, such as routes with private prefix or AS, but also identify some hidden anomalous routes, such as the routes breaking the commercial relationships and other generated network models. Even for the detecting of average bogus routes, ISP-Health can use the routing information collected from multiple monitored positions and apply the rules comprehensively to the assembled data. The capabilities of ISP-Health are summarized in Table 1. The symbol "✓" means ISP-Health has the listed capability and "*" indicates the multiple-view detecting ability.

Table 1. The Capabilities of ISP-Health

Type of Anomalous Routes	Capabilities
Containing private prefix	✓
Containing unallocated prefix	✓
Containing reserved prefix	✓
Containing private AS	✓
Containing unallocated AS	✓
Containing AS-Loop	✓
Forged routes	✓
Mismatching the Internet Hierarchy Model	✓*
Violating the ISP Commercial Relationship constrain	✓*
MOAS conflict	✓*

7 Conclusion

The inter-domain routing system is prone to human errors or malicious attacks, and it is faced with great challenges to run healthily. This paper presents a rule-based framework to detect the route anomalies and routing attacks. A rich set of detection rules are defined, and they work together with the Basic Models and the Generated Models of the Internet to find out hidden routing misbehaviors from the BGP data of monitored networks. We have implemented ISP-Health, a prototype under the proposed framework, which demonstrates several advantages over other detecting systems. In the future, we will focus on the definitions of more detection rules and try to put ISP-Health into practice.

Acknowledgement

We gratefully acknowledge the support from National Natural Science Foundation of China under Grant No. 90204005, the High-Tech Research and Development Program under Grant No. 2005AA121570 and 2003AA121510, and the National Laboratory for Modern Communications Foundation under Grant No. 51436050605KG0102.

References

1. B. Halabi. Internet Routing Architectures. Cisco Press, second edition, 2001.
2. S. Kent and C. Lynn and K. Seo. Secure Border Gateway Protocol (Secure-BGP). IEEE Journal on Selected Areas in Communications, 18(4): 582-592, April 2000.
3. S. Murphy. Border Gateway Protocol Security Analysis. IETF Internet Draft, draft-murphy-bgp-vuln-00.txt. November 2001.
4. J. Cowie, A. Ogielski, B. Premore, and Y. Yuan. Global Routing Instabilities during Code Red II and Nimda Worm Propagation. http://www.renesys.com /projects/bgp_instability

5. S. A. Misel. Wow, AS7007! NANOG mail archives. http://www.merit.edu/mail.archives/nanog/1997-04/msg00340.html.
6. R.Mahajan, et al. Understanding BGP Misconfiguration. ACM SIGCOMM' 2002
7. X. Zhao, D. Pei, L.Wang, D. Massey, A. Mankin, S.F. Wu and L. Zhang. An Analysis of BGP Multiple Origin AS (MOAS) Conflicts. ACM SIGCOMM Internet Measurement Workshop 2001.
8. G. Huston. BGP Table Statistics. http: //www.telstra.net/ops/bgp/index.html.
9. Di-Fa Chang, Ramesh Govindan, John Heidemann. Locating BGP Missing Routes Using Multiple Perspectives. ACM SIGCOMM, 2004.
10. A. Broido, E. Nemeth, K. Claffy. Internet Expansion,Rrefinement and Churn. ETT, Jan 2002
11. L. Gao. On Inferring Autonomous System Relationships in the Internet. IEEE Global Internet Symposium, 2000.
12. L. Subramanian, S. Agarwal, R. H. Katz. Characterizing the Internet Hierarchy from Multiple Vantage Points. IEEE INFOCOM, 2002.

Transaction of Web Services Based on Struts*

Gong-Xuan Zhang, Ping-Li Wang, and Wen Chen

Department of Computer Science and Technology,
Nanjing University of Science and Technology,
210094 Nanjing, China
Gongxuan@mail.njust.edu.cn

Abstract. There are many frameworks for web applications and Struts is one of them with a collection of Java code designed to help you build solid applications while saving time. It provides the basic skeleton and plumbing, and takes complex applications as a series of basic components: Views, Action classes, and Model components. Web services are new technology for next generation Internet. In this paper, Struts-enabled framework is first described and then a transaction scenario is discussed for web services applications.

1 Introduction

Internet has been applied over most of our real world. Companies use Internet for their businesses to provide customers with their products or services. With the explosion of the number and type of services, some new mechanisms and frameworks are required to meet the needs of customers with ease. Web services are new technology for us to discover, deploy, invoke and migrate web resources. There are many web application frameworks developed to improve business processes and quality of services over Internet, such as BEA WebLogic 8.x, Borland Enterprise Server 6.5, IBM Websphere Application Server 5.0, and Oracle 11. All of them support web services technology.

Struts framework, a key infrastructure or model embedded in web application servers, collaborated with any one of above concerned products, provides the basic skeleton and plumbing. With struts framework, complex applications, being compliant web applications based on the Java Servlet specification, are taken as a series of basic components: Views, Action classes, and Model components. The rests of this paper are organized as follows: MVC architecture of struts framework is described in Section 2, a new model of ISWS (Integrating Struts into Web Services) is discussed in Section 3, the transactional pool model is given in Section 4. And the last is the conclusion.

2 Struts Components

The Struts framework, first developed in 2001, combines the best of servlets and JSPs. Being a compliant Web application (or Webapp, for short) implies, among other things, that Struts application has

---

* This is partly sponsored by the Educational Fund of Jiang Su, and the project number is 04KJB520077.

J. Cao, W. Nejdl, and M. Xu (Eds.): APPT 2005, LNCS 3756, pp. 427–434, 2005.
© Springer-Verlag Berlin Heidelberg 2005

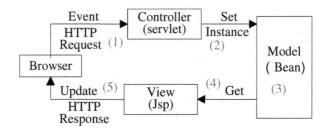

Fig. 1. MVC Model 2

• A standard directory structure
• Certain standard configuration files (web.xml and so on)
• Dynamic functionality deployed as Java classes and .jsp pages
• A standard Web Archive (.war file) format for deployment

Although it's not required reading, it would be very useful to you as a Struts developer familiar with the Java Servlet specification version 2.3 (or 2.2, depending on the application server). The Struts open source framework is based on the Model 2, or Model-View-Controller, approach to software design. The Model 2 framework evolved from the Model 1 design, which included JavaServer Page technology. With struts framework, it is easy to build flexible applications and a set of JSP custom tags for building JSP pages. The Model-View-Controller architecture of Struts simplifies building Web applications by providing a model into which you plug components. Take, for example, a simple application for updating a user's address information. In this case, the Model-View-Controller architecture might break the application into the following components [2,3,4]:

Model: A programming model that provides an internal representation of the data.
View: A view to be used to display the user's information.
Controller: A controller to assist in validating the user's entries and choosing the right view to display the results. It determines what processes to perform and what steps to take next.

For web applications, a proposed MVC architecture can be depicted in Fig.1. with five steps as following:

1. Users give their requests under HTTP;
2. The Controllers, most of them are servlets, receive the requests and put them forward to the Models, and get results from the Models;
3. The Models, encapsulate data structures and transactional logics, make actions on databases. Java Beans are Model Components;
4. After receiving results, the Controllers put them forward to the Views that are interfaces to users. View Components are made of JSPs that are embedded into HTML;
5. The Views show the results back to user's browsers.

Based on MVC model 2, all components are divided into coordinatve classes, servlets and JSP tags in Struts framework which has the characteristic of 'business logics apart from representation logics'.

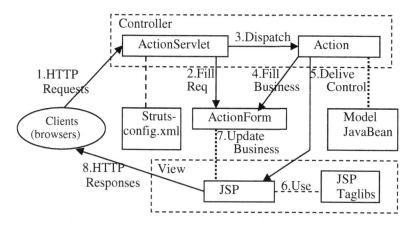

Fig. 2. Struts framework

Fig. 2 shows a Struts framework, with which the three components of MVC are fit together by struts configuration file (struts-config.xml). This means that struts framework is a series of components and most efforts, for web application developers, are to break down complex web applications into a series of simpler components. When a Struts-enabled web application is created, web.xml file is updated, struts-config.xml and tiles-def.xml deployment descriptor files are created, and then copied to the WEB-INF directory.

3 ISWS Architecture

With assumption for an Internet with the dynamic business entities can communicate within them or trade partners. This is the next generation Internet, with which web services are core techniques gradually. XML, SOAP, WSDL, UDDI and so on, are key techniques of web services [5,6,7].

Web services allow a business enterprise to publish its on-line applied businesses, other companies or applied softwares can access the on-line services. The web services technology, which can be taken as a kind of deploying the object components on the Websites, provides remote service invocation mechanism base on XML and SOAP protocols over the Internet. At first, the services provider gives his services definitions and builds the services modules interfaces, makes use of the WSDL to describe the service access entrance (URL) and remote invocation interfaces, and publishes them on Internet for requesters. And then, the services requester invokes the remote procedures (web services) by the names and parameters described in the document WSDL, the web services response his request and execute the function, and return the results to the requester.

With web services technology, some excellent Struts-based web applications can be developed [8]. A proposed model, called ISWS--Integrate Struts into Web Services, is shown in Fig. 3. It is divided into three parts of MVC model, services controller and Web services implementation. The services controller is a key component of which bridges MVC model and Web Services. A client request, for instance, is first processed in MVC model, and dispatched by services controller then. The controller invokes related services of web services implementation. Finally, the results are sent back to MVC model through the services controller and shown on the client screen with view components of MVC model.

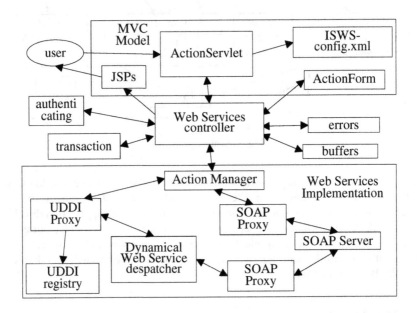

Fig. 3. ISWS Architecture

In general, web services are invoked statically by WSDL and dynamically by UDDI. The two typical invocations will be respectively implemented with SOAP proxy and UDDI proxy in web services implementation of the ISWS model.

4 TPL Transaction

Third Party Logistics (in short, TPL) is a popular, newer pattern for e-commerce. TPL focuses on goods' shipping, transporting and storing. Many resources can be optimized and saved during goods' circulating with TPL pattern. As a general example, a TPL system consists of 6 parts (called 6 roles) from booking an order to completing the shopping as fellow:

1. Customer—books orders for his/her shopping on a market website;
2. Market (shopping center)—handles orders, signs invoices and asks shipping;
3. Logistics Center—schedules shipping list according to invoices;
4. Store House—stores all goods in it;
5. Shipping Group—arranges vehicles to transport the ordered goods;
6. Manufacturer—provides goods for the others.

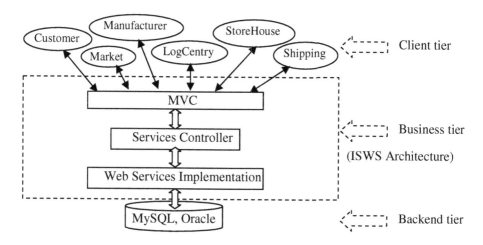

Fig. 4. ISWS application

Fig. 4 is a application model of ISWS for TPL with six roles, and each of them can be constructed into three types of components. Note that JSPs are interfaces for clients to invoke remote services, servlets are controllers with listening clients' requests, creating connection and passing messages between JSPs and EJBs, and EJBs are models to execute SQL statements on database management systems. All the components will be programmed in Java and deployed with WebLogic, or other web application servers at running time.

Just as discussed above, TPL components, which are developed as Web services and encapsulated in the YH_TPL.war (Web ARchive, created in building time), work together by the struts-config.xml of TPL system. In a struts-config.xml, many items, such as Global Forwards, ActionMappings, ActionForm beans and JDBC DataSources, can be defined. Bellow are part contents of TPL struts-config.xml.

```
<?xml version="1.0" encoding="UTF-8"?>
<!DOCTYPE struts-config PUBLIC "-//Apache Software Foun-
dation//DTD        Struts        Configuration        1.1//EN"
"http://jakarta.apache.org/struts/dtds/struts-
config_1_1.dtd">
<struts-config>
  <form-beans>
  <form-bean              name="loginActionForm"              type="
com.struts.basic.LoginActionForm"/>
  <form-bean              name="marketInfoForm"              type="
com.struts.admin.enterprise.market.marketInfoForm"/>
  <form-bean             name="marketGoodsInfoForm"             type="
com.struts.admin.enterprise.marketGoods.marketGoodsInfoFo
rm"/>
    ......
  </form-beans>
<global-forwards>
<forward name="index" path="/index.jsp" redirect="true"/>
```

```
<forward                name="noExit"                path="/in-
clude/userNotExisted.jsp"/>
<forward name="globalerror" path="/include/error.jsp"/>
<forward                name="dialogClose"                path="
/include/dialogClose.jsp"/>
<forward name="error" path="/include/error.jsp"/>
      ......
</global-forwards>
<action-mappings>
      ......
<action    name="marketInfoForm"    path="/admin/enterprise/
market/marketSaveAction"          scope="request"          type="
com.struts.admin.enterprise.market.marketSaveAction">
<forward          name="success"          path="/admin/enterprise
/market/closedialog.jsp"/>
</action>
<action  path="/admin/enterprise/market/marketChangeAction
"  scope="request"  name="marketInfoForm"  attribute=  "mar-
ketInfoForm"                                            type=
"com.struts.admin.enterprise.market.marketChangeAction">
<forward          name="success"          path="/admin/enterprise
/market/marketChange.jsp"/>
</action>
<action  path="/admin/enterprise/market/marketUpdateAction
"       scope="request"       name="marketInfoForm"       type=
"com.struts.admin.enterprise.market.marketUpdateAction">
<forward          name="success"          path="/admin/enterprise
/market/closedialog.jsp"/>
</action>
<action     path="/admin/enterprise/market/marketAddAction"
scope="request"                                         type=
"com.struts.admin.enterprise.market.marketAddAction">
<forward          name="success"          path="/admin/enterprise
/market/marketAdd.jsp"/>
</action>
<action
type="com.struts.admin.enterprise.market.marketDeleteActi
on" path="/admin/enterprise/market/marketDeleteAction">
<forward          name="success"          path="/admin/enterprise
/market/marketListAction.do"/>
</action>
      ......
</action-mappings>
<message-resources                                  parameter=
"com.struts.ApplicationResources"/>
</struts-config>
```

In a web application environment, the number of customers is unfixed and many requests, especially data change requests, are probably on the same database or different ones at same time. So there must be transaction mechanism to keep updated data coherent. In order to develop and implement the TPL system, we define some data interfaces with 2PC agreement for its business tier, adapted to WebLogic server,

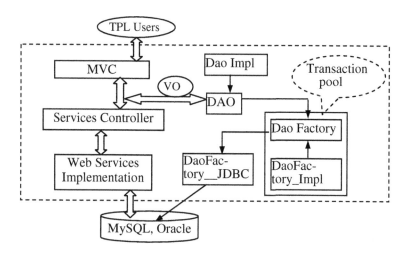

Fig. 5. TPL Transaction with ISWS Model

under ISWS architecture. The data interfaces are based on transaction and called transactional pools located in transaction component next to services controller. Some data operations are distribued in the Web Services Implementation. TPL transaction abstract interfaces model is shown In Fig.5. All TPL components access database tables through VOs (Value Objects) that pass the parameters to DAOs (Data Access Objects, interfaces). When data access invoked, WebLogic server creates instances (Dao Factory and its implementing) and the instances access related databases under DaoFactory_JDBC connections. So many instances may access different data tables over several databases in different servers. When data access finished, WebLogic server will destroy the instances and release their resources.

To keep the instances transactional, it is important to configure transaction mechanism in the file of WebLogic's server.xml as following:

```
<!-- TPL WebLogic  Server Configuration File -->
<Server>
       ......
<Context  path="/TPL"  docBase="TPL"  debug="0"  reload-
able="true" crossContext="true">
<Resource   auth="Container"   name="jdbc/TPL"   type="
javax.sql.DataSource"/>
<ResourceParams name="jdbc/TPL">
     <parameter>
     <name>factory</name>
     <value>org.objectweb.jndi.DataSourceFactory</value>
     </parameter>
</ResourceParams>
<!-- Description of the resource "UserTransaction -->
<Resource  name="UserTransaction" auth="Container" type="
javax.transaction.UserTransaction"/>
<ResourceParams name="UserTransaction">
  <parameter>
```

```
<name>factory</name>
<value>org.objectweb.jotm.UserTransactionFactory
</value>
    </parameter>
    <parameter>
    <name>jotm.timeout</name>
    <value>60</value>
    </parameter>
    </ResourceParams>
        ......
</Server>
```

At same time, the ofollowing key java statements must be added into most action components to activate the transaction mechanism.

```
Context ctx=new InitialContext();
UserTransaction                    utx=(UserTransaction)
ctx.lookup("java:comp/ UserTransaction");
utx.begin();
MarketDao                              marketDao1
=DaoAbstractFactory.getFactory().getMarketDao();
Market []market1=new Market[2];
market1[0]=new Market("22");
market1[1]=new Market("33");
Boolean bTrans= marketDao1.insert(market1);
......
if (bTrans) utx.commit();
else    utx.rollback();
```

5 Conclusion

With Struts-enabled framework, many enterprises can build their web applications quickly and easily. The reason is that most businesses can be constructed with MVC components (here Web Services), and then dynamically deployed into a web application server. You can update or add some business modules at any time.

References

1. Andrew S. Tanenbaum. Distributed Systems: Principles and Paradigms. Prentice-Hall, Inc (2002).
2. Chuck Canvaness. Programming Jakarta Struts. O'Reilly Media, Inc (2002).
3. James Goodwill. Mastering Jakarta Struts. Wiley Publishing, Inc (2003).
4. James Turner. Struts Kick Start. Sams Publishing, Inc (2003).
5. Eric Amstrong. The Java Web Services Tutorial. Higher Education Press, Pearson Eduction (2003).
6. The Website. Design a simple service-oriented J2EE application framework. http://www.javaworld.com/javaworld/jw-10-2004/jw-1004-soa.html
7. Zhao Qiang. Developing J2EE application (WebLogic+JBuilder). Publishing House of Electronics Industry (2003).
8. IBM Company. Architect Struts applications for Web services. http://www-106.ibm.com/developerworks/webservices/library/ws-arcstruts/

A Method of Aggregate Query Matching in Semantic Cache for Massive Database Applications

Jianyu Cai, Yan Jia, Shuqiang Yang, and Peng Zou

School of Computer, National University of Defense Technology, Changsha, China
jianyucai@163.com

Abstract. Aggregate queries are frequent in massive database applications. Their execution tends to be time consuming and costly. Therefore efficiently executing aggregate queries is very important. Semantic cache is a novel method for aiding query evaluation that reuses results of previously answered queries. But little work has been done on semantic cache involving aggregate queries. This is a limiting factor in its applicability. To use semantic cache in massive database applications, it is necessary to extend semantic cache to process aggregate query. In this paper, query matching is identified as a foundation for answering aggregate query by semantic caches. Firstly a formal semantic cache model for aggregate query is proposed. Based on this model, we discuss aggregate query matching. Two algorithms are presented for aggregate query matching. These two algorithms have been implemented in a massive database application project. The practice shows the algorithms are efficient.

1 Introduction

Aggregate queries are pervasive in massive database applications, whose execution tends to be time consuming and costly. Therefore promotion of their efficiency will largely improve the performance of the system. Semantic cache is a novel scheme for aiding query evaluation, which can reuse results of previously answered queries [1]. But most of existing solutions research on SPJ (Select Project Join) query and little work have been done on semantic cache involving aggregate queries [1][2][3][4][5]. This is a limiting factor in its applicability and it is mostly used in small-scale database applications. In order to utilize semantic cache in massive database applications, it is necessary to extend semantic cache to support aggregate query. Query matching is primary for answering query by semantic cache. When a query comes, it has to decide whether the query be answered by cache. In order to process aggregate queries, semantic cache requires the function of aggregate query matching.

Several works considered the problem of answering queries using views in the presence of grouping and aggregation. [6] extended the treatment of grouping and aggregation to consider mutli-block queries. They presented a matching algorithm that break the matching task into many smaller sub-matches replying on a general matching infrastructure. Their approach is too complex for semantic cache. [7] involved a set of transformations in the query rewrite phase. In this approach, the algorithm performs syntactic transformations on query until it is possible to identify a subexpression of the query that is identical to the view. However, the purely syntactic nature of this approach

J. Cao, W. Nejdl, and M. Xu (Eds.): APPT 2005, LNCS 3756, pp. 435–442, 2005.
© Springer-Verlag Berlin Heidelberg 2005

is a limiting factor in its applicability. [8] relaxed the limitation of [7] and proposed a more semantic approach that find views related to the query. But [8] only gave a sketch. [9] considered the formal aspects of answering aggregate queries using views. They discussed the problem of completion and didn't give any practical method of query matching.

Researches on SPJ query matching can't be used in aggregate query matching. Other related works have deficiencies in terms of using semantic information or haven't found their way into practice. In this paper we propose a method of aggregate query matching in semantic cache for massive application. We assume that update and insert operations have little effect on cache in the massive application.

The remainder of the paper is organized as follows. Section 2 describes formally semantic cache, which supports aggregate query. Section 3 defines containing match and overlapping match. Based on the definition, two algorithms of query matching, AQCM and AQOM are given in section 3. Section 4 concludes the paper.

2 Formal Description of Semantic Cache

From a logical point of view, a semantic cache is composed of a set of cache items. Each cache item is the result of aggregate query. In this research, we assume that both the queries and the cached items are defined by relational algebra expressions involving only projections, selections, aggregations, groups and joins. Thus we extend the model of [3] and give formal definitions of semantic cache in this section. Without a lose of generality, we ignore $\neq$ comparison and assume that the problem is in the real domain in this study. Now we make conventions as follows:

1. Compare Predicate, P, where $P = X \ op \ c$, X is an attribute of a base relation, $op \in \{\leq, \geq, <, >, =\}$, c is a domain value or a constant.
2. Join Predicate, P, where $P = (X=Y)$, X, Y are join attribute.
3. Simple Predicate, P, is either a Compare Predicate or a Join Predicate.

Definition 1. Given a database $D = \{R_i | R_i$ is a base relation, $i=1,2, ...,m\}$, a Semantic Cache item, S, is a tuple $<A,F,T,P,C>$, where $C=\pi_{A,F}(\sigma_P(R_{i1} \times R_{i2} \times, ... \times R_{in}))$; $T=\{R_{i1},R_{i2},...,R_{in}\} \subseteq D$; $A=\{a_l | a_l \in R_k, R_k \in D\}$; $F=\{f(b) | b \in R_i, R_i \in T, f \in Ag\}$, $Ag=\{$MAX, MIN,SUM,COUNT$\}$; $P=p_1 \wedge p_2 \wedge ... \wedge p_j$ where each p_j is a simple predicate.

In definition 1, P indicates the constraints that the tuples in the semantic cache item S satisfy, while C represents the actual content. T is the base relations that are involved in the cache item creation. A defines the non-aggregation attributes in S. At the same time, A represents the grouping attributes in S. F defines the aggregation columns in S. Ag gives the aggregation functions that is possible in S. Semantic cache items are the result of Select-Project-Group-Join (SPGJ) operations. For SUM and COUNT can compute AVG, we don't discuss AVG in this paper.

Example 1. Consider two base relations in a databae D:Student(Sid, Sname, Age, Class) and Course(Cid, Cname, Sid, Period). Suppose there is a query Q_1 that will generate the semantic cache item S:

Query Q_1: Select Sname, SUM(Period) From Stuent, Course Where Age>20 and Student.Sid = Course.Sid Group by Sname;

Therefore, the semantic cache item S is represented as $<A,F,T,P,C>$, where $A=\{Sname\}$; $F=\{SUM(Period)\}$; $T=\{Student,Course\}$; $P=(Age>20) \wedge (Stdent.Sid = Course.Sid)$; and C is the result of query Q_1.

Since semantic cache items are actually the result of query, we can qualify the semantic information of queries in the same way as we specify those for cache items. There we define a query just as definition 1.

Definition 2. A Query Q is represented by a virtual semantic cache item, $<A_Q,F_Q, T_Q,P_Q,C_Q>$. The fields in the tuple have the same implication as definition 1. C_Q is empty before query execution.

3 Aggregate Query Matching

This section resolves aggregate query matching based on the formal description of last section. First, discuss some definitions and techniques of aggregate query matching. Then give two algorithms of aggregate query matching.

3.1 Query Matching

Query matching decides whether the query be answered by cache item. This section discusses some definitions and techniques of query matching.

Definition 3. Consider a semantic cache item $S=<A,F,T,P,C>$ and a query $Q=<A_Q,F_Q, T_Q,P_Q,C_Q>$. We say Q matches S, if there exists a relational algebra expression E, containing only project, selection, join, aggregation and group operations, such that $E(C)\neq\varnothing$, and $E(C)\subseteq C_Q$. If $E(C)=C_Q$, we say Q can be fully contained by S. This match type is called containing match.

Group operator is an important factor of query matching. On the premise of that grouping attributes of aggregate query are contained by those of semantic cache item, we firstly discuss the conditions that aggregation functions satisfy in present of query matching. If a query matches a semantic cache item, there are some possible situations as for their aggregation functions. The equivalence of their aggregation functions is one of conditions. Otherwise, aggregation functions of query can be computed based on aggregation functions and grouping attributes in semantic cache item. Definition 4 describes these situations.

Definition 4. Consider a semantic cache item $S=<A,F,T,P,C>$ and a query $Q=<A_Q,F_Q, T_Q,P_Q,C_Q>$. Given an aggregation function $f_1(a)\in F_Q$. We say $f_1(a)$ is derivable from F and A, denoted by $f_1(a)=h(F,A)$, if there exists another aggregation function $f_2(b)\in F$, such that one of the following four conditions holds:

(1) $f_1=f_2, a=b, f_1\neq COUNT$;
(2) $f_1=f_2, f_1=COUNT$;
(3) $f_1=SUM, f_2=COUNT, a\neq b, a\in A$;
(4) $f_1=MAX|MIN, a\in A$;

If there is an aggregation function $f_l(a)$ in F_Q such that $f_l(a)=h(F,A)$, then F_Q is derivable from F and A, denoted by $F_Q \leftarrow D(F,A)$. For any aggregation function $f_l(a)$ in F_Q, if $f_l(a)=h(F,A)$, then F_Q is fully derivable from F and A, denoted by $F_Q=D(F,A)$. If $F_Q=D(F,A)$ and there exists an aggregation function $f_l(a)$ in F_Q such that $f_l(a) \neq h(F,A)$, then F_Q is partially derivable from F and A, denoted by $F_Q \approx D(F,A)$.

Theorem 1. Consider a semantic cache item $S=<A,F,T,P,C>$, and a query $Q=<A_Q,F_Q, T_Q,P_Q,C_Q>$, suppose P_A is its predicate attribute set. If it is not specified, P_A is the predicate attribute set of P in this paper. Then we have:

(1) If $T=T_Q$, $A_Q \subseteq A$, $F_Q \leftarrow D(F,A)$, $P_A \subseteq A$, and $P \wedge P_Q$ is satisfiable, then Q matches S.
(2) If $T=T_Q$, $A_Q \subseteq A$, $F_Q=D(F,A)$, $P_A \subseteq A$, $P_Q \Rightarrow P$, then Q can be fully contained by S.

Proof. Proof of 1: Suppose Q doesn't match S, there does not exist a relational algebra expression E, containing only project, selection, group and aggregation operations, such that $E(C) \neq \varnothing$, $E(C) \subseteq C_Q$.

Let us construct a query $Q'=<A_Q,F_Q,T_Q, P \wedge P_Q,C_Q>$. Since every tuple which satisfies $P \wedge P_Q$ will always satisfy P_Q, thus we have $C_Q' \subseteq C_Q$. Also $P \wedge P_Q$ is satisfiable, thus $C_Q' \neq \varnothing$.

Because $T=T_Q$, $A_Q \subseteq A$, $F_Q \leftarrow D(F,A)$, $P_A \subseteq A$, and $S=<A,F,T,P,C>$, then we have $C_Q' = \pi_{A_Q,F'} \left(\sigma_{P_Q}(C) \right)$, where F' is a new set of aggregation functions derived from $F_Q=D(F,A)$. Hence we can find a $E = \pi_{A_Q,F'} \sigma_{P_Q}$, such that $E(C)=C_Q' \subseteq C_Q$, $E(C) \neq \varnothing$. This conflicts with the assumption at the beginning, so Q matches S.

Proof of 2: Suppose Q is not be fully contained by S, there exists a tuple $t_Q \in C_Q$, and for t_Q, there does not exist a tuple $t \in C$ such that $t_Q=E(t)$, E is a relational algebra expression.

Let us construct a semantic cache item $S'=<A,F,T,P_Q,C'>$. Since $T=T_Q$, $A_Q \subseteq A$, $P_A \subseteq A$, and $P_Q \Rightarrow P$, we have $C' \subseteq C$. Also we have $A_Q \subseteq A$ and $F_Q=D(F,A)$, thus $C_Q = \pi_{A_Q,F'}(C')$, where F' is a new set of aggregation functions derived from $F_Q=D(F,A)$. For tuple $t_Q \in C_Q$, there exist at least one tuple $t \in C'$, such that $t_Q = \pi_{A_Q,F'}(t)$.

Because $C' \subseteq C$, we have $t \in C$. This conflict with the assumption made at the beginning, so Q can be fully contained by S.   □

Query Q matches semantic cache item S, but it isn't be concluded that S fully contains Q. That is to say, Q may only retrieve partial results from S. Definition 5 describes the case.

Definition 5. Consider a semantic cache item $S=<A,F,T,P,C>$ and a query $Q=<A_Q,F_Q, T_Q,P_Q,C_Q>$. We say S overlaps Q and this match type is overlapping match, if one of the following three conditions holds:

(1) $F_Q=D(F,A)$ and $P \wedge P_Q$ is satisfiable;
(2) $F_Q \approx D(F,A)$ and $P_Q \Rightarrow P$;
(3) $F_Q \approx D(F,A)$ and $P \wedge P_Q$ is satisfiable;

3.2 Query Matching Algorithms

Query matching is the process that determines whether query is answered by semantic cache. When a query is compared to a semantic cache item, there can be three different match types: containing match, overlapping match and disjoint match. Obviously, disjoint match represents that query can't retrieve any result from cache. So we focus on containing match and overlapping match based on the concepts defined before.

3.2.1 Containing Match

Implication problem is central to containing match [10]. We give some concepts of implication and show how to test containing match.

Definition 6. Consider a predicate $P=p_1 \wedge p_2 \wedge \cdots \wedge p_j$, where each p_j is a simple predicate. For all $(X<c_i)$, $(X \leq c_i)$ and $(X=c_i)$ in P, let $C_{up}^X = \min(c_i)$ for variable X and C_{up}^X satisfy all $(X = Y)$ in P. If X can be equal to C_{up}^X, C_{up}^X is closed; otherwise C_{up}^X is open. For all $(X>c_i)$, $(X \geq c_i)$ and $(X=c_i)$ in P, let $C_{low}^X = \max(c_i)$ for variable X and C_{low}^X is restricted by all $(X = Y)$ in P. If X can be equal to C_{low}^X, C_{low}^X is closed; otherwise C_{low}^X is open.

Theorem 2. Given a predicate P, which is a conjunctive of simple predicates. P is satisfiable if and only if for each variable X in P, $C_{low}^X < C_{up}^X$ or $C_{low}^X = C_{up}^X$ and both C_{low}^X and C_{up}^X are closed.

Proof. The proof is similar with [10]. We omit it. □

Theorem 3. Suppose C_{up}^X and C_{low}^X are derived from P_Q. P_Q implies P if and only if P_Q is unsatisfiable, or

(1) for any $(X = Y) \in P$ there exists $(X = Y)$ in P_Q, or $(X = Y)$ can be derived from P_Q;

(2) for any $(X<c) \in P$, $c > C_{up}^X$, or $c = C_{up}^X$ and C_{up}^X is open;

(3) for any $(X>c) \in P$, $c < C_{low}^X$, or $c = C_{low}^X$ and C_{low}^X is open;

(4) for any $(X \leq c) \in P$, $c \geq C_{up}^X$;

(5) for any $(X \geq c) \in P$, $c \leq C_{low}^X$;

(6) for any $(X=c) \in P$, $c = C_{up}^X = C_{low}^X$.

Proof. The proof is similar with [10]. We omit it. □

As a result of definition 6, theorem 2 and theorem 3, an algorithm to evaluate if P_Q implies P is provided next.

Algorithm 1 Implication (P_Q, P)

Input: P_Q and P. Both P_Q and P are conjunctive of simple predicates.
Output: If P_Q implies P, true is returned. Otherwise, false is returned.

Step1: Evaluate if P_Q is unsatisfiable.

   1.1 Construct the minimum range C_{low}^X, C_{up}^X for each X by scanning all $(X\ op\ c)$ in P_Q.

   1.2 Construct the links by scanning all $(X = Y)$ in P_Q. The links consist of equivalent variables.

   1.3 Modify C_{low}^X and C_{up}^X through the equivalent relation of links.

   1.4 If any $C_{low}^X < C_{up}^X$, or $C_{low}^X = C_{up}^X$ and both C_{low}^X and C_{up}^X are closed, P_Q is satisfiable. Otherwise, P_Q is unsatisfiable. In this situation, P_Q implies P and return true.

Step2: For any $(X = Y)$ in P, if it cannot be concluded from the links in 1.2, return false.

Step3: Process the compare predicates in P.

   3.1 For each $(X<c)$, if $c < C_{up}^X$ or $c = C_{up}^X$ and C_{up}^X is closed, then return false.

   3.2 For each $(X>c)$, if $c > C_{low}^X$ or $c = C_{low}^X$ and C_{low}^X is closed, then return false.

   3.3 For each $(X \leq c)$, if $c < C_{up}^X$, then return false;

   3.4 For each $(X \geq c)$, if $c > C_{low}^X$, then return false;

   3.5 For each $(X=c)$, if $c \neq C_{up}^X$ and $c \neq C_{low}^X$, then return false.

Step4: P_Q implies P; return true.

On the basis of theorem 1 and algorithm 1, we give algorithm AQCM (Aggregate Query Containing Match) to evaluate if S contains query Q.

Algorithm 2 AQCM(Q,S)

Input: Query Q, Semantic cache item S.

Output: If S contains Q, return true. Otherwise, return false.

Step1: If $T=T_Q$, go to Step2. Otherwise, return false.

Step2: If $A_Q \subseteq A$, go to Step3. Otherwise, return false.

Step3: For each aggregation function $f_i(a)$ in Q, evaluate the following situations.

   3.1 If f_i is MAX or MIN, then there exists $f_2(b)$ in F such that $f_1=f_2$ and $a=b$. If f_2 does not exist and a does not belong to A, then return false.

   3.2 If f_i is COUNT, then there exists $f_2(b)$ in F such that f_2=COUNT. If f_2 does not exist, then return false.

   3.3 If f_i is SUM, then there exists $f_2(b)$ in F such that f_2=SUM and a=b or f_2=COUNT and $a \in A$. If f_2 does not exist, then return false.

Step4: If $P_A \subseteq A$, use algorithm1 to evaluate if P_Q implies P.

3.2.2 Overlapping Match

Definition 5 describes overlapping match. We can modify algorithm 2 to evaluate condition (2). So we discuss overlapping match that described by condition (1) and (3). When Q and S overlap, S only provides partial result of Q. Thus Q has to be split into two subqueries: one part that retrieves the portion of Q satisfied by S, and the other part that cannot be satisfied. This process is called query trimming [3]. If there exists join predicate in a query, query trimming must process $(X \neq Y)$. For reducing the complexity of query trimming and processing, we assume that the join predicates of P_Q are equivalent to the ones of P. The algorithm AQOM (Aggregate Query Overlapping Match) for overlapping match is described next.

Algorithm 3 AQOM(S,Q)

Input: Query Q, Semantic cache item S

Output: If Q intersects with S, return true. Otherwise, return false.

Step1: If $T=T_Q$, go to Step2. Otherwise, return false.

Step2: If $A_Q \subseteq A$, go to Step3. Otherwise, return false.

Step3: For any aggregation function $f_1(a)$ in Q, if one of the following conditions holds, then go to Step5.

    3.1 If f_1 is MAX or MIN, then there exists $f_2(b)$ in F such that $f_1=f_2$ and $a=b$. If f_2 does not exist, then $a \in A$.

    3.2 If f_1 is COUNT, then there exists $f_2(b)$ in F such that $f_2=$COUNT.

    3.3 If f_1 is SUM, then there exists $f_2(b)$ in F such that $f_2=$SUM or $f_2=$COUNT and $a \in A$.

Step4: If none of condition3.1-3.3 holds, return false.

Step5: If $P_A \subseteq A$, go to Step 6. Otherwise, return false.

Step6: Construct the minimum range C_{low}^X, C_{up}^X for each X by scanning all $(X \ op \ c)$ in P_Q.

Step7: Construct the links by scanning all $(X = Y)$ in P_Q. Each link consists of equivalent variables.

Step8: Construct the minimum range C_{low}^X, C_{up}^X for each X by scanning all $(X \ op \ c)$ in P.

Step9: Construct the links by scanning all $(X = Y)$ in P. Each link consists of equivalent variables.

Step10: Compare the links of Step7 with the links of Step8. If the two link sets aren't equivalent, return false.

Step11: Modify C_{low}^X and C_{up}^X through the equivalent relations of links.

Step12: If any $C_{low}^X < C_{up}^X$, or $C_{low}^X = C_{up}^X$ and both C_{low}^X and C_{up}^X are closed, $P \wedge P_Q$ is satisfiable; return true. Otherwise, return false.

4 Conclusions

Existing researches of semantic caches emphasize SPJ query and don't investigate aggregate query. But aggregate queries are pervasive in massive database applications. We have to extend query matching to support aggregation and group operation so that semantic cache can be used in massive database applications. In this paper, we define formally semantic cache. Based on the definitions, we discuss when caches can answer an aggregate query. We focus on containing match and overlapping match. Then two match algorithms are presented.

    All the algorithms in this paper have been implemented in a massive database application StarTP. The application has a TB database and a lot of aggregate queries. The practice shows our algorithms are efficient. For the future research, we intend to work along more complex aggregate query matching, heuristic query matching and consistence maintenance.

Acknowledgments. This research has been supported by the National High-Tech Research and Development Plan of China under Grant No. 2003AA115410, No. 2003AA115210, No.2003AA111020 and No. 2004AA112020.

References

1. Dar S.,Franklin M. J.,Jonsson B. T.,Srivastava D.,Tan M.: Semantic data caching and replacement. In: Proc of 22th Int'l Conf on Very Large Data Bases. Mumbai (Bombay), India: Morgan Kaufmann, 1996. 330~341
2. Parke Godfrey ,Jarek Gryz.: Answering Queries by Semantic Caches. In: Proc of the 10th DEXA. Florence, Italy: Springer Verlag, August 1999. 485~498
3. Qun Ren, Margaret H. Dunham, Vijay Kumar.: Semantic Caching and Query Processing. IEEE Transactions on Knowledge and Data Engineering. 2003,15(1):192~210
4. Basu J.: Associative caching in client-server databases: [PhD dissertation]. Stanford University,1998
5. Dongwon Lee ,Wesley W. Chu.: Semantic caching via query matching for web sources. In: Proc of the 8th international conference on Information and knowledge management. Kansas City,Missouri: ACM Press, 1999. 77~85
6. M. Zaharioudakis, R. Cochrane, G. Lapis, H. Pirahesh, M. Urata.: Answering Complex SQL Queries Using Automatic Summary Tables. In: Proc ACM SIGMOD Int'l Conf on Management of Data. Dallas: ACM Press, 2000. 105~116
7. A. Gupta, V. Harinarayan, and D. Quass.: Aggregate-query processing in data warehousing environments. In: Proc of 21th Int'l Conf on Very Large Data Bases. Zurich: Morgan Kaufmann, 1995. 358~369
8. D. Srivastava, S. Dar, H.V. Jagadish, A. Levy.: Answering queries with aggregation using views, In: Proc of 22th Int'l Conf on Very Large Data Bases. Mumbai (Bombay),India: Morgan Kaufmann, 1996. 318~329.
9. S. Cohen, W. Nutt,s and A. Serebrenik.: Rewriting aggregate queries using views. In: Proc of 18th Symposium on Principles of Database Systems. Philadelphia: ACM Press, May 1999.
10. Sun X H.,Kamel N.,Ni L.M.: Solving implication problems in database applications. In: Proc ACM SIGMOD Int'l Conf on Management of Data. Portland: ACM Press, 1989. 185~192

A Parallel Modular Exponentiation Scheme for Transformed Exponents

Chin-Chen Chang[1] and Yeu-Pong Lai[2]

[1] Department of Information Engineering and Computer Science,
Feng Chia University, Taichung, Taiwan, 40724, R.O.C.
[2] Department of Information Engineering and Computer Science,
Chung Cheng Institute of Technology, National Defense University,
Tauyuan, Taiwan, 335

Abstract. This paper introduces an efficient method to compute modular exponentiation operations in parallel. For the parallel process of a modular exponentiation operation, the exponent is transformed into mixed radix digits first. Each digit is then an exponent for a partial result of the modular exponentiation operation. Because the computing processes for these partial results are highly independent, they can be carried out concurrently. The bases in these partial exponentiation operations can be pre-computed and used till the exponent moduli set changed. If the largest mixed radix digit is k-bits with respect to m exponent moduli, the time complexity for the proposed scheme is then $k + log_2m$. The performing complexity is very efficient, compared with other methods. Since the comparison is based on the same modular multiplication hardware, the performance is better if the fewer operations required. In the scenario of two exponent moduli, the performance improvment is approximately 40%. Finally, the proposed scheme is presented with a parallel algorithm for which the computing architecture is also illustrated in the paper.

Keywords: Mixed radix conversion, modular exponentiation, parallel computation, residue number system.

1 Introduction

One reason that modular operations are important to computer systems is that modular operations reduce the scalar of operands. Computing speed is therefore increased. Another reason for modular operations being important is the application of modular operations in cryptosystems. Most cryptosystems perform with the modular exponentiation and multi-exponentiation operations for security reasons, such as those cryptosystems based on RSA scheme[12] or ElGamal scheme[4] and their varieties[3]. However, for legal users, the massive computation required in modular exponentiation operations makes the use of these cryptosystems impractical. Thus, many researchers are dedicated to making these operations performed efficiently using hardware or software approaches[7].

The modular exponentiation operation computes $B^x \bmod N$, where the variables B, x and N are the base number, the exponent, and the modulus, respectively. The operation will be discussed in Section 2. Computation in the modular

J. Cao, W. Nejdl, and M. Xu (Eds.): APPT 2005, LNCS 3756, pp. 443–452, 2005.
© Springer-Verlag Berlin Heidelberg 2005

exponentiation operation can be divided into two separated parts: modular multiplication operation and modular square operation. In the sequential computing hardware, these modular square operations are compulsory. Most researchers focus on scanting the number of the modular multiplication operations in the modular exponentiation operation[1][6]. With fewer modular multiplication operations, the modular exponentiation operation can perform more efficiently. The optimal solution to accelerate computing is therefore to find an addition chain of the exponent[2][13]. Shorter addition chains mean that fewer modular multiplication operations are required in the modular exponentiation operation. However, finding an optimal addition chain is an NP-hard problem[6]. In other words, this problem can not be solved in polynomial time. Thus, most research is aimed at finding a feasible addition chain.

Actually, each term of an addition chain is a partial result of the modular exponentiation operation. These terms are used only once in computing, because the addition chain is for a certain exponent. For different exponents, the procedure to find addition chain should be performed again. To overcome the shortcoming of re-computing, some researchers then bring out the pre-computation concept to compute the partial results first and then store them. The most well-known method is the extended window method, abbreviated EWM[5]. In this method, the computation cost for partial results is ignored in modular exponentiation operations, since it is considered as a pre-processed procedure.

This paper proposes a parallel method to compute modular exponentiation. The exponent is presented in a mixed radix number format for certain moduli. Then, each mixed radix digit of the exponent is inputted into a process element of a parallel computing architecture. This computing architecture is regular so that it is easy to set up or be implemented. The multiple of these moduli is the largest number being presented in the mixed radix number system, so it should be larger than the scalar of the exponent.

In the next section, numerical examples are provided to explain the modular exponentiation operation. The proposed exponentiation operation is then described in Section 3. It is further explained using a parallel algorithm and a parallel architecture. Subsection 3.1 presents the MR conversion using examples. Subsection 3.2, then describes the application of the MR digits to computing the modular exponentiation in parallel. The computing performance of this parallel scheme and of two others is then presented in Section 4. Finally, the conclusions are given in Section 5.

2   Modular Exponentiation

Modular exponentiation operations are always performed using modular square operations and modular multiplication operations. In other words, an exponent is first transformed into a binary presentation. The modular exponentiation operation can then be performed using modular square operations and modular multiplication operations by referring to the bits of the exponent from left to right. The modular multiplication operation is performed when the pointed-to bit of the exponent is one. When the pointer shifts to the next bit, the modular square operation is then performed. The modular multiplication operation multiplies the temporary result by the base number and then performs the modular operation, where the temporary result is set as 1 initially. The modular square operation squares the temporary result and then performs the modular operation, too. This method can be considered to be the left-to-right method, since the pointer scanning the exponent is from left to right.

$$4^5 = 4^{101} = [(1 \times 4)^2]^2 \times 4 = 1024.$$

Besides, the modular exponentiation operation can be performed in a different way. The scan order can be from right to left. With this method, the modular multiplications are performed when the pointed-to bit of the exponent, scanned from the least significant bit to the most significant bit, is 1. The multiplier is squared when the pointer shifts further to the next bit. When the scanned bit is 1, the modular multiplication operation is performed. Otherwise, the modular multiplication operation is not performed. In the following example, the multipliers are 4^1, 4^2, and 4^4, which are obtained via a series of modular square operations.

$$4^5 = 4^{101} = (4^1)^1 \times (4^2)^0 \times (4^4)^1 = 1024.$$

Both the left-to-right and right-to-left methods performs sequentially. These instructions determine whether modular multiplication is to be carried out or not, depending on the corresponding bit value of the exponent. For instance, modular multiplication is performed when the bit value of the exponent is one. When the bit value is zero, modular multiplication is not performed. The right-to-left method scans the exponent from right to left. In other words, the scanning order is from the least significant bit (LSB) of the exponent to the most significant bit (MSB) of the exponent. The left-to-right method scans the exponent in the reverse order that is from the MSB to the LSB.

The right-to-left method can be intuitively considered as a kind of parallel computing methods, since the modular square operations can be performed on a processor distinct from the one performing the modular multiplication operation. Thus, the computing performance of the right-to-left method will be better than that of the left-to-right method if there are two processors available. The performance of the left-to-right method can be improved by using signed bits to eliminate the number of 1-bits so as to reduce the number of modular multiplication operations. For the right-to-left method, the computing performance can be improved only by improving the hardware architecture or by designing an efficient modular operation technique. The Montgomery multiplication algorithm is a good solution for accelerating the modular multiplication operation since the modulus can be transformed to a power number of 2, 2^r[10]. The modular operations are then to simply fetch the least r-bits if the operations performed in a regular computer architecture.

The left-to-right method is a "strict" sequential method. The most popular way to improve computing performance is to transform the format of the exponent, such as by introducing a signed number to representing the exponent. These techniques minimize the number of non-zero bits in the transformed exponent so that fewer modular multiplication operations are required[11]. The exponent is transformed into the non-adjacent form (NAF) or divided into windows (EWM). EWM is the extended window method.

3   The Proposed Scheme

According to the review of modular exponentiation operations in the previous section, modular exponent operations can be performed in two different ways according to the scanning order. The right-to-left method performs more efficiently than the left-to-right method, if the computing powers of the process elements are the same in the two computing architectures for these two methods. When the right-to-left method performs on two process elements computing architecture, the computing time is the time spent only in performing the modular square operations. That is because the modular multiplication operation can compute concurrently in the other processor while the modular square operation is performing.

This section presents a more efficient way to compute the modular exponentiation in parallel. Unlike the right-to-left method, the computing performance of the proposed method can be improved further when more process elements are provided. However, the exponent should be transformed into mixed radix digits (MR digits) first. Each process executes a partial modular exponentiation for a corresponding MR digit of the exponent. Finally, the partial results are multiplied with each other for the final result.

Subsection 3.1 introduces the mixed radix conversion (MRC) for transforming the exponent into several MR digits. The transformation is simple compared to the complexity of the modular multiplication (or square) operation, so that the computing cost for the transformation can be ignored. The parallel architecture, described in Subsection 3.2, is so regular that the architecture is only a reconfiguration of the existing parallel architectures.

3.1   Mixed Radix Conversion

Generally, in computing systems, a large scalar is always represented by several smaller digits. The representation should be reversible so that the original number can be reconstructed according to these smaller digits. The ways to transform numbers are the residue number transformation and the mixed radix conversion. The residue number system is used extensively and efficiently in digital signal processing. This is because the addition, subtraction, and multiplication operations on residues of two numbers are equivalent to these operations on these two numbers. The "equivalent" stands for those two numbers being the same, which are the number transformed from the operated residues and the result operated direct on the two numbers. These residues are always smaller than the original binary number so that the operations on the residues are more efficient than on the binary numbers. The other representation method, the MR system, also has this property of smaller scalar operands. Moreover, the MR system is superior to the RNS in several application areas, such as in the sign determination, magnitude comparison and overflow detection.

A binary number x can be transformed to a set of numbers $(a_1, a_2, \ldots, a_m)$ with respect to the moduli $q_1, q_2, \ldots, q_m$, where x is equal to $a_1 + a_2 \times q_1 + \ldots + a_m \times q_1 \times q_2 \times \ldots \times q_{m-1}$ and $0 \le a_i < m_i$ for $i = 1, 2, \ldots, m$. For example, the moduli set (q_1, q_2, q_3) is $(3, 5, 8)$ and the number x is 116.

Conversion Phase:

Let the moduli set (q_1, q_2, q_3) be $(3, 5, 8)$ and let the number x be 116. Therefore, the number x is divided by the moduli, q_1, q_2, q_3, sequentially to obtain the MR digits. The conversion procedure is as follows:

$$116 \div 3 = 38 \cdots 2$$
$$38 \div 5 = 7 \cdots 3$$
$$7 \div 8 = 0 \cdots 7$$

Thus, the MR presentation of x is $(2, 3, 7)$.

Reversed Phase:

The number x can be derived from $(2, 3, 7)$ with respect to $(3, 5, 8)$ by the following equation.

$$x = a_1 + a_2 \times q_1 + a_3 \times q_1 \times q_2 = 2 + 3 \times 3 + 7 \times 3 \times 5 = 2 + 9 + 105 = 116.$$

According to the description of the modular exponentiation operation in the previous section, the operation consists of modular multiplication and modular square operation. If the exponent is in the format of the MR system, the exponentiation operation can be divided into several partial exponent computations. The exponent of each partial computation is the related MR digit, and the base is the exponentiation of the original base. These varietal bases can be computed efficiently as the moduli set $(q_1, q_2, \ldots, q_m)$ is determined for representing the exponent. Therefore, these bases, $B_1, B_2, \ldots, B_m$, can be pre-computed and saved for any exponent. For the modular exponentiation $B_1^{a_1} \times B_2^{a_2} \times \ldots \times B_m^{a_m} \bmod N$, these bases are always in the interval of $[0, N-1]$. The partial exponentiation computation, $B_j^{a_j} \bmod N$, is then highly independent so that the parallel computation is used for improving computing performance.

Let the base B be 7 and the exponent be 116. The exponent can be transformed into MR digits, $(2, 3, 7)$ with respect to the moduli set $(3, 5, 8)$. For the base, the computations 7^3 and $(7^3)^5$ can be pre-computed. The computation of $(7^3)^5$ is the exponentiation of 7^3 so that the pre-computation of $(7^3)^5$ can be computed very efficiently. The two pre-computations 7^3 and $(7^3)^5$ can be considered as two "bases" for the two corresponding MR digits, 3 and 7. The computation of 7^{116} is divided into three partial exponentiation computations of three bases, 7, 7^3 and 7^{15}.

Therefore,

$$7^{116} = 7^{2+3\times3+7\times3\times5} = 7^2 \times (7^3)^3 \times (7^{15})^7 = (B_1)^2 \times (B_2)^3 \times (B_3)^7 \text{ , where}$$
$B_1 = 7, B_2 = 7^3$ and $B_3 = 7^{15}$.

The partial exponents (2,3,7) are the MR digits of the original exponent. The pre-computations can be used for different computations if there is no change in the base of the exponentiation operations and in the moduli set used in exponent transformation. Actually, the base is always fixed in the exponentiation operation for the cryptosystems based on the ElGamal signature scheme and its variations. The moduli set is used for transforming the exponent within a certain field so that the moduli may not be changed after chosen. Thus, the pre-computations can be stored in ROM for efficiently fetching. Above all, for accelerating the exponentiation operations, the partial computations can be computed concurrently. Each processing element fetches a certain exponent base from ROM memory. The parallel computing method will be introduced in the following section.

3.2 Parallel Computing

Parallel computation is a way to improve computing performance. Sometimes, parallel computation should be applied to certain computing architectures. The computing architectures are specific with special interconnections. To generally apply the parallel computation scheme, the parallel programming grammar is designed. The program can also manipulate the computation with a parallel computing hardware or a multi-thread processor. Subsection 3.2.1 proposes an algorithm for parallel processing. The next subsection describes the hardware interconnection.

Proposed Algorithm. To enable operations to perform concurrently, the algorithm should specify which computing processors operate at the same time. The commands are therefore different from the usual ones. Every computing processor also performs using a memory distinct from others. The following algorithm computes the exponent operation. The depth of this algorithm is $\lceil log_2 m \rceil$, where the variable m is the number of the MR moduli. The first concurrent level computes all the m partial results of the exponentiation operation. Thus, it consumes operating time. In other levels, the only operations left are multiplication operations. This algorithm can also be applied to modular exponent operations by replacing the multiplication operations with modular multiplication operations.

Algorithm // Parallel computing
Begin
    Forall processor p_i, $1 \le i \le m$ do in parallel
      Begin
        Processor p_i: $partial = B^{a_i}$;
        //every processor fetches B from its memory
      End
    For $d = 1$ to $\lceil log_2 m \rceil$ do
      Begin
        Forall processor p_i, $i = 1 + 2^d \times j, 0 \le j \le \lceil m/2^d \rceil - 1$ do in parallel
        Begin
          Processor p_i: SendMsg($p_{i+2^{d-1}}, partial$),
                      ReceiveMsg($p_{i+2^{d-1}}, partial'$);
          Processor $p_{i+2^{d-1}}$: SendMsg($p_i, partial$),
                      ReceiveMsg($p_i, partial'$);
        End
        Forall processor p_i, $i = 1 + 2^d \times j, 0 \le j \le \lceil m/2^d \rceil - 1$ do in parallel
        Begin
          $partial = partial \times partial'$;
        End
      End
    End

In the above algorithm, the temporary results are stored in the variables *partial* of each processor and transferred to its couple processor. In every level, the couple processor for each processor is different. The couple processor of the i-th processor is the $(i + 2^{d-1})$-th processor, where the variable d is the depth of levels. After every processor receives the desired partial result, the processors compute and update the data. This procedure is performed for $\lceil log_2 m \rceil$ times repeatedly to obtain the final result.

Actually, the exponent operations are only performed at the first level. Every process fetches a pre-computed base for a certain exponent moduli set from a distinct memory space. The exponent in the i-th processor is the corresponding MR digit a_i of the original exponent. Since these MR digits are smaller than the related moduli, the complexity of these partial exponentiation operations is much less than that of the original exponentiation operation.

Model of Parallel Computation. This section describes the parallel computing architecture. In this architecture, m processors work in parallel, except during data transmission. Each processor concurrently computes a partial result that is an exponentiation of a certain base fetched from its memory. The exponent for this partial exponentiation operation is the corresponding MR digit of the original exponent. These partial results are then multiplied together to obtain the final result of the exponentiation operation. Multiplication is performed in parallel within $\lceil log_2 m \rceil$ levels. The number of interconnection lines is $2 \times (2^{\lceil log_2 m \rceil} - 1)$. This architecture is so regular that almost all the parallel computing hardware can be reconfigured to manipulate the operations.

4   Discussions

In above, numbers can be transformed into mixed radix digits via several division operations. These division operations are not time consuming compared with the exponentiation operation. The proposed algorithm consists of two computing parts: (1) performing the partial exponentiation operations with the exponent of MR digits and (2) multiplying two partial results in parallel within $\lceil log_2 m \rceil$ levels. The time complexity of multiplication operations can be considered to be the complexity of only $\lceil log_2 m \rceil$ multiplication operations, since these operations perform concurrently in different processors. For the same reason, the number of partial exponentiation operations can be also considered to be one.

However, the computing performance is very hard to analyze, since the performance is highly dependent on the relation between the exponent and the moduli. If the MR digits are small in scalar, the operation will be more efficient. The digit corresponding to a modulus with a larger scalar may not be larger than the digit corresponding to a modulus with a smaller scalar. Although the previous claim is absolutely correct, the moduli are also sorted in the order $q_1 < q_2 < \ldots < q_m$. The MR digits therefore satisfies $a_1 < q_1, a_2 < q_2, \ldots, a_m < q_m$. Thus, the MR digit in a preceding position is in a smaller domain than those after it. In other words, the preceding digit has a higher probability of having a smaller scalar.

Let the largest MR digit a_j be in k bits. The complexity for computing partial results is then k square operations when the right-to-left scheme is applied, since the largest exponent dominates the complexity of the concurrent operations. Table 1 illustrates the performance comparison among Right-to-left scheme, Left-to-right scheme and the proposed scheme.

For further analyzing the improvement of parallel computing in computing performance, the complexity comparisons are presented for $m = 2$. If the computing powers for implementing these three methods are the same, the operation time proportions to the number of multiplication (or square) operations. To the proposed scheme, the more process elements are in the parallel architecture, the

Table 1. Performance comparison

	Multiplication	Square
Right-to-left scheme	0	$n^{\diamond}$
Left-to-right scheme	$n/2$	n
Proposed scheme	$log_2 m^{\ddagger}$	$k^{\natural}$

$\diamond$: The variable n is the number of bits to present the exponent.
$\ddagger$: The variable m is the number of moduli.
$\natural$:The variable k is the number of bits to represent the larger MR digit.

Table 2. Performance comparison for certain numbers of moduli

bits	1	2	3	4	5	6	7	8	9	10
range	2	4	8	16	32	64	128	256	512	1024
q_1	null	2	7	15	31	63	127	255	511	1023
q_2	null	3	6	14	29	62	125	254	510	1022
Max		6	42	210	899	3906	15875	64770	260610	1045506
Bits of Max.		2.58	5.39	7.71	9.81	11.93	13.95	15.98	17.99	19.99
R-to-L		3	6	8	10	12	14	16	18	20
L-to-R		5	9	12	15	18	21	24	27	30
Proposed		3	4	5	6	7	8	9	10	11

more improvement is. In the worst scenario, the MR digits are in the same bit length of the corresponding modulus. The numerical example is tabulated in Table 1. The first row is for the maximal bit length of moduli. The moduli should be within the range from 0 to the range number shown in the second row. For example, if the bit length is 10, the moduli should be smaller than 1024. The following rows are then for the co-prime moduli chosen within the range. The exponent therefore should be less than the number in the "Max" row. The bit length of the largest exponent is presented in the following row. The last three rows are the numbers of modular multiplication (or square) operations for these three methods. Table 2 shows the improvement is magnificent. For example, for two moduli of 5 bits, the modular multiplication (or square) operations perform for 6 times only. With the right-to-left method, there are 10 operations. The improvement of the proposed method compared to the right-to-left method is 40%, which is obtained from $(10 - 6)/10$. Thus, the proposed method performed very efficiently.

5   Conclusions

Cryptography, which has become more and more important recently, requires that many modular exponentiation operations be performed. Thus, the performance of a cryptosystem depends on the performance of computing the modular exponentiation. Parallel processing is a good strategy to accelerate computing in computers.

However, modular exponent operations are strictly sequential. The operations can not be computed using parallel architecture. To overcome this restriction, a transformation of the exponent was designed. The exponent should be transformed into MR digits as mentioned in Subsection 3.1. Each MR digit is used as an exponent for computing a partial exponentiation result. These computing procedures are highly independent so that parallel processing can be applied. Each process element in parallel architecture computes a partial exponentiation result. The exponent for the partial exponentiation is the corresponding MR digit. The range of the exponent for the partial results depends on the scalar of the moduli. Therefore, the moduli are in increasing order. The larger moduli are placed in the more posterior positions. This also introduces an advantage in pre-computing the bases. In addition, in computing systems, large numbers may be stored in the residue format so that conversion from residues to MR digits is also important. The conversion procedure was explained in Section 3.1. If it is assumed that the largest MR digit has k bits, the parallel algorithm then has a time complexity of $k + \lceil log_2 m \rceil$, where m presents the number of moduli. The performance improvement of the proposed method is very notable. Thus, this paper introduces a good way to accelerate the computing of modular exponentiation operations.

References

1. ARNO, S., AND WHEELER, F. S. Signed digit representations of minimal Hamming weight. *IEEE Transactions on Computers 42* (1993), 1007–1010.
2. BOS, J., AND COSTER, M. Addition chain heuristics. *In Advances in Cryptology-Proceedings of Crypto '89 435* (1990), 400–407.
3. CHANG, C. C., AND LAI, Y. A flexible data-attachment scheme on e-cash. *Computers and Security 22* (2003), 160–166.
4. ELGAMAL, T. A public key cryptosystem and a signature scheme based on discrete logarithms. *IEEE Transactions on Information Theory 31* (1985), 469–472.
5. GORDON, D. M. A survey on fast exponentiation methods. *Journal of Algorithms 27* (1998), 129–146.
6. JOYE, M., AND YEN, S. Optimal left-to-right binary signed-digit recoding. *IEEE Transactions on Computers 49* (2000), 740–748.
7. LAI, Y., AND CHANG, C. C. An efficient multi-exponentiation scheme based on modified Booth's method. *International Journal of Electronics 90* (2003), 221–233.
8. LAI, Y., AND CHANG, C. C. Parallel computational algorithm for generalized Chinese remainder theorem. *Computers and Electrical Engineering 29* (2003), 801–811.
9. MILLER, D. F., AND MCCORMICK, W. S. An arithmetic free parallel mixed-radix conversion algorithm. *IEEE Transactions on Circuits and systems–II: Analog and Digital Signal Processing 45* (1998), 158–162.
10. MONTGOMERY, P. Modular multiplication without trail division. *Mathematics of Computation 44* (1985), 519–521.
11. OKEYA, K., AND SAKURAI, K. Use of montgomery trick in precomputation of multi-scalar multiplication in elliptic curve cryptosystems. *IEICE Transactions on Fundamentals E86-A* (2003), 98–112.
12. RIVEST, R. L., SHAMIR, A., AND ADLEMAN, L. A method for obtaining digital signatures and public-key cryptosystems. *Communications of the Association for Computing Machinery 21* (1978), 120–126.
13. ROOIJ, P. Efficient exponentiation using precomputation and vector addition chains. *In Advances in Cryptology-Proceedings of Eurocrypt'94 950* (1994), 389–399.

Content Selection Model for Adaptive Content Delivery[*]

Chen Ding[1], Shutao Zhang[2], and Chi-Hung Chi[3]

[1] School of Computer Science, Ryerson University, Canada
[2] School of Computing, National University of Singapore, Singapore
[3] School of Software, Tsinghua University, Beijing, China 100084
chichihung@mail.tsinghua.edu.cn

Abstract. In order to adapt content delivery to different client capabilities and preferences, we propose a content selection model to automatically classify HTML content based on its functionality, then map client descriptions on preferences and device capabilities into our classification scheme, and finally selectively deliver the content which users want and which devices can handle. The experiment shows that our content selection model could reduce HTML object size, object latency and page latency. Therefore, it is effective in saving network resources and improving clients' access experiences.

1 Introduction

In today's web, HTML is the primary language used to create web pages and HTML traffic is estimated to represent 20% of total web traffic [2]. Most of HTML tags are designed to help web publishers present their information in a certain format, while some of them are not. For example, values of "keywords" and "description" attribute in "meta" tag are used by search engines to identify keywords in the page, while users may not care about them. Edge Side Includes (ESI) [4] language markups may be inserted into HTML pages to facilitate content caching and selection, and they are more for web intermediaries instead of end users. Usually, web clients are only interested in content which could be presented through browsers. So, if HTML content is not designed for presentation purpose, it doesn't have to be delivered to clients. But servers still need to keep them because they are of interest to other parties such as search engines or web proxies. Disregarding the content itself, sometimes, web users may also have their particular preferences when they request web content, such as language preference, interest on advertisement, privacy preservation, etc. Therefore, content should be selectively delivered based on client preferences.

In order to deliver the content based on client capabilities and preferences, many of the current solutions have proposed different transcoding algorithms on multimedia objects, either by reducing the quality of the object or converting to a different media [1] [5] [6] [10] [11] [13]. In this paper, we are trying to solve this problem in a more general way. We work on the container first, which is the HTML page itself, and then on all the embedded multimedia objects. We propose a content selection model to

---

[*] This research is supported by the funding 2004CB719400 of China.

J. Cao, W. Nejdl, and M. Xu (Eds.): APPT 2005, LNCS 3756, pp. 453 – 460, 2005.
© Springer-Verlag Berlin Heidelberg 2005

automatically classify HTML content based on its functionality and deliver only the content which users want and which devices can handle.

There are several contributions by our proposal: Its aim is to improve the content delivery performance of the HTML document, which includes both the HTML page and all the embedded objects. Most of current solutions only focus on one. The selection model is automatic. The content classification is based on heuristics instead of human-edited descriptions. There is no overhead on web publishers, both in server processing power and in manpower. It is especially suitable for those existing HTML documents. It is flexible. When necessary, web publishers could provide guidance on content selection, and for those multimedia objects, we could also use existing transcoding algorithms to further adapt the content.

2 Related Work

To enable web content publishers to provide customized content to different clients, IETF (Internet Engineering Task Force) have proposed general frameworks in which web content can be modified on the way from servers to clients, including Internet Content Adaptation Protocol (ICAP) [7] and Open Pluggable Edge Services (OPES) [12].

Besides the framework, there are extensive researches on content adaptation. Web content adaptation can be performed on web servers, web intermediaries (i.e. proxies), or web clients. Many of the solutions are proxy based [1] [5] [14]. Fox et al. [5] developed a system to distill or transcode images when they pass through a proxy, in order to reduce the bandwidth requirements. Chandra and Ellis [1] presented techniques of quantifying the quality-versus-size tradeoff characteristics for transcoding JPEG images. Spyglass Prism [14] is a commercial transcoding proxy. There are also other approaches trying to address the issue of content adaptation on the server side. They either provide external content annotations or give guidance on how to process the web content. Hori et al. [6] proposed to annotate HTML pages to guide the transcoding procedure. Mogul et al. [9] [10] proposed a server directed transcoding scheme. Most of content adaptation solutions focus on one or two types of multimedia objects without looking into the whole web document. InfoPyramid [11] [13] is a representation scheme for handling the Internet content (text, image, audio and video) hierarchically along the dimension of fidelity (in different qualities but in the same media type) and modality (in different media types).

In addition, we need descriptions about the client's preference and capabilities. W3C has proposed the Composite Capability and Preference Profile (CC/PP) [3] to achieve this goal. Wireless Application Protocol (WAP) Forum has proposed a similar approach called User Agent Profile (UAProf) [15] to handle clients' descriptions. Both CC/PP and UAProf are based on Resource Description Framework (RDF) and their goals are describing and managing software and hardware profiles. In this study, we will use CC/PP or UAProf for client side descriptions.

3 Content Selection Model

The general process of content selection can be viewed as follows. When a client wants to access certain HTML content from a server, it sends a request to the server together with client descriptions on its preferences and hardware capability constraints. Upon receiving the request, the web server retrieves the content and passes it to a content selector, together with client descriptions. The content selector then selects HTML content based on client descriptions, and passes the selected content to the client.

To do the content selection, first, HTML content should be classified based on its functionalities, and we refer to this process as HTML content classification. Since this classification scheme is not known to clients, as the second step in the content selection model, we need some mechanisms to convert clients' own descriptions on their preferences and device capabilities to our pre-defined content classes.

3.1 HTML Content Classification

HTML documents contain various types of content, and each piece of content has its own functionalities. Without loss of generality, we define 6 content classes. The definition is for HTML documents. For dynamic pages written in server-side scripts, since web servers will convert them to HTML before sending them to clients, this classification scheme can also apply.

- C_PRESENTATION – content for presentation purposes.
- C_META - meta information of a page, e.g. meta keywords which could be used by search engines, meta author to recognize page authors.
- C_DYNAMIC – dynamic content which will be executed at client side, e.g. scripts written in JavaScript, Java Applets, plug-ins.
- C_EXTENSION – extensional markup in a HTML page, e.g. ESI markups.
- C_DOCUMENTATION – comments in a page.
- C_REDUNDANT – redundant content in a page, e.g. blank spaces, new lines, tabs.

Most of HTML content can be classified into the first big class. We have further defined its sub-classes. We haven't exhausted all the situations and only list commonly used functionalities.

- C_P_HTML_STRUC – the structure information of a HTML page, e.g. <html>, <head>, <body>, <frame>.
- C_P_HTML_MM – multimedia effect of a page, e.g. background image, background sound.
- C_P_TEXT – information related to text content in a page, including text itself, format, style, layout, etc.
  - C_P_TEXT_FORMAT – format information of text content, e.g. text font, color, size or style sheet information.
  - C_P_TEXT_LAYOUT – layout and location of text content in a page, e.g. <table>, <p>, <ul>.
  - C_P_TEXT_CONTENT – actual text content enclosed by various tags, e.g. "hello world" in <p>hello world</p>.
- C_P_OBJECT – embedded objects in a HTML page.
  - C_P_IMAGE – embedded images in a page. Example sub-classes include:
    - C_P_IMAGE_JPG – images with file type .jpg.

- • C_P_IMAGE_GIF – images with file type .gif.
  - • C_P_IMAGE_PNG – images with file type .png.
- • C_P_VIDEO – embedded videos in a page. Example sub-classes include:
  - • C_P_MPEG – videos with file type .mpeg.
  - • C_P_REAL – videos with file type .ra.
- • C_P_AUDIO – embedded audios in a page.

The following table illustrates classification for some HTML content.

Table 1. HTML Content and Related Classes

HTML content	Classes (in the form of class.sub-class)
<p>Good Morning</p>	<p>, </p> -- C_PRESENTATION.C_P_TEXT.C_P_TEXT_LAYOUT , -- C_PRESENTATION.C_P_TEXT.C_P_TEXT_FORMAT "Good Morning" -- C_PRESENTATION.C_P_TEXT.C_P_TEXT_CONTENT
<meta name = "keywords" content="xml">	C_META
<object data="canyon.png" type="image/png">	C_PRESENTATION.C_P_OBJECT.C_P_IMAGE.C_P_IMAGE_PNG
<SCRIPT type="text/vbscript" src="http://someplace.com/progs/vbcalc"></SCRIPT>	C_DYNAMIC.C_DYNAMIC_JS
<!-- comments... -->	C_DOCUMENTATION
<esi:when test="!$(QUERY_STRING)"> <esi:include src="viewsource.html?rp= $(REQUEST_PATH)"/> </esi:when>	C_EXTENSION.C_EXTENSION_ESI

3.2 Preferences and Capability Mapping

We have defined a classification scheme for HTML content. However, clients may not describe their preferences and device capabilities based on this scheme. We need some mechanisms to map their descriptions into our classification scheme. First, let us introduce some definitions.

R: The set containing all the rules.

D: The set containing all the descriptions of client preferences and device capabilities.

A: The set of transformation actions, "+" for keeping the content, "-" for removing the content, and "t" for transforming the content.

C: The set of all possible content classes, including ones we might not list.

DESCRIPTION: A description d is a string of characters to describe one of client's preferences or device capabilities.

For example, "the device is image capable" is a description. Usually clients send their preferences and device capabilities in a sequence of descriptions $<d_1, d_2, ..., d_n>$, e.g. "image capable, screen size: 15x9 chars".

RULE: A rule Ru is to map a description d to a set S where $d \in D$, $S \in P(A \times C)$, and $P(A \times C)$ is power set of $A \times C$.

For example, a description d specifies that a client does not need the documentation. Then there is a corresponding rule Ru:

$$Ru(d) = \{(\text{``-''}, \ C\_DOCUMENTATION)\}$$

It is interpreted as: documentation should not be included in final delivered content. If the class has sub-classes, this rule should be applied on all sub-classes, too.

Let's look at another example. A description d specifies that a client can only process GIF images. Then the rule Ru will be:

$$Ru(d) = \{(\text{``+''}, \ C\_P\_IMAGE\_GIF), (\text{``t''}, \ C\_P\_IMAGE\_OTHER\_THAN\_GIF)\}$$

where $C\_P\_IMAGE\_OTHER\_THAN\_GIF$ should be any class which is the sub-class of C_P_IMAGE and which is not $C\_P\_IMAGE\_GIF$.

After this rule is applied, all the gif images should be kept and all other types of images should be transformed. The rule itself does not specify which kind of transformation will be performed. It depends on the actual implementation of the transformation engine. It could be transformation into a gif image or simply removing it.

In general, when a client specifies it is capable (incapable) of processing a content class or (not) interested in a particular class of content, all the content classified into this class should (not) be delivered to the client, and sometimes, transformation is a necessary step before the delivery.

The mapping procedure described above is generic and flexible. In the current stage, we only implemented two actions (keeping or removing content). In our future work, the policy engine similar to [11] could be implemented for transcoding embedded objects in HTML documents. In the remaining of the paper, we only consider these two actions.

3.3 HTML Content Selection

After the mapping, clients' requirements can be described in the context of our classification scheme. Then, we can select appropriate HTML content for clients.

Figure 1 illustrates the procedure of making the content selection. Basically, for each fragment of HTML content, we check its classes and decide whether it should be selected or removed, based on client's preferences and other parameters including the machine's local conditions and network status. After selecting all necessary fragments, we can combine them to form a new HTML document. When we define rules used in mapping procedure, we have to be careful to make sure applying rules will not affect the completeness of the remaining HTML document.

Below, we list some typical scenarios of applying the content selection model.

Scenario 1: Keep all the content except those classified into C_REDUNDANT. Typically, with enough resources, a client does not put any limit on the content he/she will accept. Only redundant content is removed.

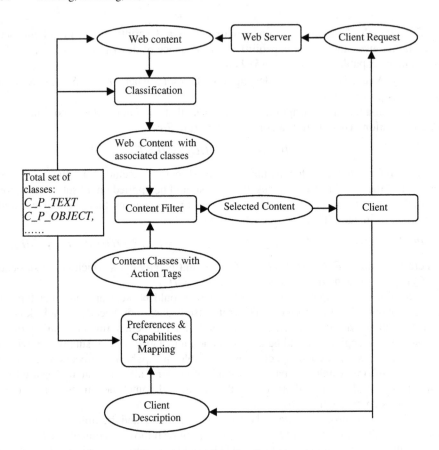

Fig. 1. Content Selection Procedure

Scenario 2: Keep content for presentation only. All the content classified into C_PRESENTATION is selected and other content removed. A client might be interested in all the content which can be displayed through a browser, but does not care about others, for example, documentation in the page source.

Scenario 3: Keep content for presentation only and accept all content except images. For example, client's device is a mobile phone whose hardware is not image capable. So, image content has to be filtered out. In this case, we should deliver content in C_PRESENTATION – C_P_IMAGE.

Scenario 4: Keep static content for presentation only. In this case, scripts or applets in the content have to be removed. We should deliver content in C_PRESENTATION – C_DYNAMIC.

Scenario 5: Keep text content only. No embedded objects (images, audio, video, etc.) and scripts are left. The client is only able to display text content due to its limited device capability or the client prefers to save cost on obtaining information from the web. In this case, we should deliver the content in C_P_TEXT+ C_P_HTML_STRUC.

4 Performance Study

In this section, we will study the impact of content selection model on HTML pages retrieval. The object URLs are from NLANR trace [8] on August 2003. Figure 2 shows the page latency in different scenarios. We observe that page latency reduction in scenario 1 and scenario 2 are less than HTML object latency reduction for objects in all size ranges. This is probably because in both scenarios, content selection model does not filter out any embedded objects. In scenario 3, it reduces the page latency much more significantly than it does for HTML object latency for objects greater than 1K bytes. There are two possible reasons. One is that in scenario 3, the selection model filters out the inline images which are the majority of embedded objects. The other reason is that when a portion of content is filtered out, it makes some of the embedded objects "nearer" to the beginning of the HTML page. Therefore, browser can download them earlier. In scenario 4, although it filters out roughly the same percentage of content from HTML document as in scenario 3, its page latency reduction is much smaller. This is because sizes of images embedded in a page are usually much larger than sizes of scripts. Scenario 5 reduces page latency the most. This is expected as scenario 5 filters out all the embedded objects.

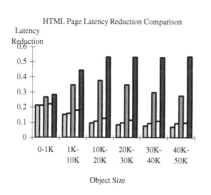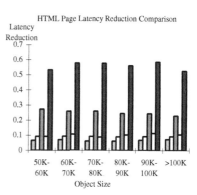

Fig. 2. HTML Page Latency

5 Conclusions and Future Work

In this paper, we propose a content selection model to address the adaptive content delivery problem. On the server side, the HTML content is classified based on its functionality. Client preferences and device capabilities are then mapped to this classification scheme by a rule engine. Finally, content selection algorithm selects appropriate content to deliver to the client. From our performance study, we can conclude that the content selection model is effective in saving network resources and improve clients' access experiences by reducing HTML object sizes and page latency. The model is especially effective in cases where the HTML object size is small and clients want to filter out some embedded objects.

References

1. S. Chandra, C. S. Ellis, JPEG Compression Metric as a Quality Aware Image Transcoding, In *Proceedings of the 2nd USENIX Symposium on Internet Technologies and Systems*, 1999.
2. ClickZ Internet Statistics and Demographics, http://www.clickz.com/stats/.
3. Composite Capability and Preference Profile, http://www.w3.org/Mobile/CCPP/.
4. ESI language specification 1.0, http://www.esi.org, 2000.
5. Fox, S. D. Gribble, E. A. Brewer, and E. Amir, Adapting to Network and Client Variability via On-Demand Dynamic Distillation, In *Proceedings of 7th International Conference on Architectural Support for Programming Languages and Operating Systems (ASPLOS)*, 1996.
6. M. Hori, G. Kondoh, K. Ono, S. Hirose, and S. Singhal, Annotation-Based Web Content Transcoding, In *Proceedings of The 9th WWW Conference*, 2000.
7. Internet Content Adaptation Protocol (I-CAP), http://www.i-cap.org.
8. IRCACHE Proxy Traces, http://ircache.nlanr.net.
9. B. Knutsson, H. H. Lu, and J. C. Mogul, Architecture and pragmatics of Server directed transcoding, In *Proceedings of the 7th International Workshop on Web Content Caching and Distribution*, 2002.
10. J. C. Mogul, Server Directed Transcoding, In *Computer Communications* 24(2):155-162, 2001.
11. R. Mohan, J. R. Smith, and C. S. Li, Adapting Multimedia Internet Content for Universal Access, In *IEEE Transactions on Multimedia*, 1(1):104–114, 1999.
12. Open Pluggable Edge Service (OPES), http://www.ietf-opes.org.
13. J. R. Smith, R. Mohan, and C. S. Li, Transcoding Internet Content for Hetero-geneous Client Devices, In *Proceedings of IEEE International Symposium on Circuits and Systems (ISCAS)*, 1998.
14. Spyglass-Prism, http://www.opentv.com/support/primer/prism.htm.
15. User Agent Profile, WAP forum, http://www.wapforum.org/what/technical/SPEC-UAProf-19991110.pdf.

Dynamic Service Provisioning
for Multiplayer Online Games

Jens Müller, Rafael Schwerdt, and Sergei Gorlatch

Westfälische Wilhelms-Universität Münster, Germany

Abstract. Multiplayer online games have become a popular class of distributed applications with an enormous amount of running Internet-based game sessions. The basic concept of how to provide game services for users has not changed for years: High-bandwidth, dedicated game servers are statically set up to continuously run game sessions, regardless of how many users actually play. This straightforward approach is inefficient, because it does not take the current user demand into account, thus wasting resources. In this paper, we present a novel system architecture for organizing dynamic, on-demand game services for single-server online games. Our system allows users to book game services for immediate play or some time in advance. The system takes the users' demands into account and dynamically sets up the required server resources in an efficient way. In contrast to the usually offered flat-rate rental of servers on at least a monthly basis, our system allows to charge users depending on the actual services usage and to incorporate new pay-per-use business models.

1 Introduction

Hundreds of thousands of users regularly play online games over the Internet using personal computers or entertainment consoles. With regard to the technical infrastructure, games can be categorized into two classes: Single-server games and multi-server games. Especially *Massively Multiplayer Online Games (MMORPG)*, which provide a persistent game world for thousands of users, require a sophisticated multi-server architecture in order to ensure a fluent and responsive game experience. The other, not less popular class of single-server online games consists of action, strategy or sports games that are played by two to one hundred users in a single game session.

The most important distinction between these two classes of games is that users can set up their own servers for single-server games while this is not possible for multi-server MMORPGs. In MMORPGs, the developer does not publish the server software: Users have to join publisher-operated game sessions and pay a monthly fee to be able to play on these servers. For single-server games, however, game developers usually provide the server software for free, and users and third-party service providers can set up their own game sessions. Third-party game hosters rent out servers for single-server action games (especially for *First Person Shooter (FPS)* games like *Counterstrike* or *Battlefield 1942*) to users on a monthly basis. This way, servers continuously run a session of a particular game independently of the actual user demand. The session location service *gamespy* [2] regularly reports over 60.000 statically running sessions for the ten most popular games, while the number of users actually playing is quite

J. Cao, W. Nejdl, and M. Xu (Eds.): APPT 2005, LNCS 3756, pp. 461–470, 2005.
© Springer-Verlag Berlin Heidelberg 2005

dynamic and ranges from about 50.000 to 250.000 users depending on weekday and daytime.

In this paper, we aim at a novel service infrastructure which allows to dynamically provide the game sessions based on current user demand. Our approach is to transfer the paradigm of *Grid Computing* [6] from scientific computing into the area of computer entertainment. In short, a computational Grid provides resources like computation power or storage space to a user who does not care about where the resources are located, as long as all requirements like computation power or security are met. In the area of multiplayer online games, a Grid built of game servers should provide game sessions depending on the current user demand in a dynamic way. We focus on the very popular genre of *First Person Shooter (FPS)* games which require Internet-based servers with high-bandwidth connections.

In our service infrastructure, the game sessions running at game service providers are dynamically managed. For each service provider, a special service accepts user requests to book and start services and schedules the game servers to fulfill this user demand. At the top level, a global directory and account service provides a central access point for users to all the service providers and lets users choose the best service to make use of. There exist several techniques for global directory services like *UDDI*, dynamic scheduling of applications and single-sign-on end user portals, known in the distributed computing area. However, a combined architecture for dynamic provisioning of online computer game services has to deal with several specific problems. Online games, as we discussed in recent work dealing with game scalability [8,9], are highly responsive, soft real-time systems that require a large amount of processing power and communication bandwidth. The novel architecture presented in this paper aims at providing these required resources to end users in a dynamic way.

The remainder of this paper is organized as follows: In Sect. 2, we present an overview of our Grid system for dynamic online game provisioning. We describe use case scenarios for users, administrators and game developers in Sect. 3. Section 4 presents implementation details of our current prototype. Finally, Sect. 5 gives an overview of possible extensions of our system, compares our approach to related work and concludes the paper.

2   System Architecture

In the current hosting approach for Internet-base, single-server game sessions, users rent a dedicated game server from a service provider on a usually monthly basis. The game server then runs day and night, regardless of whether there are actually users playing, which may result in an enormous waste of resources.

In order to make this static approach dynamic, our system runs a lightweight process called *GameService Controller (GSC)* on each game server host. A general overview of the different components of the system are shown in Fig. 1. The GSC is able to start up and shut down game server processes on its host. Usually, several game server hosts reside at a computing centre, where a special *GameService Manager (GSM)* administrates all the local GSCs, taking into account current user bookings of servers and the policies of the computer centre. The technical setup of the system in the computing centre (installed games, informations about GSCs, etc) as well as the provider-specific policies for setup and service prices are stored in a local *GameService Database (GSDB)*.

The involved computing centres and the corresponding GSM are known to a *Game-Service Web Server (GSWS)* which acts as the user portal to the complete system. Users can query, book and start game sessions via the GSWS. Additionally, a global user account database ensures a single sign-on mechanism for users: After the initial authentication, a user can access all the functionality of the system spanning several independent computing centres. Additionally, the GSWS deals with the billing issues for service usage and this way provides seamless billing for users who have access to several independent service providers.

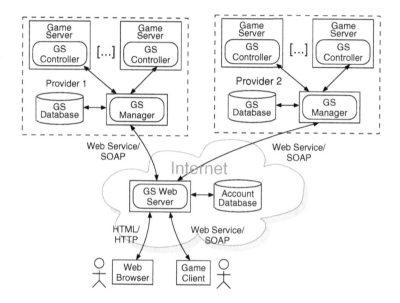

Fig. 1. System architecture

We designed this three-tier hierarchy (GSCs, GSMs, GSWS) in order to provide a good system scalability for a large amount of managed game sessions. From the top-level view, the GSWS only knows about GSMs which, in turn, manage their local GSCs. For possible user actions like searching available servers or booking a session, each tier aggregates the GameService components of the lower tier, rendering a single and comprehensive service for end users.

3   System Usage

From the user's point of view, the system provides a general access to game services. A typical use case for a single user is to set up a game session with/against known people at specific rules. Such rules define various aspects of the game like the length of the session, the number of points required to win, the particular game environment the session takes place on etc. Unlike *real-time strategy games (RTS)*, First Person Shooter (FPS) game servers require a lot of processing power and bandwidth, such that usually an end-user desktop PC can not host a game session. Additionally, a game server hosted at one of the users' machines is very problematic in terms of communication latency (the user playing on that machine has an advantage due to short inter-process communication time) and cheating (the user could manipulate the game server to gain an unfair advantage). Due to these facts, online FPS games require a dedicated Internet-based host machine.

On such a dedicated public server which already runs on the Internet and allows arbitrary users to connect to, a single user has no possibility to change the rules of the sessions. If, for example, two friends agree to play a small duel against each other with only them on a an empty public server, it will be very likely that other players join the server during their matchtime, disrupting their game experience. Using our infrastructure, these users are able to start their own, private, password-protected game session without having to rent a complete server for at least a month.

Our system provides the possibility to users to run game sessions on dedicated hosts featuring their own ruleset in an on-demand manner. The single operations a user can perform in our system are as follows:

- login to the system or register a new account
- search for available game servers to start a game session with own rules
- select a specific server based on communication quality
- set up rules before the start of the game session
- immediately start a game session or book a session for a specific time in the future
- manage account and billing information

3.1 Searching a Game Service

Consider as an example a group of users who want to play a game together. One of the users accesses the GS Web Server of our architecture, searches for an appropriate game server and defines the ruleset for the game session. Figure 2 depicts the initial user search for game servers and the corresponding system communication.

In step 1, the user logs in with a username and password, and the GSWS authenticates the user in step 2. After login, the user issues the search operation for the particular game X in step 3. Additionally, the user in this step transmits information about the other users which will participate in the actual game later on. In step 4, the GSWS forwards the search operation to the GSMs at the different service providers, which check the availability of a game server for the game X, as well as the price at which the service is offered from their local GS Database in step 5. If a game server is available, then the GSM will measure the communication quality to all the clients the user has declared to participate. One possible solution for this measurement is to measure latency and bandwidth with *ICMP echo* resp. *ping* messages (we discuss different solutions for measurement in Sect. 4). The resulting values only serve as a hint for the communication quality, because the Internet does not offer any quality-of-service guarantees, and communication quality can change abruptly and frequently. However, although this

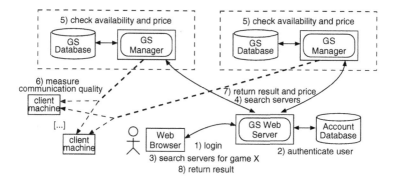

Fig. 2. A user searching and booking a game session

measurement does not provide any guarantees concerning the communication quality, it provides at least a hint about it and helps the user to differentiate services and allows providers to advertise their service at least at a "usually, but not guaranteed" communication quality. The results of the measurement as well as the price at which the service is offered are sent back to the GSWS in step 7 and forwarded to the user in step 8. The user now can choose between the offers of different service providers and immediately start the server or book the service for a period of time in the future. Figure 3 shows a screenshot of our current implementation of the user web frontend for step 3 of searching a particular game service to book.

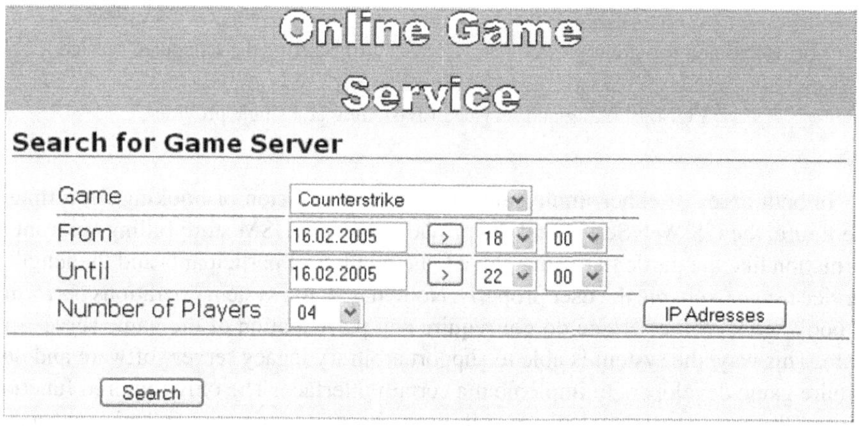

Fig. 3. Screenshot of the prototype user frontend: Search mask for a particular game service

3.2 Starting/Booking a Game Service

Resulting from the search operation, the user received the list of available services accompanied with the hint of communication quality and the cost at which each service is offered. From this list, the user chooses a particular service which can be either immediately started or booked for some time in the future if the user already knows that he wants to play in the evening or on the next day, for example. In both cases, the user has to set up the rules for the game session, like points required to win the match, the overall length of the match and the particular game environments or *maps* the session should take place on.

The user submits the service selection and session settings to the GSWS which stores informations required for billing of service's usage in the account database. Afterwards, the GSWS forwards the session settings to the particular GSM responsible for the computing centre the chosen game server resides in. Figure 4 illustrates the operations within a single service provider to store the booking from the user (Fig. 4(a)) and the actual start of the game service at the scheduled time (Fig. 4(b)).

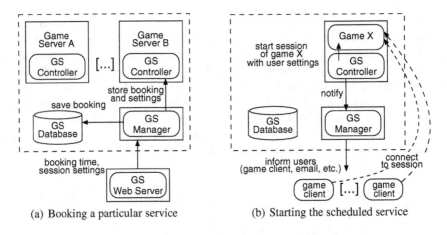

(a) Booking a particular service   (b) Starting the scheduled service

Fig. 4. Booking and service provisioning at a single provider

In both cases of either immediate starting of the session or booking for a time in the future, the GS Web Server and the provider-specific GSM store billing-relevant information like the particular game played, the number of participants and the length of service usage to charge the user properly. Note that all the system operations performed to book and start the service do not require any participation of the game server software. This way, the system is able to support arbitrary legacy server software and does require game developers to implement a certain interface. The only required functionality of the server software is the possibility to be started via a command-line call and to read session settings from a configuration file. Actually, these requirements are met by all existing single-server game software.

3.3   System Administration by Providers

Game service providers need the possibility to administrate the system on their end. They have to be able to install new games, to track bookings and usage of each game server and to specify maintenance time-frames, for which no user bookings are accepted. In our Grid infrastructure, these administrative operations are supported by the GS Manager of the provider, such that an administrator has central access to manage all the resources and set up policies and prices for the service usage.

4 Technical Design and Implementation

The prototype of our game service infrastructure has been implemented in Java. The intra-provider communication between the GS Controllers on the game server and the GS Manager is realized using Java RMI. The central interface of a service provider for the global GS Web Server is described as a web service using the Simple Object Access Protocol (SOAP). This interface allows to be used by a variety of meta-services in a platform-independent way and makes our architecture easily extensible. For example, game companies can create web portals specific for a particular game and easily integrate an on-demand booking service for game sessions in a seamless way. In the remainder of this section, we describe our approach for two important problems: how to describe game session setting in a generic way and how to measure the quality of communication to user clients during the search of suitable server resources.

4.1 Generic Description of Game-Specific Session Settings

The game server software has to be considered as a legacy application, because there is no common interface to start a game session and define session rules. Since each game software has its own syntax for describing rules, we designed a generic XML-based description to express all the specific settings. This is possible due to the fact that, although each game has its own syntax for describing the rules, the settings themselves like number of points to win, total time of the sessions etc. are the same in all of these games. For example, the servers for the games *Unreal Tournament 2003* [5] (left example below) and *Quake 3* [7] (right example) read the session settings from a configuration file at startup. The following extracts show some settings for a session.

```
FragLimit=30            fraglimit 30
TimeLimit=20            timelimit 20
MaxPlayers=8            set sv_maxclients 8
Maps=DM-Antalus         set map q3dm17
GamePassword=mygame     seta sv_privatePassword "mygame"
```

Unreal Tournament 2003   *Quake 3*
configuration file excerpt   *configuration file excerpt*

Although the syntax for the settings in the two games is different, the semantics are the same. In this example, one game round ends after 20 minutes or when one of the maximum of eight players achieves a score of 30 points. The game-specific environment is set to different map names. Both sessions require a player to provide the password mygame in order to join the session.

We analysed the sessions rules which are defined by users for several games and designed a generic XML-based description for these settings. Based on the XML description for a specific game, the GS Web Server asks the user to set up all session parameters. For the game Unreal Tournament 2003 for example, the `TimeLimit` setting description in the XML file looks like:

```
<element numeric="true">
    <name> TimeLimit </name>
    <character> = </character>
    <default-value> 0 </default-value>
    <min-constraint> 0 </min-constraint>
    <max-constraint> 60 </max-constraint>
    <eligible-as> Round Time Limit </eligible-as>
</element>
```

The GS Web Server displays a text or drop-down field for each setting and is able to check user inputs depending on the minimum and maximum constraints given in the XML file. This way, the user easily can input all required parameters and the GS Web Server does not have to know about the different configuration file characteristics. The user settings are submitted to the GS Manager, and the GS Controller at the chosen game server generates the actual configuration file describing the settings in the game-specific syntax before the startup of the game server.

4.2 Measuring Communication Quality to Clients

As discussed in Sect. 3.1, the architecture has to be able to give the user a hint about the quality of available services. Multiplayer real-time games are soft real-time systems and the game experience depends on the communication latency between the clients and the server. Especially for fast-paced First Person Shooter games, the communication latency has to be lower than about 120 ms to operate properly, as presented in [4] for the game *Unreal Tournament 2003* and in [3] for the game *Quake 3*. The user has to be able to choose a particular service taking into account not only the price but also the quality of the service.

The Internet does not provide any guarantees about communication quality and, therefore, the service providers can not guarantee a certain communication latency. Because of this, our architecture provides only a hint about the communication quality. However, if there are no unforeseeable and explosive increases in general Internet-communication, the general best effort latency of a computing centre to an end user home PC actually depends on the quality of the computing centre link-up. If a provider makes a larger investment for a better link-up, then he usually can provide a better service. Our system allows users to differentiate services not only by price, but also by the quality of communication and therefore by the technical investment of providers as well.

There are several technical solutions to measure the communication quality between hosts on the Internet: ICMP echo messages (sending a ping), the *Network Weather Service (NWS)* [11] or direct, socket-based communication between processes residing at the hosts.

In our current implementation, we use ICMP echo messages to measure the communication latency to players. The user booking the session has to input the IP-addresses of all players' home computers. Each GS Manager pings these hosts and transfers the results accompanied with the price of the service to the GS Web Server. However, this solution does not work when the players' PCs reside behind a firewall of a NAT router, because these hosts are not reachable for the providers' echo messages. In this situation, the user searching for the service can specify hosts of the Internet-communication companies which provide the general Internet link-up for the players with a firewall and thus obtain a hint about the expected latency in the game.

5   Conclusion and Related Work

There already has been some work to make the setup of the multi-server approach for MMORPG dynamic. There are two related projects to mention: The commercial *Butterfly.net* [1] provides a Grid-system for an efficient usage of game servers at a local computer centre and is specifically designed for massively multiplayer games which use several servers for a single game session. However, the Butterfly.net only manages resources which all reside at a single computing centre and, therefore, aims at a dynamic and efficient setup of game servers at a local area network rather than at a global, Internet-wide setup of servers. The second project, presented in [10], proposes an Internet-wide Grid infrastructure for a dynamic setup of single-server online game sessions comparable to our architecture. However, several important issues like the generic support of a variety of different games or the selection of the "best" resource for a particular user request are not discussed.

We implemented our architecture as a prototype and currently are testing it extensively. Users are able to search for and dynamically set up game sessions featuring their own rules, without renting a game server for a long period of time. This way, the overall efficiency of game server machines is drastically improved, because hosts which are not scheduled to run a game session at the moment remain available for other tasks. This opens the possibility to combine entertainment and scientific Grid-computing running on the same resources: Hosts without a game session running are available to run scientific applications and vice versa, leading to a general and efficient usage of high-performance hosts. The dynamic setup allows to charge users according to a pay-per-use business model, which promises to attract more customers than the current long term renting of game servers. The web services interface for accessing services of providers allow to incorporate search, start and book operations of game sessions into a variety of portals ran by game companies or independent game sites.

The architecture supports single-server online games in a generic way. Due to our XML-based description of session configuration, new games can easily be supported. The game server software itself is not required to implement a certain interface in order to be compatible with our architecture.

Measuring communication quality for the Internet is problematic due to the lack of quality-of-service guarantees. However, our architecture allows to use hints about communication latency. We designed the interface to the measurement module to be extensible, such that the technical alternatives discussed can be incorporated if required. Especially the Network Weather System is an interesting alternative to the currently used ICMP echo messages, because it traces latency regularly and can provide a history of communication quality, as well as a prediction for the future.

References

1. Butterfly.net <http://www.butterfly.net>.
2. Gamespy <http://www.gamespy.com/>.
3. G. Armitage. Sensitivity of Quake3 players to network latency, imw2001 poster, 2001.
4. T. Beigbeder, R. Coughlan, C. Lusher, J. Plunkett, E. Agu, and M. Claypool. The effects of loss and latency on user performance in unreal tournament 2003. In *Proceedings of ACM Network and System Support for Games Workshop (NetGames)*, Portland, Oregon, USA,, September 2004.
5. Epic Games. Unreal Tournament game series <http://www.unrealtournament.com/>.
6. I. Foster and C. Kesselmann, editors. *The Grid: Blueprint for a New Computing Infrastructure*. Morgan Kaufmann, 1998.
7. ID Software. Quake 3 Arena <http://www.idsoftware.com/games/quake/quake3-arena/>.
8. J. Müller, S. Fischer, S. Gorlatch, and M. Mauve. A proxy server-network for real-time computer games. In M. Danelutto, D. Laforenza, and M. Vanneschi, editors, *Euro-Par 2004 Parallel Processing*, volume 3149 of *Lecture Notes in Computer Science*, pages 606–613, Pisa, Italy, Aug. 2004. Springer-Verlag.
9. J. Müller and S. Gorlatch. GSM: A game scalability model for multiplayer real-time games. In *IEEE Infocom 2005*, Miami, Florida / USA, Mar. 2005. IEEE Communications Society.
10. A. Shaikh, S. Sahu, M. Rosu, M. Shea, and D. Saha. Implementation of a service platform for online games. In *Proceedings of ACM Network and System Support for Games Workshop (NetGames)*, Portland, Oregon, USA,, September 2004.
11. R. Wolski, N. T. Spring, and J. Hayes. The network weather service: A distributed resource performance forecasting service for metacomputing. *Journal of Future Generation Computing Systems*, 15(5-6):757–768, Oct. 1999.

Principal Component Analysis for Distributed Data Sets with Updating

Zheng-Jian Bai[1,*], Raymond H. Chan[1], and Franklin T. Luk[2]

[1] Department of Mathematics, Chinese University of Hong Kong,
Shatin, NT, Hong Kong, China
{zjbai, rchan}@math.cuhk.edu.hk
[2] Department of Computer Science, Rensselaer Polytechnic Institute, Troy,
New York 12180, USA
luk@cs.rpi.edu

Abstract. Identifying the patterns of large data sets is a key requirement in data mining. A powerful technique for this purpose is the principal component analysis (PCA). PCA-based clustering algorithms are effective when the data sets are found in the same location. In applications where the large data sets are physically far apart, moving huge amounts of data to a single location can become an impractical, or even impossible, task. A way around this problem was proposed in [10], where truncated singular value decompositions (SVDs) are computed locally and used to reduce the communication costs. Unfortunately, truncated SVDs introduce local approximation errors that could add up and would adversely affect the accuracy of the final PCA. In this paper, we introduce a new method to compute the PCA without incurring local approximation errors. In addition, we consider the situation of updating the PCA when new data arrive at the various locations.

1 Introduction

Effective clustering of large data sets is a major objective in data mining. Principal component analysis (PCA) [4,5,9] offers a popular statistical technique to analyze multivariate data by constructing a concise data representation using the dominant eigenvectors of the data covariance matrix. PCA and PCA-based clustering methods play an important role in various applications such as knowledge discovery from databases [2] and remote sensing [8]; for more applications, see [7] and the references therein.

PCA is effective for high-dimensional data analysis when the data sets are collocated. However, in present-day applications, the large data sets could be distributed over a network of distant sites, and PCA-based algorithms may no longer be applicable since these distributed data sets are often too large to send to a single location. There is a growing interest in this topic of distributed data sets and here are some relevant works in the literature: the interaction of huge data sets and the limits of computational feasibility in Wegman [12], parallel methods for spectral decomposition of nonsymmetric

---

* Current Address: Department of Mathematics, National University of Singapore, 2 Science Drive 2, Singapore 117543 (matbzj@nus.edu.sg).

J. Cao, W. Nejdl, and M. Xu (Eds.): APPT 2005, LNCS 3756, pp. 471–483, 2005.
© Springer-Verlag Berlin Heidelberg 2005

matrix on distributed memory processors in Bai *et al.* [1], an efficient out-of-core SVD algorithm in Rabani *et al.* [11], an algorithm for data distributed by blocks of columns in Kargupta *et al.* [7], and a method for massive data sets distributed by blocks of rows in Qu *et al.* [10].

In this paper, we consider the problem described in Qu *et al.* [10], where the authors use truncated singular value decompositions (SVDs) in the distributed locations to reduce communications costs. Their approach is very effective when the local data matrices have *low* ranks and can be accurately approximated via a truncated SVD (note that the savings may be nonexistent when the data matrices have high ranks). In addition, the small local approximation errors may add up substantially when the number of locations is large. We will present a new algorithm for computing a global PCA of distributed data sets. In contrast to Qu's approach [10], our method introduces no local approximation errors. At the central processor, Qu's approach works with the approximate covariance matrix while we work directly with the data matrix. Our technique will likely require less communication as well. Suppose that there is an $n_i \times p$ matrix of rank m_i at the ith local site for $i = 1, \ldots, s$. While Qu's approach [10] requires $O(p \sum_{i=1}^{s} m_i)$ communication, our procedure uses $O(p^2 \lceil \log_2 s \rceil)$ communication. When s is large, it is probable that $p \sum_{i=1}^{s} m_i > p^2 \lceil \log_2 s \rceil$ as $p \ll n$. We also consider the important problem of updating, for new data do arise all the time (for example, medical information and banking transactions), and we develop a procedure for constructing a global PCA for distributed data sets with data updating, by suitably combining the PCAs of past data and the local PCAs of new data.

This paper is organized as follows. Section 2 contains a brief review of the basic concepts. In Section 3 we present an algorithm for computing the global PCA of distributed data sets, and we include a numerical example to illustrate the advantages of our method. In Section 4 we develop a technique for computing the global PCA of distributed data sets with updating. Load balancing for communications and computation is discussed in Section 5, and Section 6 concludes the paper.

Remark 1. Throughout this paper, for simplicity, we assume that there is one processor at each location and so we will use the two words *location* and *processor* interchangeably.

2   Principal Component Analysis

Let X be an n-by-p data matrix, where rows denote the observations and columns denote the features with $n \gg p$. The data covariance matrix S is given by

$$nS = X^T(I - \tfrac{1}{n}\mathbf{e}_n\mathbf{e}_n^T)X, \tag{1}$$

where

$$\mathbf{e}_\ell \equiv (1, 1, \ldots, 1)^T$$

denotes a vector of length ℓ. The PCA of X is given by an eigenvalue decomposition [3] of nS:

$$nS = V\Sigma^2 V^T, \tag{2}$$

where
$$\Sigma^2 = \text{diag}(\sigma_1^2, \sigma_2^2, \ldots, \sigma_p^2),$$

with
$$\sigma_1^2 \geq \sigma_2^2 \geq \cdots \geq \sigma_p^2,$$

and V is an orthogonal matrix. Let
$$J \equiv I - \tfrac{1}{n}\mathbf{e}_n\mathbf{e}_n^T.$$

As the matrix J is symmetric and idempotent, we may therefore compute a singular value decomposition (SVD) [3] of the column-centered data matrix JX:

$$(I - \tfrac{1}{n}\mathbf{e}_n\mathbf{e}_n^T)X = U\Sigma V^T. \tag{3}$$

Therefore, it is not necessary to form the covariance matrix S explicitly. We save work and improve accuracy by working directly with the data matrix X. The matrices Σ and V we get in (3) are exactly the matrices we need in (2).

One application of the PCA is to reduce the dimensions of the given data matrix X. To do so, let $\tilde{V}$ denote the first m columns of V, corresponding to the m largest eigenvalues of nS. The m principal components of X is given by the n-by-m matrix

$$\tilde{X} = (I - \tfrac{1}{n}\mathbf{e}_n\mathbf{e}_n^T)X\tilde{V}.$$

It is an optimal m-dimensional approximation of $(I - \tfrac{1}{n}\mathbf{e}_n\mathbf{e}_n^T)X$ in the least squares sense [6]. The ratio η_m, given by

$$\eta_m \equiv \left(\sum_{i=1}^{m}\sigma_i^2\right)/\left(\sum_{i=1}^{p}\sigma_i^2\right),$$

reflects the total variance of $\tilde{X}$ in the original data. If $\eta_m \approx 1$ for some $m \ll p$, the n-by-p transformed data matrix JX can be well represented by the much smaller n-by-m matrix $\tilde{X}$, which forms the crux of the approach described in Qu et al. [10].

3   Distributed PCA Without Updating

We start with the case of no updating. The global data matrix X is distributed among s locations:

$$X = \begin{pmatrix} X_0 \\ X_1 \\ \vdots \\ X_{s-1} \end{pmatrix},$$

where X_i is an n_i-by-p matrix, and resides at Processor i, for $0 \leq i < s$. So,

$$n = \sum_{i=0}^{s-1} n_i$$

gives the number of rows in X. Let S be the covariance data matrix corresponding to X as given in (1). If we are to form S explicitly, then we have to move X_i across the processors, and the communication cost will be $O(np)$. In [10], Qu et al. compute the local PCA for each X_i using the SVD. They then send m_i, where $m_i < p$, singular vectors to the central processor where an approximation of S is assembled, and its PCA is computed. The communication cost of the method is thus $O(p \sum_{i=0}^{s-1} m_i)$. A drawback is that the local SVD will introduce approximation errors. In the following, we give a method of finding PCA of X exactly using the QR decomposition. For simplicity, we assume that $s = 2^\ell$ and that the global PCA is computed in location 0 (i.e., Processor 0).

Algorithm 1:

- *At Processor i, for $0 \le i < s$:* Compute the column means of X_i, i.e.,

$$\bar{\mathbf{x}}_i^T = \frac{1}{n_i} \mathbf{e}_{n_i}^T X_i.$$

Form the column-centered data matrix

$$\bar{X}_i = (I - \tfrac{1}{n_i} \mathbf{e}_{n_i} \mathbf{e}_{n_i}^T) X_i = X_i - \mathbf{e}_{n_i} \bar{\mathbf{x}}_i^T. \tag{4}$$

Then compute its QR decomposition [3]:

$$\bar{X}_i = Q_i^{(0)} R_i^{(0)}, \tag{5}$$

where $R_i^{(0)}$ are upper triangular p-by-p matrices. Send n_i and $\bar{\mathbf{x}}_i^T$ to Processor 0. If $i \ge s/2$, send $R_i^{(0)}$ to Processor $(i - s/2)$. There is no need to send any $Q_i^{(0)}$.
- *At Processor i, for $0 \le i < s/2$:* Compute the QR decomposition of $R_i^{(0)}$ and $R_{i+s/2}^{(0)}$ by using Givens' rotations:

$$\begin{pmatrix} R_i^{(0)} \\ R_{i+s/2}^{(0)} \end{pmatrix} = Q_i^{(1)} R_i^{(1)}, \tag{6}$$

where $R_i^{(1)}$ are p-by-p upper triangular matrices. If $i \ge s/4$, send $R_i^{(1)}$ to Processor $(i - s/4)$. Again, there is no need to send any $Q_i^{(1)}$.
- *Continue until we reach Processor 0 after $\ell = \lceil \log_2 s \rceil$ steps.*
- *At Processor 0:* Compute the QR decomposition of $R_0^{(\ell-1)}$ and $R_1^{(\ell-1)}$ by using Givens' rotations:

$$\begin{pmatrix} R_0^{(\ell-1)} \\ R_1^{(\ell-1)} \end{pmatrix} = Q_0^{(\ell)} R_0^{(\ell)}, \tag{7}$$

where $R_0^{(\ell)}$ is an p-by-p upper triangular matrix. Form the following $(s + p)$-by-p upper-trapezoidal matrix and compute its QR decomposition by Householder's reflections:

$$\begin{pmatrix} \sqrt{n_0}(\bar{\mathbf{x}}_0 - \bar{\mathbf{x}}) \\ \sqrt{n_1}(\bar{\mathbf{x}}_1 - \bar{\mathbf{x}}) \\ \vdots \\ \sqrt{n_{s-1}}(\bar{\mathbf{x}}_{s-1} - \bar{\mathbf{x}}) \\ R_0^{(\ell)} \end{pmatrix} = QR. \tag{8}$$

Here, R is an p-by-p upper triangular matrix and

$$\bar{\mathbf{x}} \equiv \frac{1}{n} \sum_{i=0}^{s-1} n_i \bar{\mathbf{x}}_i$$

gives the column mean of X. The PCA of S can now be obtained by computing the SVD of R:

$$R = U\Sigma V^T. \tag{9}$$

Remark 2. Algorithm 1 works for an arbitrary $s > 0$ if we replace s by s_+, where $s_+ := 2^{\lceil \log_2 s \rceil}$. For $s_+ > s$, the matrices $\{X_i\}_{i=s+1}^{s_+}$ are empty.

Lemma 1. *The covariance matrix S as defined in (1) is given by:*

$$nS = R_0^{(\ell)^T} R_0^{(\ell)} + \sum_{i=0}^{s-1} n_i(\bar{\mathbf{x}}_i - \bar{\mathbf{x}})(\bar{\mathbf{x}}_i - \bar{\mathbf{x}})^T = R^T R. \tag{10}$$

In particular, the PCA of S is given by the Σ and V computed in (9).

Proof. The last equality in (10) follow from (8). To prove the first equality, we note that

$$\bar{\mathbf{x}}^T = \tfrac{1}{n}\mathbf{e}_n^T X.$$

Hence

$$(I - \tfrac{1}{n}\mathbf{e}_n\mathbf{e}_n^T)X = (I - \tfrac{1}{n}\mathbf{e}_n\mathbf{e}_n^T)\begin{pmatrix} X_0 \\ \vdots \\ X_{s-1} \end{pmatrix} = \begin{pmatrix} X_0 - \mathbf{e}_{n_0}\bar{\mathbf{x}}^T \\ \vdots \\ X_{s-1} - \mathbf{e}_{n_{s-1}}\bar{\mathbf{x}}^T \end{pmatrix}$$

$$= \begin{pmatrix} X_0 - \mathbf{e}_{n_0}\bar{\mathbf{x}}_0^T + \mathbf{e}_{n_0}(\bar{\mathbf{x}}_0 - \bar{\mathbf{x}})^T \\ \vdots \\ X_{s-1} - \mathbf{e}_{n_{s-1}}\bar{\mathbf{x}}_1^T + \mathbf{e}_{n_{s-1}}(\bar{\mathbf{x}}_{s-1} - \bar{\mathbf{x}})^T \end{pmatrix} = \begin{pmatrix} \bar{X}_0 + \mathbf{e}_{n_0}(\bar{\mathbf{x}}_0 - \bar{\mathbf{x}})^T \\ \vdots \\ \bar{X}_{s-1} + \mathbf{e}_{n_{s-1}}(\bar{\mathbf{x}}_{s-1} - \bar{\mathbf{x}})^T \end{pmatrix},$$

where the last equality follows from the definition in (4). By (4), we see that the column sums of $\bar{X}_i$ are all zero, i.e.,

$$\mathbf{e}_{n_i}^T \bar{X}_i = \mathbf{0},$$

for $0 \le i < s$. Hence

$$nS = \left(\bar{X}_0^T + (\bar{\mathbf{x}}_0 - \bar{\mathbf{x}})\mathbf{e}_{n_0}^T \mid \cdots \mid \bar{X}_{s-1}^T + (\bar{\mathbf{x}}_{s-1} - \bar{\mathbf{x}})\mathbf{e}_{n_{s-1}}^T\right)\begin{pmatrix} \bar{X}_0 + \mathbf{e}_{n_0}(\bar{\mathbf{x}}_0 - \bar{\mathbf{x}})^T \\ \vdots \\ \bar{X}_{s-1} + \mathbf{e}_{n_{s-1}}(\bar{\mathbf{x}}_{s-1} - \bar{\mathbf{x}})^T \end{pmatrix}$$

$$= \sum_{i=0}^{s-1} \bar{X}_i^T \bar{X}_i + \sum_{i=0}^{s-1} n_i(\bar{\mathbf{x}}_i - \bar{\mathbf{x}})(\bar{\mathbf{x}}_i - \bar{\mathbf{x}})^T. \tag{11}$$

Using (5)–(7), we have

$$\sum_{i=0}^{s-1} \bar{X}_i^T \bar{X}_i = \sum_{i=0}^{s-1} R_i^{(0)T} R_i^{(0)} = \sum_{i=0}^{s/2-1} R_i^{(1)T} R_i^{(1)} = \cdots = \sum_{i=0}^{1} R_i^{(\ell-1)T} R_i^{(\ell-1)} = R_0^{(\ell)T} R_0^{(\ell)}.$$

Put this in (11) and we get (10).

To get the first m principal components of X, we broadcast $\bar{\mathbf{x}}$ and $\tilde{V}$ (the first m columns of V) to every processor. Then the m principal components of X are given by the matrix $\tilde{X}$:

$$\tilde{X} = (I - \tfrac{1}{n}\mathbf{e}_n\mathbf{e}_n^T)X\tilde{V} = (X - \mathbf{e}_n\bar{\mathbf{x}}^T)\tilde{V}. \tag{12}$$

In particular, at Processor i, we have the n_i-by-m approximation $\tilde{X}_i$:

$$\tilde{X}_i = (X_i - \mathbf{e}_{n_i}\bar{\mathbf{x}}^T)\tilde{V},$$

for $0 \le i < s$. Regarding the communication costs, note that there are $\lceil \log_2 s \rceil$ steps in the algorithm. In step j, we need to move a total number $(=s/2^j)$ of p-by-p upper triangular matrices $R_i^{(j)}$. Hence the communication cost is $O(p^2 \lceil \log_2 s \rceil)$. We state once more that the PCA (i.e., Σ and V) we obtain is exact.

We ran some numerical experiments on synthetic data using **MATLAB 7.0.1**. They simulated the scenario of distributed data sets to assess computational accuracy and communication costs. Execution times are not provided since they are not meaningful in simulations (cf. [10]).

Example (Synthetic data). The data X are generated as follows (cf. [10]). Let

$$X = GE^T + N,$$

where the n-by-d data matrix G is a d-dimensional Gaussian data, i.e., its entries are identical, independently distributed (*iid*) as $\mathcal{N}(0,1)$ (normal distribution with mean 0 and variance 1), E is a p-by-d matrix with 1's on the diagonal and zeros elsewhere, and N is a p-dimensional Gaussian noise whose entries are *iid* as $\mathcal{N}(0,\sigma^2)$. We partition the data X among s processors evenly. If the modulus r after n divided by s is not zero, let the first r processors contain $\lfloor n/s \rfloor + 1$ observations.

We took $n = 6,000$, $p = 20$, $d = 2$, and $\sigma = 0.2$, and we set the local and global PC selection thresholds to be $\sqrt{0.8}$ and 0.8, respectively. To further characterize the data X, we plot the eigenvalue distribution of the theoretical covariance matrix with these parameters in Figure 1. Ten simulations were run using the distributed principal component algorithm (DPCA) proposed by Qu *et al.* [10] (Method a) and our algorithm (Method b) for various values of s. In Table 1, we report the means and standard deviations (sd) of the quantities given as follows.

$$T_{ae} = \frac{T_a}{T_e}, \quad T_{be} = \frac{T_b}{T_e} \quad \text{and} \quad T_{ba} = \frac{T_{be}}{T_{ae}},$$

where $T_a = (\sum_{i=0}^{s-1} m_i)(p+1) + s(p+3)$ with m_i being the number of PCs selected from the ith processor, $T_b = \tfrac{1}{2}p(p+1)\ell + s(p+1)$ and $T_e = np$. The quantities T_{ae} and T_{be} provide the ratios of the communication costs.

$$d_a = \frac{\|(I - n^{-1}\mathbf{e}_n\mathbf{e}_n^T)(\hat{X} - X)\|_2}{\|(I - n^{-1}\mathbf{e}_n\mathbf{e}_n^T)X\|_2},$$

$$d_b = \frac{\|(I - n^{-1}\mathbf{e}_n\mathbf{e}_n^T)(\bar{X} - X)\|_2}{\|(I - n^{-1}\mathbf{e}_n\mathbf{e}_n^T)X\|_2}$$

and

$$d_{ba} = \frac{d_b}{d_a},$$

where $\hat{X}$ is the dimension reduced data obtained by the DPCA [10] and $\bar{X} = \tilde{X}\tilde{V}^T$. Here, $\tilde{X}$ is defined in (12) where m is the number of global PCs which is obtained based on the global PC selection threshold. d_a and d_b are the relative error between the original data X and the data approximated by Methods a and b, respectively. From

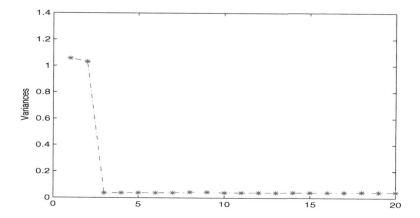

Fig. 1. Eigenvalue distribution of covariance matrix for synthetic data

Table 1. Numerical results for synthetic data

	s	1	4	8	16	32	64	128
T_{ae}	mean	.0025	.0092	.0182	.0348	.0653	.1193	.2065
	sd	.0000	.0001	.0001	.0004	.0008	.0011	.0017
T_{be}	mean	.0002	.0042	.0067	.0098	.0143	.0217	.0347
	sd	0	.0000	.0000	.0000	.0000	0	.0000
T_{ba}	mean	.0709	.4556	.3645	.2814	.2197	.1819	.1678
	sd	0	.0041	.0025	.0036	.0028	.0016	.0014
d_a	mean	.2054	.2030	.2025	.2042	.2043	.2049	.2044
	sd	.0017	.0025	.0018	.0024	.0021	.0016	.0015
d_b	mean	.1993	.1976	.1979	.1988	.1988	.1992	.1988
	sd	.0014	.0018	.0014	.0021	.0021	.0012	.0016
d_{ba}	mean	.9703	.9737	.9778	.9738	.9731	.9726	.9725
	sd	.0042	.0063	.0061	.0066	.0033	.0042	.0034

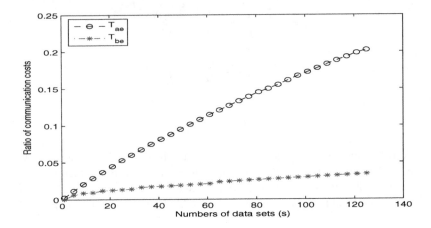

Fig. 2. Comparison of communication costs for synthetic data

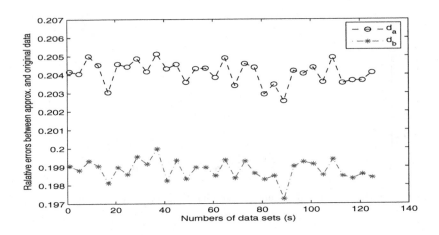

Fig. 3. Comparison of data approximate error for synthetic data

Table 1 and Figures 2 and 3, we see that our method behaves better than the DPCA in terms of both communication costs and data approximation errors.

4   Distributed PCA with Updating

In this section, we develop an algorithm for computing the PCA when new data arise in the s locations. We assume that the initial time $t_0 = 0$. In our algorithm, we use a global synchronization to keep track of the updating and the evaluation of the PCA for the global data matrix. More precisely, we fix the time instants $t_1, t_2, \ldots$, when updating is stopped and the evaluation of PCA for the global data commences.

In particular, let $X_i^{(k)}$ denote the block of data of size $n_i^{(k)}$-by-p added to Processor i between $t_{k-1} < t \leq t_k$ where $k \geq 1$. The updated matrix for this time interval is denoted by

$$
X_{n^{(k)}} = \begin{pmatrix} X_0^{(k)} \\ X_1^{(k)} \\ \vdots \\ X_{s-1}^{(k)} \end{pmatrix},
$$

where $n^{(k)} = \sum_{i=0}^{s-1} n_i^{(k)}$ is the number of rows in $X_{n^{(k)}}$. We will use $X_{n^{(0)}}$ to denote the data matrix already present at the processors at time $t_0 = 0$. For $k \geq 0$, let

$$
\bar{\mathbf{x}}_{n^{(k)}}^T = \tfrac{1}{n^{(k)}} \mathbf{e}_{n^{(k)}}^T X_{n^{(k)}} \tag{13}
$$

denote the column means of $X_{n^{(k)}}$. The p-by-p covariance matrix S_k corresponding to $X_{n^{(k)}}$ is given by

$$
n^{(k)} S_k = X_{n^{(k)}}^T (I - \tfrac{1}{n^{(k)}} \mathbf{e}_{n^{(k)}} \mathbf{e}_{n^{(k)}}^T) X_{n^{(k)}}. \tag{14}
$$

We note that for each $k \geq 0$, the PCA of S_k can be obtained by Algorithm 1 in Section 3.

We want to compute the PCA of the global data matrix collected from $t_0 = 0$ up to t_k, $k \geq 0$. Let $\mathbb{X}_{g(k)}$ denote this global data matrix

$$
\mathbb{X}_{g(k)} = \begin{pmatrix} X_{n^{(0)}} \\ X_{n^{(1)}} \\ \vdots \\ X_{n^{(k)}} \end{pmatrix}, \tag{15}
$$

where $g(k) = \sum_{j=0}^{k} n^{(j)}$ is the number of rows in $\mathbb{X}_{g(k)}$. We emphasize that the data blocks $\{X_i^{(j)}\}_{i=0}^{s-1}$ always reside on their respective processors and will not be moved. The p-by-p covariance matrix $\mathbb{S}_k$ corresponding to $\mathbb{X}_{g(k)}$ is given by

$$
g(k) \mathbb{S}_k = \mathbb{X}_{g(k)}^T (I - \tfrac{1}{g(k)} \mathbf{e}_{g(k)} \mathbf{e}_{g(k)}^T) \mathbb{X}_{g(k)}. \tag{16}
$$

We now show that $\mathbb{S}_k$ can be obtained from the covariance matrices S_j of $\{X_{n^{(j)}}\}_{j=0}^{k}$.

Theorem 1. *For any positive integer k,*

$$
g(k)\mathbb{S}_k = \sum_{j=0}^{k} n^{(j)} S_j + \sum_{j=1}^{k} \frac{g(j-1)n^{(j)}}{g(j)} \left(\bar{\mathbf{x}}_{g(j-1)} - \bar{\mathbf{x}}_{n^{(j)}} \right) \left(\bar{\mathbf{x}}_{g(j-1)} - \bar{\mathbf{x}}_{n^{(j)}} \right)^T,
$$

$$\tag{17}$$

where S_j and $\bar{\mathbf{x}}_{n^{(j)}}$ are given by (14) and (13) respectively, and

$$
\bar{\mathbf{x}}_{g(k)}^T = \tfrac{1}{g(k)} \mathbf{e}_{g(k)}^T \mathbb{X}_{g(k)}, \tag{18}
$$

is the column mean of $\mathbb{X}_{g(k)}$, which can be obtained by $\bar{\mathbf{x}}_{g(k)} = \tfrac{1}{g(k)} \sum_{j=0}^{k} n^{(j)} \bar{\mathbf{x}}_{n^{(j)}}$.

Proof. We use induction to prove the lemma. For $k = 0$, the equation (17) is obviously true. Let us assume that it is also true for $k - 1$. We first write

$$
I - \frac{1}{g(k)} \mathbf{e}_{g(k)} \mathbf{e}_{g(k)}^T = \begin{pmatrix} I - \frac{1}{g(k-1)} \mathbf{e}_{g(k-1)} \mathbf{e}_{g(k-1)}^T & 0 \\ 0 & I - \frac{1}{n^{(k)}} \mathbf{e}_{n^{(k)}} \mathbf{e}_{n^{(k)}}^T \end{pmatrix}
$$

$$
+ \begin{pmatrix} [\frac{1}{g(k-1)} - \frac{1}{g(k)}] \mathbf{e}_{g(k-1)} \mathbf{e}_{g(k-1)}^T & -\frac{1}{g(k)} \mathbf{e}_{g(k-1)} \mathbf{e}_{n^{(k)}}^T \\ -\frac{1}{g(k)} \mathbf{e}_{n^{(k)}} \mathbf{e}_{g(k-1)}^T & [\frac{1}{n^{(k)}} - \frac{1}{g(k)}] \mathbf{e}_{n^{(k)}} \mathbf{e}_{n^{(k)}}^T \end{pmatrix}
$$

$$
\equiv E_k + F_k.
$$

For E_k, using (14) and (16), we have

$$
\mathbb{X}_{g(k)}^T E_k \mathbb{X}_{g(k)} = \begin{pmatrix} \mathbb{X}_{g(k-1)} \\ X_{n^{(k)}} \end{pmatrix}^T E_k \begin{pmatrix} \mathbb{X}_{g(k-1)} \\ X_{n^{(k)}} \end{pmatrix} = g(k-1)\mathbb{S}_{k-1} + n^{(k)} S_k. \quad (19)
$$

For F_k, using (13) and (18), we have

$$
\mathbb{X}_{g(k)}^T F_k \mathbb{X}_{g(k)}
$$

$$
= \frac{1}{g(k)} \begin{pmatrix} \mathbb{X}_{g(k-1)} \\ X_{n^{(k)}} \end{pmatrix}^T \begin{pmatrix} \frac{n^{(k)}}{g(k-1)} \mathbf{e}_{g(k-1)} \mathbf{e}_{g(k-1)}^T & -\mathbf{e}_{g(k-1)} \mathbf{e}_{n^{(k)}}^T \\ -\mathbf{e}_{n^{(k)}} \mathbf{e}_{g(k-1)}^T & \frac{g(k-1)}{n^{(k)}} \mathbf{e}_{n^{(k)}} \mathbf{e}_{n^{(k)}}^T \end{pmatrix} \begin{pmatrix} \mathbb{X}_{g(k-1)} \\ X_{n^{(k)}} \end{pmatrix}
$$

$$
= \frac{1}{g(k)} \begin{pmatrix} \mathbb{X}_{g(k-1)} \\ X_{n^{(k)}} \end{pmatrix}^T \begin{pmatrix} n^{(k)} \mathbf{e}_{g(k-1)} \bar{\mathbf{x}}_{g(k-1)}^T - n^{(k)} \mathbf{e}_{g(k-1)} \bar{\mathbf{x}}_{n^{(k)}}^T \\ -g(k-1) \mathbf{e}_{n^{(k)}} \bar{\mathbf{x}}_{g(k-1)}^T + g(k-1) \mathbf{e}_{n^{(k)}} \bar{\mathbf{x}}_{n^{(k)}}^T \end{pmatrix}
$$

$$
= \frac{g(k-1)n^{(k)}}{g(k)} \left\{ \bar{\mathbf{x}}_{g(k-1)} \bar{\mathbf{x}}_{g(k-1)}^T - \bar{\mathbf{x}}_{g(k-1)} \bar{\mathbf{x}}_{n^{(k)}}^T - \bar{\mathbf{x}}_{n^{(k)}} \bar{\mathbf{x}}_{g(k-1)}^T + \bar{\mathbf{x}}_{n^{(k)}} \bar{\mathbf{x}}_{n^{(k)}}^T \right\}
$$

$$
= \frac{g(k-1)n^{(k)}}{g(k)} (\bar{\mathbf{x}}_{g(k-1)} - \bar{\mathbf{x}}_{n^{(k)}})(\bar{\mathbf{x}}_{g(k-1)} - \bar{\mathbf{x}}_{n^{(k)}})^T. \quad (20)
$$

Adding (19) and (20), and invoking the induction hypothesis, we get (17).

For each update matrix $X_{n^{(j)}}$, by Algorithm 1 and (10), its covariance matrix S_j is given by

$$
n^{(j)} S_j = R_{n^{(j)}}^T R_{n^{(j)}},
$$

where $R_{n^{(j)}}$ is p-by-p upper triangular. Hence by (17),

$$
g(k)\mathbb{S}_k = \sum_{j=0}^{k} R_{n^{(j)}}^T R_{n^{(j)}} + \sum_{j=1}^{k} \frac{g(j-1)n^{(j)}}{g(j)} (\bar{\mathbf{x}}_{g(j-1)} - \bar{\mathbf{x}}_{n^{(j)}})(\bar{\mathbf{x}}_{g(j-1)} - \bar{\mathbf{x}}_{n^{(j)}})^T.
$$

$$
(21)
$$

Let $\mathbb{R}_{g(0)} = R_{n^{(0)}}$. Using Householder's reflections, we can recursively obtain the QR decomposition of the following $(2p + 1)$-by-p matrix:

$$
\begin{pmatrix} \mathbb{R}_{g(k-1)} \\ \sqrt{\frac{g(k-1)n^{(k)}}{g(k)}} (\bar{\mathbf{x}}_{g(k-1)} - \bar{\mathbf{x}}_{n^{(k)}}) \\ R_{n^{(k)}} \end{pmatrix} = \mathbb{Q}_{g(k)} \mathbb{R}_{g(k)}, \quad (22)
$$

where $k \geq 1$ and $\mathbb{R}_{g(k)}$ is an p-by-p upper triangular matrix. It is easy to check from (21) that

$$g(k)\mathbb{S}_k = \mathbb{R}_{g(k)}^T \mathbb{R}_{g(k)}.$$

Hence the PCA of $\mathbb{S}_k$ can be obtained by computing the SVD of $\mathbb{R}_{g(k)}$:

$$\mathbb{R}_{g(k)} = \mathbb{U}\Sigma\mathbb{V}^T,$$

where Σ and $\mathbb{V}$ are p-by-p matrices. To get the first m principal components of the global data matrix $\mathbb{X}_{g(k)}$, we broadcast $\bar{\mathbf{x}}_{g(k)}$ and $\tilde{\mathbb{V}}$ (the first m columns of $\mathbb{V}$) to every processor. Then the m principal components of $\mathbb{X}_{g(k)}$ are given by the matrix $\tilde{\mathbb{X}}_{g(k)}$:

$$\tilde{\mathbb{X}}_{g(k)} = (I - \tfrac{1}{g(k)}\mathbf{e}_{g(k)}\mathbf{e}_{g(k)}^T)\mathbb{X}_{g(k)}\tilde{\mathbb{V}} = (\mathbb{X}_{g(k)} - \mathbf{e}_{g(k)}\bar{\mathbf{x}}_{g(k)}^T)\tilde{\mathbb{V}}.$$

From (21), we see that the PCA of the global data matrix $\mathbb{X}_{g(k)}$ at time t_k can be obtained from the R factors $R_{n(j)}$ of the updated matrices S_j, for $j = 0, \ldots, k$. These R factors can be computed in turn by Algorithm 1 as in (8). Once these factors are computed, they can be assembled at a particular processor to form $\mathbb{R}_{g(k)}$ as in (22) and then the PCA of $\mathbb{S}_k$ can be computed. One potential problem is that it may create bottlenecks at certain processors if the assembling are not scheduled correctly.

5 Load Balancing

In this section, we give a procedure such that the loads among the processors will be balanced provided that the size of the data blocks are more or less the same on each processor. For notational simplicity, we will denote the set of all R factors of $\bar{X}_{n(k)}$ by $\{R_{n(k)}^{(0)}\}$ (see (4) and (5)), and the subsequent set of R factors of $\{R_{n(k)}^{(i)}\}$ by $\{R_{n(k)}^{(i+1)}\}$ (see (6) and (7)). We illustrate the main idea with $s = 8$. Figure 4 gives the flowchart

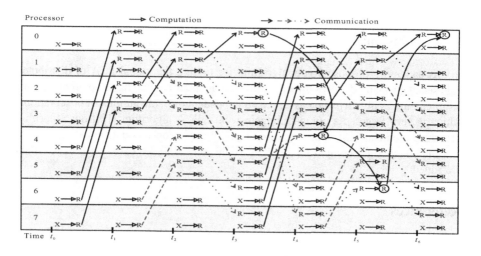

Fig. 4. Flowchart for the procedure when $s = 8$

of our algorithm when $s = 8$, i.e. $\ell = \log_2 s = 3$. In the figure, each time interval $(t_{j-1}, t_j]$ is divided into two phases: the computation phase where the QR decomposition are done, and the communication phase where the R factors are moved across the processors.

For example, in $(t_0, t_1]$, we first compute all the R factors $\{R_{n^{(0)}}^{(0)}\}$ of $\bar{X}_{n^{(0)}}$ using Algorithm 1 (marked in the figure by $X \dashrightarrow R$). There are 8 of them. Then during the communication phase, half of them will be sent to Processor i, $i < s/2 = 4$, according to Algorithm 1 (marked by the solid arrows in the figure). Then in $(t_1, t_2]$, we compute all the R factors $\{R_{n^{(1)}}^{(0)}\}$ of $\bar{X}_{n^{(1)}}$ (marked by $X \dashrightarrow R$), and the R factors $\{R_{n^{(0)}}^{(1)}\}$ of $\{R_{n^{(0)}}^{(0)}\}$ (marked by $R \dashrightarrow R$). Half of these R factors will be moved during the communication phase. However, in order to achieve load balancing, the factors $\{R_{n^{(1)}}^{(0)}\}$ should not be moved according to Algorithm 1 again, but according to the figure, i.e. to Processors 2, 3, 4, and 5 (marked by dashed arrows in the figure).

Continuing in this manner, we see that the R factor $R_{n^{(0)}}$ of the covariance matrix S_0 will be formed at Processor 0. (Recall that $R_{n^{(0)}} = \mathbb{R}_{g(0)}$ and $S_0 = \mathbb{S}_0$.) Also $R_{n^{(1)}}$ and $R_{n^{(2)}}$ will be formed at Processors 4 and 6 respectively (see the marked circles). Once $R_{n^{(k)}}$ are formed, they can be combined with previously obtained $\mathbb{R}_{g(k-1)}$ to form $\mathbb{R}_{g(k)}$ by using (22), provided that $\mathbb{R}_{g(k-1)}$ are sent there from the previous time-step (marked by curve arrows in the figure).

In this procedure, we assume that once $R_{n^{(k)}}$ is formed at time step $t_{k+\ell}$, it will be merged with $\mathbb{R}_{g(k-1)}$ to form $\mathbb{R}_{g(k)}$, see the circled-R in Figure 4. However, one can also send all these $R_{n^{(k)}}$ to a central processor, where all the $\mathbb{R}_{g(k)}$ are formed. The nice thing about this alternate approach is that if for some reasons, $R_{n^{(k)}}$ arrive to the central processor before $R_{n^{(j)}}$, for some $j < k$, then we can still form the $\mathbb{R}_{g(k)}$ at the central processor without waiting for $R_{n^{(j)}}$. Of course, $\mathbb{R}_{g(k)}$ so formed is the R factors of $\mathbb{X}_{g(k)}$ without the update block $X_{n^{(j)}}$, i.e. it is equivalent to $X_{n^{(j)}} = O$ in (15). When $R_{n^{(j)}}$ arrives at a later time, we can do the updating of $\mathbb{R}_{g(k)}$ first, and then include the contribution of $X_{n^{(j)}}$.

6   Conclusions

In this paper, we propose a new algorithm for finding the global PCA of distributed data sets. Our method works directly with the data matrices and has a communications requirement of only $O(p^2 \lceil \log_2 s \rceil)$, (i.e., independent of n, the number of observations, which is very large). As compared against the DPCA algorithm [10], our algorithm introduces no local PCA approximation errors. We also consider data updating, and we present a method for computing the PCA for the new extended data sets after new data are added.

References

1. Z. Bai, J. Demmel, J. Dongarra, A. Petitet, H. Robinson, and K. Stanley, The Spectral Decomposition of Nonsymmetric Matrices on Distributed Memory Parallel Computers, *SIAM J. Sci. Comput.*, 18(5): 1446–1461, 1997.

2. D. Boley, Principal Direction Divisive Partitioning, *Data Min. Knowl. Discov.*, 2(4): 325–344, 1998.

3. G. H. Golub and C. F. Van Loan, *Matrix Computations*, The Johns Hopkins University Press, 3rd ed., 1996.

4. H. Hotelling, Analysis of a Complex of Statistical Variables into Principal Components, *J. Educ. Psych.*, 24 (): 417–441, 498–520, 1933.

5. J. E. Jackson, *User's Guide to Principal Components*, Wiley, New York, 1991.

6. I. T. Jolliffe, *Principal Component Analysis*, Springer-Verlag, 1986.

7. H. Kargupta, W. Y. Huang, K. Sivakumar, and E. Johnson, Distributed Clustering Using Collective Principal Component Analysis, *Knowl. Inf. Syst.*, 3(4): 422–448, 2001.

8. J. B. Lee, A. S. Woodyatt, and M. Berman, Enhancement of High Spectral Resolution Remote Sending Data by a Noise-Adjusted Principal Component Transform, *IEEE Trans. Geosci. Remote Sensing*, 28(3):295-304, May 1990.

9. K. Pearson, On Lines and Planes of Closest Fit to Systems of Points in Space, *Phil. Mag.*, 2 (6): 559–572, 1901.

10. Y. M. Qu, G. Ostrouchov, N. Samatova, and A. Geist, *Principal Component Analysis for Dimension Reduction in Massive Distributed Data Sets*, Proceedings to the Second SIAM International Conference on Data Mining, April 2002.

11. E. Rabani and S. Toledo, *Out-of-Core SVD and QR Decompositions*, in Proceedings of the 10th SIAM Conference on Parallel Processing for Scientific Computing, Norfolk, Virginia, March 2001.

12. E. J. Wegman, Huge Data Sets and the Frontiers of Computational Feasibility, *J. Comput. Graph. Statist.*, 4 (4): 281–295, 1995.

Priority Conscious Transaction Routing in a Real-Time Shared Disks Cluster

Kyungoh Ohn, Sangho Lee, and Haengrae Cho

Department of Computer Engineering, Yeungnam University,
Gyungsan, Gyungbuk 712-749, Republic of Korea
{ondal, comman35, hrcho}@yumail.ac.kr

Abstract. A great deal of research indicates that the shared disks (SD) cluster is suitable to high performance transaction processing. However, the aggregation of SD cluster with real-time processing has not been investigated. By adopting cluster technology, the real-time services will be highly available and can exploit inter-node parallelism. In this paper, we propose a priority conscious transaction routing algorithm for a real-time SD cluster, which allocates real-time transactions to a node in the SD cluster. Unlike traditional routing algorithms that consider transaction affinity and load balancing only, our algorithm also considers transaction priorities inherent to real-time applications. We evaluate the performance of our algorithm under a wide variety of real-time workloads. The experiment results show that our algorithm outperforms both pure priority-based algorithms and pure affinity-based algorithms.

1 Introduction

There has been an increasing growth of real-time transaction processing applications, such as telecommunication systems, stock trading, electronic commerce, and so on. A real-time transaction has not only ACID properties of traditional transactions but also time constraints of completing its execution before deadline [5,6]. The major performance metric of real-time processing is the percentage of transactions missing their deadlines.

A cluster is a collection of interconnected computing nodes that collaborate on executing an application. A *shared disks* (SD) cluster is a representative cluster architecture for high performance transaction processing [1,13]. The SD cluster allows each node to have direct access to all disks. So it can support dynamic load balancing and seamless integration. Furthermore, the rapidly emerging technology of storage area networks (SAN) makes the SD cluster a preferred choice for reasons of higher system availability and flexible data access. The recent database systems using the SD cluster include IBM DB2 Parallel Edition [3] and Oracle Real Application Cluster [11].

Although there has been a great deal of independent research in real-time processing and SD cluster, their aggregation has very little attention [8]. The cluster technology enables highly available real-time database services, which are the core of many telecommunication services. The cluster can also achieve

J. Cao, W. Nejdl, and M. Xu (Eds.): APPT 2005, LNCS 3756, pp. 484–493, 2005.
© Springer-Verlag Berlin Heidelberg 2005

high performance real-time transaction processing by exploiting inter-node parallelism and reducing the amount of disk I/O with judicious data caching.

As a first step to the real-time SD cluster, we propose a real-time transaction routing algorithm in the SD cluster. The transaction routing is a process by a front-end router to select an execution node of an incoming transaction. The traditional transaction routing algorithms of the SD cluster have two design goals: *load balancing* and *transaction affinity* [10,14,15]. The load balancing means to avoid overloading individual node. The transaction affinity means to execute transactions with similar data access pattern on the same node (*affinity node*). To support real-time transactions, we have an additional goal of *transaction priority*. This goal means to reduce the number of transactions missing their deadlines by considering the deadline as a priority.

Our algorithm extends a well-performed traditional algorithm, named DACA (Dynamic Affinity Cluster Allocation) proposed by authors [10], to the real-time transaction processing. DACA can make an optimal balance between the affinity-based routing and indiscriminate sharing of load in the SD cluster. The contribution of this paper is to propose how the transaction priority can be incorporated into DACA. We also compare the performance of DACA and its real-time extension under a wide variety of real-time workloads.

The rest of this paper is organized as follows. Sect. 2 describes our model of the real-time SD cluster. Sect. 3 presents our algorithm in detail and Sect. 4 analyzes the experiment results. Concluding remarks appear in Sect. 5.

2   Model of Real-Time SD Cluster

Figure 1 shows our model of the real-time SD cluster. There is a router to select an execution node for each incoming transaction. The node schedules the execution of its transactions with *earliest deadline first* (EDF) policy [6]. We assume the *mixed* real-time transaction workload [7], which consists of both real-time transactions and non real-time transactions. A real-time transaction has a deadline. Executing the real-time transaction after its deadline is meaningless; hence, our model assumes *firm* real-time transactions [6].

The coupling facility of a node provides inter-node caching and global locking. Specifically, each node in the SD cluster has its own buffer and caches database pages in the buffer. Caching may substantially reduce the amount of disk I/O by utilizing the locality of reference. However, since a particular page may be simultaneously cached in different nodes, modification of the page in any buffer invalidates copies of that page in other nodes. This necessitates the use of a *cache coherency algorithm* so that the nodes always see the most recent version of database pages [1,2,9].

We assume that a record-level locking is in effect. In real-time applications, a locking protocol has to handle the *priority inversion* problem that lower priority transactions block the execution of higher priority transactions. To resolve the problem, we adopt the real-time locking protocol of [7]. Specifically, a real-time transaction aborts non real-time transactions holding locks in conflict mode.

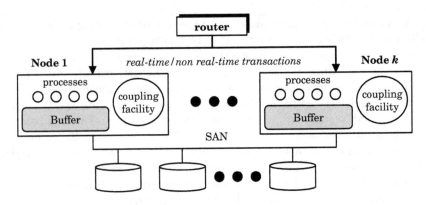

Fig. 1. Model of a real-time SD cluster

On the other hand, a non real-time transaction always waits at lock conflict. The same procedure holds between real-time transactions with different priorities. High priority transactions are always guaranteed to acquire locks without waiting.

3   Real-Time Transaction Routing

In this section, we propose a new real-time transaction routing algorithm, named *Priority conscious Dynamic Affinity Cluster Allocation* (P-DACA). We first summarize the basic idea of DACA algorithm [10]. Then we describe P-DACA algorithm that extends DACA for real-time transaction processing.

3.1   DACA Algorithm

To alleviate the routing overhead, DACA considers balancing the load of each *affinity cluster* (AC) [14]. An AC collects several transaction classes with high affinity to a given set of tables. A router maintains routing parameters. Specifically, when the router allocates a transaction of an affinity cluster AC_q to a node N_p, it increments both $\#T(AC_q)$ and $\#T(N_p)$, which means the number of active transactions at AC_q and at N_p respectively. Both counters are decremented when the transaction commits or aborts.

DACA divides the overload state into an *AC overload* and a *node overload*. The AC overload implies a state when transactions of an AC are rushed into the system. The node overload occurs when a node N_p is allocated to several ACs and $\#T(N_p)$ is over average. DACA balances the load of each node according to the overload state. If AC_q is in the AC overload state, then DACA allocates more nodes to AC_q by *node expansion* strategy. If there is no AC overload but N_p is in the node overload state, then DACA distributes some ACs assigned to N_p to other node by *AC distribution* strategy. DACA can make an optimal balance between the affinity-based routing and load balancing as follows.

- DACA tries to reduce the number of nodes allocated to the overloaded AC if the load deviation of each node is not significant. This allows DACA to reduce the frequency of inter-node cache invalidations.
- DACA prohibits allocating both an overloaded AC and other ACs to a node. As a result, DACA can achieve high buffer hit ratio for the overloaded AC. Even though several non-overloaded ACs may be allocated to a single node, efficient handling of the overloaded AC is more important to improve the overall transaction throughput.

3.2   Priority Conscious DACA (P-DACA)

Before describing the details of P-DACA, we first illustrate the problem of DACA when transactions have priorities. Example 1 shows the problem.

Example 1: Suppose there are two ACs (AC_1, AC_2) and two nodes (N_1, N_2). N_1 is an affinity node of AC_1 and executes a transaction t_1 of priority 100. N_2 is an affinity node of AC_2 and executes a transaction t_2 of priority 60. At this time, suppose a new transaction t_3 of priority 70 arrives, and t_3 belongs to AC_1. Then DACA allocates t_3 to its affinity node N_1 if N_1 is not in the node overload state. However, t_3 has lower priority than t_1 and cannot be executed until t_1 completes. So t_3 has a higher probability of missing its deadline. On the other hand, if t_3 is allocated to N_2, then it can be executed immediately.   □

The goal of P-DACA is (a) to reduce the number of transactions missing their deadlines, and (b) to take advantages of affinity clustering as DACA. To achieve this goal, P-DACA performs the following three basic steps sequentially to decide where a new real-time transaction t_r will be routed.

1. If there is an affinity node of t_r where the priority of t_r becomes the highest one, then allocates t_r to that node.
2. Else if there is a non-affinity node of t_r where the priority of t_r becomes the highest one, then allocates t_r to that node.
3. Else if there is no node that can execute t_r immediately, then allocates t_r to one of its affinity nodes in a round-robin fashion.

The underlying idea of the basic steps is to execute a real-time transaction at its affinity node only if the deadline of the transaction would not be missed (Step 1 and 3). By clustering real-time transactions to their affinity nodes, the amount of inter-node buffer invalidation can be minimized. So this idea can contribute to reduce the deadline miss ratio of following real-time transactions. Furthermore, due to the benefit of affinity clustering, it is also possible to improve the throughput of non real-time transactions.

The basic steps can be optimized if the router maintains a *priority list* for each node. The priority list is a sorted list in descending order of priorities of active real-time transactions at the node. Then a new real-time transaction t_r has to be allocated to a node where t_r can be executed faster than other nodes. P-DACA checks this condition by comparing the relative position of t_r in the

TRANSACTION_ROUTING$(t_r, P(t_r), AC_r)$

1. $\#\mathrm{T}(AC_r) = \#\mathrm{T}(AC_r) + 1$;
2. If t_r is a real-time transaction then
   (a) If there is a node $N_p \in \mathcal{S}(AC_r)$, where $\mathbf{rank}(P(t_r), N_p) < w_1$ and $\mathbf{rank}(P(t_r), N_p) < \mathbf{rank}(P(t_r), N_i)$ for all nodes $N_i \in \mathcal{S}(AC_r)$, $i \neq p$, then goto step 4;
   (b) Else if there is a node $\exists N_p \notin \mathcal{S}(AC_r)$, where $\mathbf{rank}(P(t_r), N_p) < w_2$ and $\mathbf{rank}(P(t_r), N_p) < \mathbf{rank}(P(t_r), N_i)$ for all nodes $N_i \notin \mathcal{S}(AC_r)$, $i \neq p$, then goto step 5;
   (c) Else goto step 3.
3. Select N_p, where $\#\mathrm{T}(N_p)$ is minimum for all node in $\mathcal{S}(AC_r)$;
4. If AC_r is in the AC overload state, then call **node_expansion**(AC_r);
5. Else if N_p is in the node overload state, then call **AC_distribution**(N_p);
6. $\#\mathrm{T}(N_p) = \#\mathrm{T}(N_p) + 1$; Insert $P(t_r)$ into the priority list of N_p;
7. Return N_p.

Fig. 2. Transaction routing algorithm of P-DACA

priority list of each node. Suppose $P(t_r)$ means the priority of t_r, and t_r is included in the affinity cluster AC_r. $\mathcal{S}(AC_r)$ is a set of affinity nodes of AC_r. Figure 2 summarizes the transaction routing algorithm of P-DACA.

At the step 2 of Fig. 2, the function of $\mathbf{rank}(P(t_r), N_p)$ returns the number of transactions whose priorities are higher than $P(t_r)$ at N_p. If the function returns 0, the priority of t_r will be highest at N_p. The values w_1 and w_2 are window constraints that limit the acceptable rank of t_r. w_1 is usually larger than w_2 since an affinity node of t_r could complete t_r faster. Note that if both w_1 and w_2 are set to 1, the algorithm works similar to the basic steps. A non real-time transaction is allocated to one of its affinity nodes with the lightest load (step 3). If allocating t_r would result in AC overload or node overload, then the resolution strategy of DACA has to be performed (step 4 and 5). Example 2 shows how P-DACA can resolve the problem of Example 1.

Example 2: Suppose that the information of ACs, nodes, and transactions are same to Example 1. Suppose also that both w_1 and w_2 are set to 1. Then the router allocates t_3 to N_2 that can execute t_3 immediately. This is because (a) the affinity node of t_3 is N_1 but $\mathrm{rank}(P(t_3), N_1) = 1 = w_1$, and (b) even though N_2 is not an affinity node of t_3 but $\mathrm{rank}(P(t_3), N_2) = 0 < w_2$. Note that if w_1 is set to 2, t_3 is allocated to N_1.   □

The notable features of P-DACA are two-fold. First, P-DACA allocates a real-time transaction to a node that guarantees higher probability of completing the transaction within its deadline. Even though the transaction could miss its deadline due to following transactions with higher priority, the selection strategy is the best choice at the current state. Next, P-DACA tries to allocate a transaction to its affinity node if possible. So P-DACA can achieve high buffer

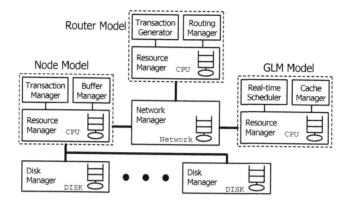

Fig. 3. Simulation model of a real-time SD cluster

hit ratio. The high buffer hit ratio in turn contributes to reduce the transaction execution time. The probability of missing deadline can be reduced as a result.

Maintaining the priority list would incur additional computing overhead. The first overhead comes from the fact that the router has to keep the priority list of each node in sorted order. An efficient implementation of ordered list with binary search can reduce the complexity of insertion, deletion, and search operations in the priority list. The next overhead comes from the fact that the priority list includes every active real-time transaction at the node. This means that a node has to report the commit of a real-time transaction to the router. Note that DACA also requires the same commit procedure for the router to maintain $\#T(AC_q)$ and $\#T(N_p)$ correctly [10].

4 Experiments

4.1 Simulation Model

To evaluate the performance of P-DACA, we have developed a simulation model of a real-time SD cluster using CSIM discrete-event simulation package [12]. Figure 3 shows the simulation model.

We model the SD cluster consisting of a single router and a global lock manager (GLM) plus a varying number of nodes, all of which are connected via a local area network. The router model consists of a *transaction generator* and a *routing manager*. The transaction generator has a role to generate transactions, each of which is modeled as a sequence of database operations. The routing manager implements three routing algorithms: P-DACA, DACA, and pure priority-driven algorithm (PRIO). PRIO does not consider the affinity and load balancing, but considers the transaction priority only.

The model of each node consists of a *buffer manager*, which manages the node buffer pool using an LRU policy, and a *resource manager*, which models CPU activity and provides access to the shared disks and the network. For

Table 1. Simulation parameters

System Parameters		
CPUSpeed	Instruction rate of CPU	1 GIPS
NetBandwidth	Network bandwidth	100 Mbps
NumNode	Number of computing nodes	8
NumDisk	Number of shared disks	20
DiskTime	Disk access time	0.01 ∼ 0.03 sec
PageSize	Size of a page	4096 bytes
ClusterSize	Number of pages in a cluster	10000
HotSize	Size of hot set in a cluster	2000 pages
DBSize	Number of clusters in database	8
BufSize	Per-node buffer size	5000 pages
FixedMsgInst	Number of instructions per messaging	20000
LockInst	Number of instructions per locking	2000
PerIOInst	Number of instructions per disk I/O	5000
PerObjInst	Number of instructions for a DB call	15000
Transaction Parameters		
TrxSize	Transaction size (# of records)	8 ∼ 12
RTPr	Probability of real-time transactions	0.2 ∼ 0.8
UpdatePr	Probability of updating a record	0.2
MPL	Number of concurrent transactions	640
ACNum	Number of affinity cluster	8
ACLocality	Probability of accessing local cluster	0.8
HotPr	Probability of accessing hot set	0.8

each transaction, the *transaction manager* forwards lock request messages and commit messages to the GLM. The disks are shared by every node.

The GLM has a role to perform the real-time concurrency control and the cache coherency control. The *real-time scheduler* implements mixed concurrency control protocol [7] for real-time locking. The *cache manager* implements ARIES/SD algorithm [9], which is a representative cache coherency algorithm in the SD cluster.

Table 1 shows the simulation parameters. Many of their values are adopted from [8,15]. Each disk has a FIFO queue of I/O requests and the disk access time is drawn from a uniform distribution between 0.01 to 0.03 seconds. The network is implemented as a FIFO server with 100 Mbps bandwidth. The CPU cost to send or to receive a message is modeled as a *FixedMsgInst* parameter.

We model that the database is logically partitioned into several *clusters*. Each database cluster has 10000 pages (40 Mbytes), and it is affiliated to a specific AC. The number of ACs is set to 8. The transaction parameter of *ACLocality* determines the probability that a transaction operation accesses a data item in its affiliated database cluster. The *HotPr* parameter models "80-20 rule", where 80% of the references to the affiliated database cluster go to the 20% of the database cluster (*HotSize*). The average number of records accessed by a transaction is determined by a uniform distribution between 8 and 12. The parameter *UpdatePr* represents the probability of updating a record. The processing associated with each record, *PerObjInst*, is assumed 15000 instructions.

The *RTPr* fraction of transactions is real-time transactions. For a real-time transaction, t, we determine its deadline (D_t) as follows [4]: $D_t = A_t + SF \times E_t$, where A_t and E_t are the arrival time and estimated execution time of t, respectively. SF is a *slack factor* that means tightness of deadlines. Its value is drawn from a uniform distribution between 1.3 and 10. E_t is computed as follows:

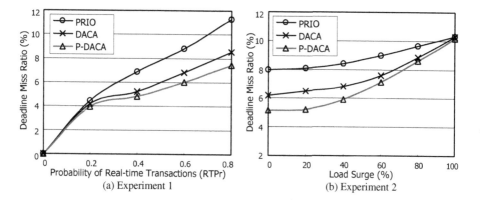

Fig. 4. Experiment results

$E_t = NumRead_t \times (PerObjInst + DiskTime) + NumWrite_t \times PerObjInst$, where $NumRead_t$ and $NumWrite_t$ are the number of read and write operations of t, respectively. Then the priority of a real-time transaction is defined as $\frac{1}{D_t}$.

The performance metric used in the experiments is a *deadline miss ratio*, which is the percentage of input transactions that the system is unable to complete before their deadlines.

4.2   Experiment Results

We first compare the performance of P-DACA with other algorithms by varying the probability of real-time transactions ($RTPr$). Figure 4(a) shows the experiment results when the input rate of each AC is equal. The system load is evenly distributed to each node as a result.

As $RTPr$ increases, the deadline miss ratio of every algorithm increases. This is due to the heavy contention between real-time transactions at high $RTPr$. There are more chances for a real-time transaction to be blocked or preempted by other real-time transactions. P-DACA performs best at every setting of $RTPr$. In particular, the performance improvement is substantial when $RTPr$ is high. P-DACA can prioritize real-time transactions with short deadlines, and thus the deadline miss ratio is reduced. PRIO also considers the priority of real-time transactions, but it performs worst. This is because PRIO suffers from frequent inter-node cache invalidations. The lower buffer hit ratio of PRIO should delay the average transaction execution time.

The next experiment determines the effect of system dynamics. Specifically, we evaluate the performance when transactions in a specific AC are surged into the system. Figure 4(b) shows the experiment results when $RTPr$ is set to 0.5. Since both $NumNode$ and $ACNum$ are set to 8 and MPL is set to 640, the steady state load per each AC is 80 transactions. A *load surge* is expressed as a fraction of its steady state load. For example, a load surge of 20% implies that the load of each non-surge AC decreases about 20% (16 transactions) and the total sum of additional load (112 transactions) goes to the surge AC.

Every algorithm performs worse as the load surge increases. This is because at high load surge there are many transactions of the surge AC, and thus the lock conflict ratio should increase. The performance difference of P-DACA and DACA becomes marginal as the load surge increases. Their performance is nearly same at the load surge of 100%. Note that P-DACA could outperform DACA by allowing real-time transactions to be executed at non-affinity nodes. However, as the load surge increases, both DACA and P-DACA allocate more affinity nodes to the surge AC. The probability of executing at non-affinity node decreases as a result. The performance of PRIO is also similar to that of P-DACA at the load surge of 100%, since there is only one AC.

5   Concluding Remarks

We proposed a new transaction routing algorithm for the real-time SD cluster, named P-DACA. The underlying idea of P-DACA is to execute a real-time transaction at its affinity node only if the deadline of the transaction would not be missed. This enables P-DACA to achieve an optimal balance between priority conscious routing and affinity-based routing. We also evaluated the performance of P-DACA under a wide variety of real-time workloads. The performance results show that P-DACA outperforms the pure priority-based or affinity-based algorithms when (a) the number of real-time transactions is large, or (b) the system load is evenly distributed.

This work only examined part of overall problem space for the real-time shared disks cluster, and many problems remain open. Developing a cache coherency algorithm for real-time transactions is part of our future work. We will then examine the hybrid effect of real-time transaction routing and real-time cache coherency algorithm. Another interesting direction of future work is a performance study between shared disks cluster and other cluster architectures, such as shared nothing, with real-time workloads.

Acknowledgements

This research was supported by Korean Ministry of Information and Communication under the University IT Research Center program supervised by the IITA (Institute of Information Technology Assessment).

References

1. Cho, H.: Cache Coherency and Concurrency Control in a Multisystem Data Sharing Environment. IEICE Trans. Information and Syst. **E82-D** (1999) 1042-1050
2. Cho, H., Park, J.: Maintaining Cache Coherency in a Multisystem Data Sharing Environment. J. Syst. Architecture **45** (1998) 285-303
3. DB2 Universal Database for OS/390 and z/OS - Data Sharing: Planning and Administration. IBM SC26-9935-01 (2001)

4. Harita, J., Carey, M., Livny, M.: Data Access Scheduling in Firm Real-Time Database Systems. J. Real-Time Syst. **4** (1994) 203-241
5. Kanitkar, V., Delis, A.: Real-Time Processing in Client-Server Databases. IEEE Trans. on Computers 51 (2002) 269-288
6. Lam, K-Y., Kuo, T-W. (ed.): Real-Time Database Systems: Architecture and Techniques. Kluwer Academic Publishers (2000)
7. Lam, K-Y., Kuo, T-W., Lee, T.: Strategies for resolving inter-class data conflicts in mixed real-time database systems. Journal of Syst. and Soft. **61** (2002) 1-14
8. Lee, S., Ohn, K., Cho, H.: Feasibility and Performance Study of a Shared Disks Cluster for Real-Time Processing. Lecture Notes in Computer Science **3397** (2005) 518-527.
9. Mohan, C., Narang, I.: Recovery and Coherency Control Protocols for Fast Intersystem Page Transfer and Fine-Granularity Locking in a Shared Disks Transaction Environment. In: Proc. 17th Int. Conf. VLDB (1991) 193-207
10. Ohn, K., Cho, H.: Cache Conscious Dynamic Transaction Routing in a Shared Disks Cluster. Lecture Notes in Computer Science **3045** (2004) 548-557
11. Vallath, M.: Oracle Real Application Clusters. Elsevier Digital Press (2004)
12. Schwetmann, H.: User's Guide of CSIM18 Simulation Engine. Mesquite Software, Inc. (1996)
13. Yousif, M.: Shared-Storage Clusters. Cluster Comp. **2** (1999) 249-257
14. Yu, P., Dan, A.: Performance Analysis Clustering on Transaction Processing Coupling Architecture. IEEE Trans. Knowledge and Data Eng. **6** (1994) 764-786
15. Yu, P., Dan, A.: Performance Evaluation of Transaction Processing Coupling Architectures for Handling System Dynamics. IEEE Trans. Parallel and Distributed Syst. **5** (1994) 139-153

Probabilistic Continuous Update Scheme in Location Dependent Continuous Queries

Song Han and Edward Chan

Department of Computer Science, City University of Hong Kong
han_song@cs.cityu.edu.hk, csedchan@cityu.edu.hk

Abstract. It is difficult to maintain the exact location of mobile objects due to the limited resources in a mobile network. A consequence of this problem is that the update cost for a location-dependent continuous query for moving objects can be quite high using traditional methods. In this paper, we propose a probabilistic update method to maintain the fidelity of the query results without incurring significant update cost. Our scheme makes use of two types of updates, one to keep the uncertainty of the mobile object's position within a specific confidence interval, and the other using probability that the moving object's location uncertainty will affect the query result as the threshold to decide whether an update should be generated or not. The effectiveness of our approach is demonstrated using a series of simulation experiments.

1 Introduction

Many mobile applications rely on the continuous tracking of mobile objects. However, due to limitations in the bandwidth of wireless networks and battery power of the mobile devices, it is difficult to maintain their exact location. This uncertainty can affect the accuracy of answers to a location-dependent continuous query (LDCQ) on these objects [1]. Researchers have proposed many dead-reckoning methods to handle the trade-off between update cost and tracking accuracy. However, a major drawback is that they do not attempt to relate the update frequency to the overall accuracy of the query. In this paper, we propose a probabilistic continuous update scheme for LDCQs. It makes use of two types of updates, a location update to keep the uncertainty of the mobile object's position within a specific confidence interval and a query accuracy update which uses the probability that the moving object's location uncertainty will affect the result as the threshold to decide whether an update should be generated. We also demonstrate how to calculate the predicted update time for mobile objects using the historical motion information stored in database.

2 Related Work

Previous research on techniques for handling LDCQs are typically based on the simplifying assumption that all moving objects know their locations and send their updates to a central database server. This line of research focuses on how the accuracy of the query can be assured given that excessive location updates consumes too much

J. Cao, W. Nejdl, and M. Xu (Eds.): APPT 2005, LNCS 3756, pp. 494–504, 2005.
© Springer-Verlag Berlin Heidelberg 2005

bandwidth, and that due to disconnections an object may not be able to continuously update its position even if ample bandwidth is available. To process LDCQs efficiently, a Moving Objects Spatio-Temporal (MOST) model was proposed in [3]. In this model, a location prediction function is defined as a dynamic attribute of a moving object to facilitate the prediction of the future locations of the object. The locations of the moving objects are tracked using some efficient dead-reckoning techniques, such as plain dead-reckoning method in which an update is generated to refresh the location of an object whenever the deviation of its current position is greater than a pre-defined threshold. In [5] the authors go beyond the object tracking problem to deal with efficient techniques in transmitting query results to clients.

Cheng et al. are the first researchers to deal with probabilistic methods in processing moving object queries [6]. The uncertainty model proposed is used in our current work, but we focus on formulating of an update strategy that will meet user fidelity requirements without incurring high update cost. As far as we know, this is the first paper to propose a probabilistic location update scheme for LDCQs.

3   System Model and Definitions

In this section we describe our system model for the support of LDCQ as well as some key definitions. Figure 1 depicts the system architecture of a mobile computing system that supports LDCQ. The system consists of a database server and a number of mobile objects connected by a mobile network. The server maintains a mobile objects database, which adopts the moving objects spatial-temporal (MOST) data model to record the location information of mobile objects dynamically. As an example, we consider a mobile object MO, whose last update is issued to the database at time t_0. At that time, the position of MO is $<x_0, y_0>$; the scalar of the speed vector and the direction are v_0 and α_0 respectively. Then the record of MO in the database will be:

Mobile Object	Update time	Position	Speed	Direction
MO	t_0	$<x_0, y_0>$	v_0	α_0

If the next update from MO is at time t_1, then at time t' in the time period $[t_0, t_1]$, the position of MO $<x', y'>$ can be predicted as: $x' = x_0 + v_0 * cos\alpha_0 * (t' - t_0)$ and $y' = y_0 + v_0 * sin\alpha_0 * (t' - t_0)$.

A mobile object in the system can be a client which issues a LDCQ with begin and end times to the location database server. At the same time, it is also a potential query target for a set of LDCQs issued by other mobile objects, in which case it generates updates to report its current location and other motion information to the database server as needed. Each update is associated with a time-stamp, which specifies the time for which the current value is valid. The query is then evaluated according to the records in the database and will be re-evaluated when there is any change in the database state during the query period. The results, in the form of a set of tuples, are collected and grouped by their begin times, each indicating the beginning of the time period for which the object specified in the tuple satisfies the condition of the query. Once the results are ready, the server may send the selected tuples to the client according to a query result transmission approach adopted by the system.

We can define the *query object* (O_Q) as an object that issues a LDCQ Q and receives the answer set S from the database server during the period [t_{begin}, t_{end}] specified by Q. For the *same* query Q, all other mobile objects are considered as moving objects (O_M^i) though not all will appear in S. Note that this distinction between query object and moving objects applies only for the same query, because a query object in Q_i may be considered as a moving object in a different query Q_j.

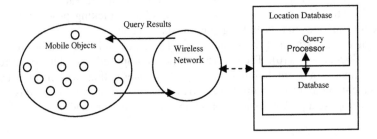

Fig. 1. The system architecture of a mobile computing system

3.1 Uncertainty Model

As the bandwidth of the wireless network is limited, it is prohibitively expensive to update the location information of all mobile objects in real-time. This leads to a discrepancy between the actual position of a mobile object and the position calculated according to information in the server database. The uncertainty in the location of a mobile object can be characterized as follows [7]:

Definition 1: An **Uncertainty Region** of mobile object *MO* at time t, $U(p, t)$, is a closed region such that *MO* can be found inside this region with a probability p.

Definition 2: Uncertainty Probability Density Function of a mobile object *MO* at time t, $f(x, y, t)$, is the probability density function of *MO*'s location at time t and

$$\int_{U(p,t)} f(x, y, t)dxdy = p$$

In this paper we only consider range queries (RQ). An RQ returns a set of tuples in the form of (O_M^i, t_i^{begin}, t_i^{end}), where [t_i^{begin}, t_i^{end}] is the period during which O_M^i will be inside the query boundary and satisfy the RQ. The relationship between a query boundary (QB) and the query is as follows: if O_Q issues a RQ at time t_0, QB(t_0) for Q is the circle whose origin is the position of O_Q at time t_0, and the radius R is specified by Q. In practice it is the movement of the QB that we are interested in.

3.2 Fidelity

In this section, we present the definition of the fidelity of a query for deviation based range query, continuous range query and probabilistic range query. Fidelity is a key user requirement and as well as a major performance metric in our simulation experiments to be described in a later section.

3.2.1 Fidelity of the Deviation-Based Range Query

Fidelity for deviation-based range query measures the deviation of the results in the database from the correct results for a range query Q. Its definition is based on the concepts of false positives and false negatives defined below [7]:

The false positives ratio of Q at time $t, f^+ (Q, t)$ and the false negatives ratio of Q at time $t, f^- (Q, t)$ are defined below respectively:

$$f^+(Q,t) = \frac{|S_{dbase}(Q,t) - S_{ideal}(Q,t)|}{|S_{dbase}(Q,t)|} \qquad f^-(Q,t) = \frac{|S_{ideal}(Q,t) - S_{dbase}(Q,t)|}{|S_{ideal}(Q,t)|}$$

where $S_{dbase} (Q, t)$ is the result set of Q at time t, evaluated using the moving object database; $S_{ideal} (Q, t)$ is the result set of Q at time t, evaluated using actual location information. $f^+ (Q, t)$ measures the fraction of objects wrongly included into the answer of Q and $f^- (Q, t)$ measures the portion of objects that are missing in the correct answer of Q.

Definition 3: Fidelity in the result of a deviation-based range query Q is maintained at time t if $E (t) = f^+ (Q, t) + f^- (Q, t) < \varepsilon$ where $E (t)$ is the error ratio of Q at time t and ε, the **Fidelity Requirement,** is a real-valued system parameter for Q.

3.2.2 Fidelity of Continuous Range Query

The fidelity defined above is only for Q at time t. For a continuous query Q, the query is active during its activation period $[t_{begin}, t_{end}]$, we need to re-define the definition of fidelity for a deviation based query to capture the overall fidelity for the full duration of the query.

Definition 4: The overall fidelity of continuous query Q over its active period is:

$$overall\_fidelity(Q) = \frac{\int_{t_{begin}}^{t_{end}} F(t)dt}{t_{end} - t_{begin}} \quad and \quad F(t) = \begin{cases} 0 & E(t) < \varepsilon \\ 1 & otherwise \end{cases}$$

3.2.3 Fidelity of the Probabilistic Range Query

In a probabilistic range query, each moving object is associated with a probability that the object will satisfy the particular query. The previous definitions for $f^+ (Q, t)$ and $f^- (Q, t)$ can be refined for probabilistic range query. First we define:

$S_{Intersection} (Q, t) = S_{dbase} (Q, t) \wedge S_{ideal} (Q, t)$
$S_{DB} (Q, t) = S_{dbase} (Q, t) - S_{Intersection} (Q, t)$
$S_I(Q, t) = S_{ideal} (Q, t) - S_{Intersection} (Q, t)$

We assume the cardinality of $S_{Intersection} (Q, t)$, $S_{DB} (Q, t)$ and $S_I(Q, t)$ are $l(Q, t)$, $m(Q, t)$ and $n(Q, t)$ respectively. Each moving object in $S_{dbase} (Q, t)$ has a relative probability p_i that it will satisfy the continuous query Q at time t, and for $S_{ideal} (Q, t)$, the relative probability p_i is 1. Now we can define:

$$f^+(Q,t) = \frac{|(\sum_{i=1}^{l(Q,t)}(1 - p_{Inter\,sec\,tion}^i)) + \sum_{j=1}^{m(Q,t)} p_{DB}^j|}{|\sum_{k=1}^{l(Q,t)+m(Q,t)} p_{dbase}^k|}, \quad f^-(Q,t) = \frac{|(\sum_{i=1}^{l(Q,t)}(1 - p_{Inter\,sec\,tion}^i)) + n(Q,t)|}{|l(Q,t) + n(Q,t)|}$$

where $p^i_{Intersection}$ is the probability for the ith object to be in $S_{Intersection}$ (Q, t), p^j_{DB} is the probability for the jth object to be in S_{DB} (Q, t) and p^k_{dbase} is the probability for the kth object to be in S_{dbase} (Q, t).

4 The Probabilistic Continuous Update Scheme

The Probabilistic Continuous Update Scheme (PCU) is used for generating location updates, and aims at maintaining high fidelity with low location update costs. In traditional time-based or distance-based schemes, an object is assigned a fixed time or distance threshold and will update once the threshold is exceeded. However, in PCU, two kinds of updates are needed to ensure that the required fidelity is maintained, namely Object Location Update and Query Accuracy Update.

Definition 5: Object Location Update (OLU) is the update issued by a query object or moving object to guarantee that at time t, its position will not be outside its uncertainty region U (p, t).

Definition 6: Query Accuracy Update (QAU) is the update issued *only* by the moving object when the change of the moving object's uncertainty region will affect the answer set for a certain Q with a probability p which is specified by the user.

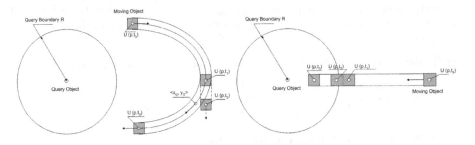

Fig. 2. Example of OLU   **Fig. 3.** Example of QAU

In Figure 2, at time t_2, the actual position of the moving object is outside the uncertainty region U (p, t_2) due to a change in its trajectory, so an OLU will be issued. In Figure 3, at time t_2, though U (p, t_2) is not exceeded, the probability that it will cross the query boundary exceeds p, so a QAU will be issued to guarantee the accuracy of the query result. It can be seen that a QAU is typically invoked as the moving object crosses the query boundary, where even minor movement may determine whether it satisfy Q and hence affect the accuracy of the query answer.

To facilitate discussion of update mechanism, we focus on two groups of moving objects. The first group, the *potential set* (PS) of moving objects, is the set of moving objects which have the chance to satisfy the query conditions during the entire query duration for a certain query. To determine which objects belong to PS, we simply consider the worst case in which the moving object and the query object are moving face to face at maximum speed. If the moving object has no chance of hitting the query boundary over the duration of the query, it will be excluded. The second group of moving objects, called the *query boundary set* (QBS), is the set of moving objects whose uncertainty region $U(p, t)$ has an intersection with the query boundary. All other moving objects need not be considered at that time because the uncertainty in their location cannot affect the accuracy of the answer to the query. Figure 5 shows two moving objects in the query boundary set at time *t*.

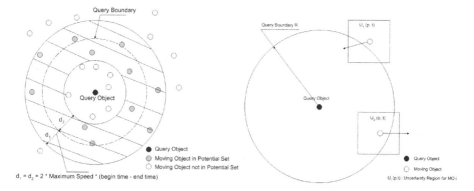

Fig. 4. Potential Set for RQ   **Fig. 5.** Query Boundary Set at time *t* for RQ

Pruning Phase: in this phase, for a certain Q, system will execute following steps:

1. Query object sends its motion information and query parameters to the database.
2. Database calculates Q's potential set and broadcasts a message to all moving objects in the set requesting them to issue the OLU in the duration of Q.
3. Based on motion information of the query object and uncertainty region for each moving object in the potential set, the database decides their initial QAU time.

Refresh Phase: in this stage, for a certain Q, the system will execute two main steps:

1. If database receives an OLU from a moving object in Q's potential set, it will re-calculate the new QAU time for that moving object, and if necessary, other objects involved in the query.
2. If a moving object is predicted to issue a QAU, database will send a message to it to request an update on its position to guarantee the accuracy of the query result.

In the following sections, we will show how to evaluate PS, OLU and QAU and the latest update records in database for O_Q and O_M are listed in Table 1.

4.1 Evaluation of the Potential Set

In Figure 6, R is the query boundary, and V_{max} is the maximum speed, and other notations are the same as those defined above.

S is the set of all moving objects in the database

$X \leftarrow \Phi$

for $i \leftarrow 1$ to $|S|$ do

$$T_i = \frac{\left| \sqrt{(x^i_M - x_Q)^2 + (y^i_M - y_Q)^2} - R \right|}{2V_{max}}$$

If $(T_i < t_{end})$ then $X \leftarrow X \cup O^i_M$

Return X

Fig. 6. Evaluation of Potential Set

Table 1. Latest update records for O_Q and O_M

Object	O_Q	O_M
Update Time	t_Q	t_M
Position	$<x_Q, y_Q>$	$<x_M, y_M>$
Speed	v_Q	v_M
Direction	α	β

4.2 Generation of OLU and QAU

4.2.1 Generation of OLU

Because the calculation of OLU is independent and same for moving objects and query objects, we just need to decide at time t, whether an OLU for a mobile object MO will be issued to the database. We assume the uncertainty probability density function for the position of MO is normally distributed and the expectation of the distribution is the predicted position of MO at time t, the variances of the distribution are parameters specified by the user. Suppose the last update (either OLU or QAU) of MO is at time t_0, and at that time, the position of MO is $<x_0, y_0>$ and the scalar and direction of the speed is v and α respectively. The predicted position at time t is $<x^P, y^P>$ (the calculation is demonstrated in Section 3). At time t, the position of MO is denoted by $<X, Y>$, where X and Y are two independent random variables and they satisfy the normal distribution. Then we have $X \sim N(x^P, \sigma_X)$, $Y \sim N(y^P, \sigma_Y)$ where σ_X and σ_Y are system parameters. MO will issue an update if its actual position at time t exceeds the predicted position's confidence interval c and we can induce that the c confidence intervals of X and Y are respectively:

$$(x^P - u_{(1-c)/2} * \sigma_X, x^P - u_{(1-c)/2} * \sigma_X) \text{ and } (y^P - u_{(1-c)/2} * \sigma_y, y^P - u_{(1-c)/2} * \sigma_y)$$

Suppose at time t, the actual position of O is $<x, y>$. MO will not issue OLU if

$$x \in (x^P - u_{(1-c)/2} * \sigma_X, x^P - u_{(1-c)/2} * \sigma_X) \text{ and } y \in (y^P - u_{(1-c)/2} * \sigma_y, y^P - u_{(1-c)/2} * \sigma_y).$$

4.2.2 Generation of QAU

At time t, if moving object O_M will cross the query boundary of the query object O_Q with a probability which is bigger than the threshold probability p, a control message will be sent by the central server to O_M to ask it generate a QAU immediately to the server. Since in a range query, all moving objects are independent, we will just consider the calculation between O_M and O_Q. Assume the positions of O_M and O_Q satisfy normal distributions, at time t ($t >= t_Q$, $t >= t_M$), the position of O_M is $<X_M, Y_M>$ and that of O_Q is $<X_Q, Y_Q>$, where X_M, Y_M, X_Q, Y_Q are independent random variables and

$$X_M \sim N(x_M^P, \sigma_x^2), Y_M \sim N(y_M^P, \sigma_y^2), X_Q \sim N(x_Q^P, \sigma_x'^2), Y_Q \sim N(y_Q^P, \sigma_y'^2)$$
$$x_M^P = x_M + v_M * (t-t_M) * cos(\alpha), y_M^P = y_M + v_M * (t-t_M) * sin(\alpha)$$

$$x_Q^{P} = x_Q + v_Q * (t-t_Q) * cos(\beta), \ y_Q^{P} = y_Q + v_Q * (t-t_Q) * sin(\beta)$$

Now we consider the relative movement of O_M to O_Q, and the relative position is $<X', Y'>$. Since X' and Y' are independent random variables, they satisfy:

$$X' = X_M - X_Q => X' \sim N (x_M^{P} - x_Q^{P}, \ \sigma_x^2 + \sigma_x'^2; \ Y' = Y_M - Y_Q => Y' \sim N (y_M^{P} - y_Q^{P}, \ \sigma_y^2 + \sigma_y'^2)$$

Based on the distribution of the X' and Y', now we can calculate the probability P_{QB} that the O_M will cross the query boundary at time t. For simplicity, we set:

$$\mu_1 = x_M^{P} - x_Q^{P}, \ \mu_2 = y_M^{P} - y_Q^{P}, \ \sigma_1^2 = \sigma_x^2 + \sigma_x'^2, \ \sigma_2^2 = \sigma_y^2 + \sigma_y'^2$$

And the probability can be calculated as:

$$P_{QB} = \iint_{\Omega} \frac{1}{2\pi\sigma_1\sigma_2} e^{-\frac{1}{2}\left[(\frac{X'-\mu_1}{\sigma_1})^2 + (\frac{Y'-\mu_2}{\sigma_2})^2\right]} dX' dY'$$

Ω is the integration area and is different depending on whether O_M is moving out or into the query boundary. Suppose the confidence interval of O_Q is q and the confidence interval of is r, then:

$$X_M \in (x_M^{P} - u_{(1-r)/2} * \sigma_X, \ x_M^{P} + u_{(1-r)/2} * \sigma_X), \ Y_M \in (y_M^{P} - u_{(1-r)/2} * \sigma_Y, \ y_M^{P} + u_{(1-r)/2} * \sigma_Y)$$
$$X_Q \in (x_Q^{P} - u_{(1-q)/2} * \sigma_X', \ x_Q^{P} + u_{(1-q)/2} * \sigma_X'), \ Y_Q \in (y_Q^{P} - u_{(1-q)/2} * \sigma_Y', \ y_Q^{P} + u_{(1-q)/2} * \sigma_Y')$$

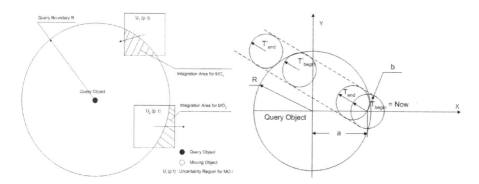

Fig. 7. Examples of the integration areas   **Fig. 8.** Calculation Period Case 2

And we can deduce:

$$X' \in (x_M^{P} - u_{(1-r)/2} * \sigma_X - (x_Q^{P} + u_{(1-q)/2} * \sigma_X'), \ x_M^{P} + u_{(1-r)/2} * \sigma_X - (x_Q^{P} - u_{(1-q)/2} * \sigma_X'))$$
$$Y' \in (y_M^{P} - u_{(1-r)/2} * \sigma_Y - (y_Q^{P} + u_{(1-q)/2} * \sigma_Y'), \ y_M^{P} + u_{(1-r)/2} * \sigma_Y - (y_Q^{P} - u_{(1-q)/2} * \sigma_Y'))$$

Two examples of the integration area Ω are shown in Figure 7.

After demonstrating how to calculate the probability that O_M will enter or leave the query boundary, now we will describe how to predict when to generate the next QAU for O_M. Suppose at time t, the expected distance between O_Q and O_M is a, the query boundary is R; the relative speed and direction are v^R and θ respectively. As it is difficult to calculate the next update time t^P directly, we need to first get the interval which bounds t^P, we call it the calculation period. Based on different relationships among a, b, R, v^R and θ, the calculation period $[t_{begin}, t_{end}]$ can be evaluated in the following table Case 2 is demonstrated in Figure 8. In case 2, there are two calculation periods $[t_{begin}, t_{end}]$ and $[t'_{begin}, t'_{end}]$. This is because O_M's uncertain region will intersect with the query boundary twice. Other cases can be explained along the same line. Based on the information of the calculation period, dichotomy is used to get the next update time t^P.

5 Simulation Results and Performance Analysis

In this section we evaluate the performance of our proposed probabilistic continuous update scheme using a number of simulation experiments. In particular we compare our scheme with a distance-based update scheme (DBU) where the moving object will issue an update to refresh its location information if the deviation of its current position is greater than the predicted value by a pre-defined threshold. We use the random waypoint mobility model [2] in our simulation. The continuous query length is set to

Table 2. Evaluation of Calculation Period

$$
\begin{cases}
R-b \le a \le b+R &
\begin{cases}
\cos\theta \ge 0 \;\;or\;\; (\cos\theta < 0 \;\;and\;\; (R-b)^2 < a^2\sin^2\theta) &
\begin{cases}
t_{begin} = now \\
t_{end} = (-a\cos\theta + \sqrt{(b+R)^2 - a^2\sin^2\theta})/v^R
\end{cases}
& (1) \\[4ex]
\cos\theta < 0 \;\;and\;\; (R-b)^2 \ge a^2\sin^2\theta &
\begin{cases}
t_{begin} = now \\
t_{end} = (-a\cos\theta - \sqrt{(R-b)^2 - a^2\sin^2\theta})/v^R \\
t'_{begin} = (-a\cos\theta + \sqrt{(R-b)^2 - a^2\sin^2\theta})/v^R \\
t'_{end} = (-a\cos\theta + \sqrt{(b+R)^2 - a^2\sin^2\theta})/v^R
\end{cases}
& (2\,*\,)
\end{cases} \\[10ex]
a > b+R &
\begin{cases}
\cos\theta > 0 \;\;or\;\; (\cos\theta \le 0 \;\;and\;\; (b+R)^2 \le a^2\sin^2\theta) &
\begin{cases}
t_{begin} = \infty \\
t_{end} = \infty
\end{cases}
& (3) \\[4ex]
\cos\theta < 0 \;\;and\;\; (b+R)^2 > a^2\sin^2\theta &
\begin{cases}
(R-b)^2 \ge a^2\sin^2\theta &
\begin{cases}
t_{begin} = (-a\cos\theta - \sqrt{(b+R)^2 - a^2\sin^2\theta})/v^R \\
t_{end} = (-a\cos\theta - \sqrt{(R-b)^2 - a^2\sin^2\theta})/v^R
\end{cases}
& (4) \\[4ex]
(R-b)^2 < a^2\sin^2\theta &
\begin{cases}
t_{begin} = (-a\cos\theta - \sqrt{(b+R)^2 - a^2\sin^2\theta})/v^R \\
t_{end} = (-a\cos\theta + \sqrt{(b+R)^2 - a^2\sin^2\theta})/v^R
\end{cases}
& (5)
\end{cases}
\end{cases} \\[10ex]
a < R-b &
\begin{cases}
t_{begin} = (-a\cos\theta + \sqrt{(R-b)^2 - a^2\sin^2\theta})/v^R \\
t_{end} = (-a\cos\theta + \sqrt{(b+R)^2 - a^2\sin^2\theta})/v^R
\end{cases}
& (6)
\end{cases}
$$

1000 sec, query boundary is 200 m. 100 moving objects roam a 1000x1000m area. Fidelity requirement and confidence level are both 95%. Speed is uniformly distributed between 12 and 60 km.

In Figure 9, we vary the object location variance (OLV) to demonstrate its effect on the fidelity of the query result. We find that our scheme can maintain the fidelity requirement while the fidelity calculated using the distance-based update scheme drops below the fidelity requirement quickly. In Figure 10 and Figure 11, we compare the number of updates in the system between PCU and DBU scheme. There are two types of updates in PCU: OLU and QAU. Figure 10 shows that OLU is around 5% less than the update number under DBU, this is because QAU in our scheme also constraint the uncertainty of the object location and decrease the number of OLU. Figure 11 shows that the total number of update of our scheme is slightly larger than DBU. This is a tradeoff for maintaining the fidelity of the query result which we think is worthwhile. In Figure 12, based on the same update number, it can be seen that PCU not only meets the fidelity requirement but also consistently provides better fidelity.

6 Conclusion

In this paper, we proposed a probabilistic continuous update scheme that goes beyond traditional location updates schemes which attempt only to maintain the uncertainty of the moving object's location to a prescribed value by trying to link the update with required fidelity in the answer to a LDCQ. Based on simulation experiments, it is shown to outperform a simple deviation based location update method for probabilistic range queries. We are currently studying the effectiveness of this scheme for other mobility models and will report the results in a future paper.

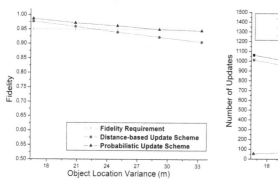

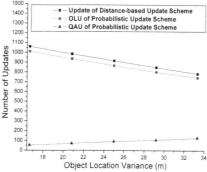

Fig. 9. Fidelity vs. Object Location Variance   **Fig. 10.** Number of updates vs. OLV

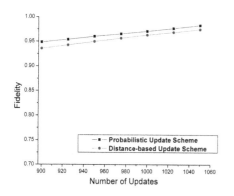

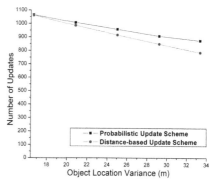

Fig. 11. Total number of updates vs. OLV   **Fig. 12.** Fidelity vs. Number of Updates

References

[1] M. H. Dunham and V. Kumar, Location Dependent Data and its Management in Mobile Database, Database and Expert Systems Applications, 1998, Proc. 9^h International Workshop on Database and Expert Systems Applications, 1998.

[2] D. Johnson and D. Maltz, Dynamic Source Routing in Ad Hoc Wireless Networks. Mobile Computing (ed. T. Imelinsky and H. Korth), Kluwer Academic Publishers, 1996.

[3] A. P. Sistla, O. Wolfson, S. Chamberlain, and S. Dao, Querying the Uncertain Position of Moving Objects, Temporal Database – Research and Practice Lecture Notes in Computer Science 1399, 1998.

[4] O. Wolfson, S. Chamberlain, S. Dao, L. Jiang and G. Mendez, Cost and Imprecision in Modeling the Position of Moving Objects, Proc. 14th International Conference on Data Engineering, 1998.

[5] H. G. Gök and Ö. Ulusoy, Transmission of continuous query results in mobile computing systems, Information Science, vol. 125, no.1 – 4, pp. 37 – 63, 2000.

[6] Reynold Cheng, Dmitri V. Kalashnikov, and Sunil Prabhakar, Querying imprecise data in moving object environments, IEEE Trans. on Knowledge and Data Engineering, Vol. 16(7), July 2004.

[7] Jinfeng Ni and C. V. Ravishankar, Probabilistic Spatial Database Operations, Proc. 8th Intl. Symposium on Spatial and Temporal Databases (SSTD), 2003.

SIP-Based Adaptive Multimedia Transmissions for Wired and Wireless Networks*

Weijia Jia and Man-Ching Yuen

Department of Computer Science, City University of Hong Kong,
83 Tat Chee Avenue, Kowloon, Hong Kong, SAR China
itjia@cityu.edu.hk

Abstract. SIP (Session Initiation Protocol) is a signaling protocol standardized by IETF, aiming to manage the multimedia transmission sessions among different parties. This paper illustrates an adaptive multimedia transmission system for wired and wireless networks based on SIP with protocol selection mechanism for a certain level of QoS guarantee. In our system, SIP is not only used for call setup signaling but also for carrying the information in the protocol selection. Using Agent Server, our system can select the most suitable protocol for adapting different situations intelligently during the connections and data buffering service is also provided for various media data flows between the end users with acceptable QoS level without any interruption and disconnection regardless of types of devices, platforms and protocols used.

1 Introduction

Mobile multimedia transmissions such as online movies, live TV, network radio and audiovisual conversation require good quality of service (QoS) anytime and anywhere. However, it is difficult to attain the targets over the integrated wired - wireless networks because the wireless devices typically have the limited resources of processing power or memory. Little work is done to enable dynamic multimedia communication, especially end-to-end multimedia transmission across wired and wireless networks. Compared with wired Internet, there are several obstacles for end-to-end wireless multimedia transmission: (1) Low capability and limited resource of terminals: A wireless terminal typically has a small display with low resolution, slow processor and small memory space. However, multimedia applications usually require high capability of graphic processing, large size of memory space and also a big screen to display pictures and videos. Obviously, wireless terminals can only support limited multimedia applications. (2) Diversity of wireless terminals: Wireless terminals often support only a limited set of data formats due to their low capabilities and limited resources. When two wireless terminals of different types communicate with each other, their supported data formats may not be compatible and thus communication cannot succeed. Although it is possible that the data sent from the sender can be

---

* This effort is partially sponsored by City University of Hong Kong strategic grants 7001587, and 7001709 and the National Basic Research Program (973) MOST of China under Grant No. 2003CB317003.

J. Cao, W. Nejdl, and M. Xu (Eds.): APPT 2005, LNCS 3756, pp. 505–514, 2005.
© Springer-Verlag Berlin Heidelberg 2005

converted into the format supported by the receiver, these conversion often costs much and is not practical or acceptable to resource-limited wireless terminals. (3) Low bandwidth of wireless networks: Due to low capability of wireless devices, multimedia data streams created by wireless devices can not be compressed very much and require high bit rate for real-time transmission. Therefore, it is difficult to develop dynamic real-time multimedia communication protocols for audiovisual conversation and videoconferencing. (4) Fluctuated bandwidth and blockouts of wireless connections: Multimedia data transmission, especially real-time transmission, requires steady high bit rate and is intolerant of package delay. On the contrary, wireless networks have fluctuated bandwidth and high probability of traffic congestion. Usually, the transmission time of multimedia data in a session is quite long, but the blackout of wireless connection may cause frequent session reconnection and data retransmission.

SIP (Session Initialization Protocol) [12] is a signaling protocol of Application Layer which is standardized by IETF (Internet Engineering Task Force), and it aims to manage multimedia sessions among different parties. The principle of SIP is to set up sessions or associations between two or more end users. SIP is not used for transmitting data, but sessions initiated with SIP can exchange various types of media data using appropriate protocols such as RTP, RSTP and so on. It can carry out bi-directional authentication and capability negotiation. SIP is simple and extensible. It accepts complementary information inserted as SIP payload for other applications. Currently, SIP is able to set up a call for a multimedia session of complex requirements by carrying more detailed information using protocols such as the Session Description Protocol (SDP) [13].

This paper proposes an adaptive protocol selection mechanism for integrated wired-wireless multimedia transmission mechanism using SIP to maintain a certain level of QoS guarantees. SIP is not only used for call setup signaling, but also carries information for protocol selection. SIP usually carries an SDP packet describing an audio or video session, indicating supported communication protocols, end terminals capabilities, QoS requirements of applications and session ID which is used for user identification of multi-parties communication. Besides selecting the most suitable protocol for adapting different situations intelligently during connection, data buffering service is provided with Agent Server such that media data flows must transmit between end users. In this way, the end users can communicate among the others at their best acceptable QoS level regardless of types of devices, platforms and protocols they are using.

The rest of this paper is organized as follows. Section 2 introduces some related works. Section 3, describes the adaptive protocol selection mechanism. Section 4 presents the multimedia transmission connections on integrated wired and wireless networks by using SIP technology. Section 5 concludes the paper.

2 Related Work

WMSTFP [2] is an end-to-end TCP-friendly multimedia streaming protocol, which is used to detect the status of the wired and wireless parties in the wireless Internet. By accurately distinguishing the packet losses due to transmission errors from the congestive losses and smoothing out the pathologic round-trip-time values caused by the highly dynamic wireless environment, higher throughput in wireless Internet can be achieved and transmission rate can be adjusted in a smooth and TCP-friendly manner.

UPnP™ Forum [3] is an industry initiative designed to enable simple and robust connectivity among stand-alone devices and PCs from different vendors. The forum members are engaged in producing standards to describe device specifications usually in XML format. iMobile [4] is a proxy-based mobile service platform designed to provide personalized services. iMobile provides a modular architecture that supports accesses from various mobile devices to various information spaces. However iMobile does not support communications between devices. Transcoding service plays a very important roll in the design of wireless multimedia system [5-8]. A video transcoding technology using intermediate data processor is proposed in [5] to enhance the quality of transcoded data. A video transcoding proxy for 3G wireless mobile Internet access and a video transcoding gateway for wireless access are proposed in [6] and [7] respectively. End-to-End Wireless Multimedia Transmission system (EEWMT) is developed and designed to transcode end-to-end data flows [8].

A lot of other related works have been engaged in wireless access of multimedia data, but most of them just consider data transmissions between wireless terminals and servers. In our proposed system, data transmission between wireless terminals is also considered using Agent Server as the intermediate party to provide a certain level of QoS guarantees. Moreover, to maintain the best acceptable QoS level, SIP is used for both call setup signaling and carrying information for processing an adaptive protocol mechanism, such that the most suitable transmission protocol is selected for transmission of various multimedia data flow adaptively during connection without interruption or disconnection.

3 Adaptive Protocol Selection

The adaptive protocol selection mechanism is used to ease the balance of transmission performance and communication interoperability among various clients and servers in wired networks and wireless mobile networks. We briefly describe its mechanism. Based on features and popularities of different existing protocols, the transmission protocols used in our system are classified as:

TCP (Transmission Control Protocol) or UDP (User Datagram Protocol),

1. Pure HTTP (HyperText Transfer Protocol) [10], and
2. A web services protocol, SOAP (Simple Object Access Protocol).

TCP/UDP may be used to provide efficient and fast data transmission. Especially, it enables real-time data transmission using UDP with the support of certain protocols like RTP (Real Time Protocol). However, it is difficult to implement using TCP/UDP because they may not be interpretable between different OS platforms. As a result, TCP/UDP implementations in a specific OS platform may not be reused in another OS platform. Pure HTTP enables communications across different platforms, and also allows communications penetrating some firewalls. However, it is not flexible for developers to support new services due to its limited number of services and commands provided. As web services protocols work based on XML [11], they have similar functionalities as HTTP. The only difference is that the web services protocols can provide a consistent and simple interface for developers to support connection services in a uniform way. Since both pure HTTP and web services protocols are used in the application layer, their data transmission rate is low and may not be suitable for the real-time multimedia transmission.

As these popular transmission protocols have different advantages and disadvantages, they are used under different situations. Table 1 shows performance comparison of different natures of data with four types of network protocols (TCP, UDP, HTTP and web services). There are many ways to select the most suitable protocol for different situations. One of the examples is to apply fuzzy membership functions [16]. There is a unique membership function associated with each input parameter. The membership functions associate with weighting factors with each input and the effective rules. These weighting factors determine the degree of influence or degree of membership (DOM) for each active rule. By computing the logical product of the membership weights for each active rule, a set of fuzzy output response magnitudes are produced. All that remains is to combine and specify these output responses.

Table 1. Performance Comparison of Different Natures of Data with Four Types of Network Protocols (TCP, UDP, pure HTTP and web services)

		TCP/IP	UDP/IP	Pure HTTP	Web Services
Text (Data size: very small)	Non-real time	Suitable (Fast data transmission rate)	Suitable (Fast data transmission rate)	Most suitable (relatively slow data transmission rate)	Most suitable (relatively slow data transmission rate)
	Real time	Suitable (Fast data transmission rate)	Suitable (Fast data transmission rate)	Most suitable (relatively slow data transmission rate)	Most suitable (relatively low data transmission rate)
Audio (Data size: small)	Non-real time	Suitable (Fast data transmission rate)	Suitable (Fast data transmission rate)	Most suitable (relatively low data transmission rate)	Most suitable (relatively low data transmission rate)
	Real time	Suitable (Fast data transmission rate)	Suitable (Fast data transmission rate)	Most suitable (relatively low data transmission rate)	Most suitable (relatively low data transmission rate)
Image (Data size: medium or large)	Non-real time	Suitable (Fast data transmission rate)	Suitable (Fast data transmission rate)	Most suitable (relatively low data transmission rate)	Most suitable (relatively low data transmission rate)
	Real time	Suitable (Fast data transmission rate)	Suitable (Fast data transmission rate)	Most suitable (relatively low data transmission rate)	Most suitable (relatively low data transmission rate)
Video (Data size: large or very large)	Non-real time	Suitable (require more network resource)	Most suitable (Fast data transmission rate)	Not suitable (Very low data transmission rate)	Not suitable (Very low data transmission rate)
	Real time	Suitable (require more network resource)	Most suitable (Fast data transmission rate)	Not suitable (Very low data transmission rate)	Not suitable (Very low data transmission rate)

To effectively communicate among the different platforms with varies devices, our adaptive protocol selection mechanism is designed possessing the following functionalities: (1) Allowing different kinds of client devices to communicate in the integrated wired and wireless networks while the communication performance is monitored at an acceptable level most of the time; (2) Allowing clients to dynamically select a suitable protocol for adapting different situations intelligently without any interruption and disconnection; (3) Providing consistent APIs for different protocols thus reducing development overhead of service modules regardless of platforms and devices, and (4) Providing simple APIs for different service modules so that the APIs are reusable and extensible for supporting new services.

We here briefly describe the implementation and design of the adaptive protocol selection system for handling the interactive communications among integrated networks and various client devices. The design issues are categorized into two parts: connection establishment and transmission protocol selection below:

Connection Establishment consists of two major steps: (1) Initialization of communication session between client devices through agent servers. A set of agent servers in the networks are responsible to provide data buffering and QoS guaranteed services. All communications between end users must pass through the agent servers. The communication session between agent server and client device can be initialized using SIP for agent server or client device. By considering different protocols supported by client devices and characteristics of transmission sessions, the way of communication between agent servers and client devices is defined during the communication initialization stage (see Section 4). (2) Notification of both protocol and platform of all communication parties to agent servers. Whenever communication is initialized either by agent servers or client devices, agent servers have to know the type of protocol of client devices, platform of client devices and QoS requirements of applications. It is because negotiations between agent servers and client devices may be required for having services of the best performance. It is also useful for selecting the most adequate transmission protocol in later step.

Transmission Protocol Selection is defined in two respects: (1) Common APIs (Application Program Interfaces) of supported transmission protocols. To support most of the services provided by transmission protocols, we have devised the common APIs of the transmission protocols available in agent servers. Each service module has a set of APIs for providing its service to client devices where the APIs are able to support different protocols, platforms and client devices. (2) Selection of a suitable transmission protocol in different situations. There exist a number of data type, such as non real-time text data and real-time video data. To balance the performance (user's point of view) and system overhead (developer's point of view), different transmission protocols are suitable for different data transmissions. To select the most suitable transmission protocol according to different situations, we have to define data transmission, the QoS requirements of session, the available protocols supported by client devices and the protocols supported by agent servers. Once agent servers have the information, they determine the most adequate transmission protocol for specific data transmission and inform client devices by providing an appropriate data transmission process using the most adequate protocol. Our system uses SIP to select the most suitable transmission protocol adaptively depending on the nature of media flows and the transmission capability [9] as detailed in next section.

4 SIP Based Multimedia Transmission Control

Our system is designed to provide services to end-to-end multimedia transmission, including both real time and non-real time transmission, with certain level of QoS guarantees. Fig. 1 shows the framework of our system for wired and wireless networks. Four main parts of our system are User Agent in client device, SIP Proxy Server, Database Server and Agent Server. We here illustrate them as below:

1. User agents (UA) are SIP endpoints that send or receive signaling messages residing on client devices and help client devices to communicate with servers. UA collect device profile of client devices, capabilities of client devices and QoS requirements of sessions to be requested, and then sends these information to the servers. UA also convert user commands and application signaling into formats that can be read by servers and also translate server responses for users and applications. Device profile is used to describe the technical specifications and capabilities of the device such as multimedia processing capability and network transmission capability. Some other device information such as manufacturer name and device model are also presented in the profile. Device profile must be in a universal format so that the servers are able to recognize all kinds of devices and provide appropriate services to them. The device profile is given as a XML formatted file listing device specifications and capabilities. Fig. 2 presents a simple example of device profile.

2. SIP proxy servers store the information of all the major SIP proxy servers and provide DNS services. Each major SIP proxy server further connects to a set of proxy servers within its network domain. The control data flow through SIP proxy servers, while all media data in communication between end users flow through agent servers only.

3. Database servers store the updated information of all agent servers in their domains and respond to the requests from either SIP proxy servers or agent servers. Examples of the information in database server are user profiles and device profiles.

4. Agent servers are application-layer routers and receive call requests from UA or another proxy, try to locate the receivers via the selected route paths defined by SIP initially, and forward the requests to another location until the given address is reached. They execute the adaptive protocol selection mechanism and keep track the change of situations during data transmission. To achieve the best level of QoS guarantees, an agent server provides many categories of services to client devices: the data buffering service prevents transmitters retransfer lost data due to network congestion or disconnection and the data transcoding service helps the heterogeneous terminals to communicate with each other seamlessly. All data flows in communications must go through agent servers until receiving terminals are reached.

Basically, SIP is a control protocol for establishing media sessions and it is used for both call setup signaling and session transmission for adaptive protocol selection mechanism. Five functionalities that support the establishment and termination of multimedia communications, for the adaptive protocol selection mechanism, are called:

1. User location detection determinates the end system to be used for communication.
2. User capability detection defines the media and media parameters to be used.
3. User availability detection decides the willingness of the called party to engage in communications.
4. Call setup establishes call parameters at both called and calling party during "ringing".
5. Call handling handles many management operations including transfer and termination of calls.

The sessions' addresses to be established are carried out in the body of the application layer message. It has two types of messages: request and response. SIP messages carry the descriptions of media sessions in their payload/header using Session Description Protocol (SDP) [13]. Some additional mechanism is needed for payload modification is defined for the servers below:

1. INVITE: This message is used to invite another participant to a session.
2. ACK: This message confirms session establishment.
3. BYE: This message is used to close a session.
4. CANCEL: This message cancels a pending INVITE message.
5. REGISTER: This message is used to register the current address of a potential participant.
6. RESPONSE: This message is used to give response to request and indicates success or failures and progress updates.

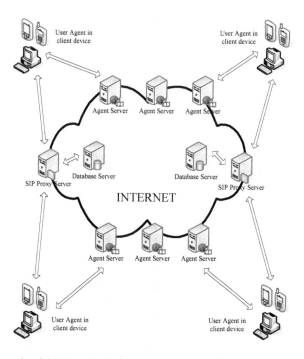

Fig. 1. Framework of SIP based end-to-exnd multimedia transmission for integrated wired and wireless networks

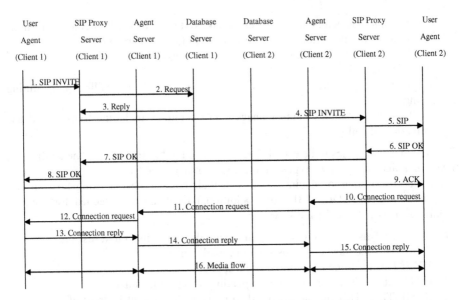

Fig. 2. Signaling scenario for connection establishment in call setup stage

To establish a connection session, a UA sends a SIP INVITE request to SIP proxy server. As mentioned before, the SIP proxy server only stores the information of all other SIP proxy servers and provides DNS services. SIP proxy server sends a request to database server which searches the information of agent servers among the end users. Once the connection path is established, the media data flows through agent servers between end users directly rather than through proxy servers. During communication, agent servers keep track the status of connection, and deploy the adaptive protocol selection mechanism. During the connection, the agent server is responsible to select the most suitable transmission protocol for efficient communications. Note that for agent server, each communication protocol will reserve a specified port which is randomly generated. Figs 2-4 illustrate the adaptive connection establishment signaling protocols used in our system for call setup stage, change of transmission protocols or connection path between end users.

Thus, our system has three advantages that support large varieties of devices and transmission environments, and also provide certain level of QoS guarantees:

(1) Heterogeneous communication: Due to the diversity of wireless terminals, data formats supported by different terminals may be not compatible with each other. In our system, data flows go through agent servers that may cause the incompatible data formats into acceptable formats according to the device profiles of the receiving terminals. As a result, terminals can send and receive data in preferred formats without concerning data format incompatibility problems.

(2) Low cost of terminal resource: Instead of terminals, agent servers are responsible to convert data formats, retransmission for lost packages, and reconnection to lost packets and saves resources for terminals more effectively and efficiently. Moreover, a certain level of QoS can be guaranteed.

(3) Faster recovery from disconnection: Connection loss is quite often for wireless connection due to signal fading, interference and path blackout. Once the connection between the receiver and its agent server is lost, the transmitter has to wait until it reconnects and retransmits the lost packages. In our system, since the terminals are connected to the specified agent servers, and data buffering in agent servers can keep receiving the data from the sender even if the receiver is disconnected. Once the receiver is reconnected, the buffered data will be delivered.

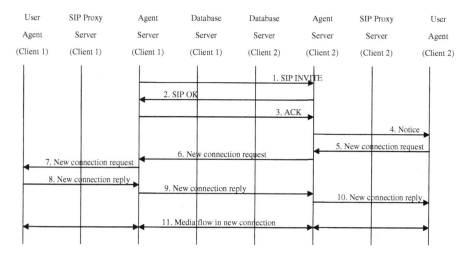

Fig. 3. Signaling scenario for adaptive change of transmission protocols during connection

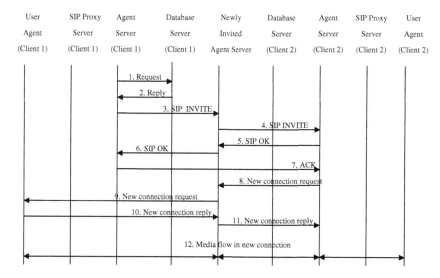

Fig. 4. Signaling scenario for change of connection path between end users during connection

5 Conclusions

We have proposed an adaptive multimedia transmission system for wired and wireless networks using SIP technology for call setup signaling, information carrying via agent servers. Based on SIP, our system can be used in various communication environments and it is extensible. Through the adaptive protocols, the most suitable connection mechanism can be selected adaptively based on the dynamic change of traffic or flows.

References

1. 3GPP, "Open services architecture", Application Programming Interface, 3G TR 29.998, http://www.3gpp.org.
2. Yang F., Zhang Q., Zhu W., Zhang Y.-Q., "Bit Allocation for Scalable Video Streaming over Mobile Wireless Internet", Proceeding of INFOCOM'2004, Hong Kong, March 2004.
3. Universal Plug and Play Forum, Understanding Universal Plug and Play, 2000, http://www.upnp.org/download/UPNP_UnderstandingUPNP.doc.
4. Chen Y.-F., Huang H., Jana R., John S., Jora S., Reibman A., Wei B., "Personalized Multimedia Services Using A Mobile Service Platform", Wireless Communications and Networking Conference, 17-21 March 2002. vol.2, pp. 918 - 925.
5. Iwasak O., Uenoyama T., Ando A., Nishitoba T., Yukitake T., Etoh M., "Video Transcoding Technology for Wireless Communication Systems", IEEE Vehicular Technology Conference Proceedings, Tokyo, 15-18 May 2000. vol.2, pp. 1577 - 1580.
6. Warabino A., Ota S., Morikawa D., Ohashi M., Nakamura H., Iwashita H., Watanabe, F., "Video Transcoding Proxy for 3Gwireless Mobile Internet Access", IEEE Communications Magazine, October 2000, Vol.: 38 , Issue: 10, pp. 66 - 71.
7. Lei Z., Georganas N. D., "Video Transcoding Gateway for Wireless Video Access", Electrical and Computer Engineering, Canadian,4-7 May 2003. vol. 3, pp. 1775 - 1778.
8. Shen J., Han B., Yuen M.-C., Jia W., "End-to-End Wireless Multimedia Transmission System", IEEE Vehicular Technology Conference Fall 2004.
9. Yuen M.-C., Cheng L., Au P.-O., Jia W., "Adaptive Generic Communications for Integrated Mobile and Internet Web-Services", The 5th International Conference on Web-Age Information Mangement 2004.
10. Berners-Lee T., Fielding R., Frystyk H., "Hypertext Transfer Protocol -- HTTP/1.0", RFC1945, May 1996.
11. Hollenbeck S., Rose M., Masinter L., "Guidelines for the Use of Extensible Markup Language (XML) within IETF Protocols", RFC3470, Mar 2002.
12. Handley M., Schulzrinne H., Schrooler E., Rosenberg J., "Session Initiation Protocol", RFC2543, IETF., March 1999.
13. Handly M. and Jacobson V., "SDP: session description protocol", RFC2327, IETF, April 1998.
14. "H.323 - Packet-based multimedia communications systems", ITU standard.
15. Schulzrinne H. and Rosenberg J., "A Comparison of SIP and H.323 for Internet Telephony", Network and Operating System Support for Digital Audio and Video (NOSSDAV), Cambridge, England, July 1998.
16. Chen G., Pham T. T., "Introduction to Fuzzy Sets, Fuzzy Logic, and Fuzzy Control Systems" CRC Press, 2000.

WM+: An Optimal Multi-pattern String Matching Algorithm Based on the WM Algorithm

Xunxun Chen, Binxing Fang, Lei Li, and Yu Jiang

Research Center of Computer Networks and Information Security Technology,
Harbin Institute of Technology, Harbin, P.R.C.
{cxx, bxfang, lilei, jy}@pact518.hit.edu.cn

Abstract. The WM algorithm, designed by Sun Wu and Udi Manber, is considered the fastest multi-pattern string matching algorithm in practice except when the pattern number is very large or the alphabet size is small[2]. Theoretically, the scanning time of WM is average-optimal (i.e. $O(n\log\sigma(rm)/m)$), but in the worst case, its scanning time can not be evaluated at all. The maximum shift of the original WM algorithm is m-B+1, where m is the minimum length of all patterns and B is the q-gram size. The tuned WM algorithm (abbreviated as WM+) can reach higher performance by improving the *shift* table building algorithm and combining the AC algorithm with the original WM algorithm. And the scanning time of the WM+ algorithm in the worst case is predictable. Experiments show that the scanning time of the WM+ algorithm is less or not great than that of the WM algorithm for varied size of m and number of patterns, especially in the worst case.

1 Introduction

Multi-pattern searching tries to solve the problem of finding all the starting positions of any occurrence of sub-strings $P=\{p, p, ..., p_r\}$ in text $T=t_1...t_n$. This is very common in the fields of information retrieve, gene comparison, virus scanning, intrusion detection and so on. A series of optimal algorithms have been developed since 1970's such as WM[1], AC[3], BM[4], SBOM[5]. Many technologies, such as native comparison, automata, filtering, n-gram[11] and bit comparison etc, have been employed and improved.

Usually, the discussion of string matching algorithms can be described as follows:

Definition 1: Suppose that $P=\{p_1, p_2, ..., p_r\}$ denotes the patterns set, $p_i=c_{i1}c_{i2}..c_{im}$, $c_{ij}\in\Sigma$, Σ the alphabet table with the size of σ, m the mean length of patterns, $M=rm$ the total length of all patterns, and $T=t_1...t_n$ the text string need to be scaned. The matching algorithm is a function $Am(P,T)$ which has two inputs P and T. The time complexity Ct and storage complexity Cs are used to evaluate the efficiency of Am. When $Ct\rightarrow O(f(n))$, in which f is a function of n and shows the relationship between searching time of Am and the text length n under the condition that contents and lengths of patterns are fixed, it means the time complexity of Am(i.e. Ct) reaches the level described by $f(n)$. And when

J. Cao, W. Nejdl, and M. Xu (Eds.): APPT 2005, LNCS 3756, pp. 515–523, 2005.
© Springer-Verlag Berlin Heidelberg 2005

$Cs \rightarrow O(g(M))$, in which g is a function of M and shows the relationship between the memory storage occupied by Am and the total length of all patterns(i.e. M), it means that the storage complexity of Am(i.e. Cs) reaches the level described by $g(M)$.

Generally, when the worst-case time complexity of an Am is proved to be $O(n)$ (i.e. $Ctw \rightarrow O(n)$), the Am is considered as a worst-case optimal algorithm; when the average-case time complexity of an Am is proved to be $O(n\log_\sigma(rm)/m)$ (i.e. $Cta \rightarrow O(n\log_\sigma(rm)/m)$), the Am is considered as an average-case optimal algorithm[6]; when the pre-processing time complexity of an Am tends to be $O(rm)$, the Am is considered to have an optimal pre-processing time; and when the storage complexity of an Am is proved to be $O(rm)$, the Am is considered to have an optimal storage[2].

There are two kinds of exact multi-pattern matching algorithms. One kind is the DFA based algorithms that are typified by using much storage to save matching time. Because of having stable time complexity $O(n)$ which is patterns-free and text-free, very easy to implement on computers, and having less instructions in each step, the DFA based algorithms are widely used as the base algorithms in many matching algorithms. The typical of them is AC[3] algorithm that is based on KMP[7]. Because the scanning time complexities are $O(n)$ in both the worst case and average case, and the pre-processing time complexity is $O(rm)$, the DFA based algorithms are regarded as the fastest algorithms when $n\log_\sigma(rm)/m \geq n$ (i.e. when the pattern lengths are very short or the pattern number is very large) [2].

The other kind is filtering based algorithms that reduce the comparison times by skipping more characters to the best of their abilities. The typical filtering based algorithm is BM[4] that puts forward the idea of from-right-to-left scanning and not comparing all the characters of the text. In practice, because the worst case is very unusual to appear in the text to be scanned, filtering based algorithms are generally faster than DFA based algorithms[2]. Presently WM[1] and SBOM[5] are considered as the two fastest exact multi-pattern matching algorithms[2] in literature, and WM is considered to be faster than SBOM in practice except that the pattern number is very large or the alphabet size is very small.

WM algorithm is a filtering based multi-pattern string matching algorithm utilizing the basic idea of BM. In fact, WM is an extending of BM for multi-pattern matching problems. WM also scans text from right to left. However, WM uses a B-size block (also called N-Gram) instead of a single character as the comparison unit, which extends the alphabet size logically. It's method can take advantage of computer word length, which is often no less than 16 bits, for increasing the scanning speed. WM also employs the idea of Horspool's[8] improvement for BM, that is, increasing the searching speed by significantly decreasing the hitting probability(i.e. the probability of the equality of text character and the last character of patterns) of BM.

There are still several parts of WM that can be improved. First, the result of pre-processing stage often makes the size of the scanning skip not to be long enough and the shorter skip size results in taking additional time in scanning step. Second, it is supposed in WM that the probability of the equality of the first two characters of two patterns combined with the equality of the B characters in middle of the two patterns is very small. Then, in some cases when this probability becomes larger, the efficiency of

WM is obviously decreased. We developed a new improved algorithm, which is called as WM+, trying to optimize the pre-processing and scanning stage of WM.

2 Algorithm: WM+

2.1 Pre-processing Stage

In the pre-processing stage of WM, the scanning time is not so easily to reduce due to the coarse *shift* table items. In this section, we enlarge the maximum skip size of scanning stage by fining SHIFT table items and thus the scanning time is reduced.

Definition 2: If a B-size string $c_1 c_2..c_B$ is completely included in the first m characters ($p_1 p_2..p_m$, m is minimum pattern length) of a pattern, we call $c_1 c_2..c_B$ a valid B-block. If the last k characters of $c_1 c_2..c_B$ appear in the head of $p_1 p_2..p_m$ (i.e. $c_{B-k+1} ..c_{B-1}c_B = p_1 p_2..p_k$), we call $c_1 c_2..c_B$ a k/B-valid B-blocks. And all B-blocks other than these two cases are called invalid B-block.

In pre-processing stage, WM only constructs indices of valid B-blocks of $p_1 p_2..p_m$ to build the *shift* table and uses shorter skip size for the k/B-valid B-blocks[1]. For example, suppose we have two patterns which are {p_1="abcd", p_2="bcdef"} and $B=2$, $m=4$, the *shift* table are $shift$("ab")=2, $shift$("bc")=1, $shift$("cd")=0, $shift$("de")=0, and the *shift* value of all other 2-size strings is $m-B+1=3$. Since B is no less than 2 and thus makes $m-B+1<m$, the average skip step in the best case is $m-1$ in scanning stage.

The optimizing method used in WM+ is described as follows. Indices of all valid and k/B-valid B-blocks are constructed to generate the *shift* table and the *shift* value of all invalid B-blocks is m in order to increase the best-case average skip size up to m. And the scanning performance will be much accelerated if m is small. Theoretically, the scanning performance of WM+ in the best case can be improved by $1/(m-1)$ times to that of WM. For the same example, the *shift* tables of WM and WM+ are shown in Table 1.

Table 1. *Shift* tables constructed by WM and WM+

B-block Indices / Algorithm	ab	bc	cd	de	?a	?b (Excluding ab)	Others
WM	2	1	0	0	3	3	3
WM+	2	1	0	0	3	3	4

2.2 Scanning Stage

WM uses knowledge of linguistics probability statistics, that is, WM supposes that the probability of the equality of the first two characters of two patterns combined with the equality of the B characters in middle of the two patterns is very small. However, the probability can be significantly affected by both the minimum pattern length and the pattern number, and it is also sensitive to the similarity of the patterns. In the worst case, because native comparison is done character by character and pattern by pattern, the most time-consumed operation in scanning stage is the comparison after the success of the coarse matching (i.e. the hit of B-block HASH). In practice, if $m<4$ or $r>5000$, the probability of the hit of B-block HASH will greatly increase, which causes the rapidly dropping of the scanning performance of WM.

WM+ uses prefix automata scanning derived from AC[3] instead of native pattern comparison after the success of the coarse matching, which decreases the uncertainty of the matching time in the worst case of scanning stage. In practice, WM+ uses filtering algorithm combined with automata based algorithm to accomplish the scanning job in scanning stage. Filtering algorithm is employed to increase the matching speed in the best case to the best of its ability by skipping the bad characters, and automata is employed to decrease the matching time in the worst case because the automata based algorithm is the optimal algorithm in the worst case.

The main body of scanning stage algorithm of WM+ with $B=2$ on zero-ended text is described as Figure 1. And the algorithm for fixed length text can be easily acquired by modifying the ending conditions of two WHILE loops in Figure 1.

```
w←m-B;  s←0;  p←0;
1:WHILE  t[w]≠0 DO
    IF shift[WORD PTR *(t+w)]>0  THEN
        w←w+shift[WORD PTR *(t+w)];
    ELSE IF p<w+B-m THEN p←w+B-m;
        WHILE  t[p]≠0 DO
            IF output[s][t[p]]>0 THEN write("Found a match.");
            ENDIF
            IF state[s][t[p]]=0    THEN
                w←p+m-B+1;  p←p+1;  s←0;  GOTO 1;
            ELSE
                s←state[s][t[p]];  p←p+1;
            ENDIF
        ENDDO
    ENDIF
ENDDO
```

Fig. 1. The main body of scanning stage algorithm of WM+ With $B=2$ on zero-ended text

In Figure 1, t is the starting address of the text. w is the pointer of the scanning algorithm, which points to the first character of the last B characters in the scanning window. s contains the current state of the prefix automata. p is the pointer of current character of the prefix automata and p can exceed w which is an important characteristic for accelerating the scanning speed. The one-dimensional array *shift* is the skipping table constructed in the pre-processing stage of WM+. The two-dimensional array *state* is the state table of the prefix automata, of which the initiating process can be found in AC[3]. For the same example as shown in section 2.1, the scanning process on the text "axyzxyzabcdefabab" is described in Figure 2.

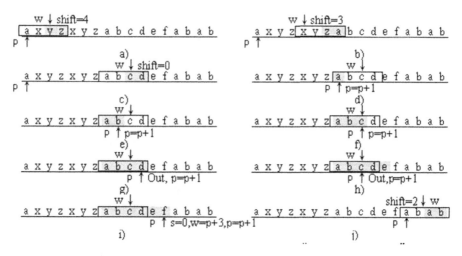

Fig. 2. The scanning process of WM+ on text *axyzxyzabcdefabab*

3 The Coarse Analysis of the Performance of WM+

In order to compare with WM, for simplicity, we suppose that there are r patterns, and one of them has a length of m, and the other $r-1$ patterns have a length of k. Let $m<k$ and $M=m+(r-1)k$, $B=\log_\sigma(rm)$. Now we will compare the performance of WM+ with that of WM in pre-processing and scanning stage respectively.

In the best case, all B-blocks in the text are invalid B-blocks. Then the WM's maximum skip length of each comparison is $m-B+1$ and the time consumed by once computing of *hash* value of a B-block is $O(B)$. So the best case time complexity of scanning algorithm of WM can achieve $O(Bn/(m-B+1))$ = $O(n\log_\sigma(rm)/(m-B+1))$, which slightly inferior than the average case optimal time complexity[2]. The WM+'s maximum skip length is m, so it can achieve the average case optimal time complexity (i.e. $O(n\log_\sigma(rm)/m)$).

For WM the worst case is that the text is a string of n same character c and the first i characters of all $r-1$ k-length patterns are c, and $m<<i<k$. In this case the *hash* function of the coarse matching will hit in every filtering operation of WM (i.e. $shift[] = 0$) and native i-length string comparison will be done r times character by character. So the scanning time complexity of WM in the worst case can achieve $O(Bnri)$ = $O(n\log_\sigma(rm)ri)$, which is greatly inferior than $O(n\log_\sigma(rm)rm)$.

In the same case, WM+ will enter automata matching stage after the hit of coarse matching and the state of the automata will be zero after matching the $i+1$ character. The coarse scanning pointer will skip $m-B+1$, and the distance from the previous pointer is $i+1+m-B+1-m = i-B+2$. So the automata will scanning all n characters, and the scanning time complexity of WM+ in this case is $O(Bn/(i\text{-}B+2))+O(n))$, greatly superior than $O(n\log_\sigma(rm)/m)+O(n)$, which is also greatly superior than $O(n\log_\sigma(rm)ri)$.

The worst case for WM+ is for $i=0$ in the above example. The scanning time complexity of WM+ is $O(Bn)+O(n) = O(n\log_\sigma(rm))+O(n)$. In this case, because WM also needs to calculate the *hash* value of the prefix B-block, to lookup the *prefix* table of the hit patterns set (suppose the size is r'), and to compare the patterns in need, the scanning time complexity of WM is $O(Bn)+O((B+r')n) = O(n\log_\sigma(rm))+ O(n(\log_\sigma(rm)+r'))$, which is inferior than $O(n\log_\sigma(rm))+O(n)$.

In average case, we also suppose the length of all patterns is m, and the text and the patterns are both random strings comply with even distribution [1]. In scanning stage, the time complexity of WM+ when coarse scanning not hit (i.e. skip size is great than zero) is $O(Bn/m)$ which is the same as that of WM[1]. Because the maximum skip size of WM+ (i.e. m) is greater than that of WM (i.e. $m\text{-}B+1$), the time consumed by WM+ is less than that of WM in practice. When hitting occurs in coarse scanning, WM needs to match prefixes and calculate the *hash* value of the prefix. The time complexity of these operations is $O(B)$, which is the same as that of the processing on the first B-characters of the text by the automata in WM+. Because the text and the patterns are both random strings complying with even distribution, the probability of finding a pattern in the text is $1/\sigma^{m\text{-}2B}$ after the B-size prefix and suffix of the pattern are both matched. The average matching times is $(m\text{-}B)/2$ in this case. After the finding of the same prefix, WM needs to execute native matching character by character for every patterns with the same prefix (suppose that the number of such patterns is r') and the time complexity of these operations is $O(r'(m\text{-}B)/2)$. For the same condition, WM+ fulfils the same job by automata which only scans the text one time and the time complexity is $O((m\text{-}B)/2)$.

Summing up the three parts of time consumed described above, we see that in average case the time consumed by WM+ is less than that of WM. Because of the low hitting probability in random cases, in practice the average case time complexities of WM and WM+ are generally the same as $O(n\log_\sigma(rm))$.

4 Experiments Result

For verifying the performance of WM+, some experiments are conducted to compare the scanning speeds of WM and WM+ in several conditions. The WM algorithm is quoted from Agrep[10]. All codes are written in ANSI C and the hardware platform is an IBM X31 with one Pentium-M 1.4GHz CPU and 512 MB DDR memory. The OS is Windows XP professional. For convenient we run Vmware 3.2.0 build-2230 on the OS and assign 192 MB memory to the virtual machine. The virtual OS is Red Hat Linux 8.0 with kernel version 2.4.18-14smp. The codes are compiled with gcc 3.2-7 for Red Hat Linux 8.0 with option -O2. The scanning time is obtained by inserting *time()* functions from begin and after the scanning code and calculating the difference of their outputs.

The first experiment is designed to test the scanning performance of WM and WM+ in the best case. The patterns are generated randomly and the minimum pattern length *m* is 5 and let *B*=2. No pattern contains character 'a' and the text is composed of 112.9 MB character 'a'. Scanning is done 10 times circularly. Theoretically, WM+ will scan 1/2 of the text and WM will scan 2/3 of the text. So the scanning time of them will be different by 25% approximately. The experiment result is shown in Table 2.

Table 2. The comparison of the scanning performances of WM and WM+ in the best case

Pattern Number	Patterns Total Length	Min. Pattern Length	Max. Pattern Length	Scanning Time of WM (in seconds)	Scanning Time of WM+ (in Seconds)
10	155	5	22	67	65
50	860	5	35	67	65
100	1693	5	39	67	65
500	8782	5	48	67	65
1000	17348	5	48	67	65
2000	34956	5	48	67	65
5000	87581	5	48	67	65
8000	141051	5	48	67	65

The result shows that the scanning performances of WM and WM+ are not very different though WM+ is slightly better than WM. In fact the difference is only about 3% which is greatly different from the theoretical estimate (i.e. 25%). It shows that the common operations in the implementations of the algorithms occupy a large proportion of operations and the time consumed by the skipping of the scanning pointer only occupies a little proportion of the total processing time, which is about 1/8 by practically and theoretically calculation.

Table 3. The comparison of the scanning performances of WM and WM+ in the average case

Pattern Number	Patterns Total Length	Min. Pattern Length	Max. Pattern Length	Scanning Time of WM (in seconds)	Scanning Time of WM+ (in Seconds)
10	172	5	25	7	7
50	851	5	34	7	7
100	1702	5	35	7	6
500	8768	5	39	6	6
1000	17686	5	41	7	7
2000	35098	5	41	7	7
5000	87386	5	52	7	7
8000	140508	5	52	10	9

The second experiment is designed to test the scanning performance of WM and WM+ in the average case. The patterns are generated randomly and the minimum pattern length m is 5 and let $B=2$ too. The text is a randomly generated string which length is 100 MB. The experiment result is shown in Table 3.

The result shows that the scanning performances of WM and WM+ are not very different though WM+ is slightly better than WM again. The reason is the same as that of the first experiment.

The last experiment is designed to test the scanning performance of WM and WM+ in the worst case. Let i be the length of the duplicated prefix (see also in section 3.2) and r' be the number of patterns having the same duplicated prefix. Two pattern sets are generated.

One set is used to test the effect on scanning time made by different values of i, which contains three parts. Part one is composed of 10 patterns in which lengths are from 5 to 22; Part two is the pattern 'bbaa'; and part three is the pattern 'aa...aZ' (the number of 'a' is i). The minimum pattern length is 4 and $B = 2$ too. The text is composed of 112.9 MB character 'a'. The experiment result is show in Figure 3a.

The other set is used to test the effect on scanning time by different values of r', which is also composed of three parts. Part one and part two are same as that of the above first two pattern sets. Part three is r' pattern 'aaaaZ' (the Z of every pattern is different from each other). The minimum pattern length is 4 and $B = 2$ too. The text is also composed of 112.9 MB character 'a'. The experiment result is shown in Figure 3b.

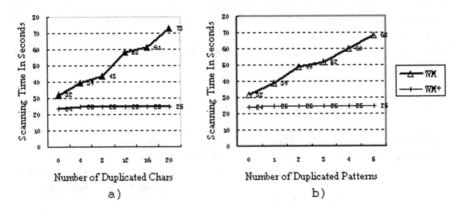

a)   b)

Fig. 3. The comparison of the scanning performances of WM and WM+ In the worst case

The experiment result shows that in the two worst cases, the scanning time of WM both increases linearly and will increase squarely if combined the two cases together. At the same time, the scanning time of WM+ keeps approximately changeless which is stable about three times greater than that7 in the average case. The results of all three cases validate the conclusion made in the coarse performance analysis.

5 Conclusions

Though it does not theoretically change the time complexity obviously, WM+ algorithm can reach higher performance in the best, worst and average case than WM by optimizing the *shift* table generation algorithm and combining the AC [3] algorithm with the original WM algorithm. And the scanning time of the WM+ algorithm in the worst case is more stable and predictable. The time complexity of WM+ in the extremely worst case when B=2 is $O(3n)$ and that of WM in the same case is in proportion to the number of overlay patterns and the overlay length. So the exact scanning time of WM in this case is not predictable and can be thousands times to that of WM+ in the extremely worst case, in which case the scanning can hardly work.

The deficiency of WM+ to WM is that the automata in it is storage consumed when patterns number is very large, and the initiating time is longer than that of WM when constructing very large automata. However, the storage complexity and the initiating time of WM+ is no more than that of DFA based algorithms (e.g. AC). For these reasons, WM+ is applicable to chronically running applications having large number of patterns and patterns not changing very often such as online virus scanning, IDS etc.

References

1. Wu, S., Manber, U.: A Fast Algorithm for Multi-pattern Searching. Report TR-94-17, Department of Computer Science, University of Arizona (1994)
2. Baeza-Yates, R., Navarro, G.: Text Searching: Theory and Practice, http://citeseer.ist.psu.edu/605426.htm (2004)
3. Aho, A., Corasick, M.: Efficient String Matching: An Aid to Bibliographic Search. Communications of the ACM, Vol 18 (1975) 333-340
4. Boyer, R., Moore, J.: A Fast String Searching Algorithm. Communications of ACM, 20 (10) (1987) 762-772
5. Allauzen, C., Raffinot, M.: Factor Oracle of a Set of Words. Technical report 99-11, Institute Gaspard-Monge, University de Marne-la-Vallee (1999)
6. Fredriksson, K., Navarro, G.,: Average-optimal Multiple Approximate String Matching. In: 14th Ann. Symp. On Combinatorial Pattern Matching(CPM'03), LNCS Vol. 2676. (2003) 109-128
7. Knuth, D., Morris, J., Pratt, V.: Fast Pattern Matching in Strings. SIAM Journal on Computing, Vol. 6(2). (1977) 323-350
8. Horspool. N.: Practical Fast Searching in Strings. Software-Practice and Experience, Vol 10(6). (1980) 501-506
9. ZHANG Xin, TAN Jianlong, CHENG Xueqi. An Improved Wu-Manber Multi-Pattern Matching Algorithm(In Chinese). Computer Application, 2003, 23(7): 29-31.
10. Wu, S., Manber, U.: Agrep — A Fast Approximate Pattern-matching Tool. In: Usenix Winter 1992 Technical Conference. San Francisco, (1992) 153-162
11. Kim, J. Y., Taylor, J. S.: Fast String Matching Using An n-gram Algorithm. Software – Practice And Experience. Vol. 24(1). (1994) 79-88

Author Index

Lecture Notes in Computer Science

For information about Vols. 1–3670

please contact your bookseller or Springer